MW00605006

Poly(ethylene glycol)
Chemistry and Biological Applications

ACS SYMPOSIUM SERIES **680**

Poly(ethylene glycol)

Chemistry and Biological Applications

J. Milton Harris, EDITOR
The University of Alabama at Huntsville

Samuel Zalipsky, EDITOR
SEQUUS Pharmaceuticals, Inc.

Developed from a symposium sponsored by the Division
of Polymer Chemistry, Inc., at the 213th National Meeting
of the American Chemical Society,
San Francisco, CA,
April 13–17, 1997

American Chemical Society, Washington, DC

This book is printed on acid-free, recycled paper.

PRINTED IN THE UNITED STATES OF AMERICA

Foreword

THE ACS SYMPOSIUM SERIES was first published in 1974 to provide a mechanism for publishing symposia quickly in book form. The purpose of the series is to publish timely, comprehensive books developed from ACS sponsored symposia based on current scientific research. Occasionally, books are developed from symposia sponsored by other organizations when the topic is of keen interest to the chemistry audience.

Before agreeing to publish a book, the proposed table of contents is reviewed for appropriate and comprehensive coverage and for interest to the audience. Some papers may be excluded in order to better focus the book; others may be added to provide comprehensiveness. When appropriate, overview or introductory chapters are added. Drafts of chapters are peer-reviewed prior to final acceptance or rejection, and manuscripts are prepared in camera-ready format.

As a rule, only original research papers and original review papers are included in the volumes. Verbatim reproductions of previously published papers are not accepted.

ACS BOOKS DEPARTMENT

Contents

CONJUGATES OF OLLGOMERIC AND LOW-
MOLECULAR WEIGHT SUBSTRATES

ix

Preface

POLY(ETHYLENE GLYCOL) (or PEG) is widely used in several areas of medicine and biological science. This book presents recent developments of these applications and the related chemistry. The book was developed from a symposium presented at the 213th National Meeting of the American Chemical Society, sponsored by the ACS Division of Polymer Chemistry, Inc., in San Francisco, CA, April 13–17, 1997. The symposium was chaired by Milton Harris and David Grainger. Most of the contributors to this volume were participants in the symposium.

During the past two decades, an increasing number of PEG applications have required covalent attachment of the polymer to a wide range of substrates, including low-molecular-weight drugs, affinity ligands, proteins, oligonucleotides, and the surfaces of biomaterials and particles such as liposomes and polystyrene latex. In this book, we have attempted to give appropriate emphasis to these contemporary areas of PEG applications, covering them as broadly as possible within the scope of a single edited volume.

Several PEG-based technologies have now appeared as commercial therapeutic, research, and diagnostic products. Examples include PEG-modified proteins and liposomes and PEG-grafted materials for use in medical and analytical devices. Research in these and related PEG applications is escalating, and is driven primarily by the many useful properties of the polymer. This trend is certain to continue, in part because of the extensive experience we now have with PEG, and in part because there are limited alternatives to PEG as a biomaterial.

The current book is truly an interdisciplinary work. The contributors and potential readers include polymer, physical, and organic chemists, biochemists, physicians, bioengineers, and separations, pharmaceutical and material scientists. We anticipate that the many professionals who are currently using PEG in their work and those who are contemplating use of PEG will find this volume beneficial. Most of the chapters are written in the form of short reviews and as such provide a useful tool for new and potential users of PEG derivatives. It will be very rewarding for those of us who have contributed to this volume if it can serve as a catalyst to kindle new ideas and applications.

The editors appreciate all the contributors through whose efforts this volume became possible. We wish all of you continued success and productivity.

SAMUEL ZALIPSKY
SEQUUS Pharmaceuticals, Inc.
960 Hamilton Court
Menlo Park, CA 94025

J. MILTON HARRIS
Department of Chemistry
Materials Science Program
The University of Alabama at Huntsville
Huntsville, AL 35899

August 5, 1997

Chapter 1

Introduction to Chemistry and Biological Applications of Poly(ethylene glycol)

Samuel Zalipsky[1] and J. Milton Harris[2]

[1]Sequus Pharmaceuticals, Inc., 960 Hamilton Court, Menlo Park, CA 94025
[2]Department of Chemistry, The University of Alabama at Huntsville, Huntsville, AL 35899

The chemistry and biological applications of polyethylene glycol (or "PEG") have been the subject of intense study both in academics and in industry. The current volume contains 28 chapters dedicated to providing a review of the major aspects of the topic. In this introductory chapter we highlight recent developments in the field, with reference to those chapters emphasizing these developments. Also we discuss those few important aspects of PEG chemistry and applications that are not described in the chapters. A number of leading references are given.

The principal focus of this book is to describe the chemistry and biological applications of polyethylene glycol (PEG), one of the most widely used biocompatible polymers (1). Traditionally, PEG has been widely used in biological research as a precipitating agent for proteins, other biological macromolecules and viruses (2). It has also been utilized since the mid 1950s for preparation of two-phase aqueous systems (3). Another widely applied property of PEG is its ability to facilitate biological cell fusion, a technique widely used in cell hybridization technology (4,5). Each of these traditional applications of PEG has its own large body of literature. The current volume does not deal with these subjects, but rather is concerned with contemporary applications of PEG, development of which has taken place over the past twenty years.

The purpose of this introductory chapter is several fold: (1) to provide a directory to the following chapters, (2) to mention some of the important recent developments in the field, and (3) to highlight some important aspects that are not covered in the following chapters.

Prime Applications of PEG Technology

The biological applications of PEG chemistry currently receiving the most attention are as follows:

1. PEG-protein conjugates for pharmaceutical applications,
2. PEG-enzyme conjugates for industrial processing,
3. surface modification with PEG to provide protein- and cell-rejecting properties,
4. surface modification with PEG to provide control of electroosmosis,
5. aqueous two-phase partitioning for protein and cell purification,
6. PEG hydrogels for cell encapsulation, drug delivery and wound covering,
7. PEG-modification of small-molecule pharmaceuticals,
8. PEG tethers for synthesis of biomolecules,
9. PEG tethering of molecules for biological targeting and signaling,
10. PEG-liposomes and micelles for drug delivery.

With the exception of two-phase partitioning, these applications are covered in this book. Two-phase partitioning was the subject of several chapters in the previous book dedicated to biological applications of PEG (6), so we have decided to forego further treatment of this subject at this time. This earlier volume also had chapters dealing with the other topics in this list, with the exception of topics seven and ten. These two subjects are represented here. The latter topic, PEG-particulates for drug delivery, was just coming to public attention at the time of the previous volume, and has since blossomed into a field of intense study, as evidenced by publication of a book dedicated to the subject (7) and by approval by the FDA of Sequus Pharmaceutical's Stealth® PEG-liposome for Doxorubicin delivery for treatment of Kaposi's sarcoma. Several chapters of the current book are dedicated to discussion of various forms of PEG-particulates (see chapters of Woodle, Lasic, La and Okumura). Attachment of PEG to the surface of lipid vesicles has had a particularly noticeable impact in the field of drug delivery since a major limitation of liposomal drug delivery (fast clearance by the liver and spleen) can be overcome by PEG attachment (7).

PEG-Proteins and Other PEG Conjugates

Since the initial description of covalent PEG-protein adducts (8-11) there has been a tremendous amount of work done on this type of conjugate. In particular, the pioneering work of Frank Davis and co-workers (10,12), discovering the enormous potential of PEGylated proteins as therapeutics, has been of central importance. A substantial portion of this book is dedicated to PEG-proteins. In addition to work dealing with development of PEG-proteins as therapeutics (see the chapters of Olson, Hershfield, Monkarsh and Sherman), several chapters are focused on unconventional applications of PEG-proteins (Mabrouk, Panza and Topchieva).

Protein conjugates constitute by far the largest group of PEG conjugates (13). They continue to enjoy tremendous popularity, and conjugation chemistries, analytical methods and biological applications have become increasingly sophisticated. FDA approval for two PEG-protein conjugates manufactured by Enzon (PEG-adenosine deaminase for treatment of ADA deficiency, and PEG-asparaginase for treatment of acute lymphoblastic leukemia), and several other PEG-proteins in clinical development by a number of companies, is a clear indication of the tremendous utility and the maturity of this technology. The chapter of Hershfield reviews the almost two decades of experience accumulated with the first approved therapeutic conjugate of this type, PEG-adenosine deaminase.

PEG-conjugates with smaller, biologically-active molecules other than proteins are also of great interest, primarily because of increased water solubility, reduced rate of kidney clearance and reduced toxicity. Although PEG conjugates of small drugs appeared in the literature earlier than their protein counterparts (*14-16*), they have not yet achieved success as FDA approved therapeutics. It is expected that this will not long remain the case. The chapters of Barany and Felix describe PEG conjugates of peptides. In Chapter 18 Jäschke has written the first comprehensive review of PEG-oligonucleotide conjugates. The topic of low molecular weight drug conjugates is discussed in the chapter by Ouchi. The chapters of Schacht and Zhao describe the synthesis of new PEGs that are well suited for delivery of small molecule drugs, and the chapters of Chen and Zhao describe the synthesis of new hydrogels of utility for this purpose.

The great popularity of PEG-conjugates is driven by the unique combination of physicochemical and biological properties of the polymer. These include excellent solubility in aqueous and most organic solutions (*17*). Among the important biological properties of PEG (of molecular weights over 2000) are favorable pharmacokinetics and tissue distribution and lack of toxicity and immunogenicity (*18-20*). The issue of safety of PEG and its conjugates is addressed in the chapters by Working and Rhee. A wide variety of biologically relevant molecules have been conjugated to PEG to take advantage of these properties (see Table I). The experience accumulated over the last two decades has shown that the useful characteristics of PEG can usually be conveyed to its conjugates. The conjugates also receive protection against protease attack and, because of their larger size, have a reduced rate of kidney clearance. Recent reviews of PEG conjugates have been provided by Hooftman, Zalipsky, Katre, and Delgado (*13,21-24*).

Chemistry for Formation of Conjugates. Side reactions. Careful execution of both PEG functionalization and coupling chemistry are essential for clean preparation of PEG-adducts. Synthesis of various functionalized derivatives of the polymer were summarized in the original review of Harris (*30*) and more recently by Zalipsky (*22,31*). One of the important recent developments in the field is identification of chemical reactivity of various PEG-reagents as well as characterization of side reactions associated with specific activated PEG derivatives. For example, mPEG-dichlorotriazine (*10*), one of the first described reactive PEG derivatives, has long been suspected to react not only with amino groups of proteins, as originally claimed, but also with other nucleophilic groups present on proteins (*22*). Recently Gotoh et al. have used NMR and amino acid analysis to show that this reagent reacts readily with phenol and imidazole groups of proteins (*32*). Synthesis of a related reagent containing two mPEG arms per chlorotriazine was recently reexamined and a new

Table I. Summary of various classes of PEG conjugates and their applications.

Conjugates of	Useful Properties & Applications	Reviews and/or Leading References
Drugs	Improved solubility, controlled permeability through biological barriers, longevity in bloodstream, controlled release.	(13) and chapter by Ouchi
Affinity ligands and cofactors	Used in aqueous two-phase partitioning systems for purification and analysis of biological macromolecules, cells. Enzymatic reactors.	(3,6)
Peptides	Improved solubility. Conformational analysis. Biologically active conjugates.	(17) and chapters by Barany and Felix
Proteins	Resistance to proteolysis, reduced immunogenicity & antigenicity, longevity in bloodstream, tolerance induction. Uses: therapeutics, organic soluble reagents, bioreactors.	(19,21,23,25, 26)
Saccharides	New biomaterials, drug carriers.	Chapter of Schacht and Hoste
Oligonucleotides	Improved solubility, resistance to nucleases, cell membrane permeability.	Chapter by Jächske
Lipids	Used for preparation of PEG-grafted liposomes.	(27)
Liposomes & particulates	Longevity in bloodstream, RES-evasion.	(7)
Biomaterials	Reduced thrombogenicity, reduced protein and cell adherence.	(6,28,29)

clean process has been introduced (*33*). The original method for preparing 2,4-bis(methoxypolyethylene glycol)-6-chloro-s-triazine was shown to yield substantial amounts of mPEG-dichlorotriazine as well as some mPEG oligomers. Since mPEG-dichlorotriazine and 2,4-mPEG-6-chlorotriazine exhibit different reactivities towards various nucleophilic groups on proteins, one should be careful in evaluating the literature published during the 1980s involving the use of the latter reagent (*25*).

Chemistry of another activated PEG derivative, the 2,2,2-trifluoroethylsulfonate or "tresylate", has undergone recent revision. Originally introduced by Nilsson and Mosbach (*34*) and adopted by several other investigators (*35-37*), mPEG-tresylate was believed to produce secondary amine attachment upon reaction with amino groups of proteins. Upon closer examination it was discovered that this reaction does not proceed via direct nucleophilic displacement of tresylate, but rather through a sequence of elimination-addition steps to give a sulfonate amide (PEG-OSO$_2$-CH$_2$-CONH-protein) (*38,39*). Additional work will have to be done on this surprising chemistry. It will be interesting to learn if coupling is completely tilted towards the elimination-addition reaction or if the direct substitution reaction is also taking place under certain conditions. If the elimination-addition mechanism in fact is dominant, then this will place serious doubt on claims made about the stability, charge and other "unique" properties of conjugates derived from use of mPEG tresylate (*21,36,37*).

The overall maturity of the field of PEG conjugation is reflected in the close attention being given to conjugation chemistry. It is our opinion that the future will reveal development of more specific and sophisticated coupling chemistries. Readers particularly interested in this subject are directed to the recent papers of Harris, Veronese and coworkers dealing with a new thiol-specific reagent, PEG-vinyl sulfone (*40*), and the chapter in this book of Zalipsky and Menon-Rudolph dealing with some unconventional conjugation methods. Specific N-terminal PEGylation of proteins with aminooxy-PEG was recently described by Gaertner and Offord (*41*). PEGylation of active-site-protected enzyme resulting in higher preservation of proteolytic activity of the conjugate was recently reported by Caliceti et al. (*42*).

Biomaterials: Surfaces and Hydrogels

Surfaces modified by covalent attachment of PEG chains are resistant to protein and cell adsorption (i.e., are "nonfouling") and consequently have received much attention in preparation of biomaterials. Several chapters of this book are concerned with methods for forming and understanding nonfouling surfaces for biomaterials applications (Caldwell, Sofia, Mrksich).

Although it may not be immediately evident, attachment of PEG to a surface acts to alter the electrical nature of a surface exposed to an aqueous environment. In essence the PEG layer forms a viscous, neutral, surface-bound layer and results in movement of the hydrodynamic plane of shear relative to the electrical double layer. Utilization of this result permits control of the extent of electroosmosis, a phenomenon of importance to a variety of applications such as capillary electrophoresis, and it provides a method for determining the thickness and density of surface-bound PEG layers (see the chapter by Emoto).

The benign nature and availability of multifunctional PEG derivatives has led to the preparation and application of crosslinked PEG networks or hydrogels (*43*). Several chapters in this book deal with PEG-polymer networks or hydrogels (Rhee, Chen, Zhao). PEG hydrogels are effective for wound covering, and the methods for forming these hydrogels can be so mild that the hydrogels can even be formed in situ (*44*). Incorporation of degradable linkages into the hydrogel can lead to slow dissolution of the material *in vivo* as discussed in the chapter in this book by Zhao. Recent work has shown that appropriately substituted PEGs can be used to prepare photochemically reversible hydrogels (*45*). The ability of these gels to adhere to other surfaces can also be controlled as shown in the chapter of Chen. The chapter of Rhee and coworkers describes use of PEG-crosslinked collagen hydrogels as benign materials for soft tissue replacement.

PEG as a Spacer Moiety

PEG is an ideal spacer molecule for many applications (*31,46,47*). Ethylene oxide repeating units are well solvated by water, binding several water molecules each (*43,48*). The length of the chain can be controlled within a desired range by the degree of polymerization of the starting polymer. Alternatively oligomers of precise length can be used. Also the high flexibility of the PEG chain contributes to the ready availability and high activity of PEG-tethered molecules. While the usual application is to link two different moieties (e.g., molecule to molecule or surface to molecule), a recent novel application is to replace chain segments within a biopolymer with PEG; this has been done with peptides (*49*) and oligonucleotides (reviewed in the chapter of Jäschke).

Molecules tethered to a surface via a PEG chain function well as targeting or binding/activating moieties. It is noteworthy that PEG-tethered proteins and proteins adsorbed to PEG-surfaces are not denatured by interaction with PEG. This "hospitality" to proteins is also revealed in the observation that enzymes tethered to plastic surfaces via a PEG linker are protected by the PEG during free radical polymerization, and they remain active in organic solvents (*50*) (see chapter by Russell).

Although it is possible and sometimes less complex to link two moieties using a homobifunctional PEG reagent (*47*), a more elegant and powerful approach is to use a heterobifunctional PEG. Recent developments in chemistries of heterobifunctional PEG derivatives allow convenient access to the essential precursors (*31*). Heterobifunctional PEGs have been utilized for grafting the polymer onto solid supports (*51*), crosslinking two different proteins (*52*), linking reporter groups to biomaterials and attaching various biologically relevant ligands to membrane-forming lipids (*53-55*). In the latter case it was demonstrated that a ligand positioned on a long PEG tether has expanded its receptor-accessible range by almost the entire length of the fully stretched PEG spacer (*55*). Incorporation of sialyl LewisX-PEG2000-lipids into unilamellar liposomes resulted in three orders of magnitude enhancement in binding to E-selectin (*54*). Small peptides linked through PEG to mPEG-grafted liposomes exhibit markedly prolonged circulating lifetimes in vivo (*53*). These examples illustrate the advantages of multifunctional PEG-tethered systems and should lead the way to more refined applications in the upcoming years.

Analysis of PEG and PEG Conjugates

One of the great challenges of PEG chemistry is analysis of PEG derivatives and their conjugates. One of the most significant recent advances in this direction has been the availability of matrix-assisted-laser-desorption-ionization mass spectrometry (MALDI-MS) (*56-58*). Several chapters in this book show examples of utilization of this powerful technique (Zhao, Olson, Felix, Zalipsky). MALDI provides the mass of the unfragmented, singly-charged molecular ion of macromolecules up to about 100,000 Daltons, and thus greatly assists determination of polydispersity and identity of PEG derivatives. Similarly, the mass of the molecular ions of PEG conjugates, such as PEG proteins, can also be determined, and the composition of PEGylation reaction products containing different numbers of PEGs ("one-mer", "two-mer", etc.) can be established. Spectra of PEG conjugates of peptides, oligosaccharides and lipids in the range below about 10,000 Daltons are characterized by a bell shaped distribution of equally spaced molecular ions 44 Daltons apart (the mass of one oxyethylene unit) (*56-58*). Higher molecular weight conjugates show a similar bell shaped distribution, but the individual peaks, 44 Daltons apart, are not well resolved.

Proteolytic cleavage of PEGylated proteins followed by HPLC separation of the peptide fragments and comparison of the peptide mapping pattern to the one obtained for the parent protein allows precise determination of PEG-modified fragments. Further application of peptide sequencing techniques provides the exact site of PEGylation (*59,60*). The power of peptide mapping can be further enhanced by MS analysis of the relevant HPLC-purified PEG-fragments (see chapter of Olson). For example, the location of PEG attachment on superoxide dismutase modified with PEG succinimidyl succinate has been determined by this approach; MS analysis of the hydrolyzed conjugate demonstrated formation of succinimidylated lysine residues at the sites of PEG cleavage (*61*).

Capillary electrophoresis is another powerful analytical method that is increasingly being used for analysis of PEG-proteins. Recent work has shown that this method can be used to determine the amount of each PEG-mer in a PEG-protein (*62,63*). CE can also be used in analysis of fragmented PEG-proteins for determination of the site of PEGylation.

Alternative Polymers for PEG replacement

What is so unique about PEG? Can other polymers match its performance in biological systems? These questions have challenged researchers over the past few years. Some insight into the nature of PEG is provided by the chapter of Karlström on the theory of PEG in solution. Various polymers have been examined as potential PEG replacements; this list includes various polysaccharides, poly(vinyl alcohol), poly(vinyl pyrrolidone), poly(acryloyl morpholine), PHPMA, and polyoxazolines.

This topic is discussed in some detail in the chapter by Veronese. Although promising results have been observed in a number of individual studies, PEG appears in many ways to be a unique polymer. It is probably safe to say that no other polymer has as wide a range of biological applications or the general acceptance of PEG.

One of the prime properties of PEG is its "exclusion effect" or "steric stabilization effect" (chapter by Lasic), which is the repulsion of other macromolecules or particles by PEG either in its free form or grafted onto a flat surface or the surface of a particle or protein (64). It is accepted that heavy hydration, good conformational flexibility and high chain mobility are principally responsible for the exclusion effect. Taking advantage of these properties, PEG grafting has been widely used as a method for reduction of various undesirable consequences of biological recognition manifested by immunogenicity and antigenicity in the case of proteins (10,13,19,21,23,26), and thrombogenicity, cell adherence, and protein adsorption in the case of artificial biomaterials (28,29).

It is interesting that PEG as well as some of the most promising alternatives act only as hydrogen bond acceptors, not as donors (65,66). It appears that hydrophilicity *per se* is of secondary importance to chain conformational flexibility and mobility (67). Evidence for this conclusion comes from various protein-polymer and particulate-polymer systems where more hydrophilic polymers, such as polysaccharides and PVA, performed only marginally compared to the PEG (68).

Effective replacement of PEG with other polymers has been achieved with poly(methyloxazoline) and poly(ethyloxazoline) for polymer-grafted liposomes (65,69), and with PVA (70), and HPMA (71) for polymer-protein and polymer-nanoparticles conjugates, respectively. Also oligopropylene sulfoxide is effective in surface protection against protein adsorption (66). Similarly, Österberg and coworkers found that surface-bound polysaccharides such as dextran and ethyl hydroxyethylcellulose were comparable to PEG for protein repulsion, although PEG was effective at a much lower molecular weight (72).

New PEG Architectures

In addition to progress in PEG applications and the key ancillary areas of synthesis of new PEG derivatives and analysis of PEG conjugates, there has also been much progress in synthesis of new PEG backbones or architectures. Until recently the only available PEGs were linear PEG diol and methoxy-PEG (or mPEG). High quality mPEG with low diol content and MWs of 20,000 and above is now available from several sources. PEG is also available in a range of MWs. One concern with regard to use of any PEG material is that long term storage can lead to peroxide formation and chain cleavage (73), so it is important to use freshly prepared materials and to store them, ideally, over nitrogen, protected from light, in the cold (1).

New PEG derivatives described in recent years include branched PEGs such as 3-armed, 4-armed, and "star" PEGs (74,75). Note that in general, as could be predicted from conformational freedom considerations, branched PEGs have smaller exclusion volumes and are less effective at protein rejection than their linear counterparts of similar molecular weight, although this difference is minimal above 2000 Daltons (75).

In the above examples the branched PEGs have reactive groups at the arm ends. Another form of branched PEG has two inert arms and a third short arm with an active functional group (*25,76*). Such materials can be made, for example, by coupling two mPEGs to cyanuric chloride or by coupling two mPEGs to the amine groups of lysine. When these branched PEGs are linked to proteins an increase in protection from proteolytic degradation is observed, even at relatively low degrees of modification (*25,76*).

Difunctional PEG can be coupled with the amine groups of lysine to provide an alternating, linear copolymer with pendant carboxyl groups separated by PEG chains (*77,78*). These carboxyl groups can be activated so that the copolymers can serve as drug carriers or starting materials for hydrogel preparation (*77,78*). Koyama and coworkers have made copolymers of ethylene oxide and glycidyl allyl ether (*79*). The resulting copolymers have pendant allyl ether groups along the polyether backbone that can be converted to useful functional groups by reaction with the appropriate thiol. The net result is a polyether with functional groups placed at random along the backbone.

Another interesting copolymer can be prepared by vinyl polymerization of mPEG allyl ether and maleic anhydride. The result is a comb polymer with PEG chains tethered onto a hydrocarbon backbone and with alternating anhydride reactive groups. This polymer has been used to modify proteins via multipoint attachment of the anhydride groups to the protein surface (*80*).

Finally, it is noteworthy that ethylene oxide-propylene oxide random, copolymers can be functionalized and used for formation of conjugates. These materials are of interest because of their temperature sensitivity (see chapter of Chen describing an example case of this phenomena). They exhibit a lower consulate solution temperature (LCST) or cloud point upon heating in aqueous solution. At the cloud point, the polymer becomes hydrophobic and is no longer water insoluble and thus can be recovered, along with any bound conjugates (*81*). Importantly, if the polymer is bound to a surface, it loses its protein rejecting ability at the cloud point (*82*).

Future Perspectives

It is expected that research and development in the above areas will continue to flourish, and commercialization of these applications will continue, probably at an accelerated pace as acceptance of the technology increases. New applications of PEG chemistry appear on a regular basis, and this process can be expected to continue. A recent example is the publication by Scott and coworkers in which PEG coatings on red blood cells are shown to hide antigenic sites and thus make possible heterologous blood donation (*83*). Certain important areas of PEG technology need more research, and given the importance of the technology, presumably this research will be done. For example, there is still little modern work published on PEG and PEG-conjugate toxicity and body clearance (see chapter of Working). Also the questions around the ability of PEG to penetrate membranes and the blood brain barrier have not been addressed in a concerted way (*84*). Anyone who has done a recent patent or literature search with PEG as a keyword can attest that interest in PEG science and technology is growing steadily, and trying to keep up with developments is a major challenge. Unless there is an unexpected and sudden reversal of this trend, PEG research, development and commercialization will continue to be an exciting discipline in which to participate.

Literature Cited

1.   Bailey, Jr., F.E.; Koleske, J.V., *Alkylene oxides and their polymers*; Marcel Dekker: New York, 1991.
2.   Polson, A. *Prep. Biochem.* **1977**, *7*, 129-154.
3.   *Partitioning in aqueous two-phase systems*; Walter, H.; Brooks, D.E.; Fisher, D., Eds.; Academic Press: San Diego, 1985.
4.   Lentz, B. *Chem. Phys. Lipids* **1994**, *73*, 91-106.
5.   Klebe, R.J.; Mancuso, M.G. *Somatic Cell Genetics* **1981**, *7*, 473-488.
6.   *Poly(ethylene Glycol) Chemistry: Biotechnical and Biomedical Applications*; Harris, J.M., Ed.; Plenum Press: New York, 1992.
7.   *Stealth® Liposomes*; Lasic, D.; Martin, F., Ed.; CRC Press: Boca Raton, FL, 1995.
8.   Pollak, A.; Witesides, G.M. *J. Am. Chem. Soc.* **1976**, *98*, 289-291.
9.   King, T.P.; Kochoumian, L.; Lichtenstein, L.M. *Arch. Biochem. Biophys.* **1977**, *178*, 442-450.
10. Abuchowski, A.; Van Es, T.; Palczuk, N.C.; Davis, F.F. *J. Biol. Chem.* **1977**, *252*, 3578-3581.
11. Rubinstein, M.; Simon, S.; Bloch, R. U.S. Patent 4,101,380 1978,
12. Davis, F.F.; Van Es, T.; Palczuk, N.C. U.S. Patent 4,179,337 1979,
13. Zalipsky, S. *Adv. Drug Delivery Rev.* **1995**, *16*, 157-182.
14. Weiner, B.-Z.; Zilkha, A. *J. Med. Chem.* **1973**, *16*, 573-574.
15. Weiner, B.-Z.; Havron, A.; Zilkha, A. *Israel J. Chem.* **1974**, *12*, 863-872.
16. Chaabouni, A.; Hubert, P.; Dellacherie, E.; Neel, J. *Makromol. Chem.* **1978**, *179*, 1135-1144.
17. Mutter, M.; Bayer, E. In *The Peptides*; E. Gross and J. Meienhofer, Ed.; Academic Press: New York, 1979; Vol. 2; pp 285-332.
18. Pang, S.N.J. *J. Amer. Coll. Toxicol.* **1993**, *12*, 429-456.
19. Dreborg, S.; Akerblom, E.B. *Crit. Rev. Ther. Drug Carrier Syst.* **1990**, *6*, 315-365.
20. Yamaoka, T.; Tabata, Y.; Ikada, Y. *J. Pharm. Sci.* **1994**, *83*, 601-606.
21. Delgado, C.; Francis, G.E.; Fisher, D. *Crit. Rev. Therap. Drug Carrier Syst.* **1992**, *9*, 249-304.
22. Zalipsky, S.; Lee, C. In *Poly(ethylene glycol) Chemistry: Biotechnical and Biomedical Applications*; J. M. Harris, Ed.; Plenum Press: New York, 1992; pp 347-370.
23. Katre, N.V. *Adv. Drug Delivery Rev.* **1993**, *10*, 91-114.

24. Hooftman, G.; Herman, S.; Schacht, E. *J. Bioact. Compatible Polym.* **1996**, *11*, 135-159.
25. Inada, Y.; Matsushima, A.; Kodera, Y.; Nishimura, H. *J. Bioact. Compatible Polym.* **1990**, *5*, 343-364.
26. Sehon, A.H. *Adv. Drug Delivery Rev.* **1991**, *6*, 203-217.
27. Zalipsky, S. In *Stealth Liposomes*; D. Lasic and F. Martin, Ed.; CRC Press: Boca Raton, FL, 1995; pp 93-102.
28. Merrill, E.W. In *Poly(Ethylene Glycol) Chemistry: Biotechnical and Biomedical Applications*; J. M. Harris, Ed.; Plenum Press: New York, 1992; pp 199-220.
29. Llanos, G.R.; Sefton, M.V. *J. Biomater. Sci. Polym. Edn.* **1993**, *4*, 381-400.
30. Harris, J.M. *J. Macromol. Sci.-Rev. Macromol. Chem. Phys.* **1985**, *C25*, 325-373.
31. Zalipsky, S. *Bioconjugate Chem.* **1995**, *6*, 150-165.
32. Gotoh, Y.; Tsukada, M.; Minoura, N. *Bioconjugate Chem.* **1993**, *4*, 554-559.
33. Ono, K.; Kai, Y.; Maeda, H.; Samizo, F.; Sakurai, K.; Nishimura, H.; Inada, Y. *J. Biomater. Sci. Polym. Ed.* **1991**, *2*, 61-65.
34. Nilsson, K.; Mosbach, K. *Methods Enzymol.* **1984**, *104*, 56-69.
35. Yoshinaga, K.; Harris, J.M. *J. Bioact. Compatible Polym.* **1989**, *4*, 17-24.
36. Delgado, C.; Patel, J.N.; Francis, G.E.; Fisher, D. *Biotechnol. Appl. Biochem.* **1990**, *12*, 119-128.
37. Knusli, C.; Delgado, C.; Malik, F.; Domine, M.; Tejedor, C.; Irvine, A.E.; Fisher, D.; Francis, G.E. *Brit. J. Haematol.* **1992**, *82*, 654-663.
38. Gais, H.-J.; Ruppert, S. *Tetrahedron Lett.* **1995**, *36*, 3837-3838.
39. King, J.F.; Gill, M.S. *J. Org. Chem.* **1996**, *61*, 7250-7255.
40. Morpurgo, M.; Veronese, F.M.; Kachensky, D.; Harris, J.M. *Bioconjugate Chem.* **1996**, *7*, 363-368.
41. Gaertner, H.F.; Offord, R.E. *Bioconjugate Chem.* **1996**, *7*, 38-44.
42. Caliceti, P.; Schiavon, O.; Sartore, L.; Monfardini, C.; Veronese, F.M. *J. Bioact. Compat. Polym.* **1993**, *8*, 41-50.
43. Graham, N.B. In *Poly(ethylene glycol) chemistry. Biotechnical and biomedical applications*; J. M. Harris, Ed.; Plenum: New York, 1992; pp 263-281.
44. Sawhney, A.S.; Pathak, C.P.; Hubbell, J.A. *Macromolecules* **1993**, *26*, 581-??
45. Andreopoulos, F.M.; Deible, C.R.; Stauffer, M.T.; Weber, S.G.; Wagner, E.; Beckman, J.; Russell, A.J. *J. Am. Chem. Soc.* **1996**, *118*, 6235-6240.
46. Akerblom, E.; Dohlsten, M.; Brynö, C.; Mastej, M.; Steringer, I.; Hedlund, G.; Lando, P.; Kalland, T. *Bioconjugate Chem.* **1993**, *4*, 455-466.
47. Harris, J.M.; Yoshinaga, K. *J. Bioact.Comp. Polym.* **1989**, *4*, 281-195.
48. Antonsen, K.P.; Hoffman, A.S. In *Poly(ethylene glycol) chemistry. Biotechnical and biomedical applications*; J. M. Harris, Ed.; Plenum: New York, 1992; pp 15-28.
49. Bouvier, M.; Wiley, D.C. *Proc. Natl. Acad. Sci. USA* **1996**, *93*, 4583-4588.
50. Yang, Z.; Mesiano, A.J.; Venkatasubramanian, S.; Gross, S.H.; Harris, J.M.; Russell, A.J. *J. Am. Chem. Soc.* **1995**, *117*, 4843-4850.

51. Zalipsky, S.; Chang, J.L.; Albericio, F.; Barany, G. *Reactive Polym.* **1994**, *22*, 243-258.

52. Paige, A.G.; Whitcomb, K.L.; Liu, J.; Kinstler, O. *Pharm. Res.* **1995**, *12*, 1883-1888.

53. Zalipsky, S.; Puntambekar, B.; Bolikas, P.; Engbers, C.M.; Woodle, M.C. *Bioconjugate Chem.* **1995**, *6*, 705-708.

54. DeFrees, S.A.; Phillips, L.; Guo, L.; Zalipsky, S. *J. Am. Chem. Soc.* **1996**, 118, 6101-6104.

55. Wong, J.Y.; Kuhl, T.L.; Israelachvili, J.N.; Mullah, N.; Zalipsky, S. *Science* **1997**, *275*, 820-822.

56. Felix, A.; Lu, Y.-A.; Campbell, R.M. *Int. J. Peptide Protein Res.* **1995**, *46*, 253-264.

57. Lee, S.; Winnik, M.A.; Whittal, R.M.; Li, L. *Macromolecules* **1996**, *29*, 3060-3072.

58. Zalipsky, S.; Mullah, N.; Harding, J.A.; Gittelman, G.; Guo, L.; DeFrees, S.A. *Bioconjugate Chem.* **1997**, *8*, 111-118.

59. Kinstler, O.; Brems, D.N.; Lauren, S.L.; Paige, A.G.; Hamburger, J.B.; Treuheit, M.J. *Pharm. Res.* **1996**, *13*, 996-1002.

60. Clark, R.; Olson, K. et al. *J. Biol. Chem.* **1996**, *271*, 21969-21977.

61. Vestling, M.M.; Murphy, C.M.; Fenselau, C.; Dedinas, J.; Doleman, M.S.; Harrsch, P.B.; Kutny, R.; Ladd, D.L.; Olsen, M.A. In *Techniques in Protein Chemistry III* ; Academic Press: New York, NY, 1992; pp 477-485.

62. Cunico, R.L.; Gruhn, V.; Kresin, L.; Nitecki, D.E. *J. Chromotogr.* **1991**, *559*, 467-477.

63. Bullock, J.; Chowdhurry, S.; Johnson, D. *Anal. Chem.* **1996**, *68*, 3258-3264.

64. Lasic, D.D.; Martin, F.J.; Gabizon, A.; Huang, S.K.; Papahadjopoulos, D. *Biochim. Biophys. Acta* **1991**, *1070*, 187-192.

65. Woodle, M.C.; Engbers, C.M.; Zalipsky, S. *Bioconjugate Chem.* **1994**, *5*, 493-496.

66. Deng, L.; Mrksich, M.; Whitesides, G.M. *J. Am. Chem. Soc.* **1996**, *118*, 5136-5137.

67. Blume, G.; Cevc, G. *Biochim. Biophys. Acta* **1993**, *1146*, 157-168.

68. Torchilin, V.P.; Omelyanenko, V.G.; Papisov, M.I.; Bogdanov, A.A.; Trubetskoy, V.S.; Herron, J.N.; Gentry, C.A. *Biochim. Biophys. Acta* **1994**, *1195*, 11-20.

69. Zalipsky, S.; Hansen, C.B.; Oaks, J.M.; Allen, T.M. *J. Pharm. Sci.* **1996**, 85, 133-137.

70. Kojima, K.; Maeda, H. *J. Bioact. Compatible Polym.* **1993**, *8*, 115-131.

71. Kamei, S.; Kopecek, J. *Pharm. Res.* **1995**, *12*, 663-668.

72. Osterberg, E.; Bergstrom, K.; Holmberg, K.; Schuman, T.P.; Riggs, J.A.; Burns, N.L.; Van Alstine, J.M.; Harris, J.M. *J. Biomed. Mat. Res.* **1995**, *29*, 741-747.

73. Yang, L.; Heatley, F.; Blease, T.; Thompson, R.I.G. *Eur. Polym.J.* **1996**, *32*, 535-547.

74. Merrill, E.W. *J. Biomater. Sci. Polym. Edn.* **1993**, *5*, 1-11.

75.   Bergstrom, K.; Osterberg, E.; Holmberg, K.; Hoffman, A.S.; Schuman, T.P.; Kozlowski, A.; Harris, J.M. *J. Biomater. Sci. Polym. Edn.* **1994**, *6*, 123-132.

76.   Monfardini, C.; Schiavon, O.; Caliceti, P.; Morpurgo, M.; Harris, J.M.; Veronese, F.M. *Bioconjugate Chem.* **1995**, *6*, 62-69.

77.   Nathan, A.; Zalipsky, S.; Erthel, S.I.; Agathos, S.N.; Yarmush, M.L.; Kohn, J. *Bioconjugate Chem.* **1993**, *4*, 54-62.

78.   Nathan, A.; Bolikal, D.; Vyavahare, N.; Zalipsky, S.; Kohn, J. *Macromolecules* **1992**, *25*, 4476-4484.

79.   Koyama, Y.; Umehara, M.; Mizuno, A.; Itaba, M.; Yasukouchi, T.; Natsume, K.; Suginaka, A. *Bioconjugate Chem.* **1996**, *7*, 298-301.

80.   Kodera, Y.; Sekine, T.; Yasukohchi, T.; Kiriu, Y.; Hiroto, M.; Matsushima, A.; Inada, Y. *Bioconjugate Chem.* **1994**, *5*, 283-286.

81.   Alred, P.A.; Tjerneld, F.; Kozlowski, A.; Harris, J.M. *Bioseparations* **1992**, *2*, 363-373.

82.   Tiberg, F.; Brink, C.; Hellsten, M.; Holmberg, K. *Colloid Polym. Sci.* **1992**, *270*, 1188-1193.

83.   Scott, M. D.; Murad, K. L.; Koumpouras, F.; Talbot, M.; and Eaton, J. W. *Proc. Nat. Acad. Sci. USA* **1997**, *94*, 0000.

84.   Harris, J. M. In *Poly(ethylene glycol) Chemistry: Biotechnical and Biomedical Applications*; J. M. Harris, Ed.; Plenum Press: New York, 1992; pp 11-12.

FUNDAMENTAL PROPERTIES
OF POLY(ETHYLENE GLYCOL)

Chapter 2

Theory of Poly(ethylene glycol) in Solution

Gunnar Karlström and Ola Engkvist

Department of Theoretical Chemistry, University of Lund,
P.O. Box 124, S-221 00 Lund, Sweden

The behavior of the poly(ethylene glycol) molecule in aqueous
solutions is reviewed, using a comparison of experimental and
theoretical means. It is stressed that the understanding of the
conformational equilibria for the poly(ethylene glycol) molecule is of
large importance both for the description of the phase behavior of
poly(ethylene glycol) - water system, as well as for the properties of
poly(ethylene glycol) solutions. In particular the influence of additives
to the solution is discussed, together with the properties of
poly(ethylene glycol) coated surfaces and the use of poly(ethylene
glycol) for purifying biochemical compounds.

Poly(ethylene glycol) (PEG) or as it is often called poly(ethylene oxide) (PEO) is a
polymer of considerable technical importance. PEG has perhaps its largest use as an
additive to control the viscosity of paint and as an additive when paper is produced. It
is also used as a head group in nonionic surfactants and to cover surfaces in order to
prevent other polymers (e.g. proteins) from binding to the surface. Naturally, a large
amount of both physical and chemical information has been collected about PEG and
PEG solutions in general and about the PEG-water system in particular *(1,2)*. The
purpose of this chapter is to try to establish a relationship between the macroscopically
observed phase behavior and a microscopic description of a PEG molecule in a water
solution.

It has long been known that, while completely soluble in water at low
temperature, PEG loses its solubility at elevated temperatures. This process is
normally given the name clouding. Complete phase diagrams for the PEG-water
system were first determined by Sakei and coworkers *(3)*. A similar behavior has been
observed for ethylhydroxyethylcellulose (EHEC) in both water and formamide by
Samii et. al. *(4)* (EHEC is a cellulose molecule substituted with ethyl and ethylene
oxide groups so that it is effectively covered with ethylene oxide (EO) groups). In a
similar way surfactants with ethylene oxide head groups phase-separate from a water

or formamide solution at elevated temperatures *(5)*. These observations are consistent with a model suggesting that the EO segments become more hydrophobic at higher temperatures.

Another line of research of relevance to this discussion starts with the investigations performed by Podo and coworkers *(6,7)*, who began in the 1970s to investigate the conformational equilibrium of 1,2-dimethoxyethane (DME), which can be regarded as a very short PEG chain, as a function of the solvent. Their main conclusion was that the polarity of the PEG chain was changed with the solvent polarity. Thus in polar solvents the PEG chain preferred polar conformations and in non polar solvents non polar structures were favored. These observations have later been verified by other investigations *(8,9)*. Recently Raman and IR investigations by Matsuura and coworkers *(10,11)* have indicated that for short PEG chains this trend does not hold in a very dilute water solution, where more non-polar conformations again start to be more populated. From these experiments it is also clear that this last trend is most pronounced for chains that contain 3 EO units and becomes weaker for longer chains. For a PEG molecule containing more than 10 EO units, this effect would hardly be observed.

The outline of this chapter will be that first we will briefly introduce some concepts of general polymer solution theory, with focus on the PEG-water system. This will be followed by a section dealing with molecular modeling of the PEG-water system. The knowledge, that we have presented in these two sections, will then be used to discuss the effect of different additives to a water solution of PEG, the behavior of a surface coated with PEG and finally the interaction between a protein and a solution containing PEG.

General Polymer Solution Theory.

The objective of this chapter is to see what information about PEG in general and about the PEG - water system in particular can be obtained from comparison of results from theoretical modeling of PEG solutions (and similar systems) with experimental data. Despite the fact that the focus should be placed on the physical properties and behavior of the PEG molecules, it is necessary to start by a short introduction into general polymer - solution modeling. The purpose of this is not to give a full account of what has been done within the field but rather to introduce the reader to the benefits and shortcomings of the three main strategies used to study polymers in solution.

Virial Expansion Techniques. The starting point for this type of modeling is to disregard the solvent and focus only on the effective interactions between the polymer molecules *(12)*. The concept of virial coefficients was originally developed to describe the deviation from ideal behavior for gases. For one mole of an ideal gas we may write

$$pV = RT \tag{1}$$

and for a real gas the following relation holds to a good approximation

$$pV = RT\left(1 + \frac{B}{V} + \frac{C}{V^2} + \dots \right) \tag{2}$$

In these equations p is the pressure, V the volume, R the gas constant, and T the absolute temperature. The constants B and C are called the second and third virial coefficients and they measure the deviation from ideal behavior. As we will see below the constants B and C are only constants at a fixed temperature, and in reality they vary quite strongly with temperature. To get some idea about the physics behind these constants, it is convenient to start with van der Waal's gas law for one mole of gas.

$$\left(p + \frac{a}{V^2} \right)(V - b) = RT \tag{3}$$

In this equation b is a correction to the ideal gas law due to the fact that each molecule has a volume and that consequently only the volume (V - b) is available for the molecules. Thus b measures the volume of one mole of molecules. The other correction term (a) originates from the facts that the molecules attract each other and that this reduces the pressure. In order to obtain the pressure of the ideal gas law this reduction of the pressure must be added to the experimentally observed pressure. Consequently a is a measure of the attraction between molecules. From what has been said above it is clear that both a and b are positive constants. If van der Waal's gas law is transformed to a form similar to the virial expansion one obtains

$$pV = RT\left(1 + \frac{pb}{RT} - \frac{a}{VRT} - \frac{ab}{V^2RT} \right) \tag{4}$$

In the second term in the parentheses p/RT can be substituted by V using the ideal gas law and one obtains

$$pV = RT\left(1 + \frac{b}{V} - \frac{a}{VRT} + \dots \right) \tag{5}$$

Comparing equations 2 and 5 we see that B = b - a/RT. Here we can identify two limiting cases. At sufficiently low temperatures a will dominate and the second virial coefficient will be negative, whereas at sufficiently high temperatures b will dominate, resulting in a positive second virial coefficient. A negative virial coefficient means that there is an effective attraction between the molecules and a positive virial coefficient that there is an effective repulsion between the molecules. In terms of the phase behavior this means that at low temperatures, where the second virial coefficient is negative one observes a liquid phase in equilibrium with a gas phase, and at high temperatures where the second virial coefficient is positive one observes only a gas phase.

The link between virial expansions for gases and for polymers in solution is straight forward. In this type of model one can neglect the solvent, since it only is a medium in which the polymer molecules are moving, in a similar way to the movements of gas molecules in a vacuum. There are however also some differences.

For the gas molecules we know that there is always a long-range attraction between the molecules, i.e. a is positive. For the polymer solution this is not necessarily true. In principle a may be negative. In general it is true, however, that at sufficiently short distances polymer molecules will repel each other and this means that at sufficiently high temperatures the polymers will dissolve in the solvent, (provided that the solvent is still a liquid and the polymer does not disintegrate). From this we may conclude that one would expect that the solubility of polymers in a solvent should increase with increasing temperature and that the second virial coefficient should increase with increasing temperature. This type of analysis has been performed for the water - dextran - PEG system by Edmond and Ogston *(13,14)*, and an analysis of the PEG - water system reveals that this system behaves abnormally. The effective interaction between PEG molecules, as measured by the second virial coefficient, becomes more attractive at higher temperatures, even when it is measured in units of RT. A similar type of thermodynamic analysis of the PEG -water system have been made by Kjellander and Florin *(15,16)*.

The important message from this type of analysis is that there are some anomalies in the PEG-water system. The analysis as such is based on macroscopic concepts and provides very little information about the mechanism on a molecular level. It was observed by Kjellander and Florin-Robertsson, however, that the thermodynamics of the PEG-water system was similar to that observed in water solutions of hydrocarbons and other non polar molecules. For systems of that type one normally observes a solubility minimum in water at temperatures close to 20 $^{\circ}$ C. The phenomenon is normally referred to as the hydrophobic effect, and is well described in the book with the same name by Tanford *(17)*, and more recently by Israelachvili and Wennerström *(18)*. The basic idea is that at low temperatures, the solubility of non-polar substances is increased due to a formation of a relatively ordered water layer around the non polar substance. The structure is preserved in order to maintain as many hydrogen bonds between the water molecules as possible, despite the presence of the solute. This is energetically favorably compared to a non-structured solvation of the hydrophobic substance, but associated with an entropic cost. At higher temperatures where entropy becomes more important the solubility decreases, or increases less than could be expected from the temperature increase. There are two weak points when this model is used to explain the observed behavior of the PEG - water system. First of all the clouding occurs in a temperature range (100 - 200 $^{\circ}$C) where an increased solubility is observed for ordinary non-polar substances. Second the concept of "structured water" is not uniquely defined on a molecular level, and consequently it is not possible to design microscopic molecular models that can reproduce what is experimentally observed, or one may say that the model can not be proved false since it does not allow for any predictions.

Microscopic Models of Clouding in the PEG - Water System.

From the previous section we learned that the phase separation observed at higher temperatures is a manifestation of an increased effective attraction between the EO segments. This feature must be handled by any model designed to describe the observed phase behavior. The ideal entropy of mixing always favors the formation of a one-phase system and this effect will always be larger when the temperature is increased. Thus there must be another effect, with entropic origin, that is stronger than the mixing entropy. In the literature there exist two different microscopical explanations that will be reviewed here. The older one is based on the work by Hirschfelder, Stevenson and Eyring from 1937 *(19)*, where the authors show that if the intermolecular potential between two species has small, strongly attractive regions and large repulsive ones, then the effective interaction between two substances may change from being attractive to being repulsive at higher temperatures. It is rather easy to associate the attractive regions with hydrogen bonds and the other regions with non hydrogen bonded interaction. An explanation along these lines was first given, for the PEG - water system, by Lang *(20)* and later Goldstein *(21)* gave it a mathematical formulation.

The other microscopic model for the phase separation at elevated temperatures focuses on the conformational degrees of freedom of the PEG chain *(22)*. Based on the notation that some of the conformations are more polar than others, one realizes that the different conformations must interact with the solvent in different ways. (The notation more polar means here, that the local dipole density is higher than for a less polar conformation.) A weak point in this latter model is that it requires that there are more non-polar conformations than there are polar ones. Quantum chemical calculations on DME show that most of the relevant conformations have a dipole moment of 1.6 to 1.8 D, and the only conformation with low energy that lacks dipole moment is the one which is anti around the two C-O bonds as well as around the C-C bond.

Today there exist two distinct ways to model a polymer solution. The most straight forward and in some sense the most accurate of these is to perform all molecular Monte Carlo (MC) or Molecular Dynamics (MD) simulations of the liquid polymer solution *(23)*. The advantage of this type of modeling is that it gives a full description of the probability distributions for the studied system, provided that the intermolecular potentials employed are sufficiently accurate enough. The drawback is that it is not possible at present to study systems that are so large that accurate thermodynamic properties can be determined and phase diagrams calculated.

Since this is exactly what we initially are focused on in this chapter we are forced to turn to the other type of thermodynamical modeling based on lattice theory. This type of model is normally given the name Flory - Huggins theory after Flory and Huggins, who independently developed the model *(24)*. The model gives an expression for the entropy and energy of mixing between a solvent and a solute and is derived by looking for possible conformations of a polymer chain on a lattice. For a full derivation see reference *(25)*. The basic assumptions behind the theory are that a polymer segment and a solvent molecule have the same size, that they are randomly mixed, apart from the connectivity of the polymer chain, and that the enthalpic part of the polymer - solvent interaction depends only on the solvent and solute concentrations. The following expressions are obtained

$$A_{mix} = U_{mix} - T S_{mix} \tag{6}$$

$$U_{mix} = n_{tot} \, \psi_1 \, \psi_2 \, W_{12} \tag{7}$$

$$S_{mix} = - R \, n_{tot} \left(\psi_1 \ln \psi_1 + \frac{\psi_2}{M} \ln \psi_2 \right) \tag{8}$$

In these equations A_{mix} is Helmholtz free energy of mixing for the polymer - water system. In the same way U and S are the internal energy and entropy. n_{tot} means the total number of moles of entities, where an entity is either a solvent molecule or a polymer segment. φ_1 and φ_2 are the volume fractions of the solvent and the polymer, M the degree of polymerization of the polymer and finally w_{12} measures the effective interaction between the solvent and polymer molecules. The model, as was discussed above, is based on a lattice, and consequently it reflects the behavior at constant volume, but frequently U is replaced with H (the enthalpy) and A with G (Gibbs free energy). The difference is extremely small for condensed phases. The crucial parameter, which controls the phase behavior is w_{12}. For negative or not too positive values, the polymer and the solvent are completely miscible at all concentrations, but for larger positive values a two phase-system represents the most stable solution. (Negative w_{12} means effective attraction between the polymer and the solvent). The absolute value for w_{12} where the system changes from being a one-phase system to a two-phase system at some concentrations depends on the degree of polymerization M. For very long polymer chains it approaches 0.5 RT, and in the limit when M equals 1 it takes the value 2 RT. As is indicated in the text w_{12} is an effective interaction. This means that the interaction is averaged over all relative orientations and positions. Further it is a difference according to

$$U_{mix} = U_{solution} - U_{solvent} - U_{polymer} \tag{9}$$

where

$$U_{solution} = 0.5 \, (\, n_1 \, \psi_1 \, w'_{11} + n_2 \, \psi_2 \, w'_{22} + 2 \, n_1 \, \psi_2 \, w'_{12} \,) \tag{10}$$

$$U_{solvent} = 0.5 \, n_1 \, w'_{11} \tag{11}$$

$$U_{polymer} = 0.5 \, n_2 \, w'_{22} \tag{12}$$

Using equation 4 − 7 and $\psi_1 + \psi_2 = 1$ one easily obtains

$$U_{mix} = 0.5 \, n_1 \, \psi_2 \, (\, 2 \, w'_{12} - w'_{11} - w'_{22} \,) \tag{13}$$

or

$$w_{12} = w'_{12} - 0.5 \, (\, w'_{11} + w'_{22} \,) \tag{14}$$

From equation 14 it is clear that w_{12} depends on the solvent - solvent interaction, w'_{11} and the solute - solute w'_{22} interaction, as well as on the direct solvent - solute interaction w'_{12}. If equations 6 - 8 are analyzed it is seen that w_{12} must increase more than linear with T in order for phase separation to occur as the temperature is increased. This is a rather unusual situation, since normally one would expect the interaction parameters to decrease at elevated temperatures. This fairly lengthy discussion about the interaction parameter may seem a bit academic, but it is necessary in order to illuminate the physics behind the clouding. We will conclude this section by analyzing the existing explanations in terms of the interaction parameters. Obviously the hydrogen bond explanation by Goldstein focuses on w'_{12} *(21)*. On the other hand the water structure models are centered around w'_{11} describing the water - water interaction *(15,16)*, whereas in the conformational model it is suggested that w'_{12} and w'_{22} are mainly responsible for the observed phase behavior *(22)*. In order to define a scheme to calculate the actual temperature dependence of the interaction parameter w_{12}, a molecular model is needed. Two such models exists in the literature. Below the formalism and the physics in these models will be described.

The Hydrogen Bond Model. The basic ingredient in this model is that there may or may not be a hydrogen bond between a water molecule and a segment of the PEG chain *(21)*. These two types of interaction are characterized by two interaction parameters w^{hb}_{12} and w^{nhb}_{12}, with the superscripts designating hydrogen bond and non-hydrogen bond, respectively. The probability at infinite temperature for a hydrogen bond and the probability for a non-hydrogen bond P_{inf}^{hb} and P_{inf}^{nhb} must also be specified. The sum of these two probabilities is 1 at all temperatures. The ratio of these two probabilities, we will denote g. If we introduce P as a variable describing the actual probability that contact at a given temperature between a water molecule and a PEG segment is of hydrogen bonded type, Then we may write the following equations for the free energy of the system:

$$U_{mix} = n_{tot}\, \varphi_1\, \varphi_2\, (\, P\, w^{hb}_{12} + (1 - P)\, w^{nhb}_{12}\,) \tag{15}$$

$$S_{mix} = - R\, n_{tot} \left(\varphi_1 \ln \varphi_1 + \frac{\varphi_2}{M} \ln \varphi_2 \right)$$
$$- R\, n_{tot}\, \varphi_1\, \varphi_2 \left(P \ln P + (1 - P) \ln \frac{1 - P}{g} \right) \tag{16}$$

The total free energy (A) is defined according to equation 6, and the variable P is chosen in such a way that the total free energy is minimized. When A is minimized with respect to P one obtains a minimum for larger P values at lower temperatures provided that w^{hb}_{12} is smaller (more negative or less positive) than w^{nhb}_{12}. This means that the effective interaction parameter between water and the solute becomes more repulsive at larger temperatures, and if f is large enough (larger than 2) phase separation may occur at elevated temperatures.   A model of this type was first suggested as an explanation to the clouding in the PEG-water system by Goldstein in 1984 *(21)*.   The mathematical formalism may look different in Goldstein´s work from what is described here, but the physics and the phase behavior are equivalent.

The Polymer Conformational Model.   In the conformational model it is recognized that there must exist a conformational equilibrium for the segments in the polymer chain, and that different local structures interact with the solvent (water) in different ways *(22)*. In some sense it is similar to the water structure model which assumes that different water molecules may be ordered relative each other in at least two different ways, icebergs and disordered.   In a similar way the conformational model divides up the local polymer conformations in polar and non-polar (unpolar) conformations.   It is further assumed that the polar conformations interact more favorably with water than the non-polar ones, but that the number of non-polar conformations is more numerous than the polar ones.   If we define the six interaction parameters w'_{11}, w'^{pw}_{12}, w'^{uw}_{12}, w'^{pp}_{22}, w'^{pu}_{22} and w'^{uu}_{22}, where the superscripts designate means p =polar, w = water, and u = unpolar and use the notation f for the ratio of the number of non polar to polar conformations, then the following expression is obtained for the energy   of mixing.

$$U_{mix} = n_{tot} (0.5\ \varphi_1\varphi_1 w'_{11} + \varphi_1\varphi_2 (P\ w'^{wp}_{12} + (1 - P)\ w'^{wu}_{12}) +$$
$$0.5\ \varphi_2\varphi_2 (P^2\ w'^{pp}_{22} + 2\,P\,(1 - P)\ w'^{pu}_{22} + (1 - P)^2\ w'^{uu}_{22})) \qquad (17)$$

In this equation we have used the interaction parameters, which describes the interaction between the different species.   This expression can be simplified in the same way as was used in equations 9 to 14 in the classical Flory - Huggins theory. When this is done one realizes that only one of the two diagonal terms w'^{pp}_{22} and w'^{uu}_{22} can be removed.   The other term is needed to describe the energetic difference between the polar and non polar conformers of the polymer chain.   Using some algebra equation 17 simplifies to

$$U_{mix} = n_{tot} [\ \varphi_1\varphi_2 \{ P\ w^{wp}_{12} + (1 - P)\ w^{wu}_{12} \} +$$
$$\varphi_2\varphi_2 \{\ P\,(1 - P)\ w^{pu}_{12} + 0.5\,(1 - P)^2\ w^{uu}_{22} \}] \qquad (18)$$

In equation 18 the primed interaction parameters have been replaced with non primed parameters defined according to equation 14.   The interaction parameter w^{uu}_{22} has been kept in order to describe the energetic contribution to the equilibrium between the polar and non-polar conformations.   The entropic contribution to the free energy is almost similar to that in the hydrogen bond model (equation 16) and it differs only by the lack of a factor φ_1 in the last term.

$$S_{mix} = -R\,n_{tot}\left[\left(\,\phi_1\ln\phi_1 + \frac{\phi_2}{M}\ln\phi_2\,\right)\right.$$
$$\left. + \phi_2\left(\,P\ln P + (1-P)\ln\frac{1-P}{f}\,\right)\right]$$
                                                                        (19)

Both the conformational and the hydrogen bond models predict that the effective interaction between the polymer and the solvent is temperature dependent and more repulsive at higher temperatures. The conformational model also suggests that the polymer solvent interaction becomes more repulsive at higher polymer concentrations. A weak point in the conformational model however is that the number of non-polar conformations must be as least twice as large as the number of non-polar ones (f must be at least 2) in order that a reasonable phase diagram could be calculated from the model. If the simplest possible model for PEG (DME) is analyzed one finds that out of the 27 different conformations that are in principle possible only 23 can be expected to be of any importance for the description of the system. But only three of these 23 conformations have a dipole moment that is significantly different from the others. These are the anti-anti-anti (aaa) conformation and two gauche anti gauche conformations (gag',g'ag), which lack dipole moment. The label a means that there is an anti conformation around that particular bond and the label g indicates a gauche conformation. The label g' indicates that this gauche angle is rotated in opposite direction to the previous gauche angle. Here we have used the notation to specify a conformation with labels for the dihedral angles. The dihedral angles are ordered consecutive starting from one end of the molecule. This means that the first label refers to the first C-O bond, and that the second label specifies the conformation around the C-C bond. All other conformations have a dipole moment that is close to 1.6 D. Naturally the situation will be different for longer chains but this observation, discussed above, indicates a problem with the mechanism.

Today a large amount of experimental data are available and they indicate that the polarity of the PEG polymer decreases at increased temperature as well as in less polar solvents *(26,27)*. This was briefly discussed in the introduction.

First Principle Modeling of 1,2-Dimethoxyethane in Solution.

From the preceding section it is clear that knowledge of the conformational equilibrium of a PEG chain in solution is of vital importance for our understanding of the solution properties of PEG. The first observation we must make in this context is that only very short chains, such as the dimer DME and the trimer diglyme can be treated by theoretical means if the relevant parts of the phase space should be sampled. The purpose of this section is to give a short account of the findings of different first principle approaches to the problem. We will divide this discussion into two parts; the first of these will deal with so called continuum models, where the studied molecule is placed in a cavity in a dielectric medium and a set of quantum chemical equations describing the molecular system are solved together with a set of equations describing the polarization of the dielectric medium. In the second approach statistic mechanical means are used to model DME in a molecular solvent (water). In the first type of approach the focus is placed on the solute molecule and only a very crude description of the medium is included. The second approach is more balanced in that both the solute and the solvent are treated in an equal manner. Before starting these discussions we must briefly discuss the simpler problem of the conformational behavior of DME in vacuum.

Previously it was mentioned that 27 different conformations of DME may exist. These conformations can be divided into 10 classes. Both from experimental and theoretical considerations it is clear that there are four different classes of conformations that are more important than others. These are the aaa, aga, aag, and the agg' conformations. We further note that the multiplicity of the different conformations differ. Thus there are only one aaa conformation, two aga conformations, four aag conformations and also four agg' conformations. This causes no problem regarding the stability of the different conformations in the gas phase, since this is normally characterized by the energy of each conformation. To determine the relative stability of the conformers in a condensed phase one normally uses the free energy of the conformation and now it is necessary to differ between the free energy of the conformation and the free energy of the class of conformations. Typically there are differences of a factor RTln 2, and RTln 4, between the free energy for a conformation and the free energy for a class of conformations, depending on the number of conformations that exists in a class of conformations. Nevertheless electron diffraction measurements indicate that several different conformations are present in the gas phase *(28)*. Gas phase NMR studies are a bit more conclusive and they suggests that the aga conformation has the lowest energy, followed by the aaa and the agg' conformations *(29)*. Gas phase IR measurements however, indicate that aaa is the most stable one, followed by the agg' and the aga conformations *(30)*. In fact the determination of the lowest energy conformation of DME in the gas phase is a problem where theoretical *ab initio* calculations have proved to be more reliable than experiment. Basically all high quality calculations predict the aaa conformations to be the most stable one, followed by the agg' and the aga conformations *(31,32,33)*. The relative energies are that the agg' has an energy of 0.1 kcal/mole, the aga 0.2 kcal/mole and finally the aag has an energy of 1.4 kcal/mole according to the most accurate calculations performed so far *(31)*. Since the energy difference is so small and there are more of the agg' and the aga conformations, this means that the agg' is the most probable type of conformation followed by the aga conformation, and that the aaa conformation is only slightly more probable at room temperature than the aag conformation.

The theoretical description of conformational equilibria in condensed phases is more difficult. In the simplest approach, the effect of the medium is modeled as a dielectric continuum. This type of study on DME, where the focus is on the conformation around the central C - C bond has been performed by several groups *(34,35,36)*. The main conclusion from these types of studies is that a polar medium favors a gauche conformation around the C - C bond. In physical terms this is easy to understand. When DME has a gauche conformation around the central bond then the molecule has a dipole moment and this leads to an increased stabilization when the dipole is allowed to polarize the dielectric surrounding. The actual results obtained from this type of modeling are of limited value, since the models contain parameters, and the results depend fairly strongly on these. In this context it should also be noted that this type of model, always fails to reproduce a decreased solubility at higher temperatures.

Potentially more interesting are the statistical mechanical studies of DME and PEG in a molecular aqueous solution. Several reports of this type have been published, one deals with a DME molecule in water *(37)*, and two deal with a PEG molecule in water *(37,39)*. Two other reports treat the head group of a nonionic surfactant of PEG type in water *(40,41)*. The quality of the different studies of PEG is poor and not all conformations have been covered in the simulations. In practice this means that the results cannot be trusted. The first study *(37)*, which is focused on the conformation around the central C - C bond in the molecule, shows that the aqueous solution stabilizes the aga conformation with 0.9 kcal/mole more than the aaa conformation. In our laboratory we have also studied the conformational equilibrium of DME in water *(42)*. The investigation clearly shows that the stabilization of the polar conformers in aqueous solution has an energetic origin and that the non-polar aaa conformation is entropically favored. In practice this means that it is only at room temperature that the polar conformations are favored by the solvent. Our simulations indicate that the free energy is lowered by 0.5 and 1.8 kcal/mol for the aga and aag conformations, respectively, relative to the aaa conformation at 298 K. At 398 K, close to the clouding temperature of the PEG - water system, the corresponding figures are -0.3 and 1.1 kcal/mol. This means that at this temperature the medium favors the aaa conformation relative the aga conformations, and that only an aag conformation is still more stable than the aaa conformation. Moreover, the specific heats for the different conformations indicate that the relative amount of the aaa conformation is rapidly increasing with temperature. In order to get further insight into the basis of this increase in the aaa conformation we have analyzed the energetics and found that the energetic stabilization of the polar conformations mainly are due to an improved water - water interaction, and only to a smaller extent (• 25%) due to a direct interaction between the solvent and the solute. We have also found that the effect of polarization of the electrons in the solvent and the solute are of almost no importance for the difference in solvation energies between the different conformations.

A conclusion that can be drawn from the statistic mechanical simulations of DME seems to be that the clouding behavior of the PEG - water system is closely linked to the conformational equilibrium of the PEG chain.

Influence of Additives to the PEG - Water System.

In the previous section we have seen that there seems to be a link between the polymer conformation and the behavior of the polymer in aqueous solution, but we have also seen that the clouding process could equally well be modeled as due to a hydrogen bond model. The mathematical formulation of these two models are almost the same, and many kinds of behavior that could be modeled by one of these schemes could be modeled by the other. The discussion that will be carried through below will be formulated in terms of the conformational model. Most predictions, however, apart from the polymer conformation, could be obtained from either of the models.

It has long been well known that addition of salt and other additives affects the phase diagram for the PEG-water system. Perhaps the most extreme effects are obtained if a salt like potassium phosphate is added to the PEG-water system *(43)*. For this type of system a two phases occur already at room temperature. One might say that the clouding temperature is lowered down to this temperature. An alternative way to interpret the experimental data is to say that there is a repulsion between the PEG molecules and the ions and that this causes a phase separation into two phases in the same way as addition of a second polymer, e.g. dextran would do *(43)*. If a general polymer - solvent system is analyzed according to the Flory-Huggins model, one can show that provided that $-0.5RT < w_{sa} - w_{pa} < 1.5\ RT$, then addition of a monomeric substance will stabilize the one-phase region. If on the other hand $w_{sa} - w_{pa} < -0.5RT$ or $1.5\ RT < w_{sa} - w_{pa}$, addition of the substance will stabilize the two-phase solution relative to the one-phase solution *(44)*. In these inequalities, the subscript s means solvent, p polymer and a additive. In physical terms this means that if the additive prefers the polymer with 0.5 RT relative to the solvent or the additive prefers the solvent with 1.5 RT relative to the polymer, then the added substance will favor phase separation, and the added substance will be found mainly in the phase for which it has an energetic preference. We thus see that clouding may be induced either by adding substances that are more polar than the water or by adding substances that are less polar than the polymer *(44)*. If we further adopt the two-conformation model used to describe the PEG-water system, it is easy to deduce that addition of a polar substance like NaCl will favor the polar conformations and that additives that are less polar promote the non-polar conformations and lower the clouding temperature. These predictions, for the conformation of the PEG-chain, have been verified by ^{13}C NMR measurements, which show that the polarity of the PEG chain decreases with decreasing polarity of the solvent and increasing temperature *(26,27)*.

PEG—coated surfaces.

Quite frequently PEG is used to coat surfaces in order to prevent proteins or other macromolecular material from depositing on the surface. The general mechanism behind the effective repulsion of the macromolecule from the coated surface is that if the macromolecule comes close to the surface, then the conformational degrees of freedom for the polymer are drastically reduced, and this causes an entropic repulsion between the surface and the macromolecule. The theoretical description of polymer molecules outside a surface, within the Flory - Huggins concept, has been developed by Scheeutjens and Fleer (SF) *(45,46)*. The essence of the SF model is that the space is divided into layers parallel to the surface and that the Flory - Huggins model is applied to each layer. The SF model is easily merged with the conformational model *(47)*. The outcome of this type of modeling is quite interesting and explains to some extent why PEG has been found to be useful for coating surfaces. The first and most important property of the PEG chain is that it is flexible and that the entropic repulsion consequently will be large. The PEG chain does however have some additional nice features which will be of importance for hydrophobic surfaces. The ability to change polarity of the chain, implies that close to a hydrophobic surface, the polymer will have more non polar conformations, and this leads to an increased coverage of the surface, whereas far from the surface out in the bulk, the polymer will be more polar. This will have as a consequence that the PEG chains penetrate far out into the water, implying a stronger repulsion of macromolecules. This type of behavior is very difficult to verify experimentally. However, Ahlnäs et al. *(48)* showed that for micelles formed from nonionic surfactants of PEG-type forming micelles, ^{13}C NMR measurements indicate that the conformation preferred by the chain changes from being less polar close to the hydrocarbon core to more polar for the segments closest to the hydrophilic end of the molecule.

Protein - PEG interactions in aqueous solution.

One of the main applications of PEG in biochemistry is for purification using aqueous two-phase systems. This was first suggested and tested by Albertsson in 1956 *(43,49)*. By now the literature on the subject is extensive and it is not the purpose of this work to review it. The purpose is instead to analyze the effective interaction between the protein and the polymer on a molecular level and investigate to what extent it is possible to understand the mechanisms governing the partitioning of the protein between the two phases. We will use the machinery that has been established in the previous sections. The starting point for the discussion is that we have two polymers which we can call P1 and P2. When these polymers are dissolved in water they form two phases. The first phase contains most of P1 and a small amount of P2, whereas the second phase contains the major part of P2 and some P1. Before distributing the protein between the two phases, we must understand how the water is distributed. First we note that both energy and entropy effects are of importance for the partitioning of the water. Obviously water prefers to be in the phase that contains the polymer with which it interacts most favorably, provided that the entropic contribution can be neglected. In the Flory-Huggins model this is equivalent to saying that the water prefers the polymer with the smallest interaction parameter w (least positive or largest negative). The entropic contribution on the other hand will be most favorable for the polymer with the lowest degree of polymerization. In the Flory-Huggins model this means the polymer with the smallest value of M.

To proceed we will initially assume that the protein lacks net charge. This means that it can be moved between the two phases without moving any counter-ions. With this assumption it is easily shown that the entropic contribution to the partitioning is such that it always favors partitioning of the protein to the water richer of the two polymer phases. The energetic contribution is more complex, since it depends on the difference in protein-polymer interaction between the two phases, but also on the difference in protein-water interaction compared to the protein-polymer interactions. This first contribution obviously favors the phase with the least repulsive protein-polymer interaction. Normally the energetic interaction between the water and the proteins are more repulsive than the polymer-protein interactions. This has the effect that the protein is directed to the water-poor phase. The outcome of these three effects may vary from protein to protein. Normally, for hydrophilic proteins both energetic and entropic contributions favors partitioning to the phase with the more polar polymer. If, however, the molecular weight of the more polar polymer is much higher than that of the less polar polymer then the entropic contributions may favor the other polymer phase. For more hydrophobic proteins the energetic contribution normally favors the less polar polymer phase.

Normally a protein is charged and this complicates the prediction of the partitioning of the protein, but it also introduces a possibility to control it. The basic mechanisms for this originates from a difference in affinity of an ion for the different phases. To illustrate this, we will consider a two phase system, where phase 1 contains a polar polymer and a lot of water. The other phase, phase 2, contains less water and a less polar polymer. Small hydrophilic ions prefer phase 1 and large less hydrophilic ions prefer phase 2. Thus if tetramethylammonium fluoride and sodium perchlorate are dissolved in these two phases, sodium fluoride will accumulate in phase 1 and tetramethylammonium perchlorate will be found mainly in phase 2. If one wishes to direct the protein to one particular phase, one should ensure that the counter-ion to the protein has a preference for that phase and that the co-ion has a preference for the other phase. These questions have been discussed by Johansson *(50)*.

Literature Cited

1. Bailey, F. E.; Koleske, J. V. *Poly(ethylene oxide);* Academic Press: New York, 1976.
2. *Poly(ethylene glycol) Chemistry, Biotechnical and Biomedical Applications* Harris, J. M., Ed.; Plenum: New York, 1992.
3. Saeki, S.; Kuwahara, N.; Nakata, M.; Kaneko, M. *Polymer* **1976**, *17*, 685.
4. Samii, A.; Karlström, G.; Lindman, B. *Langmuir* **1991**, *7*, 653.
5. Wärnheim, T.; Bokström, J.; Williams, Y. *Colloid Polym. Sci.* **1988**, *266*, 562.
6. Podo, F.; Nemethy, G; Indovina, P. L.; Radics, L.; Viti, V. *Mol. Phys.* **1974**, *27*, 521.
7. Viti, V.; Indovina, P. L.; Podo, F.; Radics, L.; Nemethy, G. *Mol. Phys.* **1974**, *27*, 541.
8. Tasaki, K.; Abe, A. *Polymer J.* **1985**, *17*, 641.
9. Matsuura, H.; Fukuhara, K. *J. Mol. Struct.* **1985**, *126*, 251.
10. Matsuura, H.; Sagawa, T. *J. Mol. Liq.* **1995**, *65/66*, 313.
11. Masatoki, S.; Takamura, M.; Matsuura, H.; Kamogawa, K.; Kitagawa, T. *Chem. Lett.* **1995**, 991.
12. Richards, E. G. *An Introduction to Physical Properties of Large Molecules in Solution;* Cambridge University Press: Cambridge, 1980.
13. Edmond, E.; Ogston, A. G. *Biochem. J.* **1968**, *109*, 659.
14. Ogston, A. G. In *Chemistry and Technology of Water Soluble Polymers;* Editor, Finch; C.A., Plenum Press: New York, 1983, p 659.
15. Kjellander, R.; Florin Robertsson, E. *J. Chem. Soc. Faraday Trans. 1* **1981**, *77*, 2053.
16. Florin-Robertsson, E. *Theoretical and experimental investigations of aqueous poly(ethylene oxide) solutions;* Thesis, The Royal Institute of Technology, Stockholm, 1983.
17. Tanford, C. *The Hydrophobic Effect;* John Wiley & Sons: New York, 1973.
18. Israelachvili, J.; Wennerström, H. *Nature* **1996**, *379*, 219.
19. Hirschfelder, J.; Stevenson, D.; Eyring, H. *J. Chem. Phys.* **1937**, *5*, 896.
20. Lang, J. C.; Morgan, R. D. *J. Chem. Phys.* **1980**, *73*, 5849.
21. Goldstein, R. E. *J. Chem. Phys.* **1984**, *80*, 5340.
22. Karlström, G. *J. Phys. Chem.* **1985**, *89*, 4962.
23. Allen, M. P.; Tildesey, D. J. *Computer Simulation of Liquids;* Clarendon: Oxford, 1987.

24. Flory, P.J. *J.Chem.Phys.* **1942**, *10*, 51.
25. Hill, T. L. *An Introduction to Statistical Thermodynamics;* Dover: New York, 1986.
26. Björling, M.; Karlström, G.; Linse, P. *J. Phys. Chem.* **1991**, *95*, 6706.
27. Björling, M. *Polymers at Interfaces;* Thesis: University of Lund, Lund, Sweden 1991.
28. Astrup, E. E. *Acta Chem. Scand.* **1979**, *A33*, 655.
29. Inomata, K.; Abe, A. *J. Phys. Chem.* **1992**, *96*, 7935.
30. Yoshida, H.; Kaneko, I.; Matsuura, H.; Ogawa, Y.; Tasumi, M. *Chem. Phys. Lett.* **192**, *196*, 601.
31. Jaffe, R. L.; Smith, G. D.; Yoon, D. Y. *J. Phys. Chem.* **1993**, *97*, 12745.
32. Tsuzuki, S.; Uchimaru, T.; Tanabe, K.; Hirano, T. *J. Phys. Chem.* **1993**, *97*, 1346.
33. Murcko, M. A.; DiPaola, R. A. *J. Am. Chem. Soc.* **1992**, *114*, 1010.
34. Andersson, M.; Karlström, G. *J. Phys. Chem.* **1985**, *89*, 4957.
35. Müller-Plathe, F.; van Gunsteren, W. F. *Macromolecules* **1994**, *27*, 6040.
36. Williams, D. J.; Hall, K. B. *J. Phys. Chem.* **1996**, *100*, 8224.
37. Liu, H.; Müller-Plathe, F.; van Gunsteren, W. F. *J. Chem. Phys.* **1995**, *102*, 1723.
38. Depner, M.; Schürmann, B. L.; Auriemma, F. *Mol. Phys.* **1991**, *74*, 715.
39. Tasaki, K. *J. Am. Chem. Soc.* **1996**, *118*, 8459.
40. Kong, Y. C.; Nicholson, D.; Parsonage, N. G.; Thompson, L. *J. Chem. Soc. Faraday Trans.* **1994**, *90*, 2375.
41. Kong, Y. C.; Nicholson, D.; Parsonage, N. G.; Thompson, L. *J. Chem. Soc. Faraday Trans.* **1995**, *91*, 4261.
42. Engkvist, O.; Karlström, G. *J. Chem. Phys.* 1997, *106*, 0000.
43. Albertsson, P-Å. *Partition of Cell Particles and Macromolecules, 2nd Ed.;* Almqvist & Wiksell: Stockholm, 1971.
44. Johansson, H-O.; Karlström, G.; Tjerneld, F. *Macromolecules* **1993**, *26*, 4478.
45. Scheutjens, J. M. H. M.; Fleer, G. J. *J. Phys. Chem.* **1979**, *83*, 1619.
46. Scheutjens, J. M. H. M.; Fleer, G. J. *J. Phys. Chem.* **1980**, *84*, 1882.
47. Björling, M.; Karlström, G.; Linse, P. *J. Phys. Chem.* **1990**, *94*, 471.
48. Ahlnäs, T.; Karlström, G.; Lindman, B. *J. Phys. Chem.* **1987**, *91*, 4030.
49. Albertsson, P-Å. *Nature* **1956**, *177*, 771.
50. Johansson, H-O.; Karlström, G.; Mattiasson, B.; Tjerneld, F. *Bioseparation* **1995**, *5*, 269.

Chapter 3

The Conformation of Polymers at Interfaces

Danilo D. Lasic

Liposome Consultations, 7512 Birkdale Drive, Newark, CA 94560

Coating of various surfaces and colloidal particles with flexible polymers greatly reduces adsorption and increases biological compatibility of surfaces and stability of colloidal particles. In addition, the immunogenicity and antigenicity of such particles decreases drastically. These findings are extremely important in the preparation of non-fouling and biocompatible surfaces. In the case of colloidal solutions, PEGylated proteins and liposomes are commercially available as novel medical formulations due to improved biological stability.
The behavior of flexible polymers attached to the surface can be explained theoretically by the scaling theory. In this article this theoretical model as well as some experiments using osmotic stress technique and surface force apparatus are reviewed. Experiments show that this simplified theory fits experimental data as well as more complex treatments, indicating that general properties of such systems can be modeled by a rather simple theoretical approach.

Polymer coating of various surfaces and colloidal particles has a long history of thousand years in applications such as in preparation of paints, inks, food, as well as pharmaceutical and cosmetic preparations. After some empirical explanations at the beginning of this century only recently theoretical understanding of increased stability and improved dispersivity of these systems started to emerge. Here we shall briefly review scaling theory and only refer to more complex mean field models.

The problem can be especially critical in medical applications in which either larger surfaces, such as needles, patches, pumps, or artificial colloidal particles come into contact with body fluids, especially blood. The majority of materials with appropriate mechanical properties lack biocompatibility and surface/particle coating with appropriate polymer coatings which reduce protein adsorption, toxicity and immunogenicity.

Poly(ethylene glycol), which is also frequently referred to as poly(ethylene oxide), is the most widely used polymer for surface modification. In addition to the coating of surfaces and colloidal particles, technology was also developed to bind PEG chains on some proteins and encapsulate them in a PEG coating without substantial loss of activity. This treatment, does not only reduce antigenicity of these proteins but due to increase in the size eliminates their secretion by kidneys, which resembles if the size of the particles is compared to globular proteins - dialysis with a molecular cut off of ca 60 000 Da. Here we shall very briefly review the theoretical origin of surface polymer stabilization of surfaces and colloidal particles and briefly present some experimental data.

Polymers at Interfaces: Theory and Experiment

In this article we shall discuss the behavior of polymers at interfaces. We shall mostly concentrate on nonionic polymers grafted to the surface of colloidal particles and discuss some properties of sterically stabilized vesicles coated with polyethylene glycol. In the range of distances from 1-100 nm a relatively simple scaling arguments can be used to describe such systems. We shall try to explain the polymer behavior by using scaling laws which are independent of molecular details for polymers in a good solvent. For detailed introduction to scaling concepts an interested reader is referred to ref. *1-3*.

Colloidal particles and other surfaces immersed in polymer solution can either attract or repel polymers and adsorption and depletion layers can be formed on the surface, depending on the sign of interaction of polymer with the surface (*3*). Interaction at the surface can be changed by coating the surface with a polymer. Because it is difficult to control polymer adsorption, surface grafting is preferred. In the case of liposomes, self-organized colloidal particles composed from a shell of amphiphilic molecules, polymer coating can be achieved by titrating polymer-lipids into bilayers.

Scaling analysis (*1-3*) starts by a representation of a self-similar grid, in which the mesh size ξ is a decreasing function of polymer volume fraction Φ,

$$\xi = a \, \Phi^{-3/4} \tag{1}$$

where a is the size of a monomer. Typically a is around 0.3 to 0.4 nm and ξ about 0-10 nm. Exponents are normally determined or postulated empirically rather than based on a rigorous theory. For the prediction of properties of a bulk solution we must define the number of monomers, g, in the subunit of size ξ.

$$g = \Phi^{-5/4} \tag{2}$$

Using these relations many properties of polymer solutions can be calculated. For instance, osmotic pressure is simply

$$\pi = kT/\xi^3 \tag{3}$$

where k is Boltzmann constant and T temperature.

We can use this concept also to calculate the repulsive pressure, F, above surfaces with grafted polymer chains. We are looking for a model which can express F as a function of polymer grafting density (D, average distance between grafting points) and degree of polymerization (N). In the case of low grafting densities the polymer extends up to the Flory radius of the chain, given by

$$R_F = a\, N^{3/5}. \tag{4}$$

Such regime is normally referred to as polymer mushroom conformation. Increasing the polymer concentration and/or length, the polymer chains start to interact at $D \sim R_F$. This causes their extension into the extended conformation, the so-called brush. To calculate the thickness of the brush we define that the mesh size equals D:

$$\xi\,(\Phi') = D \tag{5}$$

where Φ' is concentration of polymer in the brush layer. In the mesh size D there is g_D of monomers, and

$$g_D = \Phi^{-5/4} = (D/a)^{5/3} \tag{6}$$

The overall thickness of the brush using volume fraction of polymer Na^3/LD^2, is simply

$$L = N\, D/g_D = N\ a\, (\, a/D)^{2/3} \tag{7}$$

Using the above arguments we can estimate the repulsive force between two surfaces coated with the polymer brushes. They come in contact at the distance $h = h_c = 2L$. Upon compression the polymer concentration in the brush increases as $\Phi = \Phi(h_c/h)$ resulting in increased osmotic pressure inside the brush and decrease of elastic restoring forces (entropy prefers non-stretched chains). While in an uncompressed brush the osmotic and elastic terms balance each other at L, the compressed brush exhibits repulsion

$$F = (kT/D^3)\, [\, (h_c/h)^{9/4} - (h/h_c)^{3/4}\,]. \tag{8}$$

In a mushroom regime there is no elastic restoring force and h_c scales with $N^{3/5}$. The excluded volume interaction can be calculated from

$$F_v = kT\ \Phi^2\, LD^2 \tag{9}$$

where LD^2 is volume occupied by one chain and Φ can be expressed as N/LD^2. Using L from eq. 7, one gets:

$$F = N\, kT\, (a/D)^{4/3} \tag{10}$$

Scaling to $\xi = D$, taking into account number of monomers per blob, $g = (D/a)^{5/3}$, one gets

$$F = N\, kT\, (a/h)^{5/3}. \tag{11}$$

Repulsive pressure can be calculated from the free energy of chain by derivative against h and dividing by D^2 to normalize per unit area:

$$F' = (kT/D^2 \, h_c) \, (h_c/h)^{8/3}. \tag{12}$$

It is interesting to study the colloidal particles with grafted polymers in a solution of the same polymer. At lower concentrations of free polymer the depletion layer forms between grafted and free polymers. If two depletion layers overlap the osmotic pressure pushes the particles together and the so-called depletion flocculation occurs. Further addition of free polymer, however, eliminates depletion layers because repulsive potential becomes uniform throughout the solution and depletion layer reforms and the floc redisperses (*3*).

Another important feature is the stability under washing. In the case of adsorbed polymers they can be desorbed by washing and in the case of lipid anchored polymers in liposomes they can dissociate from liposomes due to their higher aqueous solubility. Usually this desorption follows first order kinetics and constant is proportional to the Boltzmann factor exp-$\{ (G_b - G_f)/kT \}$ where G indicates the energy difference between membrane associated and free state. In the case of PEGylated lipids, this factor is proportional to the cmc of the PEG-lipid.

$$G = kT \ln cmc \tag{13}$$

Scaling predictions are basically a random walk model of polymer chains. More complex theories, such as mean field lattice theory include segment interactions and polymer can be modeled as a self-avoiding random path in a potential field. For equilibrium properties it is reasonable to use mean field which can be calculated by self-consistence, i.e. the potential generated by the distribution of the segments must produce the very same distribution. Mean field lattice models are normally used for calculation of shorter chains which constrict random walk to accessible lattice sites. For longer chains continuum theories are used which use diffusion in order to calculate potential (*4*).

These theoretical predictions were confirmed in several experiments. We shall briefly discuss only colloidal particles with attached chains. While simple stability observations of colloidal suspension with and without grafted polymers revealed the presence of additional repulsive force, first experimental proof of the brushes grafted on colloidal particles were shown for silica particles with grafted polydimethylsiloxane. Observed thickness of the brush h $\approx$ N $a^{5/3}$ /$D^{2/3}$ by neutron scattering was the first experimental confirmation of the "brush" regime (*5*).

The first equivalent experiment with poly(ethylene) glycol coated particles in aqueous solution was done by using liposomes containing poly(ethylene glycol) conjugated lipids, PEG-lipids. After a brief introduction of liposomes we shall show that the behavior of polymer coated liposomes is also in agreement with scaling laws and that their stability can be predicted theoretically.

In parallel, a great amount of work was devoted to prepare, characterize and theoretically explain larger surfaces coated with polymers. The problem of many surfaces, especially the ones which come into contact with blood, is protein adsorption.

Coating such surfaces with PEG induces protein resistance, as well as decreases toxicity and immunogenicity. Several methods to graft such surfaces, including metals, silica, plastics (polyurethane), rubber and similar with PEG were developed (*6*). In addition to quasi irreversible adsorption, they include photo chemical curing of PEG incorporated in a polymerizable resin (PEG hydrogels), and chemical binding of functionalized polymer onto functionalized surfaces. A typical reaction is the Schiff base reaction between PEG-CHO and aminated (-NH$_2$) surfaces (*6*).

In addition to theoretical descriptions discussed, computer simulations were also used to describe the behavior of PEG on the surfaces (*7*). Methods of molecular mechanics can yield some information on static and dynamic structure of polymers and their surroundings. Parameters, such as bond distances and angles, torsion angles and interaction constants, used in these calculations are obtained from experimental techniques, or, when this is difficult, from ab initio quantum mechanical calculations. These parameters are used to calculate the potential energy function of all the atoms in a closed system (to define entropy) of i atoms:

$$V = 0.5 \{ \Sigma_i K_b (b\text{-}b_o)^2 + \Sigma_\theta (\theta - \theta)^2 + \Sigma_\Phi [1 + s \cos(n\Phi)]\}$$
$$+ \Sigma_i [A/r_i^{12} - C/r_i^6] + \Sigma_{ij} q_i q_j/Dr_{ij} \qquad (14)$$

where terms describe potential associated with bond lengths, valence angle, torsion angle, van der Waals interaction (as Lennard Jones potential in this case), and electrostatic interaction, respectively. Energy minimization is used to describe the static structure, while dynamic information is obtained by using methods of molecular dynamics and Monte Carlo simulation. In the latter methods atoms are moved along random trajectories. They move into new position only if the energy there is lower and in such a way structure development in time can be described. In molecular dynamics, Newton equations of motion for each atom in the system are solved

$$\mathbf{F} = m \, \mathbf{a} \qquad (15)$$

and from potential and force,

$$\mathbf{F} = - \partial V / \partial \mathbf{r}_i = m_i \, d^2 \, \mathbf{r}_i/dt^2 \qquad (16)$$

acceleration of each atom can be calculated and this defines the system in time. These are very laborious calculations and with present supercomputers, systems containing several thousand atoms can be simulated over a time period below a nanosecond. Obviously, the addition of solvent and protein reduces the scale and simulations of polymer or polymer-protein interaction *in vacuo* and in real conditions may differ. Limited number of atoms require introduction of boundary which strains the system additionally (some simulations use no boundaries but in such case entropy will eventually remove molecules from the picture). In order to simplify calculations other approximations are normally made and because the potential function itself does not contain all the contributions, such simulations may yield sometimes questionable results if parameters are not properly adjusted. Nevertheless, carefully performed simulations can give valuable information on static and dynamic behavior of such systems.

Polymer grafted liposomes

Liposomes are colloidal particles, in which self-assembled lipid bilayer(s) encapsulate(s) solvent, in which they freely diffuse, into their interior. Due to their properties, such as colloidal size, encapsulation ability, and control over bilayer and interface properties, several applications are being extensively studied (*8*). Many of them, however, are hampered by non-specific reactivity of liposomes and in order to by-pass that, liposomes were coated with inert polymers. The so-called sterically stabilized liposomes were shown to be rather stable, not only colloidally, but also in liposomicidal environments and below we shall review their stability. Typically, such liposomes contain in their membrane PEGylated lipids and empirical work has established that 5 mol% of lipid containing covalently linked ^{2000}PEG gave rise to optimal stabilization (*9*).

Using the above introduced scaling theory, it is possible to understand the stability of sterically stabilized systems as a function of two controlling parameters: polymer chain grafting density (D) and their length (degree of polymerization N) (*1-3*), as will be shown below.

Qualitatively the increased colloidal stability of liposomes with grafted PEG chains was observed and attributed to reduced protein adsorption due to steric stabilization (*10*) before quantitative measurements of repulsive pressure above surface with grafted polymer were performed. Corona of attached polymer causes repulsion because the entropy of chains is reduced upon approach (loss of configurational entropy) as well as excluded volume and osmotic repulsion. Depending on the size of the polymer (R) and distance between attachment/grafting points (D) polymer can have different conformations. At very low surface coverages (D >> R) the polymer forms either pancake like structures or inverse droplet-like structures, depending if it forms adsorption or depletion layer on the surface. At D > R the so-called mushroom conformation is present while for D ≤ R polymer chain start to interact, forcing their extension into the so-called brush conformation (*1-3*). Figure 1 shows these conformations schematically.

Repulsive pressure was measured between large multilamellar vesicles, with and without grafted polymers, by the osmotic stress technique. The effect of curvature, which is in the case of large multilamellar vesicles, in contrast to small vesicles, negligible, does not change scaling law which has undefined proportionality constants. At 4.5 mol% ^{2000}PEG-lipid in the bilayer D = 3.4 nm and R_F = 3.4 nm indicating that the system is in the mushroom conformation. Repulsive pressure follows scaling law as (*11*)

$$P = (5/2)\, kTN/D^2/a \,(a/(h/2))^{8/3} \tag{17}$$

This analysis of the polymer layer thickness gave h ≈ 5 nm, slightly higher than expected. Very similar results were obtained also by surface force apparatus measurements (*12*) as shown in Fig. 2. At three different grafting densities the following fits were found to describe the observed force = f (distance) profiles:

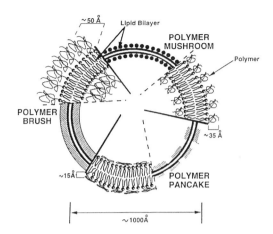

Fig. 1. Schematic presentation of various polymer conformations on the liposome surface.

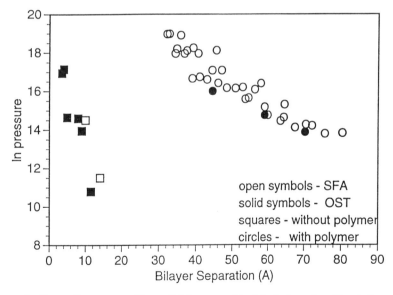

Figure 2. Force-distance profiles of bilayers with (circles) and without (squares) polymer coating as measured by osmotic stress (solid) and surface force apparatus (open) techniques. 4.5 mol% of [2000]PEG was grafted onto large multilamellar liposomes (osmotic stress method) or into supported bilayers (SFA) by titrating the bilayer with PEG-lipid. Adapted from ref. 11 and 12.

At 1.3 mol%

$$F_c(h)/R = 1.6 \ (2 \ kT/D^2) \ [\ (R_F/h)^{5/3} - 1]$$ (18)

where the numerical pre-factor is close to expected unity (12).
At 4.5 and 9 mol% of ^{2000}PEG-lipid the observed profiles could be fit by eq 19:

$$F_c(h)/R = (16 \ kT \ \pi \ h_c)/ \ (35D^3) \ [\ 7(2h_c/h)^{5/4} + 5 \ (h/2h_c)^{7/4} - 12]$$ (19)

where

$$h_c = D \ (R_F/D)^{5/3}$$ (20)

and fits are shown in Figure 3. Force between two cylindrical surfaces (F_c) and repulsive pressure (P), as measured by the osmotic stress technique, can be calculated using Derjaguin approximation

$$F_c(h) \ /R = 2\pi \ P \ (h) \ dh.$$ (21)

The thicknesses of the polymer were 3.6, 3.6-3.8 and 7-7.5 nm, for 1.3, 4.5 and 9 mol%, respectively, in a rather good agreement with the theory which predicted 3.5 , 3.5-4 and 6.5-7.5 nm (12). In still unpublished work, the polymer thickness, as measured by small angle neutron scattering of vesicles containing three different polymer chain lengths (molecular weights 750, 2000, and 5000) at several grafting densities, has confirmed expectations from theory. The same group also extended these principles to create preparations of extremely homogeneous spontaneous vesicles (13). Polymer thickness as determined by electrophoretic mobility (zetapotential) measurements were also in agreement with scaling predictions (14,15).

Conformational changes in monolayers of PEG-lipids and their mixtures with phospholipids were measured by film balance experiments and fluorescence microscopy: two phase transitions were observed - at low coverages pancake - mushroom and at higher mushroom   - brush (16). In the pancake state phase segregation was observed. Only when a depletion layer was formed in the mushroom regime the lipid molecules could intermix with PEG-lipids. With other words, PEG, which is surface active by itself, has a collapse pressure of about 8-15 mN/m (increasing with N) and can spread on the surface before the pressure due to increasing lipid concentration pushes it into the subphase (17).

Composite polymer-lipid films were deposited on solid substrates and studied by ellipsometry. Disjoining pressures as a function of humidity were studied and it was found that distance vs pressure curves are governed by steric and electrostatic forces and were in good agreement with surface force apparatus and osmotic stress technique measurements (16). Recently, phase diagram of PEG-lipid systems were published (18,19). While it was shown that fractions of ^{2000}PEG-lipid in the bilayer > 10-18% cause liposome solubilization and formation of mixed micelles (19,20) and that pure PEG-lipids form small micelles (18-22), lipid monolayers of pure PEG-lipids were

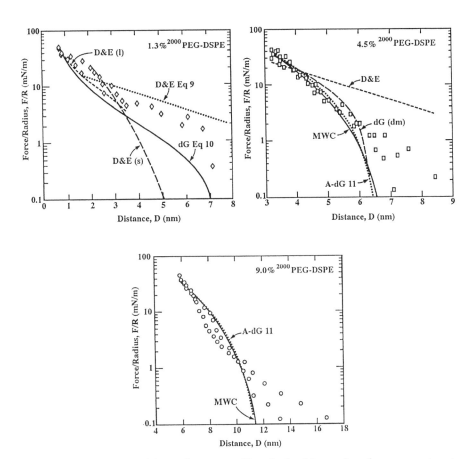

Fig. 3. Theoretical fit of force distance profiles obtained by surface force apparatus at three different surface densities of PEG. Adapted from [12]. Abbreviations refer to models by Dolan-Edwards, Alexander-deGennes and Milner-Witten-Cates.

shown to be stable up to >45 mN/m at area of ca 0.75 nm^2 per molecule (*12,16*). Some anecdotal evidence showed also that spherical vesicles could be prepared from pure PEG-lipids. Obviously, more work is needed to reconcile these observations. It is possible that PEG-lipid undergoes phase transformation from random coil into helix. This transition may be exploited for changing very stable liposomes into fusogenic upon triggered conformational transition of PEG chains. Indeed, it was reported that liposomes containing large fractions of ^{600}PEG-lipids were fusogenic (*23*).

Experimental measurements using osmotic stress technique (*11*), zeta-potential measurements (*14*) and surface force apparatus (*12*) have shown that polymer extends, in agreement with theoretical calculation, about 5 nm above the bilayer in the case of 5 mol% of ^{2000}PEGylated lipid in any membrane.

The thickness of the coating can be expressed in a good solvent as $h_c = DN(a/D)^{3/5}$ where N is degree of polymerization, a size of a monomer and D the mean distance between the grafting points. Using scaling arguments we can estimate polymer thickness for effective stabilization: a general criterion becomes then A (b/h_c) < kT. Although scaling theory does not provide numerical prefactor in h_c/b = A/kT, one can estimate from the fact that typically, A/kT is of order 1/10, the coating of approximately 5-10% of the particle diameter can give rise to effective steric stabilization (*24*). This simple analysis may explain why long circulation was achieved only with small vesicles; larger vesicles would require longer PEG chains which are unstable in the bilayer and would require preparation of more stable bilayers. Such experiment, however, has not been performed yet.

First published studies of PEGylated liposomes did not elaborate on the mechanism of increased biological stability. The vague claim of increased surface hydrophilicity (*25*) was shown to be incorrect (*26*). Obviously, surface attached polyelectrolytes would, according to such an explanation, yield the best biological stabilization, what is, not the case. The qualitative idea of protein adsorption mediated instability (*10*), which postulated that increased stability of PEGylated liposomes is due to reduced adsorption of proteins due to steric repulsion, was tried to be encompassed into a different theoretical model (*27*). Authors simulated the polymer conformation in a random flight model and simply stated that when the liposome surface is coated with polymer protein adsorption is prevented. This model does not take into account any conformational change of polymer as a function of D and N. Furthermore it predicts isotropic polymer cloud which cannot exert any repulsive pressure. Simply coating the surface with polymer does not reduce any adsorption. It is only the change of polymer conformation upon approach of a protein or other particle/macromolecule which gives rise to the repulsive pressure (see eq. 8 and 12). The authors basically calculated Flory radius of polymer in a complicated way including in the cases where this is not correct (D < R_F). Therefore those equations have no physical meaning.

While theoretical physics does not distinguish between various polymers (till they are in a good solvent), experiments do. Poly (ethylene) glycol (PEG) is by far the most widely used polymer to impart steric stabilization (*8*). This is probably due to well established chemistry of this polymer, good solubility in many solvents as well as high anisotropy (i.e. length/thickness ratio) which enables high flexibility and motility. The majority of researchers link polymer to the amino group of phosphatidylethanolamines because of its reactivity. PEG can be, however, coupled also on diacyl phosphatidyl glycerol, to propyl (backbone) group and some other, especially non-ionic, lipids.

The fact that polymer stabilization of liposomes is a general phenomenon was confirmed when other polymers were discovered which rendered similar biological stability. Poly-(2-methyl-2-oxazolidine and poly-(2-ethyl-2-oxazolidine) with $N = 50$ attached to DSPE and incorporated into liposome bilayers at 5 mol% showed similar stability profile as PEG (*28*). Poly(acryl amide) and poly (vinyl pyrrolidone) also increased blood circulation times of liposomes (*29*). The inferior results, as compared to PEG and (*28*) are probably due to a weak hydrophobic anchor which caused release of these molecules from liposomes.

Aqueous solubility of PEG-lipids, which has pronounced effects on liposome stability is not considered in any of these theoretical models. It is very likely that finite liposome stability may be due to dissociation of PEG-lipid molecules from the liposomes. Indeed, exchange rates of these lipids from donor to acceptor vesicles were shown to depend on the hydrophobic anchor and were found to be in the range of exchange halflives from 20 min to 12 hours, depending on the acyl chains (*30*). As expected, aqueous solubility critically depends on hydrophobic anchors and follows a scaling relation $\alpha \ N^{3/5}$ with respect to the PEG size (*30*). The cmc of ^{2000}PEG-DSPE has been recently reported to be between 3 to 5 μM (*31*).

Normally distearoyl chains are used to increase the anchoring effect, preferably in the mechanically strongest (most cohesive) DSPC/Chol bilayers (*32*). Sphingomyelin, which can also make mechanically very strong bilayers, with attached PEG chains, and the stability of ensuing liposomes have not been reported yet. Neither were arachidoyl ($C_{20:0}$) and behenoyl ($C_{22:0}$), or longer fatty acid bilayers, nor polymerized liposomes coated with PEG.

With the development of monoclonal antibodies and long-circulating liposomes, their targeting to particular cells become feasible. In addition to classical techniques to link antibodies, lectins, or other ligands to PE, coupling to the far end of PEG chain was developed. Inert terminal methoxy group is replaced with a reactive functional group suitable for conjugation after liposomes were prepared containing such reactive polymer-lipid. Several different ways to couple antibodies onto liposomes are discussed in ref. *33*.

Liposomes with grafted chains have shown successful applications due to increased stability. While these systems are very useful models to test various theories, in practice, liposomes with grafted polymer chains have shown increased stability in vitro and in vivo. Blood circulation times in rodents, a standard test for liposome characterization and their biological stability, were prolonged up to a hundred fold. It was shown that antitumor activity of systemically administered anticancer drugs is proportional to blood circulation times (*8,11*). Anti-cancer agents epirubicin and doxorubicin encapsulated in sterically stabilized liposomes were very effective in pre-clinical and clinical studies resulting in the first FDA approved liposomal formulation (*34*).

Following these breakthroughs some further developments are forthcoming. These include targeted liposomes and programmable liposome systems. Inclusion of lipids with time dependent conjugated charge or polymer can add further ways to regulate colloidal and mechanical stability and phase behavior of liposomes. For instance, bilayers composed from micelle and inverse micelle forming molecules, such as PEG-lipids and dioleoyl phosphatidylethanolamine, form stable bilayers. However, if PEG polymer is cleaved (by hydrolysis or thyolysis (*35*) of a bond with known reaction kinetics, by photo- or pH-induced trigger) or simply by the dissociation of the whole molecule from the bilayer (due to shorter acyl chain(s),for instance), phase transition occurs and encapsulated material is released (*35*). In addition to lamellar - hexagonal phase transition, one can induce lamellar - micellar by cleaving acyl chains of the lipids (*36*). Fusion or liposome dissolution (lamellar - micellar phase transition) can be induced also by surface grafted (*37*) or free polymers in solution (*38*) which change their conformation as a function of pH. Very recent studies of phase behavior of lipid monolayers containing PEG-lipids using neutron reflectometry have shown consistent picture with above findings. It was shown that rather short PEG chains (N=50) tethered to the air0water interface follow scaling behavior and exhibit mushroom - brush transition (*39*). A depletion layer could be also observed.

In conclusion, synergistic combination of specific properties of colloidal particles and polymers led to systems with novel characteristics and applications. An example are sterically stabilized liposomes and their applications as drug delivery vehicles in chemotherapy (*34,40*). Continuing efforts of many scientists from diverse areas are making novel breakthroughs likely.

References

1. deGennes, P.G. *Scaling concepts in polymer physics*, Cornell Univ. Press, **1979.**
2. Auvray, L. *Structure des polymeres aux Interfaces*, Laboratoire L. Brillouin report, CEN Saclay, July **1991.**
3. deGennes, P.G. *Adv.Colloid Interface Sci.* **1987**, *27*, 189-209.
4. Milner, S. *Science* **1991**, *251*, 205-10.
5. Auroy,P; Auvray, L; Leger L. *Phys. Rev. Lett.* **1991**, *66*, 719-22.
6. Golander C,G; Herron, J.N; Claesson P; Stenius, P; Andrade, J.D. in *Poly(ethylene glycol) Chemistry*, Harris, J.M.; Ed.; Plenum Press, NY, **1992,** pp. 221-45.
7. Lim, K., Herron, J.N. in *Poly(ethylene glycol) Chemistry*, Harris, J.M.; Ed.; Plenum Press, NY, **1992,** pp. 29-56.
8. Lasic, D,D; Martin, F. (Eds.) *Stealth liposomes*, CRC Press, Boca Raton, **1995,** 1-289.
9. Papahadjopoulos, D; Allen, T.M; Gabizon, A; Mayhew, E; Huang, S.K; Woodle, M.C; Lasic, D.D; Redemann, C; Martin, F.J. *Proc. Natl. Acad. Sci. USA* **1991**, *88*, 11460-4.
10. Lasic, D.D; Gabizon, A; Huang, K; Martin, F; Papahadjopoulos, D. *Biochim Biophys. Acta* **1991**, *1070*,187-92.

11. Needham, D; Hristova, K; McIntosh, TJ; Dewhirst, D; Wu, N; Lasic, D.D. *J. Liposome Res.* **1992**, *2*, 411-39.

12. Kuhl, T; Leckband, D;, Lasic, D.D; Israelachvili, J.N. *Biophys.J.* **1994**, *66*, 1479-86.

13. Joannic, R; Auvray, L; Lasic, D.D. Phys. Rev. Lett., in the press

14. Woodle, M.C.; Collins, L.R; Sponsler, E; Kossovski, N; Papahadjopoulos, D; Martin F.J. *Biophys J.* **1992**, *61*,902-8.

15. Shimada, K; Miyagashima, A; Sadzuoka, Y; Nozawa, Y; Mochizuki, Y; Ohshima, H; Hirota, S. J. *Drug Targ.* **1995**, *3*, 283-9.

16. Baekmark, T.R; Elander, G; Lasic, D.D; Sackmann, E. *Langmuir* **1995**, *11*, 3975-87.

17. Winterhalter, M; Marzinka, S; Benz, R; Kasianowicz, J. *Biophys. J.* **1995**, *69*, 1372-81.

18. Kenworthy, A.K; Hristova, K; Needham, D; McIntosh TJ. *Biophys. J.* **1995**, *68*, 1920-36.

19. Kenworthy, A.K; Simon, S.A; McIntosh, T.J. *Biophys. J.* **1995**, *68*, 1903-1920.

20. Lasic, D.D; Woodle, M.C; Martin, F; Valentincic, T. *Period. Biol.* **1991**, *27*, 287-91.

21. Frederik, P.M; Stuart, M; Bomans, P.H; Lasic, D.D in . *Nonmedical Applications of Liposomes*, Lasic,D.D; Barenholz, Y; Eds.; CRC Press, Boca Raton, **1996**, *Vol.I* (*Theory and Basic Science*), pp.302-309.

22. Lasic, D.D. *Liposomes: from Physics to Applications*, Elsevier, Amsterdam, New York, Tokyo, **1993**, p. 460 (ref.51).

23. Okumura, Y; Yanauchi, M; Yamamoto, M; Sunamoto, J. *Proc. Japan Acad.* **1993**, *69 B*, 45-50.

24. Bruinsma,R; Pincus, P. *Curr. Op. Solid State & Mat. Sci.* **1996**, *1*, 401-406.

25. Senior, J.H; Delgado, C; Fisher, D; Tilcock, C; Gregoriadis, G. *Biochim. Biophys. Acta* **1991**, *1062*, 77-82.

26. Blume, G., Cevc, G. *Biochim. Biophys. Acta* **1993**, *1146*, 157-62.

27. Torchilin, V.P; Omelyanenko, V.G; Papisov, M.I; Bogdanov, A.A; Trubetskoy, V.S; Herron, J.N; Gentry, C.A. *Biochim. Biophys. Acta* **1994**, *1195*, 11-21.

28. Woodle, M.C; Engbers, C.M; Zalipsky, S. *Bioconjug.Chem.* **1994**, *5*, 493-6.

29. Torchilin, V.P; Shtilman, M.I; Trubetskoy, V.S; Whiteman, K; Milstein, A.M.: *Biochim.Biophys.Acta* **1994**, *1195*, 181-4.

30. Silvius, J.R; Zuckermann, M.J. *Biochemistry* **1993**, *32*, 3153-9.

31. Uster, P.S; Allen, T.M; Daniel, B.E; Mendez, C.J; Newman, M.S; Zhu, G.Z. *FEBS Lett.* **1996**, *386*, 243-246.

32. Lasic, D.D; Needham, D. *Chem. Rev.* **1995**, *95*, 1601-24.

33. Hansen, C; Kao, G; Moase, E; Zalipsky, S; Allen, T.M. *Biochim. Biophys. Acta* **1995**, *1239*,133-144.

34. Lasic, D.D. *Nature* **1996**, *380*, 561-562.

35. Kirpotin, D; Hong,K; Mullah, N; Papahadjopoulos, D; Zalipsky,S. *FEBS Lett.* **1996**: *386*, 115-9.

36. Anderson, V.C; Thompson,D.H. *Biochim. Biophys. Acta* **1992**, *1109*, 33-42.

37. Thomas, L.J; Tirrell,D.A. *Acc. Chem. Res.* **1992**, *25*, 336-342.
38. Parente, R.A; Nir, S; Szoka, F. *Biochemistry* **1990**, *29*, 1720-28.
39. Majewski, J.; Kuhl T.L.; Gerstenberg, M.C.; Israelachvili, J.N.; Smith, G.S. *J. Phys. Chem.*, in the press
40. Lasic, D.D; Papahadjopoulos, D. *Science* **1995**, *267*, 1275-1276.

Chapter 4

Safety of Poly(ethylene glycol) and Poly(ethylene glycol) Derivatives

Peter K. Working[1], Mary S. Newman[1], Judy Johnson[2], and Joel B. Cornacoff[2]

[1]SEQUUS Pharmaceuticals, Inc., 960 Hamilton Court, Menlo Park, CA 94025
[2]Nycomed, Inc., Wayne, PA 19087

The safety of the PEGs has been studied and reported for over fifty years. Despite the fact that PEGS are generally recognized as safe, there is a paucity of recent literature on this topic. In this chapter we review the literature from early studies, and present new information on the safety of two different PEG polymers, one delivered subcutaneously as a component of an imaging agent and the other intravenously as a component of long-circulating liposomes encapsulating an oncologic product. These recent studies confirm the relative safety of PEG, reporting no significant adverse effects related to PEG administration.

The purpose of this chapter is to review and summarize studies of the safety of the PEGs. The chapter begins with a brief overview of the extensive studies of the toxicity, absorption, distribution, metabolism and excretion of various PEG polymers, completed several years to several decades ago. We also present heretofore unpublished information on the safety of PEG polymers as utilized in two of their major roles today: as vehicles for parenteral imaging agents and as components of intravenously infused, long circulating liposomes containing cytotoxic anti-cancer agents. The emphasis of the review presented here is on the use and safety of PEG polymers in pharmaceutical products. Readers interested in the cosmetic uses of the PEG polymers are referred to a review published several years ago (1).

Safety of Polyethylene Glycol

The majority of the studies conducted with the PEGs have utilized the oral route of administration, with relatively fewer employing intravenous or intraperitoneal routes.

Acute Toxicity. Fatal poisonings with PEG have been reported in human burn patients who had been treated with a PEG-based antimicrobial cream containing a mixture of three fractions of PEG with average molecular weights of 300, 1000 and

4000, with PEG-300 comprising more than 95% of the total PEG in the preparation (2). The poisoning had clinical features similar, but not identical, to those seen in ethylene glycol poisonings. A subsequent investigative study revealed that small PEGs of molecular weights less than 400 may be oxidized in vivo into toxic diacid and hydroxy acid metabolites, which are found in the urine of PEG-treated patients, via sequential oxidations by alcohol dehydrogenase and aldehyde dehydrogenase (3). The reaction rate of the oxidation decreased markedly with increasing molecular weight of the PEG, suggesting that larger molecular weight PEGs will not be significantly metabolized via this route. In fact, the toxicity of larger PEGs (molecular weights > 1000 daltons) has long been known to be very low when given orally (4). Extensive toxicity studies have shown that PEG-4000 can be safely administered in 10% solutions to rats, guinea pigs, rabbit and monkeys at dose levels up to 16 g/kg (5).

Intravenous administration of PEG-400 in a 0.9% NaCl solution to mice and rats determined median lethal doses of 7.6 mL/kg and 7.1 mL/kg, respectively (6). Another study in rats determined that the intravenous LD_{50} of PEG-200 was more than 10 mL/kg (7). No toxicity was observed in rabbits receiving a total dose of 10 g/kg of PEG-1450, -3350 or -6000 in 5% solution given as a slow (2.5 mL/min) intravenous injection (8). Intravenous studies in non-rodent species have been rarely reported. However, Shideman and Procita studied the safety of poly(propylene glycol) in dogs (9). Decreased blood pressure and reversible depression in respiration were seen in dogs given intravenous injections of PEG-400 and PEG-3350 at doses of 2 to 3 g/kg. These symptoms increased in dogs that received doses of 3 g/kg or greater and eventually resulted in complete respiratory arrest. At necropsy, the dogs were found to have pulmonary edema and small infarcts in the lungs, but no changes in the heart or kidneys.

Subchronic Toxicity. When repeated intravenous injections of 1g of PEG-300, -400, -1450, -3350 or-6000 were administered 6 days a week for 5 weeks to rabbits (n=5-9/group), there were no deaths in groups that received PEG-300 or PEG-1450, but one death each in groups treated with PEG-400, -3350 and -6000 (8). When nine beagle dogs were given intravenous injections of saline or 10% PEG-3350 in 0.85% NaCl at doses of 10, 30, or 90 mg/kg/day for up to 178 days, no changes were observed in appetite or body weight and there were no clinical signs of toxicity (10). There were no gross or microscopic lesions found at necropsies conducted on days 43, 99 or 178, and there were no significant differences in organ weight, clinical chemistry or hematology values between control and PEG-treated dogs. Three of eight chronically epileptic monkeys that received intravenous of 60% PEG-400 in water for 4 weeks had significantly reduced seizure frequency relative to baseline levels. A separate group of 8 similarly treated monkeys produced similar results, although 3 monkeys were removed from the study due to observed toxicity (11). These animals exhibited a variety of relatively minor clinical signs of toxicity, as well tissue damage at the site of injection. Similar responses occurred in other epileptic monkeys and normal monkeys with similar treatment regimens (12).

Other Toxicities. No adverse reproductive or teratogenic effects were seen in subchronic and chronic oral toxicity studies of PEG at doses of 1 or 10 g PEG-200/kg in rats (*13*). PEG-400 was negative in the Chinese hamster ovary cell mutation test, the sister chromatid exchange test and the unscheduled DNA synthesis assay (*14*). PEG was not mutagenic in the mouse TK+/- forward mutation assay (*15*). PEG was not carcinogenic when administered orally, intraperitoneally, or subcutaneously to various test animals (*1*).

Studies to evaluate the hemolytic potential of PEG have shown that the EC_{50} value (concentration at which 50% of the erythrocytes are lysed) for the lysis of human erythrocytes by PEG-400 was 30% (total volume percent of the cosolvent in whole blood) (*16*). Others have reported that the hemolytic potential of PEG-400 was reduced when combined with various combinations of ethanol, polypropylene glycol, water, and/or saline (*17,18*). PEG-3350 caused crenation and clumping of rabbit erythrocytes at concentrations of 10% or greater. This finding was confirmed in vivo when 10% PEG-3350 was intravenously infused to rabbits. Blood samples obtained from animals that died from pulmonary hemorrhage contained numerous clumps of cellular elements. Rabbits that received 5% PEG-6000 solutions did not have this reaction (*8*).

Subcutaneous and intramuscular administration of PEGs to rats caused increased vascularization and fibroblastic proliferation at the site of administration that gradually disappeared with time, but otherwise caused no serious tissue damage (*19*).

PEG Metabolism and Disposition

Studies of the metabolism of PEG after parenteral administration are limited. Metabolism studies have shown that high molecular weight PEGs (> 1000 daltons) are minimally absorbed from the gastrointestinal tract. Renal clearance studies in dogs showed that PEGs up to molecular weight of 4000 are filtered at a rate comparable to creatinine, suggesting that it was excreted by glomerular filtration without participation of the renal tubules (*20*). As noted above, lower molecular weight PEGs may be oxidized to diacid and hydroxy acid metabolites by alcohol and aldehyde dehydrogenase (*3*).

Recovery of PEG in urine and feces ranges from 60 to 80% after intravenous administration in most species. In rats that received 10 mg of radiolabeled PEG 4000 (approximately 70 mg/kg), mean cumulative recovery of label was 81% at 7 days, with 61% in the urine and 20% in the feces (*10*). In rabbits given 0.4 or 0.75 g of PEG 6000, 47% and 67% of the total administered dose was recovered in urine after 24 hours (*21*). Studies in humans provided similar results, showing that the PEGs are excreted mainly via the kidney. In 6 men injected intravenously with PEG 6000, approximately 63% of the dose was recovered in urine after 1 hour, and 96% was recovered after 12 hours. PEG 800, though, was excreted at a slower rate and with less recovery (*22*). The authors speculated that this was the result of relatively reduced bioavailability of the lower molecular weight PEG for renal excretion, owing to its ability to diffuse more rapidly into the tissues and out of the bloodstream. Studies in cats administered intravenous PEG 900 showed that the biliary clearance of this PEG was significant, however, being more than 90-fold than of mannitol (*23*). Nearly 75% of the PEG was excreted unchanged, but 26% was excreted in the bile after oxidation to carboxylic acids, demonstrating that significant metabolism of PEGs occurs even in this molecular weight range. The bile was relatively enriched with the larger sizes of both the unchanged PEG and the oxidized molecules, suggesting that metabolism does not have a major role in the biliary excretion of PEG. These authors speculated that PEG moves passively with the water flux into the bile canaliculi, and that the smaller PEG molecule are reabsorbed with the water more distally. PEG molecules with molecular weight around 1000 daltons are not reabsorbed and are consequently concentrated in the bile.

Yamaoka and coworkers conducted an extensive study of the distribution and tissue uptake of ^{125}I-labeled PEG of different molecular weights (range, 6000 to 190,000 daltons) in mice (24). High molecular weight PEGs were retained in he blood for a longer period than low molecular weight PEGs, and urinary clearance decreased with increasing molecular weight. PEG tended to accumulate in muscle, skin, bone and liver to a higher extent than other organs, with no relationship to molecular weight, with the rate of accumulation directly correlated with vascular permeability. Using isolated mouse peritoneal macrophages or rat Kupffer cells, these investigators showed that the amount of phagocytosed PEG was enhanced at molecular weights of 50,000 and above; smaller molecular weight PEGs, in contrast, were more readily translocated to the extravascular compartment.

Recently, a new class of PEG polymers has been developed, which is a strictly alternating copolymer of PEG and lysine in which PEGs of variable molecular weights are linked to the α- and ε-amino groups of lysine (25). The increased chemical versatility of the polymer will permit much less restrictive attachment of ligands and increased utility. The biodistribution and toxicity of the copolymer appears similar to those reported for PEG. Studies with ^{14}C-labeled PEG-Lys copolymer in mice showed no excess uptake by liver, spleen or kidney (26). Most of the label was confined to the circulating blood, with extravascular label distributed uniformly among the tissues. Excretion of the polymer, like that of PEG, was primarily via the kidney and biliary tract. The polymer was not acutely toxic in mice or rats, as judged by clinical observations and histopathological evaluation of tissues, when administered intravenously or intraperitoneally at doses up to 10 g/kg.

Safety and Disposition Studies with "Free" Polyethylene Glycol

PEG is often found as an excipient in topical and parenteral drug products, e.g., SANDRIL (Eli Lilly), ROBAXIN (Robins), and VEPESID (Bristol-Myers). Despite the fact that PEGS are generally recognized as safe, there is a paucity of recent literature on their safety. In the studies described here, the safety of a 15% w/v sterile solution of PEG-1450 (Union Carbide Corporation, Danbury, CT) was assessed in a one-month subcutaneous administration studies in rats and dogs. The absorption, excretion and pharmacokinetics of [^{14}C$_2$]-PEG-1450 terminally tagged with a labeled ethyleneoxy unit were also evaluated. The subcutaneous route was chosen as it was the intended route for clinical administration of a drug product under development. These findings have not been published elsewhere. This PEG derivative is intended for use as a component of a prolonged circulation imaging agent in humans.

Safety Studies with PEG-1450.   Sprague-Dawley rats (10/sex/group) were administered PEG-1450 in distilled water at dosages of 375, 750 or 1500 mg/kg which corresponded to dose volumes of 2.5, 5.0 or 10.0 mL/kg. Control animals received a single subcutaneous dose of 0.9% sterile sodium chloride solution at a dosage of 10.0 mL/kg. Injections were made in the scapular and flank regions once per day for 28 days. At necropsy (24 hours following the final dose) blood samples were taken for hematological and clinical chemistry assessments and a complete necropsy was performed on all animals.

There were no treatment-related adverse effects on the clinical appearance or behavior of the animals throughout the study or on body weight gain. The only statistically significant clinical chemistry effects were a slight decrease in plasma alanine aminotransferase (13-23%) and aspartate aminotransferase (3-18%) in all groups of both sexes. In addition, blood urea concentrations were decreased (10-11%) in both sexes at a dosage of 1500 mg/kg. The changes were not dosage-related and consequently were not considered to be biologically significant. At Day 29 there was a small, statistically significant decrease in mean cell hemoglobin concentration (2%) in both sexes at the high dose. There were also small, statistically significant, increases in prothrombin time (1-4%) in both sexes at the mid and high dose. Statistically significant increases were also observed in the number of segmented neutrophils (40-85%) in both sexes at 375 mg/kg. None of these changes were considered to be of biological significance. There were no significant compound-related macroscopic or histomorphologic findings observed in any animal at necropsy.

In a second study, Beagle dogs (6/sex/group) were administered PEG-1450 subcutaneously once daily for 29 days at doses of 37.5, 150 or 600 mg/kg, corresponding to 0.25, 1 or 4 mL/kg. An additional group of 6 dogs of each sex received 0.9% sterile sodium chloride solution at a volume of 4 mL/kg. Doses were administered at eight injection sites of the shaved back. As in the case of the rat, the test article was well tolerated, and no treatment-related adverse effects were noted. Although there were minor alterations in a number of hematologic and clinical chemistry parameters, none were considered of biological significance. No significant macroscopic or histomorphologic findings were noted at necropsy.

The no-observed-adverse-effect-level (NOAEL) of PEG-1450 was greater than 1500 and 600 mg/kg in the rat and dog, respectively. The most consistent finding between the two studies was minor hemorrhage, inflammation and degenerative lesions at the site of injection.

Disposition Studies with PEG-1450. An aqueous solution of $[^{14}C_2]$-PEG-1450 (15% w/v) was administered subcutaneously to male Sprague-Dawley rats at doses of 375, 750 or 1500 mg/kg (equivalent to 2.5, 5.0 or 10.0 mL/kg). Blood and plasma samples were generated and assayed for radioactivity by liquid scintillation counting to determine the relationship between dose and exposure. Radioactivity rapidly appeared in the circulation after subcutaneous administration. Dose proportional increases in exposure as measured by $AUC_{(0-\infty)}$ were demonstrated in both plasma and blood over the dose range. The terminal elimination half-life of plasma radioactivity was 3-4 days, and $AUC_{(0-t)}$ (t=216 hr for low dose; 360 hr for mid and high doses) accounted for 95-100% of the $AUC_{(0-\infty)}$ values. Blood levels of radioactivity were consistently lower than the corresponding plasma concentrations throughout the absorption and distribution phases, and the terminal elimination half-life in blood was somewhat longer (10-12 days) than in plasma. $AUC_{(0-t)}$ in blood accounted for only 59-65% of the total $AUC_{(0-)}$ value, suggesting preferential association with the cellular components in blood. This relatively longer elimination phase represents a minor fraction of administered dose. In an excretion/balance study wherein an aqueous solution of $[^{14}C_2]$-PEG-1450 (15% w/v) was administered subcutaneously to rats (4/sex) at 326 mg/kg, the most of the administered radiolabel (63.5 ± 7.9%) was recovered in urine in the first 12 hours post-dose, with an additional 11.4 ± 3.2% of the dose excreted in the feces over the seven-day study period. Most of the remaining radioactivity (~10%) was found in the cage rinsings with an additional ~2 % of the dose in the carcass. Unchanged PEG appeared to be the single radioactive component present in metabolic profiles of male or female 12 hour rat urine obtained by size exclusion HPLC.

A similar toxicokinetic profile was observed in beagle dogs that were dosed subcutaneously with $[^{14}C_2]$-PEG-1450 (15% w/v) at doses of 37.5, 150 or 600 mg/kg (equivalent to 0.25, 1.0 or 4.0 mL/kg). Like the rat, dose proportional increases in $AUC_{(0-\infty)}$ were observed in both blood and plasma over the dose range. The terminal elimination half-life was 8-10 days, and $AUC_{(0-t)}$ (t=672 hr) accounted for 92-98% of the total $AUC_{(0-\infty)}$. The terminal elimination half-life in blood was 15-20 days, and $AUC_{(0-t)}$ accounted for only 70-73% of the total $AUC_{(0-\infty)}$ value.

Safety and Kinetic Studies with Lipid-Derivatized Polyethylene Glycol

DOXIL (STEALTH liposomal doxorubicin HCl) is formulated in liposomes that contain a PEGylated lipid, N-(carbamoyl-methoxypolyethylene glycol 2000)-1,2-distearoyl-sn-glycero-3-phosphoethanolamine sodium salt (MPEG-DSPE), another phospholipid, hydrogenated soy phosphatidylcholine, and cholesterol. The total lipid content of the DOXIL formulation is approximately 16 mg/mL, and the doxorubicin HCl concentration is 2 mg/mL. DOXIL is approved in the U.S. for treatment of Kaposi's sarcoma in patients who are intolerant to or have failed treatment with standard chemotherapeutic regimens for this disease. A typical dose regimen for DOXIL is a dose of 20 mg/m^2 of body surface area (BSA), or approximately 0.5 mg/kg body weight, assuming a BSA area of 1.7 m^2 and a body weight of 70 kg, given once every two to three weeks. Considering the lipid dose administered at this level of DOXIL treatment, the typical patient would receive approximately 32 mg/m^2, or 0.71 mg/kg, of MPEG-DSPE with each DOXIL treatment.

Use of MPEG in Liposomes. MPEG-DSPE is used as a component of the DOXIL liposomes to increase their blood circulation time and, as a result, increase the tumor concentration of the active ingredient of DOXIL, doxorubicin, a potent anti-tumor drug. Liposomes are microscopic vesicles, usually in the sub-micron size range, consisting of one or more lipid bilayers enclosing an internal aqueous space. Liposomes form spontaneously when certain amphipathic lipids, including phospholipids and mixtures of phospholipids and cholesterol, are hydrated during agitation (27). When administered parenterally, non-PEGylated liposomes are rapidly phagocytosed and removed from circulation, largely through the scavenger function of cells of the mononuclear phagocytic system (MPS) (*28*). The MPS consists of resident macrophages in connective tissue (histiocytes), liver (Kupffer cells), lung (alveolar macrophages), free and fixed macrophages in the spleen and lymph nodes, macrophages in the bone marrow, peritoneal macrophages in the serous cavity, osteoclasts in the bone tissue and microglia in nerve tissue. Depending upon the route of administration, liposome uptake can occur at some or all of these sites. Following intravenous administration, non-PEGylated liposomes are primarily removed from circulation by macrophages in the liver, spleen and bone marrow.

The incorporation of MPEG, which forms a hydrophilic layer of approximately 5 nm thickness at the surface of the liposomes, reduces uptake of the liposomes by the macrophages of the MPS, resulting in a dose-independent long blood circulation time compared to conventional, non-PEGylated liposomes. Uptake of STEALTH liposomes is sharply reduced compared to conventional liposomes, with approximately 10% of the administered dose of STEALTH liposomes generally found in the liver and spleen by 24 hours post-dose (*29,30*). In contrast, combined liver/spleen levels of the best conventional liposomes ranged from 15 to 31% of administered dose by one day after treatment (Table I).

The designation "Stealth" is a reference to the ability of PEGylated liposomes, which are also often referred to as sterically stabilized liposomes, to evade detection by the MPS. The term "sterically stabilized" is derived in analogy to stabilization of inorganic colloid particles, in which steric modifications of the particle surface result in reductions in particle-to-particle interactions that could lead to aggregation or fusion. In colloids, steric stabilization is accomplished by addition of surface charge or via coating of the surface with various molecules, including starch and PEG (*31*). In the case of liposomes in biological environments, steric stabilization not only reduces particle-to-particle interactions, but also decreases adsorption of various macromolecules onto the liposome surface, the loss of liposomal components to other particles, and interactions of the liposomes with cells, all of which provide greater liposome stability (*32,33*).

Table I. Effect of the MPEG Polymer on Tissue Distribution of Liposomes in Rats

| Lipid Composition | *% of Administered Dose at 24 hours Post-Dose* | | |
	Blood	*Liver*	*Spleen*
MPEG-DSPE/PC/Chol	32.3	8.4	2.7
PG/PC/Chol	0.2	13.8	1.3
PC/Chol	0.8	28.4	3.3

MPEG: N-carbamoyl-methoxypolyethylene glycol 2000; DSPE: PC: phosphatidyl choline; 1,2-distearoyl-sn-glycero-3-phosphoethanolamine; PG: phosphatidyl glycerol; Chol: cholesterol

SOURCE: Adapted from reference 30.

Owing to its MPEG component, DOXIL also has altered tissue disposition, with lower peak concentrations in most organs, but higher concentration in tumors than after treatment with conventionally formulated, non-liposomal doxorubicin HCl. The increased tumor concentration of doxorubicin after DOXIL treatment is due to the small size of the liposomes (mean diameter = 100 nm) and, probably more importantly, the leakiness of tumor microvasculature to small molecules and even colloidal particles (*34,35*). By virtue of their extended lifetime in blood circulation, STEALTH liposome extravasation is likely to occur through the leaky vessel walls of tumors. STEALTH liposomes are even more permeable than non-PEGylated liposomes of the same size and nearly identical lipid composition, with 3- to 4-fold higher levels of accumulation in tumors (*36*). The preferential accumulation of STEALTH liposomes is apparently attributable not only to their slower plasma clearance, but also to their higher vascular permeability relative to non-PEGylated liposomes, probably because of reduced interactions with the vascular endothelium due to the MPEG coating.

Acute Toxicity of MPEG-DSPE. The MPEG used in STEALTH liposomes of DOXIL is a synthetic derivative of polyethylene glycol with a molecular weight of approximately 2000 daltons, and could potentially have intrinsic toxicity. Liposomes cannot be prepared with MPEG-DSPE as their sole component because of its physicochemical characteristics, but the chemical freely forms a micellar mixture in aqueous solutions, and the toxicity of such micelles can be directly assessed. In such a study, male ICR mice (n = 10 per group) received intravenous injections of 14 or 284 mg/kg of MPEG-DSPE micelles in saline. Mice were observed for signs of toxicity, including effects on body weights and clinical observations for up to 22 days post-treatment. No necropsy was performed. There were no deaths or clinical signs of toxicity in either dose group, and body weight gain in the two groups was equivalent. The high-dose of MPEG-DSPE administered in the study (284 mg/kg) was nearly 15-fold the MPEG-DSPE dose delivered at a DOXIL dose that caused doxorubicin-related toxicity in mice (19.1 mg/kg MPEG-DSPE at a dose of 12 mg/kg DOXIL), suggesting that no toxicity will be associated with the administration of MPEG-DSPE at the levels in DOXIL. Moreover, as noted, in humans a dose of 20 mg/m^2 DOXIL is equivalent to an MPEG-DSPE dose of 31.9 mg/ m^2 or 0.77 mg/kg. Using a dose scaling method for extrapolation to humans (*37*) to express the MPEG-DSPE dose in mice as an equivalent surface area dose in humans, the non-toxic high dose in mice (284 mg/kg) is equivalent to a human dose of 23.7 mg/kg MPEG-DSPE, or approximately 30-fold the amount of MPEG-DSPE administered with a 20 mg/m^2 dose of DOXIL. Thus, MPEG-DSPE component of the liposome likely does not pose a significant risk of toxicity in humans, particularly considering the well known significant toxicity of doxorubicin itself.

Safety of MPEG-DSPE in Liposomes. Incorporation of MPEG into long-circulating liposomes could, of course, result in the development of some significant unique toxicity, and numerous studies using empty MPEG-DSPE-containing liposomes have been conducted to examine this possibility. These studies have the advantage of preserving the disposition of the MPEG-DSPE that occurs when it is a liposome component and permitting the assessment of toxicity in the absence of the generally toxic drug encapsulated in the liposome. In these studies, MPEG-DSPE-containing STEALTH placebo liposomes were used as a control group in single and repeated-dose studies, as well as tested directly in specific pharmacology and mutagenicity studies.

The acute toxicity of MPEG-DPSE in STEALTH liposomes has been evaluated in mice, rats, rabbits, dogs and monkeys, with no evidence of toxicity related to the lipid components. For example, in an acute safety study conducted as part of the evaluation of a liposomal cytotoxic drug in early development, male and female cynomolgus monkeys (n = 4 per sex) received single intravenous injections of MPEG-DPSE-containing liposomes at a dosage of 3.6 g lipid/kg (215 mg/kg MPEG-DPSE). There were no clinical signs of toxicity, changes in clinical chemistry or hematology parameters or histopathological findings indicative of MPEG-related toxicity.

In repeated dose studies in rats and dogs, lower dosages were utilized, for as many as 10 treatments over a 30-week period. Dogs received ten intravenous infusions of 16 mg/kg total lipid (3.2 mg/kg MPEG-DSPE), administered once every third week. With the exception of minimal clinical observations that occurred during the first few infusions, which are described in more detail below, there were no signs of treatment-related toxicity. There was no pathological evidence of storage of lipids (e.g., foamy macrophages), even in the longest of the multi-dose toxicity tests, nor were there features suggesting disturbance of cellular membranes, both hypothetical possibilities if the liposomes containing MPEG were incorporated into the membranes. Neither was there any evidence suggestive of emboli formation, as determined by extensive microscopic evaluation of tissues. In a standard Segment II (teratogenicity and embryotoxicity) test, rats received dosages of STEALTH placebo liposomes of 8 mg/kg lipid (1.6 mg/kg MPEG-DSPE). No maternal or fetal toxicity was observed, and there was no evidence of developmental toxicity in treated animals.

Despite some evidence that PEG-1450 has minor irritant potential when injected subcutaneously, bolus subcutaneous or intravenous injection of MPEG-containing STEALTH liposomes at doses up to 16 mg lipid (3.2 mg MPEG-DPSE) in New Zealand white rabbits (n = 3 per group) exhibited no evidence of test material-related irritation. Incubation of STEALTH liposomes in vitro in human blood at concentrations up to 8 mg/mL of lipid (1.6 mg/mL MPEG-DSPE) was not hemolytic and was compatible with human serum and plasma when administered in a standard Segment II test (*38*).

As part of pre-approval testing of DOXIL for the European Union, a standard battery of mutagenicity tests were performed, using placebo liposomes at the highest dosage technically feasible. Either with or without an exogenous mammalian microsomal metabolic activation system, MPEG-containing liposomes showed no evidence of mutagenicity in vitro in the *Salmonella* reverse mutation assay (100-3200 μg/plate of MPEG-DSPE), did not induce forward mutations at the thymidine kinase (TK) locus in mouse lymphoma L5178Y cells (100-3200 μg/mL MPEG-DSPE), nor caused chromosomal aberrations or polyploidy in Chinese hamster ovary (CHO) cells (320-3200 μg/mL MPEG-DSPE). They also did not induce micronuclei in bone marrow polychromatic erythrocytes (PCEs) of mice in vivo. Male and female CD-1 mice (n = 15 per sex) received a single intravenous injection of undiluted STEALTH placebo liposomes, a 1:2 dilution or a 1:4 dilution of placebo liposomes in saline. Dose volume was 20 mL/kg. The doses of liposomes delivered corresponded in lipid dose to MPEG-DSPE doses of 16, 32 and 64 mg/kg. Twenty-four, 48 and 72 hours after treatment, bone marrow was extracted and processed for evaluation of PCE micronuclei. There was no increase in micronucleated PCEs over levels observed in the vehicle control (saline) in either sex at any time point.

Although there was no evidence of capillary obstruction, as judged by the lack of respiratory system, renal or cardiovascular system dysfunction, a study of the potential neurobehavioral effects of MPEG-containing liposomes was conducted in Sprague-Dawley rats. Ten rats/group received single bolus intravenous injections of undiluted placebo liposomes, a 1:2 dilution of the stock placebo liposome solution with physiological saline, a 1:10 saline dilution (48, 24 and 4.8 mg/kg MPEG-DSPE, respectively) or saline only. The STEALTH liposomes were dosed undiluted at the maximum recommended dose volume in rats (15 mL/kg), so that the lipid dose, and the MPEG-DSPE dose, delivered was the highest possible. MPEG-DSPE doses per kg were in excess of the dose administered at the LD_{50} of DOXIL in rats (approximately 2.2 mg/kg). Neurological signs and symptoms were evaluated approximately 5 minutes post-dose using a functional observation battery (FOB), a motor activity assessment and a startle response test. The neurobehavioral evaluation included distinct categories of responses that represented a wide range of behavioral and functional endpoints. The FOB included indices of autonomic function, muscle tone and equilibrium, sensorimotor reactivity, thermal sensitivity and central nervous system function. No evidence of any behavioral or functional effects related to treatment with placebo liposomes was seen, nor were there any animal deaths. Responses across dose groups were within normal limits and were consistent with behavioral profiles reported for the Sprague-Dawley rat.

As noted above, dogs treated with STEALTH liposomes exhibited transient acute effects associated with the intravenous infusion. These took the form of mild acute reversible cardiovascular effects limited to the infusion period and clinical signs (prostration, hypersalivation, lethargy), which were presumed to be related to the relatively rapid intravenous infusion of a large amount of lipid. These reactions peaked at the second administration of liposomes and decreased in both incidence and severity thereafter, demonstrating that they were not anaphylactic. To further evaluate this response, a cardiovascular function study was conducted in dogs implanted with radiotelemetry transmitters to monitor arterial pressures, heart rate and respiration. Treatment occurred in four phases. In the first phase, all dogs received saline to establish baseline values for the effect of the infusion procedure on the parameters being monitored. In the second phase, one day later, all dogs received 1.0 mL/kg placebo liposomes (3.2 mg/kg of MPEG-DSPE) at an infusion rate of 1.0 mL/minute and in the third, one week later, groups of three dogs each were treated with 1.0 mL/kg placebo liposomes at 0.25, 0.5 or 1.0 mL/minute. In the final phase, four dogs were pretreated with the antihistamines diphenhydramine HCl (BENADRYL, 4 mg/kg) and ranitidine HCl (ZANTAC, 2 mg/kg) 30 minutes prior to treatment with the MPEG-containing liposomes (1.0 mL/kg, 1.0 mL/minute). The remaining dogs were pretreated with saline. Clinical signs related to test article infusion included hypoactivity, flushing, diarrhea and emesis began shortly after the start of the infusion and resolved within 1-2 hours after treatment ended. The infusion was characterized by a statistically and physiologically significant decrease in blood pressures (19-70%) in 8 of 9 dogs that began immediately after the start of treatment and resolved rapidly after the end of treatment (within 2 hours). Respiration rate and ECGs were unaffected. Heart rates were generally unaffected, with no consistent compensatory acceleration. The extent of the hypotension was greater after the second treatment with placebo liposomes than after the first and was not affected by dose rate, and the slower rate of infusion resulted in a longer duration of hypotension. Pretreatment with antihistamines decreased, but did not eliminate the drop in blood pressure and had no effect on the incidence or severity of clinical signs of toxicity. The biological significance of this finding is not readily apparent, but may be related to histamine release in response to the infusion of a relatively large amount of lipid (the average dog in this study received a total lipid dose of 160 mg lipid in a volume of 10 mL infused over 10-40 minutes). Similar responses have been reported in dogs given other colloidal suspensions, and they do not appear to be related to the MPEG content of the liposomes. No such response has been noted in mice, rats, rabbits or monkeys administered MPEG-containing STEALTH placebo liposomes, further suggesting that the severity of this response may be unique to dogs.

In summary, this extensive series of studies, although conducted to evaluate the safety of liposomes as part of the drug safety studies done for DOXIL, also provided some measure of the relative safety of parenterally administered MPEG. The lipid components of the liposomes are all ubiquitous dietary lipids and constituents of the normal mammalian plasma membrane that are administered at a relative small fraction of their endogenous levels. Thus, the absence of adverse effects associated with the lipids is not surprising, and the negative findings of these studies are a demonstration of the safety of the MPEG-2000 polymer used in STEALTH liposomes.

Summary and Final Conclusions

A review of the literature of the safety of the PEG polymers suggests that they are safe for use in parenterally administered pharmaceuticals. Recent studies reviewed here of the safety of PEG used in imaging agents and as a component of long-circulating liposomes supports this conclusion, with no significant findings of toxicity in several species. Though the intended uses of the two new PEGylated formulations discussed are different, their lack of toxicity is consistent with the long reported findings of other and earlier studies. However, as with any new chemical or drug, safety is dependent upon exact chemical nature (e.g., PEG or a PEG derivative, PEG chain length, etc.), dose, mode of administration and numerous other components. Clearly, each new PEG and PEG derivative will require an independent and comprehensive review of its safety before its use.

Acknowledgments

The authors acknowledge Mark Graham and Peter Hoffman of Sanofi Research Division, Alnwick, Northumberland, UK and Dijon, France for their work on PEG-

1450, and Robert Abra and Anthony Huang of SEQUUS Pharmaceuticals, Menlo Park, California, USA, for their work on pegylated liposomes.

References

1. Pang, S.N.J.; *J. Amer. Coll. Toxicol.* **1993**, *12*, 429-456.
2. Bruns, D.E.; Herold, D.A.; Rodeheaver, G.T.; Edlich. R.F. *Burns*, **1982**, *9*, 49-52.
3. Herold, D.A.; Keil, K.; Bruns, D.E. *Biochem. Pharmacol.* **1989**, *38*, 73-76.
4. Smyth, H.F.; Carpenter, C.P.; Weil, C.S. *J. Am. Pharmacol. Assoc.* **1950**, *44*, 27-30.
5. Johnson, A.H.; Darpatkin, H.H.; Newman, J. *Br. J. Hematol.* **1971**, *21*, 21-41.
6. Bartsch, W.; Sponer, G.; Dietmann, K.; et al. *Arzneimittelforsch* **1976**, *26*,1581-1583.
7. Crook, J.W.; Hott, P.; Weimer, J.T.; et al. US Army Chemical Systems Laboratory Tech. Rept. ARCSL-TR-81058., 1981.
8. Smyth, H.F., Jr.; Carpenter, C.P.; Shaffer, C.B. *J. Amer. Pharmaceu. Assoc.* **1947**, *36*, 157-160.
9. Shideman, F.E.; Procita, I. *Pharmacol. Exp. Ther.* **1951**,*103*, 293-305.
10. Carpenter, C.P.; Woodside, M.D.; Kinkead, E.R.; et al. *Toxicol. Appl. Pharmacol.* **1971**, *18*, 35-40.
11. Lockard, J.S.; Levy, R.H. *Life Sci.* **1978**, *23*, 2499-2502.
12. Lockard, J.S.; Levy R.H.; Congdon, W.C.; et al. *Epilepsia* **1979**, *20*, 77-84.
13. Starke, W.C.; Pellerin, R.J. US Army Chemical Systems Tech. Rept. ARCSL-TR-81029, 1981.
14. Bushy Run Research Center, (1980). Cited in (1).
15. Wangenheim, I., Bolcsfoldi, G. *Mutagenesis* **1988**, *3*, 193-206.
16. Reed, K.W.; Yalkowsky, S.H. *J. Parenter. Sci.Technol.* **1985**, *39*, 64-69.
17. Fort, F.L.; Heyman, I.A.; Kesterson, J.W. *J. Parenter. Sci. Technol.* **1984**, *38*, 82-87.
18. Smith, B.L.; Cadwaller, D.E. *J. Pharm. Sci.* **1967**, *56*, 351-356.
19. Carpenter, C.P.; Shaffer, C.B. *J. Am. Pharm. Assoc. Sci. Ed.* **1952**, *41*, 27-29.
20. Shaffer, C.B.; Critchfield, F.H.; Carpenter, C.P. *Am. J. Physiol.* **1948**, *152*, 93-99.
21. Shaffer, C.B.; Critchfield, F.H.; Nair, J.H. III. *J. Am. Pharm. Assoc. Sci. Ed.* **1950,** *39*, 340-344.
22. Shaffer, C.B.; Critchfield, F.H. *J. Am. Pharm. Assoc. Sci. Ed.* **1947**, *36*, 152-157.
23. Friman, S.; Egestad, B.; Sjovall, J.; Svanvik, J. *J. Hepatol.* **1993**, *17*, 48-55.
24. Yamaoka, T.; Tabata, Y.; Ikada, Y. *J. Pharm. Sci.* **1994**, *83*, 601-605.
25. Topcheiva, I.N. *Polym. Sci. USSR* **1990**, *32*, 833-851.
26. Nathan, A.; Zalipsky, S.; Ertel, S.I., et al. *Bioconjugate Chem.* **1993**, *4*, 54-62.
27. Szoka, F.; Papahadjopoulos, D. *Ann. Rev. Biophys. Bioeng.* **1980**, *9*, 467-508.
28. Senior, J.H. *CRC Crit. Rev. Ther. Drug Carrier Sys.* **1987**, *3*, 121-193.

29.   Woodle, M.C.; Matthay, K.K.; Newman, M.S.; et al. *Biochim. Biophys Acta.* **1992**, *1105*, 192-200.

30.   Woodle, M.C.; Newman, M.S.; Working, P.K. *Stealth Liposomes*; Lasic, D.D.; Martin, F., Ed.; CRC Press: Boca Raton, FL, 1995; pp. 103-117.

31.   Napper, D.H. *Polymeric Stabilization of Colloidal Dispersions*; Academic Press: New York, NY, 1983.

32.   Lasic, D.D.; Martin, F.J.; Gabizon, A.; et al. *Biochim. Biophys. Acta* **1991**, *1070*, 187-192.

33.   Woodle, M.C.; Lasic, D.D. *Biochim. Biophys. Acta* **199**, *1113*, 171-199.

34.   Dvorak, H.F.; Nagy, J.A.; Dvorak, J.T.; et al. *Amer. J. Pathol.* **1988**, *133*, 95-109.

35.   Wu, N.Z.; Dewhirst, M.W. *Microvascular Res.* **1993**, *46*, 231-53.

36.   Wu, N.Z., Da, D., Rudoll, T.L.; et al. *Cancer Res.* **1993**, *53*, 3765-3770.

37.   Freireich EJ et al. *Cancer. Chemother. Rep.* **1966**, *50*, 219-244.

38.   Working, P.K.; Dayan, A.D. *Human Exp. Tox.* **1996**, *15*, 752-785.

Poly(ethylene glycol)-Modified Liposomes and Particulates

Chapter 5

Poly(ethylene glycol)-Grafted Liposome Therapeutics

Martin C. Woodle

Genetic Therapy, Inc., 938 Clopper Road, Gaithersburg, MD 20878

PEG-lipid conjugates have been synthesized and incorporated into liposomes to form a steric polymer surface barrier that enhances drug delivery applications. The PEG steric coating reduces protein binding, cellular recognition, and uptake. Thus these liposomes persist in the blood permitting extravasation into tumors, infections, and sites of inflammation. Thus PEG-grafted liposome formulations can deliver encapsulated drugs to these pathological sites, shown with doxorubicin treatment of tumors and Kaposi's Sarcoma. Other drugs are being evaluated as PEG-grafted liposome drug formulations to take advantage of this form of "passive" targeting. Additionally, efforts are being applied to obtain ligand-mediated targeting or ligand presentation through chemical conjugation to the exterior surface of the PEG coating. Improved drug targeting and use for gene therapy, which requires intracellular delivery, are limited by a need to control tissue distribution separately from drug release or cellular interactions.

Liposomal Drug Delivery

Pharmaceutical products are often enhanced by tissue and cell specific localization of drugs, in particular biological macromolecules such as proteins. Of the drug delivery systems in development, only PEG-grafted long circulating liposomes are considered here. Other forms of liposomes have been reviewed elsewhere (*1-5*).

A key achievement in liposome drug delivery has been introduction of forms exhibiting prolonged circulation. These new forms overcome a major obstacle common to colloidal drug delivery systems: rapid detection and uptake by cells of the mononuclear phagocytic system (MPS). Such uptake reduces beneficial effects of encapsulated drugs and poses a potential for toxicity to these cells (*6,7*). Long circulating liposomes were achieved initially with cohesive lipid bilayers combined with a small particle size and negligible surface charge and then extended with 1) fluorocarbon based phospholipids or 2) inclusion of specific glycolipids (*1,8,9*). These forms enhanced drug delivery but remained dependent on "rigid" liposome bilayers imposing a significant limitation.

Subsequent studies found that inclusion of PEG-lipid conjugates, poly(ethylene glycol)-phosphatidylethanolamine (PEG-PE), also reduces MPS uptake of liposomes but without a requirement for rigid bilayers, reviewed elsewhere (*10*). For drug delivery, PEG coated liposomes have several advantages: a) prolonged circulation independent of the bilayer physicochemical properties thereby permitting separate control of drug loading/leakage and blood circulation, b) plasma pharmacokinetics independent of lipid dose thereby simplifying dose escalation, and c) compatibility with ligands or other chemical functionalities at the outer surface of the PEG coating.

Another key achievement in drug delivery has been realization that simply by persisting in the blood stream small liposomes and other colloids can localize into several important pathological tissues, including solid tumors, infections, and apparently most sites of active inflammation. For these kinds of pathologies, such agents overcome another challenge to drug delivery systems: the difficulty of large macromolecules and particles to pass through the endothelial cell layer and basement membrane of blood vessels limiting access to tissues outside the vascular compartment (*5,8,11-13*). Consequently, PEG-grafted liposomes not only can keep drugs in the blood but they also can deliver them to these selected but important disease areas which can result in greatly enhanced efficacy of encapsulated drugs.

The subject of review is the chemical, physical, and biological properties of PEG-grafted, or sterically stabilized, liposomes as related to applications for drug delivery. The physicochemical and biological properties of PEG-grafted liposomes are described followed by a discussion of the consequences to drug delivery. Results obtained with existing therapeutic agents are described. Finally, their potential drug delivery under active investigation including ligand targeting is discussed.

Properties of PEG-Grafted Liposomes

The biological properties of PEG coated liposomes advantageous for drug delivery are a consequence of physicochemical interactions of the polymer with proteins and cells.

Physico-Chemical Properties. A steric barrier is thought to be formed on the surface of PEG-grafted liposomes which inhibits protein adsorption, particle opsonization, and concomitant MPS uptake (*1,14*). Physical measurements are beginning to substantiate this mechanism but further investigation and refinement are needed (*14-20*). PEG-grafted liposomes also are referred to here as sterically stabilized liposomes (SSL).

Liposomes with surface bound PEG have been obtained both by chemical conjugation onto preformed liposomes or by addition of PEG-lipid conjugates to the lipid mixtures used for liposome formation. A number of PEG-lipid conjugates have been prepared but many exhibit similar effects despite chemical differences. The effects are attributed to specifics of the physicochemical properties.

Physico-Chemical Properties of PEG-Lipid Conjugates. PEG was one of the first synthetic polymers examined to graft onto the surface of liposomes for the purpose of prolonging blood circulation. It is still preferred despite early expectations that other hydrophilic polymers would give similar effects (*4,21,22*). Attempts to use dextran for a similar increase in blood circulation apparently were not successful while some success was reported with polyvinylpyrolidone (23) and polyglycerols (24). These polymers showed similar effects but none proved to be the equal of PEG. More recently, other polymers equal to PEG have been identified: two forms of polyoxazolines which represent a very different class of water soluble polymers (25,26). Thus, the properties of PEG-grafted liposomes may be more firmly described as a physical phenomena rather than a specific unique biochemical effect of PEG.

A number of methods have been used to conjugate PEG to lipids, described in recent reviews (27,28). A common approach has been to react activated PEG with the primary amine of PE. Mono-functional methoxy-PEG reactive at only one end reduces the number of potential products. PEG-PE containing a carbamate or urethane linkage has good stability characteristics (29) and is commercially available. The chemical structure of this form of PEG-PE is shown in Figure 1. Note that chemical differences can exist between molecules all referred to as PEG-PE yet the overall nature of these molecules is quite similar. For example, several methods can be used to conjugate PEG to PE but most eliminate protonation of the amine in aqueous solution. Thus these result in different chemical species but with similar pK properties. For liposomes containing them, the similarities appear more important than the chemical differences.

The carbamate linkage provides a stable PEG-PE both for storage and in vivo (29). The greatest chemical instabilities of this PEG-PE are probably due to fatty acid unsaturation and the esters within the lipid portion. It is not clear whether oxidation of the PEG or degradation of the carbamate linkage represent stability problems. The biological compatibility of the chemical nature of PEG-PE may become a concern for materials intended as components in pharmaceutical products. Nonetheless, recent US FDA approval has been given to a liposome formulation containing PEG-PE and doxorubicin (Dox).

An alternative view of potential PEG linkage instability is as a method to achieve intentional loss of the PEG-coating in specific biological environments. An example of this concept has been illustrated with PEG-lipid conjugates with different alkyl components which result in gradual loss of the PEG-lipid from the liposome (30). Another report described an approach using a disulfide linkage which is sensitive to reduction (31). Other specific chemical conditions of interest for loss of the PEG coat include acid pH and target binding (3,32-35). At the present transformation of liposomes due to these environments have been examined largely with conventional liposomes or glycolipid forms of long-circulation (36). Unfortunately, differences in pH, redox conditions, or other chemical parameters between healthy and target tissues may be small and thus will probably require considerable optimization of the chemical nature of the PEG-lipid conjugate.

Introduction of PEG-PE replacing the distal methoxy with chemical functionalities useful for conjugation is an important advance (27,37-40). A number of heterobifunctional PEG-lipid conjugates have been reported; some shown in Figure 2. A substantial range of functionalities has been prepared such that conjugation of most biological molecules to the surface of the PEG coating on liposomes is possible through standard conjugation reactions. Consequently, these heterobifunctional PEG-lipid conjugates represent important methods to provide ligand-mediated targeting,

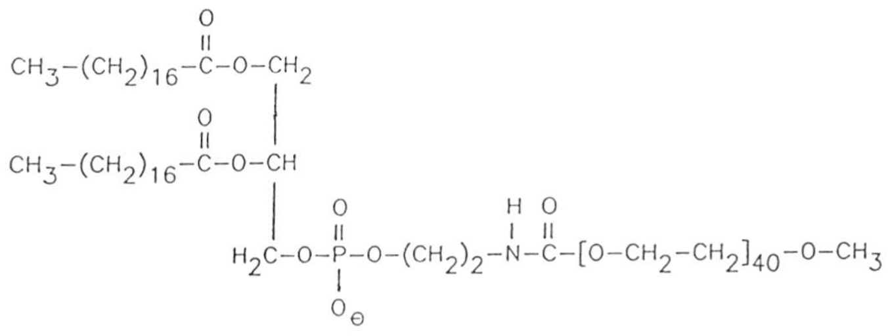

Figure 1. The chemical structure of PEG-PE prepared with 2 kd PEG and a carbamate linkage to DSPE.

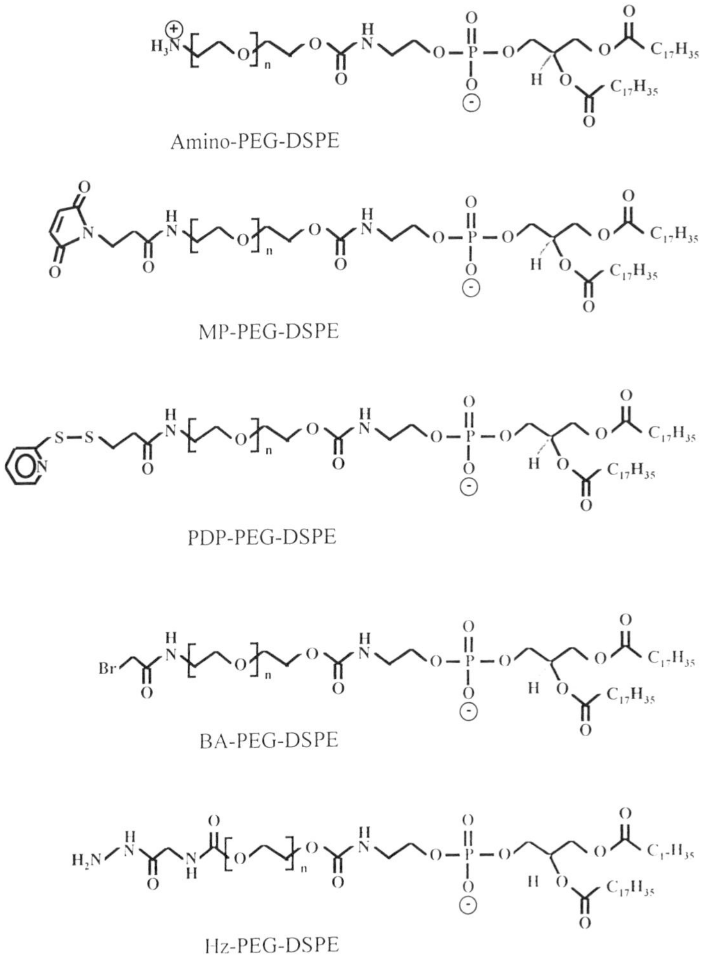

Amino-PEG-DSPE

MP-PEG-DSPE

PDP-PEG-DSPE

BA-PEG-DSPE

Hz-PEG-DSPE

Figure 2.   The chemical structure of several heterobifunctional PEG-DSPE conjugates with functional groups at the distal end of the PEG: primary amine, MP - maleimide, PDP - disulfide or sulfhydryl, Br - bromoacetate, and Hz - hydrazide. Adapted from reference (27).

ligand presentation, and other therapeutic advances to liposome drug delivery systems and can be expected to form the basis of considerable ongoing efforts. The physical properties of PEG-lipid conjugates are distinct from those of the polymer and lipid components and in some interesting ways (29,41-43). The combination of hydrophilic, water soluble PEG with hydrophobic lipids results in surfactant properties. Studies of PEG-PE show that it can form micelles giving clear solutions in water even up to concentrations as high as several mg/ml (41). This means that PEG-PE has the potential to destabilize bilayer formation when added at a high mole content relative to the other lipids. It also means that PEG-PE has the potential to disassociate from the liposome and form micelles in equilibrium with bilayers (18,30). In fact, this "problem" has been taken advantage of to achieve a slow loss of the PEG-coating by selection of the lipid anchor. A reversal of this process, insertion of PEG-PE from micelles into preformed liposomes, also has been used as a means to prepare liposomes with a PEG coating (44). The physical properties of PEG-lipid conjugates and liposomes containing them are not fully understood.

Colloidal Surface Steric Stabilization. A steric barrier on the surface of liposomes and other PEG coated particles inhibiting interactions with proteins and cells is a working hypothesis to explain their in vivo properties (1,45). Some physical measurements have been reported substantiating this mechanism (14-20,46,47). Perhaps the most extensive efforts include surface repulsion, steric pressure, and zeta potential measurements to estimate polymer thickness on the liposome surface (15,17,18,20,46). These studies are in fairly good agreement; the coating formed by 2000 kd PEG is reported to be on the order of about 5 nm in thickness (almost 4 nm by zeta potential (20) and 6 nm by steric pressure (18)). Note that steric stabilization of liposomes refers to a protection from destructive biological interactions: extraction and loss of lipids from the liposome, protein adsorption onto and into the bilayer, and cellular uptake followed by digestion. Further evidence of the PEG action as a steric barrier comes from findings that gradual loss from liposomes by disassociation restores liposome-cellular interactions, e.g. endocytosis and fusion (48).

The covalent attachment of PEG to the surface of liposomes results in anchoring the water-soluble polymer to the bilayer surface. The grafting density and PEG molecular weight determines the extent of interaction between the PEG molecules. Also, the hydrophilic and dynamic properties of the polymer could result in exclusion from the liposome surface, i.e. the polymer may try to keep as far from the surface as permitted by the covalent attachment. This concept is taken here to suggest that the space occupied by the PEG at low grafting densities may resemble that of a pear with its stem attached to the liposome surface. When the PEG surface grafting density is low each molecule is isolated. This regime also is referred to as the "mushroom" domain (1). Note that evidence in support of the exclusion of PEG from the liposome surface is lacking, in particular direct measurement of a depletion layer. The exact location of the PEG varies over time; it is a dynamic rather than static shape and thus barrier. The polymer mobility should be reduced at the anchoring site and increase as the distance from the lipid anchor increases; it is proposed to play a significant role in the steric stabilization mechanism (14). The exact nature of polymer shape and dynamics needs further investigation.

If the grafting density is increased, at some point each PEG molecule begins to encounter its neighbors. Beyond this point, the polymers are forced together into a more extended shape in a regime referred to as the "brush" domain in reference to the structure of straight bristles in a brush. With liposomes, the PEG-PE content of bilayers eventually reaches a limit as discussed above which means that it apparently forms micelles mixed with the liposomes rather than very extended brush forms of the PEG polymer chains. The actual state depends upon the surface pressure between PEG chains relative to the energetics of micelle versus bilayer formation. There may be a preferred grafting density that provides the greatest barrier to in vivo interactions, such as regions at or near the transition between mushroom and brush (*10,20*).

Zeta potential measurements of PEG-PE containing liposomes may provide some support for the importance of the transition between mushroom and brush (*20*). Note that PEG-PE, as shown in Fig 1, retains a phosphate which is ionized at physiological pH and thus it contributes a negative liposome surface charge providing electrophoretic mobility and zeta potential. Zeta potential measurements have been used to determine the thickness of the PEG layer on the liposome surface due to reduction in hydrodynamic movement. The transition between mushroom and brush regimes can be estimated as a function of both PEG molecular weight and grafting density and used to compare these surface thickness measurements with biological measurements. Both prolonged blood circulation and surface thickness consistent with the transition region are obtained with 2 to 5 kd PEG at a mol% of around 5% but not with 2 kd PEG at < 2 mol% or with 0.75 kd PEG at up to 30 mol% (*20*). These data suggest a correlation of a coating thickness proximal to the mushroom-brush transition with prolonged circulation. If this correlation is correct it suggests that 1 kd PEG-PE at 10 mol% or higher should also be effective. It also suggests that higher molecular weight PEG at lower mol% may also fall on the line. These conditions need to be tested to verify the correlation. There appears to be a lower limit to the PEG coating thickness and an upper limit, perhaps representative of a shift of the equilibrium toward micelle formation.

Another unanswered question is the result of multiple liposome surface attachments of the polymer on the biological properties. Such multiple attachments should reduce the mobility of the polymer and thus may be deleterious. A simple method to approach such mechanistic studies would be to work with PEG conjugated to lipids at both ends. However, a material of this type might be difficult to incorporate before liposome formation since the lipid at each end of the PEG could permit connections between lipid bilayers. This may be a case best served by adding the PEG-lipid to preformed liposomes (*44*).

In summary, the exact natures of the polymer surface density, length, dynamics, and other properties determinant of in vivo properties need further investigation. It appears that the effect may be firmly described as a physical phenomena rather than a specific and unique biochemical property of PEG.

Biological Properties. Prolonged circulation of liposomes was first observed through use of highly cohesive bilayers, subsequently through incorporation of specific glycolipids with small headgroups, and now with surface-grafted PEG (*1,2,4,5*). The initial findings with PEG-grafted liposomes were obtained in two rodent species;

results in rats are shown in Figure 3 (*49*). The effect has been confirmed in human clinical studies (*10,50*). Typically PEG-grafted liposomes exhibit a circulation half-life of 12-20 hours in rats or mice and 40-60 hours in humans. The biological properties of liposomes containing PEG-PE have been examined but primarily the focus has been on pharmacokinetics and effects on encapsulated drugs.

Avoidance of MPS Uptake And Its Consequences. The current understanding of liposome biological interactions is based primarily on in vitro studies with cell cultures and to a lesser extent from in vivo administration. The generally accepted interactions fall into four categories: (1) exchange of materials such as lipids and proteins between liposomes and the biological environment, (2) adsorption or binding of liposomes to cells, (3) cellular internalization by endocytosis or phagocytosis, and (4) rarely fusion of bound liposomes with cell membranes. The determination of which interactions will occur depends strongly on lipid composition, type of cell, presence of specific receptors, and many other parameters.

In vivo, the majority of liposomes are taken up by MPS cells through endocytosis and phagocytosis. The few escaping distribute into tissues and eventually disintegrate into individual components. Thus the fate of encapsulated drugs is to be degraded within endosomes and lysosomes although a few can survive and be released back into the blood, i.e. doxorubicin and amphotericin B. Thus MPS uptake has proven to be a major barrier to liposomal drug delivery.

Sterically stabilized liposomes exhibit reduced macrophage uptake in tissue culture studies. The actual mechanism appears to involve reduction in protein adsorption and reduced cellular interactions (*45,51-54*). In fact, combination of PEG-PE and antibodies covalently attached to the surface of liposomes reduces antibody binding to cell surfaces expressing the specific antigen according to the PEG molecular weight and mol% as compared with similar liposomes lacking the PEG-PE (*55,56*). With reduced MPS uptake and prolonged circulation, sterically stabilized liposomes have overcome the rapid inactivation of encapsulated drugs found with earlier types of liposomes. They also reduce the potential for toxicity of MPS tissues by encapsulated drugs. Consequently, SSL provide opportunities to bypass rapid degradation or excretion of therapeutic agents from the blood.

Pharmacokinetics and Tissue Distribution. Upon iv administration, liposomal drug formulations are diluted extensively into the blood and other biological fluids. These fluids, especially the blood, contain isotonic concentrations of salts and various proteins which can alter the liposome physical properties. Of particular importance are the opsonin proteins, lipoproteins, phospholipases, divalent ions, extent of dilution, and specific liposome properties such as size, lamellarity, lipid composition, and surface charge (*1,57*). Liposomes are thought to adsorb a number of different proteins as well as small lipophilic molecules. However, the interactions of liposomes with in vivo fluids does not usually result in instantaneous liposome destruction. With highly cohesive bilayers, in vivo changes can have minimal influence on liposome integrity. Cellular interactions, though, are influenced by adsorbed plasma proteins, i.e. opsonin proteins leading to recognition and cellular uptake of liposomes and other particles by the MPS.

Liposome uptake and clearance from blood generally is thought to occur by stepwise processes: opsonization or marking by blood proteins followed by macrophage uptake of the marked liposomes. Unfortunately, the opsonization process is poorly defined both in terms of the specific plasma factors and in how they promote phagocytosis. Nevertheless, clearance occurs with multiple kinetic components and the overall process can be saturated at high lipid dose. Pharmacokinetic (PK) studies of liposome blood clearance revealed an apparently special liposome composition showing prolonged circulation at high saturating lipid dose: distearylphosphatidylcholine (DSPC) mixed with cholesterol (1). However, the usefulness of this formulation for drug delivery is constrained by narrow limits on particle size, lipid composition, lipid dose, and net surface charge.

Originally it was thought that MPS uptake would at least allow liposomes to be useful for drug delivery to macrophages. However, following iv administration they are rapidly removed from blood by macrophages primarily in liver and to a lesser extent in spleen and bone marrow at the expense of macrophages in other tissues, especially the lymph nodes. Note that SSL exhibit reduced, but not eliminated, macrophage recognition and phagocytosis. Thus an interesting application of PEG-grafted liposomes which may seem somewhat like a contradiction is to increase drug exposure to MPS but applies to tissues outside liver and spleen (58-60).

PEG-grafted liposomes exhibit prolonged blood circulation comparable to that attained at saturating doses of small neutral SUVs composed of DSPC:Chol. Blood circulation measurements in rats using an encapsulated aqueous radioactive metal tracer are shown in Figure 3 (49). Similar results of PEG-grafted liposomes have been reported by a number of laboratories. Blood clearance studies of entrapped drugs show similar results (46,61). Overall, the results indicate that their in vivo behavior is not substantially effected by drug encapsulation. Importantly, recent clinical studies have confirmed prolonged circulation largely independent of lipid dose in humans (62,63). Prolonged circulation independent of lipid dose and other liposome properties, e.g. bilayer lipid composition and surface charge, is very important for drug delivery. PEG-grafted liposomes are useful for drug delivery but also represent a means to further elucidate colloid recognition and uptake mechanisms in vivo.

Studies of tissue distribution have focused on MPS organs while other tissues have not received as much attention. With the liver and spleen, the traditional site of liposome uptake, the combined hepatosplenic uptake is reduced consistently in all reports with liposomes containing 2 kd PEG-PE. Small liposomes, mean diameters < 100 nm, have been examined in most reports. Prolonged circulation also can be retained by larger liposomes, up to 200 or even 300 nm, although the tissue distribution shifts towards the spleen (64). This may prove useful for vaccines and other applications. Within the liver, a study of distribution between parenchymal and non-parenchymal cells indicated that about 80% of small SSL are associated with hepatocytes where as this dropped to 50% with a larger size (65). A careful evaluation of the differences in uptake by these two tissues as a function of chemical and physical composition is helpful in gaining a better understanding of the differences in these two important MPS tissues and the consequences for liposome delivery of drugs (66).

In general, uptake of PEG-grafted liposomes by healthy tissues other than MPS is not well characterized. A fairly large amount of the injected dose clears from the blood and distributes to skin, "carcass", and as yet unidentified tissues or locations and remains there for prolonged periods (46,67). This phenomena provides a reservoir in tissues other than the MPS from which drugs can be released over time. It also suggests that encapsulated drugs distribute to these unknown tissues and locations posing potential new efficacy or toxicity due to drug exposure to these tissues.

Pathological Tissue Localization. One of the most important roles a drug delivery system can provide is selective localization to desired pathological sites, a limited form of the magic bullet concept. PEG-grafted liposomes exhibit significant accumulation into several important pathologies: solid tumors, infections, and inflammations (10,12,68). In one example shown in Figure 4, increased levels of Dox were shown to accumulate in an implanted tumor when the drug was encapsulated in PEG-grafted liposomes compared with free drug (69). In this example, tumor accumulation occurs over the prolonged period of liposome circulation. Poorly controlled tumor angiogensis resulting in increased vascular leakage may explain this phenomenon (13). Despite significant differences in the types of pathological tissues into which PEG-grafted liposome localize, the results may be indicative of a common mechanism: increased extravasation into areas of leaky vasculature as a result of prolonged circulation (12,70). Evidence of increased permeability of PEG-grafted liposomes has also been reported (71). Regardless of mechanism, up to 10% of the injected dose can localize in the pathological sites. This suggests a potential for substantial improvements in efficacy of encapsulated therapeutic agents active toward these pathologies. Localization derived enhancement of encapsulated drugs has been demonstrated for treatment of tumors and infections, discussed in the following section, but remains to be demonstrated for inflammations. It should be stressed that successful efficacy improvements require achieving a balance between retention of the agent by the liposome during circulation and accumulation in a pathological site and its release from the liposome to exert activity subsequent to liposome accumulation.

PEG-Grafted Liposome Drug Formulations

Doxorubicin. The development of PEG-grafted liposomes has been largely in combination with doxorubicin (Dox) and it represents the best understood therapeutic application of long circulating liposomes (72,73). Also pH and ion gradient loading methods were developed with Dox, achieving high encapsulation efficiencies regardless of lipid composition permitting new formulations to be prepared and tested easily (74-76). The result of Dox formulation with PEG-grafted liposomes is enhanced efficacy in preclinical animal tumor models (5,10,69) and in clinical studies (62,63,77-79). An example of implanted tumor treatment by PEG-grafted liposome formulations of Dox and another oncology agent are compared with effects of the free drugs in Figure 5. This data shows that animals treated with a saline control and the free drugs at the maximum tolerated dose show unabated tumor growth whereas animals treated with the PEG-grafted liposome formulations show tumor inhibition and, more importantly, regression. This effect of the PEG-grafted liposomes is probably substantially increased activity of the Dox at the tumor. This formulation of

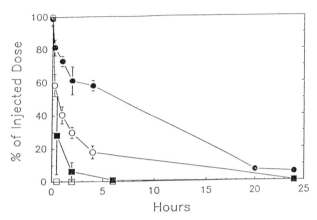

Figure 3. Blood circulation of liposomes following iv administration in rats: Ga-desferal labeled liposomes containing PEG-PE:PC:Chol, closed circles, PC:Chol, open circles, PG:PC:Chol, closed squares, and free Ga-desferal label, open squares (*49*). Copyright CRC Press, used with permission.

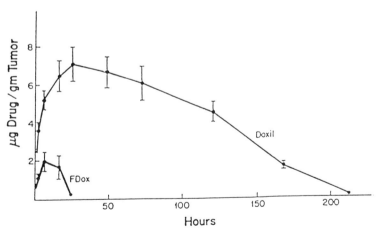

Figure 4.   Localization of doxorubicin in implanted tumors determined by microfluorimetry following 0.9 mg/kg Doxil or free Dox given by iv administration at time zero (*69*). Copyright CRC Press, used with permission.

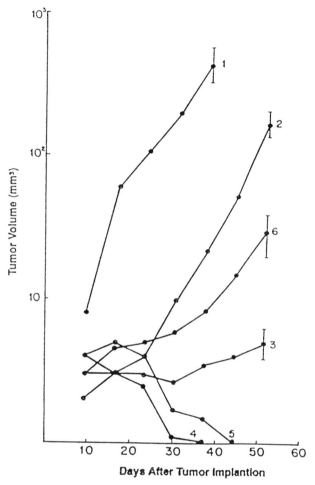

Figure 5. Efficacy of Dox and vincristine in PEG-grafted liposome formulations for the treatment of MC2 implanted tumors in mice (*69*). Treatment groups are as follows: 1) saline control; 2) 1 mg/kg Vcr liposome given on days 3, 10, and 17; 3) 6 mg/kg Dox liposome given on days 3, 10, and 17; 4) alternating treatment equivalent to groups two plus three where Vcr liposome given on days 3, 10, and 17 and Dox liposome given on days 6, 13, and 20; 5) alternating treatment equivalent to groups three plus two where Dox liposome given on days 3, 10, and 17 and Vcr liposome given on days 6, 13, and 20; and 6) simultaneous treatment equivalent to groups two plus three where both liposome drugs given on days 3, 10, and 17. Copyright CRC Press, used with permission.

Dox has been developed (Doxil®) and has recently been approved by the US FDA for treatment of Kaposi's sarcoma patients. Clinical considerations of long circulating liposome formulation therapeutics are reviewed in a recent book (*50*).

Plasma pharmacokinetic studies of Doxil in rats, dogs, and humans show prolonged circulation essentially insensitive to dose. The half-life in humans is 40 to 60 hours resulting in about 10% of the injected dose still in the plasma after one week (*62,63*). The initial plasma levels and area-under-the-curve (AUC) were linear with on dose. Liposomes circulating for a week in humans creates the requirement, and challenge, to retain drug for this period of time until they reach the tumor. Apparently this requirement is fulfilled with Doxil since unencapsulated "free" Dox in the plasma is quite low and the majority of the drug in the blood is bound to liposomes. The formulation consists of fully hydrogenated lipids with cholesterol forming a rigid liposome bilayer and the Dox forms a gel inside the liposomes, perhaps together minimizing leakage (*74*). Once at the tumor, the observed efficacy of Doxil means that drug release occurs even from this stable formulation. The efficacy of Doxil strongly supports the need for prolonged circulation to obtain tumor localization and stability during circulation yet balanced with release at the tumor.

An interesting question is whether optimization of the Doxil formulation for balance between stability and release can further improve the therapeutic index. The use of temperature sensitive liposomes and application of heat to the region of the tumor has been studied (*80-86*). The concept is to selectively enhance tumor release. In practice it turns out that extravasation is also enhanced at the heated site and thus both localization and release are influenced. The actual extent of efficacy improvement which can be obtained by this or other manners has yet to be fully determined. It may prove to be difficult to accurately control release at desired sites.

Other Oncology Agents. PEG-grafted liposome formulations which localize in tumors promise broad applicability to many anticancer agents. Conceptually, most current oncology agents could benefit from better delivery to solid tumors, or even just reduced distribution to healthy tissues, and a number have been considered for encapsulation in PEG-grafted liposomes. However, success will be limited to those that, like Dox, can be encapsulated with an appropriate balance between stability and release. Given the wide range of oncology agents with ranging physical properties and relatively few means to adjust liposome-drug interactions, it can be a formidable challenge to identify appropriate formulations. Nonetheless, what has been learned about liposome interactions of existing agents from earlier studies with conventional liposomes appears applicable to PEG-grafted liposomes.

One approach has been to focus on liposomal formulations of agents applicable to pH or ion gradient "after-loading" methods. This method generally provides high efficiency and stable encapsulation (*1*). Improved efficacy has been reported with long circulating liposome formulations prepared in this manner using epirubicin, vincristine, or mitoxantrone (*69,75,87-89*). Improved therapy of animal tumor models with vincristine in PEG-grafted liposomes alone or in a combination protocol with Doxil is shown in Figure 5. The results suggest enhancement of the individual drugs and even greater enhancement by combination in an alternating regimen. This demonstrates broad applicability of PEG-grafted liposomes for oncology drug delivery and particularly for drugs used in combination protocols, but the applicability will be limited to those drugs capable of "meta-stable" liposome loading.

Other loading methods include passive encapsulation and lipophilic derivatives or prodrugs. Passive encapsulation is applicable to most water soluble molecules but the loading efficiency tends to be quite low, a serious limitation of such formulations for use as a pharmaceutical product. This approach has been applied to encapsulation of cytosine arabinoside (araC) in PEG-grafted liposomes which also gave improvements in efficacy (*90*). Note that existing drugs are active at intracellular compartments and they probably accomplish this by diffusion through the cell membrane. Consequently, many tend to leak out of liposomes, such as 5-fluorouracil (5-FU) and taxol (*91,92*). However, if an agent can leak out, it also should be able to diffuse into liposomes. In this case, encapsulation of traps inside the liposomes can be considered as a means to stably entrap the drug inside liposomes. Also, derivatization of agents to make them lipophilic, e.g. addition of acyl groups, can provide liposome binding as studied with 5-FU (*91*). These approaches should further expand the therapeutic application of PEG-grafted liposomes for oncology.

Other Therapeutics and Diagnostics. The ability of PEG-grafted liposomes to persist in the blood provides applications to drug delivery outside oncology. Localization occurs at pathologies other than tumors, including infections and inflammation. Development of antibiotics has been examined using both antibacterial and antifungal agents. The results have shown localization at severe lung infection and demonstrated improvements for gentamicin and amphotericin B (*93-95*). Alternatively, blood persistence alone can be used to provide controlled systemic release of therapeutics such as biomacromolecules or simply to act as a platform for prolonged display of ligands to the blood (*96,97*). One study reported using a peptide hormone, vasopressin (*98*). Other examples can be found in recent evaluations for cytokines (*99*), and radiopharmaceuticals (*100-104*). Importantly, the properties of PEG-grafted liposomes suggests consideration for diagnostics (*105*). PEG-grafted liposomes provide multifaceted capabilities applicable to many agents and indications.

Potential Future Developments: Targeting & Intracellular Delivery?

Ligand Driven Localization. Despite the demonstrated success of PEG-grafted liposomes to passively localize therapeutic agents into a few pathological sites, e.g. tumors & infections, the addition of ligands to their surface may have important benefits. One concept is the use of surface ligands to provide targeting, or increase it, to specific pathological tissues. Note that ligands targeted to cells located within the blood or tissues in contact with the blood represent the lowest barrier to binding. Another application is presentation of ligands on the surface of the liposomes to increase their exposure to the blood or other tissues where the PEG-grafted liposomes are prevalent (*96,106*). A specific example is to deliver an enzymatic activity on the surface of the liposomes which converts a widely distributed small molecule substrate to an active therapeutic agent but only where the liposome has distributed (*107,108*).

Many therapeutic applications can be devised relying on presentation of cell surface receptors, their ligands, inhibitors, or other molecules on the liposome surface.

Unfortunately, an adverse consequence of steric stabilization is inhibition of liposome surface interactions which diminishes the value of surface bound ligands. Nonetheless, this problem can be overcome by a couple of strategies. One successful strategy has been to place the ligand at the exterior of the PEG coating by conjugation to the distal end of the PEG. This is shown schematically in Figure 6 for liposomes containing Hz-PEG-PE and a ligand with aldehyde functional groups. Liposomes incorporating the modified polymer-lipid conjugate are incubated with the aldehyde containing ligand and the resulting conjugate is formed at the exterior of the PEG coating. In this method, the conjugation can be converted to a more stable form by a subsequent reduction step. A number of alternative chemistries are possible for conjugation to the distal end of the PEG, see Figure 2. Another strategy is to use unstable PEG conjugation methods which result in a loss of PEG-coating and restoration of ligand interactions (*31,48*). This is an important future area of study.

The addition of a hydrazide to the end of PEG-PE (Hz-PEG-PE) has proven especially useful for conjugation of proteins and peptides to the surface of PEG-grafted liposomes (*37*). The method is based on conjugation to ligands with an aldehyde as shown in Figure 6. In the case of glycosylated proteins, the aldehyde can be formed by periodate oxidation of the glycosylation site. With certain peptide sequences an N-terminal threonine amino acid residue can be similarly oxidized (*96*). This method has some important advantages over most other methods developed for conjugation of antibodies to liposomes, in particular a greater selectivity of orientation compared with methods using random derivatives of primary amino groups (*109*). Also, it uses gentle oxidation conditions and can be performed as a two step, one vessel reaction procedure. Interesting features of this method are 1) determination of conjugation site and orientation by glycosylation site, 2) an ability to quench excess periodate with N-acetylmethionine at the end of the first reaction without producing substances that interfere with the hydrazide conjugation reaction (or thought to be troubling to pharmaceutical application), and 3) the choice of either leaving the hydrazone as a reversible linkage or reducing it to an irreversible linkage (*37,96,110,111*). Consequently, this method is one of the best approaches developed to date for conjugation of peptides and glycoproteins to PEG-grafted liposomes.

Any ligand-directed targeting method is dependent upon identification of selective ligands. A major interest is to enhance cell selective targeting and antibodies are an obvious selective ligand. Antibody targeted liposomes, or immunoliposomes, have been studied extensively and combination with PEG-grafted liposomes can be referred to as sterically stabilized immunoliposomes (SSIL) (*67,112*). A number of studies have evaluated SSIL (*55,67,110-118*). One of the first findings was of an adverse effect of the PEG coat on liposome surface bound antibody binding to its ligand. Subsequently, increased recognition of antibodies on SSIL was achieved by reducing the polymer coating thickness or moving antibody attachment to the exterior of the PEG. The studies with antibodies conjugated to SSL provide evidence for value but also complications of cell targeting by SSIL (*111*). So far, targeting to solid tumors has proven difficult to increase beyond that attained by simple PEG-grafted liposomes (*67*) and thus other ligands or strategies may prove to be of greater value.

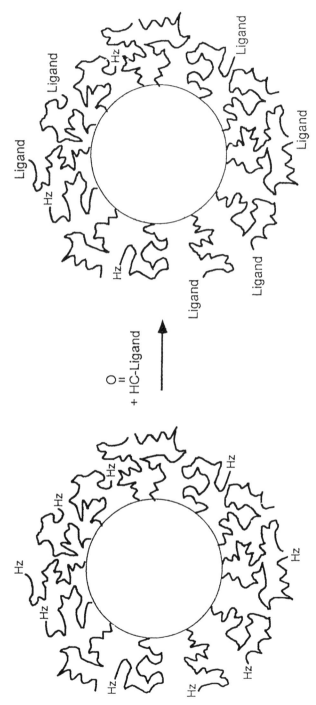

Figure 6. Scheme for ligand presentation on the external surface of PEG-grafted liposomes by conjugation of aldehyde containing ligands to the distal end of Hz-PEG-PE containing liposomes.

For several reasons, antibodies and other large proteins may not be the best ligand for drug delivery. For example, most conjugation methods developed for antibodies and proteins, and in particular those used to prepare immunoliposomes, rely on derivatization of antibody side chains without specificity and thus give many orientations on the liposome surface. Also, the nature of the chemistry used for conjugation influences antibody function, as found for conjugation of chelators to antibodies (*109,119*). Strategies to address this are use of glycosylation sites for conjugation and conjugation to antibody fragments through naturally occurring free sulfhydryl groups on cysteine residues. In the former case oxidized glycosylation sites can be coupled with hydrazide-PEG-PE where as in the latter case coupling can be made with maleimido-PEG-PE (*40,120,121*) or the more conventional maleimido-PE (*56*). Recently studies with antibody fragments coupled through maleimido-PEG-PE indicated that the ligand resulted in target-specific endocytosis equal to that on conventional immunoliposomes (*121*). Consequently these two coupling chemistries may prove to be preferred to maintain antibody function. Another significant problem identified with SSIL is reintroduction of nonspecific in vivo interactions leading to increased recognition and blood clearance. Successful development of SSIL will require adjustment of many parameters to address these and other issues. The efforts are beginning to yield information on possible advantages of SSIL.

The use of small molecules such as peptides and oligosaccharides as ligands on the surface of PEG-grafted liposomes may prove more useful than antibodies, their fragments, and other large proteins. For example, small peptides or oligosaccharides can provide similar binding constants for cell targeting while generally requiring specific chemistries and thus have defined orientations. Perhaps most importantly, they also provide better opportunities to control nonspecific in vivo interactions. For example, a study with amino-PEG-PE in liposomes demonstrated that low levels of cationic charge can be accommodated at the end of the polymer (*38*). This indicates that positive charge can be accommodated at the surface of the polymer as may be required for many small molecule ligands. Also, a range of conjugation methods have now been developed applicable to many small molecules intended either for targeting or for other purposes. Unfortunately, active small molecules are more difficult to identify and in some cases more difficult to prepare, e.g. many oligosaccharides. When they are identified and produced, though, they should help to overcome many disadvantages of antibodies and large proteins. In particular, the extent of immune responses generated by antibodies on PEG-grafted liposomes may be reduced or perhaps completely eliminated with a small peptide. Clearly, the end of the PEG coating is the preferred location for such targeting, as demonstrated using folate to provide selective recognition and uptake by cells expressing its receptor (*122*). Also, efforts have been reported using a small peptide fragment of laminin, YIGSR, with strong binding to laminin receptor for the purpose of inhibiting metastasis formation (*96*). Likewise, similar activities have been reported with small oligosaccharide ligands (*123,124*). This area of investigation is only just beginning, but a rapid expansion of results in this area can be expected.

Intracellular Delivery. Recently considerable effort has been reported to create non-viral intracellular delivery systems. Many efforts are focused upon cationic lipid complexes of nucleic acid agents (*125,126*). As for liposomes, the original suggestion that they could fuse with cells to deliver drugs directly to the cytoplasm has not been realized. Fusion of cell membranes is a highly regulated and controlled essential cellular process such as exocytosis and endocytosis. It does not occur readily with liposomal bilayers except when induced such as with reconstituted viral proteins or inclusion of pH sensitive lipids that destabilize and disrupt the membrane at low pH (*1,127-130*). To be successful, liposomal non-viral intracellular delivery may depend on the capabilities of PEG-grafted liposomes: reduced non-specific interactions, blood persistence, and extravasation through endothelial barriers to reach target tissues. In addition, they will demand other capabilities: membrane fusion with target cells to delivery agents intracellularly. Therefore intracellular delivery will depend on separation and control of these mutually exclusive requirements. Another important challenge is to emulate viral assembly by efficient encapsulation processes. Thus non-viral gene therapy represents a significant challenge for drug delivery.

Two forms of gene function therapeutic agents have been considered in general: large double stranded DNA, plasmids, for conveyance of new gene expression to cells and small oligonucleotides, antisense, ribozyme, and triplex oligos, for selective interactions with gene function (*131*). Both kinds seem to require intracellular delivery with only a few exceptions (*132,133*). Nonetheless, some therapeutic strategies using oligonucleotides are independent of intracellular delivery, i.e. agents self-sufficient in intracellular localization or extracellular activity, and thus traditional liposomal transport of oligonucleotides has received interest (*134*). Ultimately, though, both types of these therapeutic agents will be enhanced by intracellular delivery systems with some of the PEG-grafted liposome capabilities.

The development of long circulating liposomes for non-viral gene delivery has only just begun but some studies are encouraging, e.g. (*35,122*). As with ligand targeting, an adverse consequence of steric stabilization is inhibition of liposome surface interactions and cellular uptake, although evidence already exists that the barriers can be overcome (*121*). Thus, intracellular delivery, dependent upon protein and cellular interactions, with PEG-grafted liposomes represents a significant challenge. This problem may prove less significant given results with folate where the surface PEG was not prohibitive of endocytosis and cellular uptake (*122*). Since PEG is prohibitive of intracellular delivery, methods to remove it selectively are being developed. As mentioned above, loss of the PEG coating over time and chemical sensitive linkers have been described as approaches (*31*). Another potential problem is that simple addition of fusogenic materials to the surface of PEG-grafted liposomes may simply destabilize the liposomes. Therefore tissue or binding sensitive materials may be required as reported using target sensitive and pH sensitive immunoliposomes (*3,32-34*). It may be possible to prepare long circulating liposome formulations which bind and then are triggered much like a virus. However, this approach may be limited to target cells that take up the liposomes by endocytosis. These strategies represent efforts to develop synthetic viral particles, a formidable but interesting challenge.

Acknowledgments

Thanks to Samuel Zalipsky, Demetrios Papahadjopoulos, Joel Cohen, Gert Storm, Dan Lasic, John Jaeger, Irma Bakker-Woudenberg, Chesi Barenholz, Peter Working, Ken Huang, Alberto Gabizon, Terry Allen, Hans Schrier, Frank Szoka, Pieter Cullis, Mary Newman, Paul Uster, Frank Martin, Milton Harris, and many others for helpful discussions and comments.

References

1. Lasic, D. D. *Liposomes: from physics to applications;* ; Elsevier: Amsterdam, 1993.
2. Oku, N.; Namba, Y. *Crit. Rev. Ther. Drug Carrier Sys.* **1994**, *11*, 231-270.
3. Collins, D. *Liposomes as tools in basic research and industry*, Philippot, J. R. and Schuber, F., Ed.; CRC Press: Boca Raton, 1994, pp 201-214.
4. Allen, T. *Adv. Drug Deliv. Rev.* **1994**, *13*, 285-309.
5. Gabizon, A. A. *Adv. Drug Del. Rev.* **1995**, *16*, 285-294.
6. Storm, G.; Oussoren, C.; Peeters, P. A. M.; Barenholz, Y. *Liposome Technology*, Gregoriadis, G., Ed.; CRC Press: Boca Raton, 1993; Vol. 3, pp 345-383.
7. Daemen, T.; Regts, J.; Meesters, M.; TenKate, M. T.; Bakker-Woudenberg, I. A. J. M.; Scherphof, G. L. *J. Control. Rel.* **1997**, *44*, 65-76.
8. Forssen, E. A.; Male-Brune, R.; Adler-Moore, J. P.; Lee, M. J.; Schmidt, P. G.; Krasieva, T. B.; Shimizu, S.; Tromberg, B. *J. Cancer Res.* **1996**, *56*, 2066-2075.
9. Guillod, F.; Greiner, J.; Riess, J. G. *Biochim. Biophys. Acta* **1996**, *1282*, 283-292.
10. Lasic, D. D.; Martin, F. J. *Stealth Liposomes;* ; CRC Press: Boca Raton, 1995.
11. Thorpe, P. E.; Burrows, F. J. *Breast Can. Res. Treat.* **1995**, *36*, 237-251.
12. Longman, S. A.; Tardi, P. G.; Parr, M. J.; Choi, L.; Cullis, P. R.; Bally, M. B. *J. Pharmacol. Exp. Ther.* **1995**, *275*, 1177-1184.
13. Jain, R. K. *Science* **1996**, *271*, 1079-1080.
14. Torchilin, V. P.; Omelyanenko, V. G.; Papisov, M. I.; Bogdanov, A. A.; Trubetskoy, V. S.; Herron, J. N.; Gentry, C. A. *Biochim. Biophys. Acta* **1994**, *1195*, 11-20.
15. Needham, D.; Hristova, K.; McIntosh, T. J.; Dewhirst, M.; Wu, N.; Lasic, D. D. *J. Lipo. Res.* **1992**, *2*, 411-430.
16. Blume, G.; Cevc, G. *Biochim. Biophys. Acta* **1993**, *1146*, 157-168.
17. Khul, T. L.; Leckband, D. E.; Lasic, D. D.; Israelachvili, J. N. *Biophys. J.* **1994**, *66*, 1479-1488.
18. Kenworthy, A. K.; Hristova, K.; Needham, D.; McIntosh, T. J. *Biophys. J.* **1995**, *68*, 1921-1936.
19. Jansen, J.; Song, X.; Brook, D. E. *Biophys. J.* **1996**, *70*, 313-320.
20. Cohen, J.; Khorosheva, V.; Woodle, M. *Biophys. J.* **1996**, *70*, A223.
21. Senior, J.; Delgado, C.; Fisher, D.; Tilcock, C.; Gregoriadis, G. *Biochim. Biophys. Acta* **1991**, *1062*, 77-82.
22. Torchillin, V.; Papisov, M. I. *J. Lipo. Res.* **1994**, *4*, 725-739.
23. Torchilin, V. P.; Shtilman, M. I.; Trubetskoy, V. S.; Whiteman, K.; Milstein, A. M. *Biochim. Biophys. Acta* **1994**, *1195*, 181-184.
24. Maruyama, K.; Okuizumi, S.; Ishida, O.; Yamauchi, H.; Kikuchi, H.; Iwatsuru, M. *Int. J. Pharm.* **1994**, *111*, 103-107.
25. Woodle, M. C.; Engbers, C. M.; Zalipsky, S. *Bioconj. Chem.* **1994**, *5*, 493-496.

26. Zalipsky, S.; Hansen, C. B.; Oaks, J. M.; Allen, T. M. *J. Pharm. Sci.* **1996**, *85*, 133-1137.
27. Zalipsky, S. *Bioconj. Chem.* **1995**, *6*, 150-165.
28. Zalipsky, S. *Stealth Liposomes*, Lasic, D. and Martin, F., Ed.; CRC Press: Boca Raton, FL, 1995, pp 93-102.
29. Parr, M. J.; Ansell, S. M.; Choi, L. S.; Cullis, P. R. *Biochim. Biophys. Acta* **1994**, *1195*, 21-30.
30. Holland, J. W.; Cullis, P. R.; Madden, T. D. *Biochem.* **1996**, *35*, 2610-2617.
31. Kirpotin, D.; Hong, K.; Mullah, N.; Papahadjopoulos, D.; Zalipsky, S. *FEBS Lett.* **1996**, *388*, 115-118.
32. Babbitt, B.; Burtis, L.; Dentinger, P.; Constantinides, P.; Hillis, L.; McGirl, B.; Huang, L. *Bioconj. Chem.* **1993**, *4*, 199-205.
33. Tari, A. M.; Fuller, N.; Boni, L. T.; Collins, D.; Rand, P.; Huang, L. *Biochim. Biophys. Acta* **1994**, *1192*, 253-262.
34. Zhou, F.; Huang, L. *Immunomethods* **1994**, *4*, 229-235.
35. Slepushkin, V. A.; Simoes, S.; Dazin, P.; Newman, M. S.; Guo, L. S.; deLima, M. C. P.; Duzgunes, N. *J. Biol. Chem.* **1997**, *272*, 2382-2388.
36. Pinnaduwage, P.; Huang, L. *Biochem.* **1992**, *31*, 2850-2855.
37. Zalipsky, S. *Bioconj. Chem.* **1993**, *4*, 296-299.
38. Zalipsky, S.; Brandeis, E.; Newman, M. S.; Woodle, M. C. *FEBS Lett.* **1994**, *353*, 71-74.
39. Haselgrubler, T.; Amerstorfer, A.; Schindler, H.; Gruber, H. J. *Bioconj. Chem.* **1995**, *6*, 242-248.
40. Shahinian, S.; Silvius, J. R. *Biochim. Biophys. Acta* **1995**, *1239*, 157-167.
41. Lasic, D. D.; Woodle, M. C.; Martin, F. J.; Valentincic, T. *Period. Biol.* **1991**, *93*, 287-290.
42. Hristova, K.; Kenworthy, A. K.; McIntosh, T. J. *Macromol.* **1995**, *28*, 7693-7699.
43. Kenworthy, A. K.; Simon, S. A.; McIntosh, T. J. *Biophys. J.* **1995**, *68*, 1903-1920.
44. Uster, P. S.; Allen, T. M.; Daniel, B. E.; Mendez, C. J.; Newman, M. S.; Zhu, G. Z. *FEBS Lett.* **1996**, *386*, 243-246.
45. Yoshioka, H. *Biomater.* **1991**, *12*, 861-864.
46. Woodle, M. C. *Chem. Phys. Lipids* **1993**, *64*, 249-262.
47. Oku, N.; Tokudome, Y.; Namba, Y.; Saito, N.; Endo, M.; Hasegawa, Y.; Kawai, M.; Tsukada, H.; Okada, S. *Biochim. Biophys. Acta* **1996**, *1280*, 149-154.
48. Holland, J. W.; Hui, C.; Cullis, P. R.; Madden, T. D. *Biochem.* **1996**, *35*, 2618-2624.
49. Woodle, M. C.; Newman, M. S.; Working, P. K. *Stealth Liposomes*, Lasic, D. and Martin, F., Ed.; CRC Press: Boca Raton, 1995, pp 103-118.
50. Woodle, M. C.; Storm, G. *Long circulating liposomes: old drugs, new therapeutics* ; R.G. Landes: Austin, Texas, 1997.
51. Papahadjopoulos, D.; Gabizon, A. A. *Liposomes as tools in basic research and industry*, Philippot, J. R. and Schuber, F., Ed.; CRC Press: Boca Raton, 1994, pp 177-188.
52. Harasym, T. O.; Tardi, P.; Longman, S. A.; Ansell, S. M.; Bally, M. B.; Cullis, P. R.; Lewis, S. L. C. *Bioconj. Chem.* **1995**, *6*, 187-194.
53. Oja, C. D.; Semple, S. C.; Chonn, A.; Cullis, P. R. *Biochim. Biophys. Acta* **1996**, *1281*, 31-37.
54. Semple, S. C.; Chonn, A.; Cullis, P. R. *Biochem.* **1996**, *35*, 2521-2525.
55. Mori, A.; Klibanov, A. L.; Torchilin, V. P.; Huang, L. *FEBS Lett.* **1991**, *284*, 263-266.

56.   Park, J. W.; Hong, K.; Carter, P.; Asgar, H.; Guo, L. Y.; Keller, G. A.; Wirth, C.; Shalaby, R.; Knotts, C.; Wood, W. I.; Papahadjopoulos, D.; Benz, C. C. *Proc. Natl. Acad. Sci. (USA)* **1995**, *92*, 1327-1331.
57.   Amselem, S.; Cohen, R.; Barenholz, Y. *Chemistry and Physics of Lipids* **1993**, *64*, 219-237.
58.   Huang, S. K.; Martin, F. J.; Jay, G.; Vogel, J.; Papahadjopoulos, D.; Friend, D. S. *Am. J. Pathol.* **1993**, *143*, 10-14.
59.   Trubetskoy, V. S.; Torchilin, V. P. *J. Lipo. Res.* **1994**, *4*, 961-980.
60.   Moghimi, S. M.; Hawley, A. E.; Christy, N. M.; Gray, T.; Illum, L.; Davis, S. S. *FEBS Lett.* **1994**, *344*, 25-30.
61.   Allen, T. M.; Newman, M. S.; Woodle, M. C.; Mayhew, E.; Uster, P. S. *Int. J. Cancer* **1995**, *62*, 199-204.
62.   Gabizon, A.; Catane, R.; Uziely, B.; Kaufman, B.; Safra, T.; Cohen, R.; Martin, F.; Huang, A.; Barenholz, Y. *Cancer Res.* **1994**, *54*, 987-992.
63.   Northfelt, D. W.; J, M. F.; Working, P.; Kaplan, L. D.; Russell, J.; Amantea, M. A.; Newman, M.; Volberding, P. A. *J. Clin. Pharmacol.* **1996**, *36*, 55-63.
64.   Litzinger, D. C.; Huang, L. *Biochim. Biophys. Acta* **1992**, *1127*, 249-254.
65.   Scherphof, G. L.; Morselt, H.; Allen, T. M. *J. Lipo. Res.* **1994**, *4*, 213-228.
66.   Moghimi, S. M.; Davis, S. S. *Crit. Rev. Ther. Drug Carrier Syst.* **1994**, *11*, 31-59.
67.   Goren, D.; Horowitz, A. T.; Zalipsky, S.; Woodle, M. C.; Yarden, Y.; Gabizon, A. *Br. J. Cancer* **1996**, *74*, 1749-1756.
68.   Oku, N.; Tokudome, Y.; Tsukada, H.; Kosugi, T.; Namba, Y.; Okada, S. *Biopharm. Drug Dispos.* **1996**, *17*, 435-441.
69.   Vaage, J.; Barbera, E. *Stealth Liposomes*, Lasic, D. and Martin, F., Ed.; CRC Press: Boca Raton, 1995, pp 149-171.
70.   Huang, S. K.; Martin, F. J.; Friend, D. S.; Papahadjopoulos, D. *Stealth Liposomes*, Lasic, D. and Martin, F., Ed.; CRC Press: Boca Raton, 1995, pp 119-126.
71.   Dewhirst, M.; Needham, D. *Stealth Liposomes*, Lasic, D. and Martin, F., Ed.; CRC Press: Boca Raton, 1995, pp 127-138.
72.   Gabizon, A.; Pappo, O.; Goren, D.; Chemla, M.; Tzemach, D.; Horowitz, D. *J. Lipo. Res.* **1993**, *3*, 517-528.
73.   Oku, N.; Doi, K.; Namba, Y.; Okada, S. *Int. J. Cancer* **1994**, *58*, 415-419.
74.   Lasic, D. D.; Frederik, P. M.; Stuart, M. C. A.; Barenholz, Y.; McIntosh, T. J. *FEBS Lett.* **1992**, *312*, 255-258.
75.   Harrigan, P. R.; Wong, K. F.; Redelmeier, T. E.; Wheeler, J. J.; Cullis, P. R. *Biochim. Biophys. Acta* **1993**, *1149*, 329-338.
76.   Haran, G.; Cohen, R.; Bar, L. K.; Barenholz, Y. *Biochim. Biophys. Acta* **1993**, *1151*, 201-215.
77.   Harrison, M.; Tomlinson, D.; Stewart, S. *J. Clin. Oncol.* **1995**, *13*, 914-920.
78.   Uziely, B.; Jeffers, S.; Isacson, R.; Kutsch, K.; Wei-Tsao, D.; Yehoshua, Z.; Libson, E.; Muggia, F. M.; Gabizon, A. *J. Clin. Oncol.* **1995**, *13*, 1777-1785.
79.   Muggia, F.; Hainsworth, J.; Jeffers, S.; Miller, P.; Groshen, S.; Tan, M.; Roman, L.; Uziely, B.; Muderspach, L.; Garcia, A.; Burnett, A.; Greco, F. A.; Morrow, C. P.; Paradiso, L. J.; Liang, L.-J. *J. Clin. Oncol.* **1997**, *15*, 987-993.
80.   Iga, K.; Ogawa, Y.; Toguchi, H. *Pharm. Res.* **1992**, *9*, 658-662.
81.   Huang, S. K.; Stauffer, P. R.; Hong, K.; Guo, J. W.; Phillips, T. L.; Huang, A.; Papahadjopoulos, D. *Cancer Res.* **1994**, *54*, 2186-2191.
82.   Ning, S.; MacLeod, K.; Abra, R. M.; Huang, A. H.; Hahn, G. M. *Int. J. Radiat. Oncol. Biol. Phys.* **1994**, *29*, 827-834.
83.   Unezaki, S.; Maruyama, K.; Takahashi, N.; Koyama, M.; Yuda, T.; Suginaka, A.; Iwatsuru, M. *Pharm. Res.* **1994**, *11*, 1180-1185.
84.   Oku, N.; Naruse, R.; Doi, K.; Okada, S. *Biochim. Biophys. Acta* **1994**, *1191*, 389-391.

85. Gaber, M. H.; Hong, K.; Huang, S. K.; Papahadjopoulos, D. *Pharm. Res.* **1995**, *12*, 1407-1416.
86. Gaber, M. H.; Wu, N. Z.; Hong, K.; Huang, S. K.; Dewhirst, M. W.; Papahadjopoulos, D. *Int. J. Rad. Oncol. Biol. Phys.* **1996**, *36*, 1177-1187.
87. Mayhew, E. G.; Lasic, D. D.; Babbar, S.; Martin, F. J. *Int. J. Cancer* **1992**, *51*, 302-309.
88. Genne, P.; Olsson, N. O.; Gutierrez, G.; Duchamp, O.; Chauffert, B. *Anticancer Drug Des.* **1994**, *9*, 73-84.
89. Chang, C. W.; Barber, L.; Ouyang, C.; Masin, D.; Bally, M. B.; Madden, T. D. *Br. J. Cancer* **1997**, *75*, 169-177.
90. Allen, T. M.; Mehra, T.; Hansen, C.; Chin, Y.-C. *Cancer Res.* **1992**, *52*, 2431-2439.
91. Doi, K.; Oku, N.; Toyota, T.; Shuto, S.; Sakai, A.; Itoh, H.; Okada, S. *Biol. Pharm. Bull.* **1994**, *17*, 1414-1416.
92. Sharma, A.; Straubinger, R. M. *Pharm. Res.* **1994**, *11*, 889-896.
93. Bakker-Woudenberg, I. A. J. M.; Lokerse, A. F.; ten Kate, M. T.; Mouton, J. W.; Woodle, M. C.; Storm, G. *J. Infect. Dis.* **1993**, *168*, 164-171.
94. Bakker-Woudenberg, I. A. J. M.; ten Kate, M. T.; Stearne-Cullen, L. E. T.; Woodle, M. C. *J. Infect. Dis.* **1995**, *171*, 938-947.
95. van Etten, E. W. M.; van Vianen, W.; Tijhuis, R. H. G.; Storm, G.; Bakker-Woudenberg, I. A. J. M. *J. Control. Rel.* **1995**, *37*, 123-129.
96. Zalipsky, S.; Puntambekar, B.; Boulikas, P.; Engbers, C. M.; Woodle, M. C. *Bioconj. Chem.* **1995**, *6*, 705-708.
97. Storm, G.; Koppenhagen, F. J.; Heeremans, A. L. M.; Vingerhoeds, M. H.; Woodle, M. C.; Crommelin, D. J. A. *J. Control. Rel.* **1995**, *36*, 19-24.
98. Woodle, M. C.; Storm, G.; Newman, M. S.; Jekot, J. J.; Collins, L. R.; Martin, F. J.; F.C. Szoka, J. *Pharm. Res.* **1992**, *9*, 260-265.
99. Kedar, E.; Braun, E.; Rutkowski, Y.; Emanuel, N.; Barenholz, Y. *J. Immunther. Emphasis Tumor Immunol.* **1994**, *16*, 115-124.
100. Woodle, M. C. *Nucl. Med. Biol.* **1993**, *20*, 149-155.
101. Tilcock, C.; Ahkong, Q. F.; Fisher, D. *Biochim. Biophys. Acta* **1993**, *1148*, 77-84.
102. Oku, N.; Namba, Y.; Takeda, A.; Okada, S. *Nucl. Med. Biol.* **1993**, *20*, 407-412.
103. Kumar, S.; Singh, T.; Khar, R. K.; Sharma, S. N.; Chauhan, U. P. *Pharmazie* **1993**, *48*, 613-616.
104. Tilcock, C. *Liposomes as tools in basic research and industry*, Philippot, J. R. and Schuber, F., Ed.; CRC Press: Boca Raton, 1994, pp 225-240.
105. Oyen, W. J. G.; Boerman, O. C.; Storm, G.; Van Bloois, L.; Koenders, E. M.; Claessens, R. A. M. J.; Perenboom, R. M.; Crommelin, D. J. A.; Van de Meer, J. W. M.; Corstens, F. H. M. *J. Nucl. Med.* **1996**, *37*, 1392-1397.
106. Oku, N.; Tokudome, Y.; Koike, C.; Nishikawa, N.; Mori, H.; Saiki, I.; Okada, S. *Life Sci.* **1996**, *58*, 2263-2270.
107. Vingerhoeds, M. H.; Haisma, H. J.; van Muigen, M.; van de Rijt, R. B. J.; Crommelin, D. J. A.; Storm, G. *FEBS Lett.* **1993**, *336*, 485-490.
108. Blume, G.; Cevc, G.; Crommelin, M. D. J. A.; Bakker-Woudenberg, I. A. J. M.; Kluft, C.; Storm, G. *Biochim. Biophys. Acta* **1993**, *1149*, 180-184.
109. Zara, J.; Pomato, N.; McCabe, R. P.; Bredehorst, R.; Vogel, C.-W. *Bioconj. Chem.* **1995**, *6*, 367-372.
110. Zalipsky, S.; Newman, M. S.; Puntambekar, B.; Woodle, M. C. *Polym. Mater. Sci. Eng.* **1993**, *67*, 519-520.
111. Allen, T. M.; Agrawal, A. K.; Ahmad, I.; Hansen, C. B.; Zalipsky, S. *J. Lipo. Res.* **1994**, *4*, 1-25.
112. Emanuel, N.; Kedar, E.; Bolotin, E.; Smorodinsky, N. I.; Barenholz, Y. *Pharm. Res.* **1996**, *13*, 861-868.

113. Maruyama, K.; Kennel, S. J.; Huang, L. *Proc. Natl. Acad. Sci. (USA)* **1990**, *87*, 5744-5748.
114. Klibanov, A. L.; Maruyama, K.; Beckerleg, A. M.; Torchilin, V. P.; Huang, L. *Biochim. Biophys. Acta* **1991**, *1062*, 142-148.
115. Torchilin, V. P.; Klibanov, A. L.; Huang, L.; O'Donnell, S.; Nossiff, N. D.; Khaw, B. A. *FASEB J.* **1992**, *6*, 2716-2719.
116. Mori, A.; Huang, L. *Liposome Technology*, Gregoriadis, G., Ed.; CRC Press: Boca Raton, 1993; Vol. 3, pp 153-162.
117. Mori, A.; Kennel, S. J.; Huang, L. *Pharm. Res.* **1993**, *10*, 507-514.
118. Ahmad, I.; Longenecker, M.; Samuel, J.; Allen, T. M. *Cancer Res.* **1993**, *53*, 1484-1488.
119. Trubetskoy, V. S.; Narula, J.; Khaw, B. A.; Torchilin, V. P. *Bioconj. Chem.* **1993**, *4*, 251-255.
120. Huwyler, J.; Wu, D.; Pardridge, W. M. *Proc. Natl. Acad. Sci. (USA)* **1996**, *93*, 14164-14169.
121. Kirpotin, D.; Park, J. W.; Hong, K.; Zalipsky, S.; Li, W.-L.; Carter, P.; Benz, C. C.; Mullah, N.; Papahadjopoulos, D. *Biochem.* **1997**, *36*, 66-75.
122. Lee, R. J.; Low, P. S. *J. Biol. Chem.* **1994**, *269*, 3198-3204.
123. Merwin, J. R.; Noell, G. S.; Thomas, W. L.; Chiou, H. C.; DeRome, M. E.; McKee, T. D.; Spitalny, G. L.; Findeis, M. A. *Bioconj. Chem.* **1994**, *5*, 612-620.
124. DeFrees, S. A.; Phillips, L.; Guo, L.; Zalipsky, S. *J. Am. Chem. Soc.* **1996**, *118*, 6101-6104.
125. Remy, J.-S.; Sirlin, C.; Behr, J.-P. *Liposomes as tools in basic research and industry*, Philippot, J. R. and Schuber, F., Ed.; CRC Press: Boca Raton, 1994, pp 159-170.
126. Lasic, D. D. *Liposomes in Gene Delivery;* ; CRC Press: Boca Raton, 1997.
127. Wang, C. Y.; Huang, L. *Biochem.* **1989**, *28*, 9508-9514.
128. Wilschut, J.; Hoekstra, D. *Membrane fusion* ; Marcel Dekker, Inc.: New York, 1991.
129. de Lima, M. C. P.; Hoekstra, D. *Liposomes as tools in basic research and industry*, Philippot, J. R. and Schuber, F., Ed.; CRC Press: Boca Raton, 1994, pp 137-156.
130. Duzgunes, N.; Nir, S. *Liposomes as tools in basic research and industry*, Philippot, J. R. and Schuber, F., Ed.; CRC Press: Boca Raton, 1994, pp 103-136.
131. Sobol, R. E.; Scanlon, K. J. *The internet book of gene therapy* ; Appleton & Lange: Stamford, Connecticut, 1995.
132. Vlassov, V. V.; Balakireva, L. A.; Yakubov, L. A. *Biochim. Biophys. Acta* **1994**, *1197*, 95-108.
133. Nestle, F. O.; Mitra, R. S.; Bennett, C. F.; Chan, H.; Nickoloff, B. J. *J. Invest. Dermatol.* **1994**, *103*, 569-575.
134. Leserman, L. *Liposomes as tools in basic research and industry*, Philippot, J. R. and Schuber, F., Ed.; CRC Press: Boca Raton, 1994, pp 215-223.

Chapter 6

Poly(ethylene oxide)-Bearing Lipids and Interaction of Functionalized Liposomes with Intact Cells

Y. Okumura[1,4], N. Higashi[2,5], V. Rosilio[3], and J. Sunamoto[1,2]

[1]Department of Synthetic Chemistry and Biological Chemistry, Graduate School of Engineering, Kyoto University, Yoshida-Hommachi, Sakyo-ku, Kyoto 606–01, Japan
[2]Supermolecules Project, Japan Science and Technology Corporation, Keihanna Plaza, Seika-cho 1–7, Soraku-gun, Kyoto 619–02, Japan
[3]Physico-Chimie des Surfaces, Unité de Recherche Associée, Centre National de la Recherche Scientifique 1218, Université Paris-Sud, 92296 Châtenay-Malabry, France

The aggregation behavior of poly(ethylene oxide)-bearing lipids (PEO-lipids) in aqueous media or at the air-water interface was investigated. The PEO-lipids have a PEO moiety (average number of PEO units, 5 to 31) on the hydroxy group of 1,3-didodecyloxy-2-propanol. The outer surface of lecithin liposome was functionalized by reconstitution of the PEO-lipids, and the interaction of intact cells with the functionalized liposome was examined. Using fluorescent and biological markers, fusion of the PEO-lipid reconstituted liposome with various intact cells, such as carrot protoplast, Hela, B16 melanoma, and Jurkat cells was demonstrated.

Liposome is composed of lipid bilayer membrane (a review of liposome (*1*)). The same bilayer membrane structure also constitutes the fundamental part of cytoplasm membrane (a review of biomembranes (*2*)). The bilayer membrane is a sort of supramolecular assembly of lipids and therefore can self-organize. As a consequence, a unique mode of interaction, membrane fusion, is observed between liposome and intact cells.

The cell-liposome fusion has a potential application in medicine or cell engineering. The restricted passage of water soluble substances through bilayer membrane allows the liposome to hold these substances in its interior water pool. By the fusion between intact cell and liposome, the liposomal contents are directly transported into cytosol. This feature can be useful, for example, to obtain a local physicochemical information of cytosol such as pH, viscosity and polarity by site specific introduction of probe molecules. On the other hand, the fusion can be particularly valuable for introduction of substances which are susceptible to degradation with lysosomal

[4]Current address: Department of Chemistry and Material Engineering, Faculty of Engineering, Shinshu University, 500 Wakasato, Nagano 380, Japan

[5]Note: On leave from the Kao Institute for Fundamental Research, 2606 Akabane, Ichikaimachi, Haga, Tochigi 321–34, Japan

enzymes.   The introduction of such the substances, genes, antisense oligonucleotides, enzymes or peptides, is often hampered by lysosomal degradation if internalized *via* endocytosis.

Several methods other than endocytosis have been proposed for transportation of various substances into cell; e.g. micro-injection (*3*), electroporation (*4*), utilization of erythrocyte ghosts (*5*), lipofection (*6*), and DEAE dextran (*7*).   However, some of these techniques require conditions rather harsh to target cell.   The direct fusion between liposome and intact cell would be more moderate and convenient.

Although there are many possible applications of the cell-liposome fusion, the technique has one major problem: The fusion of conventional liposome with intact cell does hardly occur.   Therefore, an effective fusogen is required.   The cell-liposome fusion using lysozyme (*8,9*) and, most notably, that using virus proteins was first reported.   Okada and his coworkers developed a liposome that is modified with a fusogenic protein of Sendai virus. Using this liposome, they demonstrated successful introduction of biologically active materials such as diphtheria toxin A fragment (DTA) and a plasmid encoding DTA into a target cell (*10-13*).   However, this method requires careful handling of the virus.

Poly(ethylene oxide) (PEO) has been commonly used for induction of cell-cell fusion of protoplasts (*14*) or several mammalian cells (*15*).   PEO has been applied also to fusion between monolayer culture of eucaryotic cells and glycolipid-containing liposome which was adhered to the cell by lectin (*16*).   However, the fusion with PEO is accompanied by such problems as cytotoxicity of PEO or undesirable homo-fusion between cells or between liposomes.

For a quest of simple and well-regulated cell-liposome fusion under conditions mild to cell, we investigate artificial lipids that bear a PEO moiety as the hydrophilic head group (PEO-lipids, Figure 1) (*17-22*).   With the hydrophobic anchor, the PEO moiety is to be localized near the outer surface of liposome, where the fusion is supposed to take place.   In this article, we would like to describe function of the PEO-lipid functionalized liposome paying attention to the physicochemical characteristics of the PEO-lipids.

Synthesis of PEO-lipids

One can think many possible "PEO-bearing lipid" structures for anchoring a PEO moiety to liposomal membrane.   Without exact knowledge of the fusion mechanism, it is hardly possible to pick the one which is the best as the fusogen.   Nonetheless, several factors were in our consideration in the initial designing of our PEO-lipid molecules.

Membrane fusion involves reorganization of membrane lipids.   Therefore, we first assumed that liposomal membrane which is more prone to the reorganization should be in favor of the fusion.   Presumably, this could be achieved by loosening the packing of the lipid molecules in bilayer membrane.   On the other hand, after reconstitution of PEO-bearing lipid, the liposomal membrane still has to provide a barrier to the leakage of encapsulated substances.   This, however, favors tight lipid packing of the liposomal membrane.   As a compromise, our PEO-lipids have two C_n alkyl moieties. A double alkyl anchorage provides the better barrier function than a single alkyl one which significantly disturbs the integrity of bilayer membrane and often results in loss of the barrier function.   Meanwhile, we selected C_{12} as the hydrophobic leg shorter than that of liposomal phosphatidylcholine so that it can cause perturbation to the lipid packing.   We have little foresight about how the chain length of the PEO moiety might affect the hydrophilic-lipophilic balance of the lipid and, subsequently, the possible induction of the fusion.   Thus the adequate PEO chain length needs investigation.

We noticed that the PEO-bearing lipids shown in Figure 1, which were first synthesized by Kuwamura in a Japanese journal in 1961 (*23,24*) (later the procedure was described by Okahata et al. in an English literature (*25*)), satisfies the above criteria and can be obtained through a simple synthetic route.   In the procedure, the PEO lipid(12,a22.5) and PEO-lipid(12,a31), were synthesized and fractionated in accordance with Kuwamura's procedure.   The successive addition of ethylene oxide leaves ambiguity in the number of the oxyethylene units in the PEO-lipids.   In our case, the average length of the PEO moiety was estimated based on the integrals in ^{1}H-NMR, and the prefix "a" signifies this.

Recently, in order to circumvent the ambiguity, we have developed another way to obtain PEO-lipids by condensation of 1,3-dialkyloxy-2-propanol with PEO that has an exact number of oxyethylene units.   For example, in the case of PEO-lipid(12,15), PEO of precisely 5 oxyethylene units was separated by precision distillation from a commercially available PEO mixture.   Three molecules of the PEO were condensed to obtain a longer PEO of exactly 15 oxyethylene units.   Then further condensation of the PEO with 1,3-didodecyloxy-2-propanol yielded PEO-lipid(12,15).

Characterization of Aggregates Containing PEO-lipids in Aqueous Media

PEO-lipids in Aqueous Media.   An early study of Kuwamura concerns the solution property of the PEO-lipids as surfactant.   His study determined some of the basic properties of the PEO-lipids in aqueous media, including clouding point and surface tension-concentration relationship (*23*).   Also, Okahata *et al.* investigated self-aggregation of several different PEO-lipid(m,n) (m = 8 ~ 18 and n = 6 ~ 30) in water (*25*).   All of the PEO-lipids they examined, except PEO-lipids(8,n), yielded aggregates.   Different morphologies among those aggregates were shown by either static light scattering or electron microscopy with negative staining.   The critical aggregate concentration (*cac*) of PEO-lipid determined from the surface tension measurement was in the range of 10^{-5}-10^{-6} M and decreased with an increase in the alkyl chain length.

Also in our case, the PEO-lipids formed a self-aggregate in an aqueous medium (Okumura, Y.; Morone, N.; Sunamoto, J., manuscript in preparation).   Laser dynamic light scattering study of the aggregate indicated that PEO-lipid(12,a5) forms large aggregates, probably, of lamellar structure, while PEO-lipid(12,a13) forms vesicles with diameter of approximately 100 nm.   PEO-lipid(12,a31) showed relatively weaker light scattering in an aqueous medium than the other two.   Dynamic light scattering measurement using a powerful argon laser (488 nm, 1 W) revealed formation of an aggregate smaller than 10 nm in the diameter.   The aggregate of PEO-lipid(12,a31) is thus likely to be an aqueous micelle.   This result is conceivable from the proposed relationship (*26*) between the morphology of aggregate and relative area occupied by the head and the tail of surfactant.   With a longer PEO chain, the PEO moiety occupies more space or surface area (*vide infra*) (*27, 28*).   This makes PEO-lipid(12,a31) prefer to form micelles, which allows large head groups.   In the previous report (*25*), Okahata *et al.* described that PEO-lipid(12,a10) forms irregular vesicles and that PEO-lipid(12,a15) forms vesicles and lamellar phase.   This seems to be consistent with the result of our PEO-lipid(12,a13).   For PEO-lipid(12,a28), they found a lamellar structure in the electron microscopic observation.

PEO-lipid in Monolayers at Air/water Interface.   Using surface pressure and surface potential measurements, Rosilio *et al.* investigated the behavior of monolayers composed of a single PEO-lipid or a PEO-lipid/DMPC mixture formed at the air/water interface (*27, 28*).   In the case of a single PEO-lipid (Figure 2), the length of the PEO moieties affected the interaction among the PEO-lipids.   A PEO-lipid(12,a5) monolayer had a steep surface pressure-surface density curve, which is similar to that of a DMPC monolayer, indicating that PEO-lipid(12,a5) molecules interact each other at their aliphatic moieties as DMPC molecules do.   With PEO-lipid(12,a31) monolayer, in contrast, expanded film type behavior with larger limiting molecular area and higher

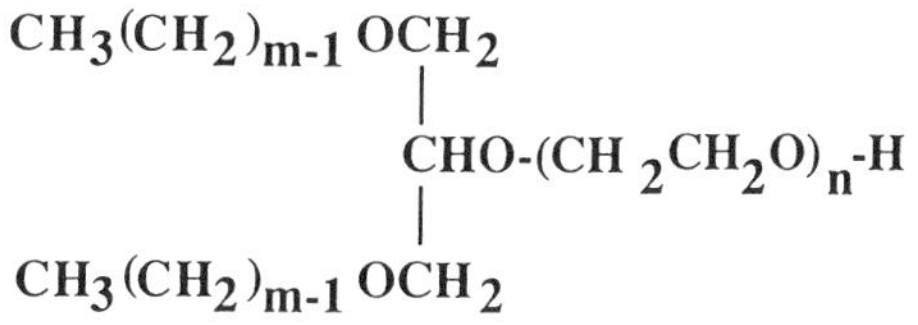

$$CH_3(CH_2)_{m-1} OCH_2$$
$$|$$
$$CHO\text{-}(CH_2CH_2O)_n\text{-}H$$
$$|$$
$$CH_3(CH_2)_{m-1} OCH_2$$

Figure 1. The structure of PEO-lipid(m,n).

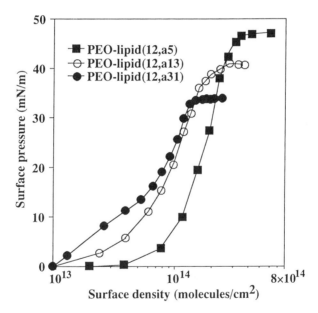

Figure 2. Surface pressure-surface density plots of PEO-lipid monolayers (adapted from ref. 27).

compressibility was observed. This revealed that interaction of the head group becomes predominant for the larger PEO moieties.

In mixed monolayers of PEO-lipid and DMPC (Figure 3), less compressible DMPC mainly controlled the collapse of the mixed monolayer, and the effect of the PEO moiety on the collapse was less significant. The collapse of the mixed lipid monolayer occurs at the pressure close to the collapse pressure of DMPC. Here, the interaction in the aliphatic groups between PEO-lipid and DMPC was certainly present, and the both components behave in unison. However, with high content of PEO-lipid(12,a31) in the mixed monolayer, the collapse pressure became close to that of PEO-lipid(12,a31). A small kink appeared in the collapse region of the curve, which indicates that PEO-lipid(12,a31) and DMPC collapsed independently. This behavior suggests lack of the interaction at the aliphatic moieties between the two lipids, possibly because the steric hindrance due to the large PEO head group could hamper the close contact at the aliphatic part.

Incorporation of PEO-lipids into Lecithin Liposome

Liposome was preformed from phosphatidylcholines (typically eggPC, that is extracted and purified from egg yolk) usually using reverse-phase evaporation method (29). Liposome thus prepared typically has the diameter of 100-200 nm. In order to achieve a high loading of water soluble substances, we used a modified reverse-phase evaporation procedure devised by Nakanishi and his coworkers (20). This was especially required for encapsulation of substances with relatively high molecular weight such as peptides or DNA. For example, using this procedure, FITC-dextran was introduced into egg PC liposome with encapsulation efficiency of 17%, while diphtheria toxin fragment A protein (DTA) was with 1.3% (20).

PEO-lipid was then spontaneously reconstituted into liposome by injecting an ethanol solution of the PEO-lipid to a liposome suspension preformed. With PEO-lipid(12,a31), the PEO-lipid-reconstituted liposome can be separated from free PEO-lipid by centrifugation. Hence, for example, PEO-lipid(12,a31) could be stably reconstituted into the liposome up to 30 mol% to liposomal eggPC (21).

Even after the incorporation of PEO-lipids, the liposome kept encapsulated materials for a reasonable period of time. For the reconstitution of 30 mol% PEO-lipids, the leakage of water soluble FITC-dextran (molecular weight 19600) from the liposome was less than 20% 1 hr (Okumura, Y.; Yamauchi, M.; Sunamoto, J., manuscript in preparation). However, in the reconstitution of 60 mol% PEO-lipid(12,a13) or PEO-lipid(12,a31), drastic leakage of FITC-dextran was observed, suggesting extensive damage of the liposomal membrane.

Kodama et al. examined behavior of PEO-lipid(12,a11.5)-functionalized DMPC liposome with high sensitivity differential scanning calorimetry (DSC) (30). Incorporation of PEO-lipid into the liposome up to 20 mol% caused a decrease in the gel-liquid crystalline phase transition temperature from 24.5°C of simple DMPC liposome to 21.4°C. At the same time, the cooperativity in the transition decreased, and the transition enthalpy became larger. A sharp transition peak was observed for the incorporation of up to 10 mol% PEO-lipid. At the PEO contents higher than 10 mol%, the DSC peak becomes broader and asymmetric, indicating partial phase separation in the liposomal membrane.

Interaction of PEO-lipid-functionalized Liposome with Plant and Animal Cells

Evaluation of Fusion Efficiency. For evaluation of cell-liposome fusion efficiency, there are the following four techniques.

Fluorescent Dye Introduction.   Liposome is double-labeled with two different types of fluorescent probes, water-soluble and hydrophobic ones. The fusion causes the mixing of both interior and lipid membrane contents between the cell and the liposome.   The water soluble probe originally encapsulated in the liposome will appear in the cytosol.   At the same time, the hydrophobic probe initially in the liposomal membrane will relocate only to the cytoplasma membrane, not to the cytosol. With this technique, it is possible to distinguish fusion from endocytosis.   If the liposome undergoes endocytosis, both of the probes will appear together in the cytosol.

Resonance Energy Transfer.   The resonance energy transfer (RET) technique has been frequently used for qualitative evaluation of liposome-liposome fusion (*31-33*).   We can prepare a fusogenic liposome containing the two membrane fluorescent probes, (7-nitrobenz-2-oxa-1,3-diazol-4-yl)-phosphatidylethanolamine (NBD-PE) and (Lissamine rhodamine B sulfonyl)-phosphatidylethanolamine (Rh-PE). The two probes are in close contact in the liposomal membrane so that energy transfer between the two probes easily occurs.   After the fusion with cell, the two probes are far apart each other in the cell membrane by dilution with cell lipids.   This causes diminished energy transfer between the two probes.   However, possible quench of the fluorescence by cellular components makes the evaluation of the cell-liposome fusion by this method tricky.   In any event, this method leaves many caveats unaddressed.

Protein or Plasmid Introduction.   If the efficiency of the fusion is extremely low, we may not reliably observe the fluorescence from the probes by usual photometric measurement.   For the fusion with mammalian cell, the sensitivity in the detection of the fusion can be improved by using an active agent introduction technique. Nakanishi *et al.* have established a method which uses DTA (*vide supra*) as the marker (*10,11,34,35*).   DTA fully keeps the original activity of native diphtheria toxin, *i.e.*, strong cytotoxic activity by inactivating elongation factor 2 in cytoplasm of target cells. One molecule of DTA kills a target cell, if it is introduced properly into cytoplasm of the target cell (*34*).   However, lack of the receptor-binding domain of native diphtheria toxin prevents DTA from passing through cell membrane, and makes DTA nontoxic even it is given at high concentration.   If DTA is delivered by endocytosis, the cells are not killed because of the complete digestion of DTA by lysozomal enzymes (*10*).   Therefore, the cytotoxicity of DTA-containing liposome can be a good evidence for the direct transfer of DTA by the fusion.

For plant protoplast, we examined the expression of GUS (β-glucuronidase) gene encapsulated in liposome (Zheng, J.; Tanaka, K.; Sunamoto, J., manuscript in preparation).   The original GUS activity in the protoplast could be suppressed by addition of methanol to the cell culture.

Direct Observation by Electron Microscopy.   In order to directly observe the fusion process, the electron microscopic observation is useful.

Interaction between PEO-lipid-bearing Liposome and Cells.   Hence, we employed the four techniques in combination for evaluation of the PEO-lipid-dependent cell-liposome fusion.

Carrot Protoplast.   First, we studied the interaction between the PEO-lipid-functionalized liposome and carrot protoplast (*17,18*).   Generally, plant cells are well-protected by cellulose or cell wall.   To remove this protection, plant cells were first treated with cellulase.   The carrot protoplast (*Daucus carota L.*) was used within 2 hr after the cellulase treatment.

PEO-lipid-functionalized liposome, containing both the hydrophobic phospholipid probes (NBD-PE and Rh-PE) in the membrane and the hydrophilic probe (FITC-dextran) in the interior of the same liposome, was coincubated with the protoplast. After washing, the protoplast was observed under a fluorescence microscope. The hydrophobic probes were localized only on the plasma membrane of the protoplast while the hydrophilic probe was distributed in the cytoplasm. This observation strongly suggests the direct fusion between the protoplast membrane and the PEO-lipid-reconstituted liposome.

The interaction was further examined by the RET technique. The RET donor and acceptor, NBD-PE (1.75 mol%) and Rh-PE (0.25 mol%), were embedded in the eggPC liposomal membrane, and the labeled liposome was incubated with carrot protoplast for 30 min at 30°C. From this RET effect (Table I), we understand that there are two factors for optimal cell-liposome fusion. First, the fusion depends on the PEO-lipid content in the liposome. The fusion was at its optimum with 23 mol% PEO-lipid. Second, the length of the PEO-moiety affected the fusion efficiency. The PEO-lipid with the shorter PEO moiety (PEO-lipid(12,a11.5)) showed the more fusion. However, we must once again notice that the RET % does not correspond to the actual fusion efficiency. The result shows only the tendency of the fusion efficiency.

Table I. Interaction of PEO-lipid-reconstituted EggPC Liposome with Protoplast

PEO-lipid content	RET %	
(mol %)	PEO-lipid(12,a11.5)	PEO-lipid(12,a22.5)
0	0	1.0
5	1.2	2.0
9	2.7	2.7
23	18.1	5.1
33	1.0	6.1
50	0	4.0

Finally, we proved the fusion between the liposome and the protoplast by GUS gene introduction (Zheng, J.; Tanaka, K.; Sunamoto, J., manuscript in preparation). Carrot protoplast was treated with the liposome encapsulating GUS gene, cultured for a several days and then homogenized. Production of 4-methylumbelliferone (4-MU) from 4-methylumbelliferyl glucuronide catalyzed by the supernatant of the homogenate was monitored using fluorescence of 4-MU at 455 nm. Figure 4 clearly shows that only the PEO-lipid-reconstituted liposome can express the GUS gene.

HeLa Cells. HeLa cells, cervicarcinoma and also known as adherent cells, were chosen to investigate the liposome-cell fusion (*19*). HeLa cells were separately cultured as single clones on a plastic dish. After washed thoroughly with balanced salt solution, the cells were coincubated with the PEO-lipid(12,a13)-reconstituted liposome that encapsulates a plasmid encoding DTA. After the complete removal of free liposome, the cells were cultured for a week, and the colony-forming efficiency was determined. As shown in Table II, significantly retarded colony formation was observed for the cells treated with the PEO-lipid-reconstituted liposome that contains the plasmid, suggesting the fusion of the liposome with HeLa cells.

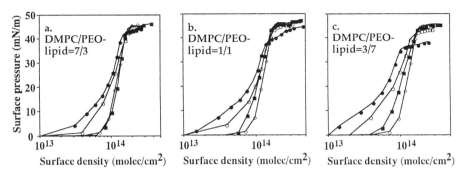

Figure 3. Surface pressure-surface density relationships for mixed DMPC/PEO-lipid monolayers. DMPC/PEO-lipid(12,a5) (B), DMPC/PEO-lipid(12,a13) (É), DMPC/PEO-lipid(12,a31) (J) and DMPC only (Å) (adapted from ref. 27).

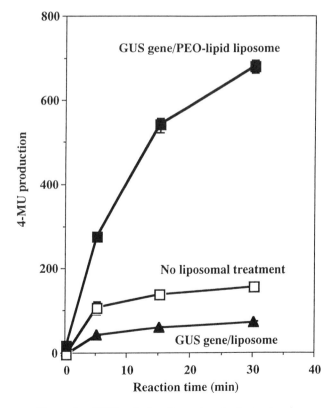

Figure 4. GUS activity of carrot protoplast cytosol.

Table II.   Colony Formation by HeLa

Treatment	PEO-lipid(12,a13) (mol%)	Number of colonies [a]
balanced salt solution only	none	360.8 ± 16.9
liposomes + PEO-lipid	none	362.8 ± 14.6
	30	346.5 ± 15.8
	60	307.5 ± 4.8
plasmid only	none	356.3 ± 19.0
plasmid in liposome + PEO-lipid	none	352.5 ± 13.7
	30	223.2 ± 59.3
	60	244.0 ± 10.5

[a] Mean ± S.E. from four measurements.

After the adequate coincubation, the cell-liposome mixture was quickly frozen, fixed and observed by a transmission electron microscope.   As shown in Figure 5, some PEO-lipid-reconstituted liposomes were in close contact with plasma membrane of HeLa cell.   In addition, the boundary between cell and liposome turned to be invisible.   This suggests the direct fusion between the cell and the liposome.

Jurkat Cell.   The interaction between the PEO-lipid- reconstituted liposome and Jurkat cell (human lymphoblastoma) was investigated (20,21).   Our aim was to compare the difference in the fusion between an adherent cell (HeLa in the above section) and a suspended one (Jurkat).   Cells of lymphoid-origin scarcely undergo endocytosis (36) are convenient to distinguish the difference between fusion and endocytosis.

In the initial trial, Jurkat cells were cultured in RPMI1640 containing 10% FCS. After washed thoroughly with PBS, Jurkat cells were coincubated with the PEO-lipid(12,a13)-reconstituted liposome that contained FITC-dextran in its interior.   The fluorescent dots were clearly observed in the cytoplasm.   This apparently suggests that the liposomes localized lysosomes.   This liposome introduction was completely inhibited by the pretreatment of the cells with cytochalasin B, an inhibitor of endocytosis (37-39).   These observations strongly suggest unexpected endocytosis of the PEO-lipid(12,a13)-reconstituted liposome by the Jurkat cell.

To achieve the fusion, we tried to reduce conceivable fusion-inhibiting factors: For example, the Jurkat cells were precultured in a medium containing less FCS (<1 %) for 16 h before the exposure to the liposome in order to suppress possible adsorption of the serum proteins.   Then, the cells were thoroughly washed with PBS and coincubated with the PEO-lipid-reconstituted liposome that contained DTA.   As shown in Table III, the Jurkat cells was significantly killed by the treatment with the PEO-lipid(12,a31)-reconstituted liposome.   Neither liposomes modified with the PEO-lipid of the shorter PEO chain nor a simple mixture of free DTA with the PEO-lipid-reconstituted liposome showed the DTA-dependent cytotoxicity.   This eliminates a possibility that free DTA would be introduced through damage in the cell membrane. The cytotoxicity was still observed even in the presence of cytochalasin B or at 4.0℃. These findings prove that the PEO-lipid(12,a31)- reconstituted liposome certainly fuses with the Jurkat cell.   The optimal length of the PEO-moiety for evoking the effective fusion seems to be different between HeLa and Jurkat.

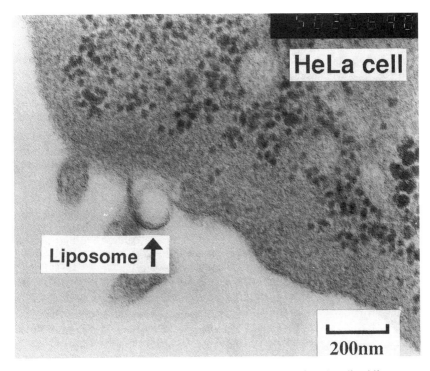

Figure 5. Electron micrographic image of PEO-lipid-functionalized liposome fusing with a HeLa cell (× 50,000).

Table III. Viability of Jurkat Cell after Interaction with Liposome

| PEO-lipid in liposome | Viability (%) [a] | |
	Without DTA	With DTA
none	90.4 ± 0.9	90.0 ± 1.4
(12,a5), 20 %	92.1 ± 0.9	91.3 ± 0.8
(12,a13), 20 %	82.0 ± 0.8	84.4 ± 1.9
(12,a13), 33 %	78.4 ± 2.1	79.6 ± 1.1
(12,a13), 50 %	66.6 ± 1.6	60.6 ± 0.4
(12,a31), 20 %	76.3 ± 0.5	37.1 ± 6.3

[a] Cell viability was determined by trypan blue staining test.   Data represent mean ± S.E. from three measurements.

B16 Melanoma Cell.   As the third example of mammalian cells, B16 melanoma cell was investigated (22).   Monolayer culture of B16 cells was treated with the liposome that contained FITC-dextran as the water-soluble marker.   A small extent of the fluorescence was observed in the cells when liposome was unmodified with PEO-lipid.   This cell uptake was completely inhibited by cytochalasin B, and we concluded that this was endocytosis, not fusion.   No detectable introduction of the fluorescence was seen with the PEO-lipid-reconstituted liposome.

By using colcemid, a cell-cycle arresting reagent (40, 41), B16 cells were arrested during mitotic phase.   The cells after the colcemid treatment could be easily detached from the wall of the culture bottle.   The suspended cells thus obtained were then treated with the PEO-lipid- reconstituted liposome that contained both FITC-dextran and octadecyl rhodamine B (Figure 6).   FITC-dextran was observed as spread in the cytosol, while octadecyl rhodamine B was seen only on the cytoplasm membrane. The independent localization of the two fluorescent probes was still observed even when the cells were treated with cytochalasin B and D.   This suggests that the present cell-liposome interaction is not endocytosis but fusion.   We further investigated the liposome-cell fusion using the DTA-embedded liposome and found that the cells were certainly killed by the DTA-dependent cytotoxicity.

Discussion

Several interesting points can be noted in our studies of the PEO-lipid-reconstituted liposome.   As seen in the case of Jurkat cells, the PEO-lipid-reconstituted liposome takes two different routes, fusion and endocytosis, depending on the PEO-lipid structure.   This fact could be a valuable for us to study intracellular transport system in biological events, e.g., intracellular metabolism of drugs transferred into target cells, processing of foreign substances for antigen presentation, etc.   Further, we would be able to draw out valuable information about cytosol such as microscopic pH, viscosity, polarity and so on in the cytosol if we could directly transport an appropriate probe into the cytosol to monitor these parameters.

One can control the fusion efficiency by modifying the incubation conditions. This is not only important for improving the fusion efficiency in practical application of the cell-liposome fusion, but also it should provide insight into the mechanism of the fusion.   The length of the PEO chain optimal for the cell-liposome fusion may be greatly dependent on the target cells.   For suspension cells like Jurkat cells or B16 cells those were arrested in mitotic phase, PEO-lipid(12,a31) was the best of the three PEO-lipids tested.   For HeLa cells or carrot protoplasts, on the other hand, PEO-lipid with shorter PEO-chain (PEO-lipid(12,a13) or PEO-lipid(12,a11.5)) showed better fusogenicity.

A

Figure 6. Confocal microscopic images of synchronized B16 cells in mitotic phase treated with the PEO-lipid functionalized liposome (PEO-lipid(12,a31), 20 mol%) at 4°C for 30 min.   A: transmission image, B: fluorescence image excited at 488 and 568 nm

Continued on next page.

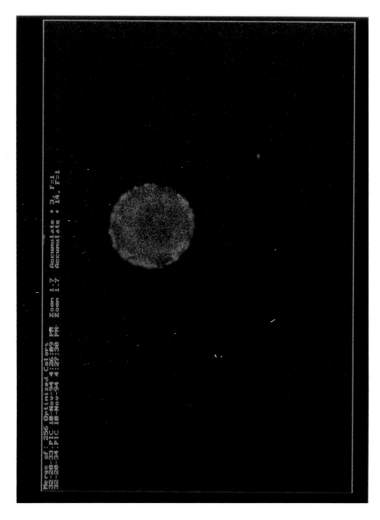

B

Figure 6. *Continued.*

In addition, cell surface proteins may affect susceptibility of the cell to the fusion with the PEO-lipid-reconstituted liposome.   In the case of Jurkat cells, the fusion efficiency is largely enhanced when the cells are pretreated with a serum-depleted medium.   The pretreatment of Jurkat cells decreases the amount of serum proteins adsorbed on the cell surface, and this may lead to closer contact between the cell and the liposome. Also in the case of B16 cells, the efficiency is enhanced when the cells are synchronized in mitotic phase.   The serum depletion induces synchronization of the cell cycle and thus enhances the fusion efficiency.   Regarding the superiority of mitotic phase, two research groups have previously shown that the cells in mitotic phase possess high membrane fluidity (*42, 43*), and this may have relation to the efficiency of the cell-liposome fusion.

Compared with free PEO, the PEO-lipid requires a far less amount of the PEO-moiety for induction of the fusion.   With free PEO, the total concentration of PEO in a fusion system is 20~60% PEO (by weight) while, with PEO-lipid, it can be less than 0.004%.   This can be attributed to the localization of the PEO moiety only to the place in which PEO works for the fusion; presumably, the proximity between the cell membrane and the liposomal membrane.   In the case of the cell-liposome fusion, the optimal length of free PEO in PEO-lipid is longer than 15 units, which corresponds to approximately 700 or more of the molecular weight.   This is similar to the optimal molecular weight of PEO for evoking fusion (*44*).   However, the kinetics of the two types of fusion are not identical.   Generally, the free PEO-induced fusion completes within 5 min (*44*) while the PEO-lipid-induced fusion requires a period longer than 15 min.   The mechanisms of the two types of fusion would be somewhat different.

In spite of extensive usage of PEO in cell biology, medicine and biotechnology today, the interaction of PEO with intact cells is still controversial.   Modification of liposomal surface with PEO derivatives prevents cell uptake in reticuloendothelial system and subsequently extends circulation time of the liposome in blood (*45-50*). On the other hand, several liposomes modified with PEO-conjugated phosphatidylethanolamines accumulate in the spleen (*51, 52*).   We have also found that the reconstitution of PEO-lipid(12, a13) into the liposome induces endocytic uptake of Jurkat cell (*20*).   PEO-bearing surfactants show a spermicidal activity (*53, 54*) and hemolysis (*55*).   These findings tell us that further studies are needed to understand the interaction of PEO and/or PEO-modified lipids with intact cells.

Distearoylphosphatidylethanolamine-conjugated PEO (DSPE-PEO) has been most commonly used for the extension of the circulation time (*56, 57*).   The effect is attributed to the steric repulsion caused by the PEO moieties on the surface of the liposomes.   Recently, in order to understand the interaction involved, the DSPE-PEO system has been examined using several physicochemical procedures.   The aggregation behavior of DSPE-PEOs in aqueous media was studied using DSC, X-ray diffraction and NMR (*58, 59*), and surface pressure (*60*) and surface force (*61*) of the monolayer of DSPE-PEOs was also investigated.

The detailed mechanism of the membrane fusion is yet to be clarified at least at the molecular level.   In the case of PEO-induced fusion, PEO raises dramatic changes on the surface of lipid membrane; it causes dehydration, modification of membrane fluidity and microviscosity, increase of surface tension, aggregation of membrane proteins and lateral phase separation (*62-67*).   These events are considered to be followed by bilayer destabilization and subsequent membrane fusion.   The most significant difference between our PEO-lipids and DSPE-PEOs exists in the structure of the lipophilic anchors.   Possibly, the 1,3-didodecyloxy-2-propanol backbone may provide destabilization of the liposomal membrane, which could be one of the keys to the fusion induction.

Acknowledgment

The authors are grateful to Mr. Koji Tanaka, Mr. Masahiro Yamauchi, Dr. Mamoru Haratake, Dr. Ji Zheng, Dr. Mahito Nakanishi, Dr. Michiko Kodama, and Dr. Adam Baszkin for the fruitful collaborations. The authors would also like to acknowledge financial supports from the Ministry of Education, Science, Culture and Sports of Japan, the Japan Science and Technology Corporation, and the Termo Science and Technology Foundation.

Literature Cited

1   Lasic, D. D. *Liposomes, from Physics to Applications*; Elsevier: Amsterdam; 1993.
2   Gennis, R. B. *Biomembranes, Molecular Structure and Function*; Springer-Verlag: New York, NY; 1989.
3   Gurdon, J. B.; Lana, H. R.; Marbaix, G. *Nature* **1971**, *233*, 177-179.
4   Zimmermann, U. *Biochim. Biophys. Acta.* **1982**, *694*, 227-235.
5   Furusawa, M.; Nishimura, T.; Yamaizumi, M.; Okada, Y. *Nature* **1974**, *249*, 449-450.
6   Felgner. P. L.; Gadek. T. R.; Holm. M.; Roman, R.; Chan, H. W.; Wenz, M.; Northrop, J. P.; Ringold, G. M.; Danielsen, M. *Proc. Natl. Acad. Sci. U.S.A.* **1987**, *84*, 7413-7417.
7   Vaheri, A.; Pagano, J. S. *Virology* **1965** *27*, 434-439.
8   Arvinte, T.; Hildenbrand, K.; Wahl, P.; Nicolau, C. *Proc. Natl. Acad. Sci. U.S.A.* **1986**,*83*, 962-966.
9   Arvinte, T.; Wahl, P.; Nicolau, C. *Biochim. Biophys. Acta* **1987**, *899*,143-150.
10   Uchida, T.; Kim, J.; Yamaizumi, M.; Miyake, Y.; Okada, Y. *J. Cell Biol.* **1979**, *80*, 10-20.
11   Nakanishi, M.; Uchida, T.; Sugawa, H.; Ishiura, M.; Okada, Y. *Exp. Cell Res.* **1985**, *159*, 399-409.
12   Kato, K.; Nakanishi, M.; Kaneda, Y.; Uchida, T.; Okada, Y. *J. Biol. Chem.* **1991**, *266*, 3361-3364.
13   Nakanishi, M.; Ashihara, K.; Senda, T.; Kondo, T.; Kato, K.; Mayumi, T. In *Trends and Future Perspectives in Peptide and Protein Drug Delivery;* Lee, V. H. L.; Hashida, M.; Mizushima, Y., Eds.; Harwood Academic Publishers: Amsterdam; in press.
14   Kao, K. N.; Michayluk, M. R. *Planta* **1974**, *115*, 355-367.
15   Ahkong, Q. F.; Howell, J. I.; Lucy, J. A.; Safwat, F.; Davey, M. R.; Cocking, E. C. *Nature* **1975**, 255, 66-67.
16   Szoka, F.; Magnisson, K. -E.; Wojcieszyn, J.; Hou, Y.; Derzko, A.; Jacobson, K. *Proc. Natl. Acad. Sci. U.S.A.* **1981**, *78*, 1685-1689.
17   Sunamoto, J.; Tanaka, K.; Akiyoshi, K.; Sato, T. *Polymer Preprints* **1990**, *31*, 155-156.
18   Sato, T.; Sunamoto, J. *Prog. Lipid Res.* **1992**, *31*, 345-372.
19   Okumura, Y.; Yamauchi, M.; Yamamoto, M.; Sunamoto, J. *Proc. Japan Acad.* **1993**, *69(B)*, 45-50.
20   Higashi, N.; Sunamoto, J. *Biochim. Biophys. Acta* **1995**, *1243*, 386-392.
21   Higashi, N.; Okumura, Y.; Yamauchi, M.; Nakanishi, M.; Sunamoto, J. *Biochim. Biophys. Acta,* **1996**, *1285*, 183-191.
22   Haratake, M.; Sunamoto, J. In *Abstract of 5th Iketani Conference--- International Symposium on Biomedical Polymers*: Kagoshima, 1995; pp 210-211.
23   Kuwamura, T. *Kogyokagakuzasshi*, **1961**, *64*, 1965-1972.

24   Kuwamura, T. *Kogyokagakuzasshi*, **1961**, *64*, 1958-1964.
25   Okahata, Y.; Tanamochi, S.; Nagai, M.; Kunitake, T. *J. Colloid Interface Sci.*
      1981, *82*, 401-417.
26   Israelachvili, J. N.; Marcelja, S.; Horn, R. G., *Quart. Rev. Biophys.*, **1980**,
      13, 121-200.
27   Rosilio, V.; Albrecht, G.; Baszkin, A.; Okumura, Y.; Sunamoto, J. *Chem.
      Lett.* **1996**, 657-658.
28   Rosilio, V.; Albrecht, G.; Okumura, Y.; Sunamoto, J.; Baszkin, A. *Langmuir*
      1996, *12*, 2544-2550.
29   Szoka, F.; Papahadjopoulos, D. *Proc. Natl. Acad. Sci. U.S.A.* **1978**, *75*,
      4194-4198.
30   Kodama, M.; Tsuchiya, S.; Nakayama, K.; Takaichi, Y.; Sakiyama, M.,
      Akiyoshi, K.; Tanaka, K.; Sunamoto, J. *Thermochimica Acta* **1990**, *163*,
      81-88.
31   Struck, D.K.; Hoekstra, D.; Pagano, R.E. *Biochemistry* **1981**, *20*, 4093-
      4099.
32   Hoekstra, D. *Biochim. Biophys. Acta* **1982**, *692*, 171-175.
33   Hoekstra, D. *Biochemistry* **1982**, *21*, 2833-2840.
34   Yamaizumi, M.; Mekada, E.; Uchida, T.; Okada, Y. *Cell* **1978**, *15*, 245-250.
35   Nakanishi, M.; Okada, Y. *Liposome Technology* ; Gregoriadis, G., Ed.;
      CRC Press: Boca Raton, FL, 1992; pp 249-260.
36   Renau-Piqueras, J.; Miragall, F.; Cervera, J. *Cell Tissue Res.* **1985**, *240*,
      743-746.
37   Mimura, N.; Asano, A. *Nature* **1976**, *261*, 319-321.
38   Takigawa, M.; Danno, K.; Furukawa, F. *Arch. Dermatol. Res.* **1987**, *279*,
      392-397.
39   Finbloom, D.S.; Martin, J.; Gordon, R. K. *Clin. Exp. Immunol.* **1987**, *67*,
      205-210.
40   Bhuyan, B. K.; Adams, E. G.; Badiner, G. J.; Trzos, R. J. *J. Cell. Physiol.*
      1987, *132*, 237-245.
41   Rieder, C. L.; Palazzo, R. E. *J. Cell. Sci.* **1992**, *102*, 387-392.
42   Lai, C. -S.; Hopwood, L. E.; Swartz, H. M. *Biochim. Biophys. Acta* **1980**,
      602, 117-126.
43   Collard, J. G.; de Wildt, A.; Oomen-Meulemans, E. P. M.; Smeekens, J.;
      Emmelot, P.; Inbar, M. *FEBS Lett.* **1977**, *77*, 173-178.
44   Blow, A. M. J.; Botham, G. M.; Fisher, D.; Goodall, A. H.; Tilcock, C. P.
      S.; Lucy, J. A. *FEBS Lett.* **1978**, *94*, 305-310.
45   Senior, J.; Delgado, C.; Fisher, D.; Tilcock, C.; Gregoriadis, G. *Biochim.
      Biophys. Acta* **1991**, *1062*, 77-82.
46   Blume, G.; Cevc, G. *Biochim. Biophys. Acta* **1990**, *1029*, 91-97.
47   Klibanov, A. L.; Maruyama, K.; Torchilin, V.; Huang, L. *FEBS Lett.* **1990**,
      268, 235-237.
48   Allen, T. M.; Hansen, C.; Martin, F.; Redemann, C.; Yau-Young, A.
      Biochim. Biophys. Acta **1991**, *1066*, 29-36.
49   Lee, K. -D.; Hong, K.; Papahadjopoulos, D. *Biochim. Biophys. Acta* **1992**,
      1103, 185-197.
50   Mori, A.; Klivanov, A. L.; Torchilin, V. P.; Huang, L. *FEBS Lett.* **1991**,
      284, 263-266.
51   Klibanov, A. L.; Maruyama, K.; Beckerlag, A. M.; Torchilin, V. P.; Huang,
      L. *Biochim. Biophys. Acta* **1991**, *1062*, 142-148.
52   Litzinger, D. C.; Huang, L. *Biochim. Biophys. Acta* **1992**, *1127*, 249-254.
53   Sunamoto, J.; Iwamoto, K.; Ikeda, H.; Furuse, K. *Chem. Pharm. Bull.*
      1983, *31*, 4230-4235.
54   Sunamoto, J.; Iwamoto, K.; Uesugi, T.; Kojima, K.; Furuse, K. *Chem.
      Pharm. Bull.* **1984**, *32*, 2891-2897.

55   Zaslavsky, B. Y.; Ossipov, N. N.; Krivich, V. S.; Baholdina, L. P.;
      Rogozhin, S. V. *Biochim. Biophys. Acta* **1978**, *507*, 1-7.
56   Woodle, M. C.; Lasic, D. D. *Biophys. Biochim. Acta* **1992**, *1113*, 171-199.
57   Zalipsky, S.; Hansen, C. B.; Lopes de Menezes, D. E.; Allen, T. M.; *J.
      Control. Release* **1996**, *39*, 153-161.
58   Needham, D.; McIntosh, T. J.; Lasic, D. D. *Biophys. Biochim. Acta* **1992**,
      1108, 40-48.
59   Kenworthy, K. K.; Simon, S. A.; McIntosh, T. J. *Biophys. J.* **1995**, *68*,
      1903-1920.
60   Baekmark, T. R.; Elender, G.; Lasic, D. D.; Sackmann, E. *Langmuir* **1995**,
      11, 3975-3987.
61   Kuhl, T. L.; Leckband, D. E.; Lasic, D. D.; Israelachvili, J. N. *Biophys. J.*
      1994, *66*, 1479-1488.
62   Lehtonen, J. Y.A.; Kinnunen, P. K. J. *Biophys. J.* **1994**, *66*, 1981-1990.
63   MacDonald, R. I. *Biochemistry* **1985**, *24*, 4058-4066.
64   Burgess, S.W.; McIntosh, T.J.; Lentz, B.R. *Biochemistry* **1992**, *31*, 2653-
      2661.
65   Ahkong, Q. F.; Lucy, J. A. *Biochim. Biophys. Acta* **1986**, *858*, 206-216.
66   Knutton, S. *J. Cell Sci.* **1979**, *36*, 61-72.
67   Tilcock, C. P. S.; Fisher, D. *Biochim. Biophys. Acta* **1979**, *577*, 53-61.

Chapter 7

Poly(ethylene glycol)-Based Micelles for Drug Delivery

Sung Bum La[1], Yukio Nagasaki[2], and Kazunori Kataoka[1,2]

[1]International Center for Biomaterials Science, Research Institute for Biosciences,
Science University of Tokyo, Yamazaki 2669, Noda, Chiba 278, Japan
[2]Department of Materials Science and Technology and Research Institute
for Biosciences, Science University of Tokyo, Yamazaki 2641, Noda,
Chiba 278, Japan

An AB-type block copolymer micelle composed of poly(ethylene glycol)
(PEG) as the hydrophilic segment and poly(amino acid) (PAA) as the
hydrophobic segment has been utilized as a drug carrier in the field of
drug delivery systems.   In this chapter, we discuss the promising
feasibility of PEG-based micelles as nanoscopic carriers for targeting of
selective drugs for cancer chemotherapy (e.g., doxorubicin and cisplatin).
In addition, the novel possibility of PEG-poly(β-benzyl-L-aspartate) (PEG-
PBLA) polymeric micelles for oral delivery of selective drugs (e.g.,
indomethacin) is also discussed based on recent *in vitro* results. Finally,
new heterobifunctional PEGs with a formyl or a primary amino group at
one end are used to prepare novel polymeric micelles with available
groups on the surface, and polyion-complex micelles are introduced as
candidate novel carriers for drug delivery.

Tremendous research effort has been expended for selective drug delivery using
polymeric devices to increase therapeutic effects and to decrease undesirable side
effects of drugs. Recently, colloidal carriers using biocompatible block copolymers to
form micelles have been studied extensively for selective drug delivery. It is well
established that block copolymers composed of poly(ethylene glycol) (PEG) as the
hydrophilic part and the appropriate hydrophobic part including polystyrene (*1*),
poly(β-benzyl-L-aspartate) (PBLA) (*2*), and polyisoprene (*3*) have been shown to self-
associate in an aqueous medium to form spherical micelles that have a dense core
consisting of hydrophobic blocks and a corona surrounded by hydrated outer PEG
segments, Figure 1.   The driving force for micellization of block copolymers is
generally considered to be attractive forces (for example, van der Waals interactions)
between collapsed hydrophobic blocks and the repulsive interactions between one of
the blocks and the solvent. Micelle-forming block copolymers with hydrophilic and
hydrophobic segments can be utilized as drug carriers because most hydrophobic drugs

are easily incorporated into the inner core of the micelles by covalent bonding or non-covalent bonding (such as hydrophobic and ionic interactions). From the pharmaceutical point of view, it is very important to combine components such as a commercially established drug or a drug that has pharmaceutical problems (e.g., low solubility in water, side effects or non-specific drug action) and a polymer with suitable properties for controlled drug delivery.

As one component of amphiphilic blocks, PEGs have been extensively used in drug delivery systems (DDS) due to their ideal properties such as non-toxicity approval for internal use, high water solubility, non-immunogenicity, and non-thrombogenicity. PEG has been used in protein modification to decrease the antigenicity of the intact protein (4) and to prolong half-life in the blood stream (5, 6). Derivatives of PEG have also been applied as intermediates for the chemical modification of proteins and peptides (7, 8), conjugation with collagen (9), inclusion compounds (10), and as prodrugs for synthetic drugs with low molecular weight (11, 12). PEG and its derivatives have been studied and reviewed extensively not only for fundamental properties but also for their chemical, biomedical and pharmaceutical applications (13-16).

This chapter is focused on our recent research with AB block copolymer micelles composed of PEG, which is suitable for the hydrophilic segment of the block copolymer to form the outer shell of the micelle, and poly(amino acids) (PAA) as drug carriers for selective drug delivery.

Polymeric Carriers for Drug Delivery

Since Ehrlich's magic bullets that were the initial strategy for delivery of selective drugs by carriers consisting of haptophore (binding moiety to the target site) and toxophore (cytotoxic portion), much study has been done on drug targeting using drug-conjugates with polymers, primarily for anticancer drugs. As criteria in selection of anticancer drugs for targeting, drugs that cannot be administered in large enough doses to kill cancer cells, or the clinical uses of the drugs are limited owing to the indiscriminate attack of drugs to both targeted cells and normal cells. By using drug-conjugates, however, it has not always been found possible to significantly accumulate drug-conjugates to target site due to their non-specific excretion and enzymatic or non-enzymatic degradation. After intravenous injection, *in vivo* fate of drug-carrier conjugates as vehicle is that the vehicles usually have a high-incidence of excretion through kidneys if their molecular weights are not large enough to avoid glomerular filtration (urinary excretion) as well as of hepatic elimination. It is noteworthy that vehicles administered in the bloodstream are rapidly and efficiently eliminated by the reticuloendothelial systems (RES), located at such organs as liver, spleen, and lung. The RES is known to be the main reason for removal of particulate vehicles from the bloodstream. Thus, it is necessary to develop polymeric drug-carriers that achieve drug targeting in the bloodstream and avoid RES recognition (17).

For drug targeting, polymeric carriers have several advantageous features in pharmaceutical application. Of prime importance is the effect reported by Maeda et al. (18, 19) of "enhanced permeability and retention (EPR)" in the target organ by using macromolecular substances.   Macromolecular drugs and drug vehicles accumulate more effectively and remain in tumor tissue for a much longer period than in normal tissue with little drainage via lymphatic clearance, following both direct injection into the tumor and indirect accumulation from the bloodstream.   The mechanism of indirect accumulation is believed to be the synergism of increased tumor vascular permeability and decreased tissue drainage.   Maeda *et al.* proposed that passive accumulation of macromolecular drugs for targeting to tumors is a general phenomenon (*20*), and this new concept has now become one of the major guiding principles in drug targeting using polymeric carriers.

Micelle-Forming Block Copolymer

For development of micelle-forming polymeric drugs, Ringsdorf et al. (*21*) first investigated the idea of using polymeric micelles for the sustained release of drugs. The A-block consisted of hydrophilic units while the B-block contained hydrophobic side-chain units and attached drug.   In this paper, AB-block copolymer micelles based on poly(ethylene oxide) and poly(L-lysine) were synthesized, and drugs such as sulfido derivatives of cyclophosphamide (CP) and an alkylating antitumor agent were fixed in a polymer analog reaction.   The CP-derivatives, which normally hydrolyze rapidly to the active metabolite 4-hydroxy-CP, can be stabilized owing to their assembly into the hydrophobic portion of the block copolymer, which was suggested by the delay in the half-life for release from minutes to several hours.   In addition, they suggested that this polymer-drug conjugate was forming a micellar system based on data for dye solubilization using Sudan red 7B (*22*).

We have proposed a concept of polymeric micelle with amphiphilic character composed of PEG and poly(aspartic acid) block copolymer (PEG-Pasp) as novel drug carrier which would be useful for targeting owing to avoiding RES recognition.   This "stealth property" is based on size as well as the steric repulsion effect of PEG.   We have carried out a systematic study to prove its pharmaceutical advantages for application of selective drug delivery.   Our results related to polymeric micelle containing selective drug (e.g., doxorubicin or DOX) for targeting were reviewed previously (*17, 23, 24*).   Thus, here we briefly discuss PEG-based, polymer-drug conjugates for drug targeting, and PEG-poly(lactide) and polyion complex micelles as candidates of novel drug carriers.   Also the novel feasibility of PEG-PBLA micelles for oral delivery are described in this chapter.

DOX-Conjugated PEG-Pasp Block Copolymer Micelle, PEG-Pasp(DOX)

Synthesis of micelle-forming PEG-Pasp block copolymers and PEG-Pasp(DOX), which was prepared by conjugating DOX to carboxyl groups of PEG-Pasp block copolymers via amide linkages, was reported and the scheme is shown in Figure 2 (*25-27*).

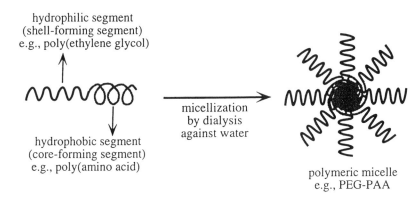

hydrophilic segment
(shell-forming segment)
e.g., poly(ethylene glycol)

micellization
by dialysis
against water

hydrophobic segment
(core-forming segment)
e.g., poly(amino acid)

polymeric micelle
e.g., PEG-PAA

Figure 1. Concept of multimolecular micellization of amphiphilic block copolymers in a selective solvent.

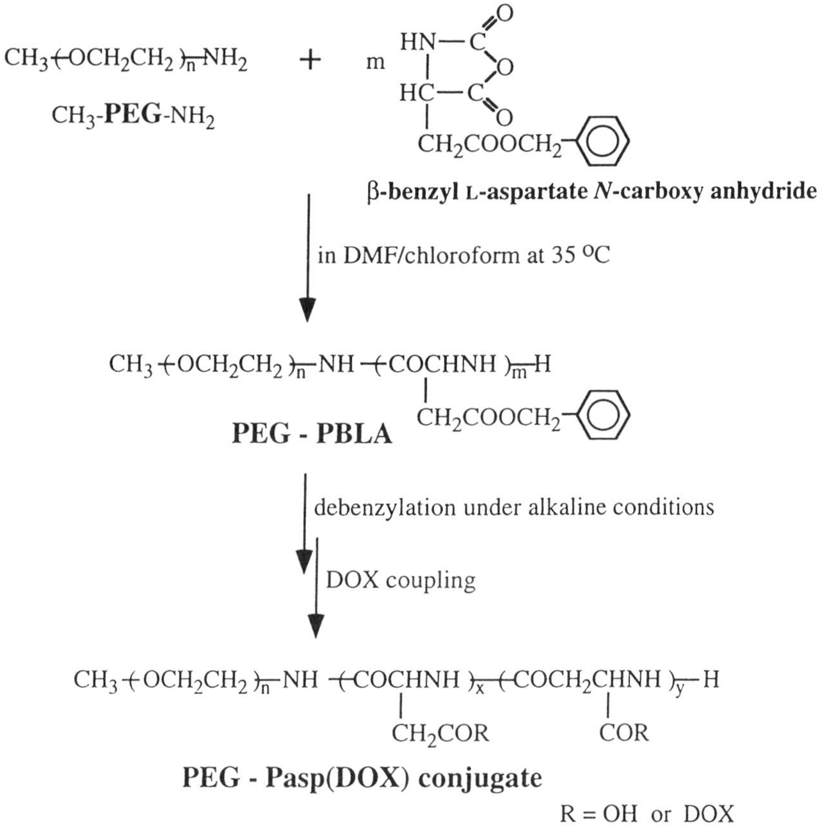

$CH_3 \text{-} (OCH_2CH_2)_n \text{-} NH_2$ +   m

$CH_3\text{-}\mathbf{PEG}\text{-}NH_2$

β-benzyl L-aspartate *N*-carboxy anhydride

in DMF/chloroform at 35 °C

$CH_3 \text{-} (OCH_2CH_2)_n \text{-} NH \text{-} (COCHNH)_m \text{-} H$

PEG - PBLA CH_2COOCH_2

debenzylation under alkaline conditions

DOX coupling

$CH_3 \text{-} (OCH_2CH_2)_n \text{-} NH \text{-} (COCHNH)_x \text{-} (COCH_2CHNH)_y \text{-} H$

CH_2COR COR

PEG - Pasp(DOX) conjugate

R = OH or DOX

Figure 2. Synthesis of the polymeric drug, PEG-Pasp(DOX).

To obtain the AB block copolymers of PEG and PBLA, β-benzyl-L-aspartate N-carboxy anhydride (BLA-NCA) was polymerized from the terminal amino group of α-methoxy-ω-amino PEG as an initiator. The block lengths of the A or B segment could be adjusted by control of the PEG molecular weights as starting materials and by the amount of BLA-NCA.   The PEG-PBLA block copolymers obtained were debenzylated under alkaline conditions to form the PEG-Pasp block copolymers without detectable Pasp chain cleavage.   Subsequently, DOX was coupled to PEG-Pasp by amide bonds between the amino group of the DOX and the carboxyl group of an aspartic acid residue in the Pasp chain using 1-ethyl-3-(3-dimethylaminopropyl)-carbodiimide as a coupling agent.

Micelle formation of PEG-Pasp(DOX) conjugates thus prepared was confirmed by dynamic light scattering (DLS) (*26-28*) and gel permeation chromatography (GPC) (*27, 29, 30*).   In this system, we have suggested that micelle formation is mainly driven through hydrophobicity and the cohesive force of the conjugated DOX itself, because both high loading capacity and high stability can simultaneously be achieved.   The size distribution of the micelles was narrow, with diameters observed in the range of 15 to 60 nm on the number average scale with various compositions of the conjugates, including the molecular weight of both segments and the DOX content.   The appropriate size for long half-life in blood of the micellar structure of the PEG-Pasp(DOX) with a dense core and shell corresponds to that of a virus; a typical example is adenovirus that is known to be stable from the outer environment owing to its supramolecular structure and its small size (70-80 nm), which may contribute to avoidance of RES uptake.   Thus, the micelle-forming PEG-Pasp(DOX) conjugates with appropriate size are large enough to avoid urinary excretion, and they may have a low incidence of uptake by the RES due to the low interfacial energy and high steric repulsion of the hydrophilic palisades surrounding the core of the micelle.

In addition, GPC study, which is a suitable method to get the micelle properties that depend on the composition, also gave evidence of micelle formation.   In the case of PEG-Pasp(DOX) composed of various lengths of PEG and Pasp segment ranging from 1000 to 12000 in molecular weight and from 10 to 80 units, respectively, the conjugates were eluted as a peak at the gel exclusion volume in phosphate-buffered solution (pH 7.4), indicating that the conjugates formed polymeric micelles because the exclusion molecular weight of the pullulan standard is over 300 000.   Micelle-forming behavior of the conjugates and their stability were found to be dependent on the composition of the conjugates.   Polymeric micelles are considered to be more stable than those of low molecular weight surfactants, and they have much lower critical micelle concentration (cmc) compared to surfactant micelles.   Such properties of the polymeric micelles can be utilized to realize stable micelle formation under *in vivo* conditions, and the micelle formation of the PEG-Pasp(DOX) conjugates in the bloodstream also was confirmed by GPC (*31*).

To evaluate *in vivo* anti-cancer activity of DOX-conjugated micelles, PEG-Pasp(DOX) having different chain length of poly(aspartic acid) were injected intraperitoneally to CDF1 female mice that were intraperitoneally inoculated with P 338 mouse leukemia (26, 28). The anti-cancer activity of PEG-Pasp(DOX) was evaluated by the ratio of the median survival period of the treated mice to that of the control mice (T/C %). The PEG-Pasp(DOX) showed higher *in vivo* anti-cancer activity than that of free DOX, and lowered toxic side effects, which were judged from the body weight change of the mice inoculated with free DOX or PEG-Pasp(DOX). The low toxicity of PEG-Pasp(DOX), as approximately 1/20 of the toxicity of intact DOX, was also observed with body weight change in DOX equivalents (31). In addition, extended *in vivo* anti-cancer activities of PEG-Pasp(DOX) against solid tumors were evaluated with three murine tumors (C 26, C 38, and mouse fibrosarcoma M 5076) and two human tumors (gastric cancer MKN-45 and mammary cancer MX-1). Superior anti-cancer activities of PEG-Pasp(DOX) were observed relative to that of free DOX in all of tumors investigated, except MKN-45 tumor, and almost the same activity was revealed in the MKN-45 tumor.

In the mechanism of anti-cancer action of micelle-forming polymeric drug PEG-Pasp(DOX), we have assumed the possibility of three major mechanisms; drug-conjugated polymeric micelles will possess the drug action by direct interaction with target cells, by slow release of free drug from the micelle (it should be noted that there exists a definite amount of physically-entrapped DOX in the core of PEG-Pasp(DOX) micelles), and by equilibrium control, in which polymer with conjugated drug dissociated from the micelle (17). The anti-cancer activity may reveal by the mechanism of one or two combined in the three major mechanisms. Our recent study revealed that sustained release of physically-entrapped DOX in the micelle at the tumor site seems to have a major role in the anti-tumor effect of the PEG-Pasp(DOX) system (32). Furthermore, we speculate that one of distinctive properties, the EPR effect, for macromolecules such as polymeric drug will play a key role in expressing superior anti-cancer activity of PEG-Pasp(DOX) relative to that of free DOX, because the micelles with a hydrated PEG outer shell may be preferentially taken up and retained by the tumor. This suggestion was supported by following studies.

The pharmacokinetics and biodistribution of the PEG-Pasp(DOX) were investigated by intravenous injection in mice of micelle which had been radiolabeled with ^{125}I (31) and ^{14}C-benzylamine (33, 34). In these studies, PEG-Pasp(DOX) was found to afford a significantly larger initial concentration in blood with a longer half-life than free DOX, indicating that DOX circulated in micellar form in blood without rapid excretion, metabolism, or absorption by tissues. By comparison, free DOX was known to be rapidly absorbed by tissues and revealed low initial concentrations with a short half-life (within 5 min).

We have hypothesized that PEG-Pasp(DOX) with prolonged blood circulation and low RES uptake will lead to enhanced accumulation at tumor sites. It is worth notice that our hypothesis is supported because the results of biodistribution of radiolabeled PEG-Pasp(DOX) injected intravenously in tumor-bearing mice substantiate our suggestion. It can be concluded that micelle-forming PEG-Pasp(DOX) conjugates have promise for targeting of DOX.

Recently, cisplatin (CDDP) was introduced into PEG-Pasp block copolymer and thus CDDP-conjugated PEG-Pasp micelles, PEG-Pasp(CDDP), were prepared by ligand substitution reaction at platinum of CDDP to aspartic acid residues of PEG-Pasp in distilled water (*35*). The micelles from PEG-Pasp(CDDP) conjugates were confirmed by DLS and GPC, and they had small diameters of ca. 16 nm based on weight average. The micelle was found to be stable in an aqueous medium as well as in the presence of 1 % sodium dodecyl sulfate, suggesting that the micelle formed an intermolecular complex by bridge formation through Pt. *In vitro* cytotoxicity of free CDDP and PEG-Pasp(CDDP) against murine B 16 melanoma cells shows that the polymeric micelles had reduced toxicity as compared with free CDDP, indicating slow release of Pt complexes from the micelles. However, extended studies, such as a prolonged incubation period for cytotoxicity and *in vivo* anti-cancer activity, are necessary to evaluate its efficacy.

PEG-PBLA block copolymer micelles

We have recently extended our research to the design of micelle-forming block copolymers with amphiphilic characteristics for installation of drugs into the inner core of the micelle by means of non-covalent interactions (i.e., physical entrapment) (Figure 3) (*36*). In this model, drugs with hydrophobic properties do not need to have appropriate chemical functional sites for covalent attachment to the block copolymers to be incorporated into the polymeric micelle. For this purpose, polymeric micelles having different compositions based on AB block copolymers of PEG and PBLA were prepared because of the suitable properties of PEG aforementioned in the former parts and, in particular, PBLA's biodegradable nature with lower toxicity. Characterization of micellar solutions of PEG-PBLA block copolymers was achieved by static and dynamic light scattering and ^{1}H-NMR and photophysical means (*37*).

Distribution of particle sizes of the PEG-PBLA micelles in water, determined by light scattering, suggested a bimodal distribution; this was assumed to be due to individual micelles with secondary aggregates. Concerning the secondary aggregates, the basic nature is still unclear. However, PEG-PBLA micelles without any secondary aggregates have been obtained recently by careful control of dialysis (*38*): drop-wise addition of water to N,N-dimethylacetamide (DMAC) solution of PEG-PBLA to form micelles with swelled core, followed by dialysis against water. This methodology of selective solvent addition has been utilized by Munk et al. (*39*) to prepare polystyrene-block-poly(tert-butyl methacrylate) with unimodal size distribution. PEG-PBLA micelles prepared by the above-mentioned method of modulated dialysis have an average diameter of 69 nm on a gamma-averaged scale with narrow size distribution. The polydispersity index, characterized by cumulant method, is calculated to be 0.1, which indicates that the micelle has an essentially monodispersive nature. In static light scattering studies, an association number of 200 a micellar molecular weight of 3.1 x 10^6 g/mol were calculated.

For physical entrapment of drugs into PEG-PBLA micelles, pyrene was first investigated as a model hydrophobic molecule (*36, 37*). Pyrene is a widely used fluorescence probe because it is one of the few condensed aromatic hydrocarbons that shows significant vibrational structure sensitive to polarity (*40*). Thus, it preferentially partitions into the hydrophobic core of the micelle with a concurrent change in photophysical properties of the molecule. It is worth note that the location of the pyrene within the PBLA core of the micelle can be determined from the fluorescence vibrational fine structure, demonstrating the ability of the PEG-PBLA polymeric micelles to solubilize drugs with hydrophobic character. Using the intensity change of vibrational fine structure of pyrene, the cmc of PEG-PBLA block copolymer in distilled water was obtained; the cmc value was less than 10 mg/L, indicating that the micelles are very stable.

In addition, the conformation of the PBLA segment in the PEG-PBLA block copolymers was studied in organic solvents (*41*). While PBLA homopolymer, with the same molecular weight as the PBLA segment in the PEG-PBLA, showed no evidence of α-helix formation in chloroform, the PBLA segments adopted a left-handed α-helix conformation in the same conditions, thus indicating that the PEG segment in the PEG-PBLA is essential to allow PBLA to adopt an α-helix structure. On the contrary, the PBLA segments had a random-coil conformation in dimethylsulfoxide (DMSO). Thus, the conformational structure of the PBLA segments in the PEG-PBLA copolymers depends on the solvents used, and the results may play an important role in controlling the properties of the resulting micelles.

For pharmaceutical applications, DOX was incorporated into the core of the PEG-PBLA micelles by physical entrapment; the loading process involved transfer of DOX and PEG-PBLA into an aqueous medium from N,N-dimethylformamide (DMF) through dialysis (*42*). In this case, we hypothesized that by bringing DOX, in its unionized form, and PEG-PBLA into an aqueous medium from a good solvent for both species, DOX could be entrapped in PEG-PBLA micelles. As the DMF is removed through dialysis, PEG-PBLA micelles form and have a prospect of incorporating DOX. Evidence for the preparation of the PEG-PBLA(DOX) polymeric micelles was derived from fluorescence spectroscopy and GPC. From the fluorescence emission spectra, DOX displayed its characteristic spectrum, whereas PEG-PBLA(DOX) had a low total fluorescence intensity, suggesting that the drug was self-associated in the micelles. In GPC studies, PEG-PBLA(DOX) eluted at the gel exclusion volume, consistent with the elevated molecular weights of the micelles. The entrapment and retention of DOX in the micelles may be highly constrained and ionization of DOX may be inhibited by the hydrophobicity of the cores and by DOX self-association. This research was focused on the possibility of preparation of PEG-PBLA(DOX) micelles by physical attachment for targeting of DOX. Subsequent studies of the micelles containing DOX are required to obtain evidence for targeting of DOX.

Recently, as an advanced step for drug targeting using PEG-PBLA micelles, PEG-PBLA block copolymers having a functional group at the end of the PEG chains were synthesized to introduce the targeting moiety on the outer shell of the PEG-PBLA micelles; these are α-methoxy- and α-hydroxy-PEG-PBLA (*43*). The micelles thus prepared were characterized by DLS and fluorescence spectroscopy. Both PEG-PBLA micelles have a small diameter (< 50 nm) and a low cmc (< 20 mg/L) in water. The effect of such a high density of hydroxy functions at the outer shell of the micelles on the biological activity (e.g., as interactions with living cells) is now under investigation.

PEG-PBLA polymeric micelles were recently investigated as carriers for oral delivery of hydrophobic drugs; indomethacin (IMC), one of the most powerful non-steroidal anti-inflammatory drugs, was incorporated into the hydrophobic inner core of the micelles by dialysis and an oil/water (O/W) emulsion method, Figure 4 (*44, 45*). Furthermore, since the solvent selected drastically affects the stability of polymeric micelles, PEG-PBLA micelles were prepared by dialysis against water using different solvents. In this case, DMAC was found to be the best solvent to form stable polymeric micelles with a narrow size distribution (average particle size based on the number distribution of the micelles: 19 nm) and avoidance of aggregation (less than 0.01 % of the total amount) (*45*). After formation of PEG-PBLA(IMC) micelles, the actual amount of IMC entrapped in the hydrophobic core of the micelles was determined by UV to be 20.4 w/w % (dialysis method) and 22.1 w/w % (O/W emulsion method), respectively.

An *in vitro* release study of IMC from these micelles was carried out in various buffer solutions at pH ranging from 1.2 to 7.4 at 37 °C (Figure 5). Under acidic conditions such as pH 1.2, the release profile of IMC was consistent with a relatively constant slow release. On the other hand, the release rate of IMC from the micelles in pH 6.5 or 7.4 was remarkably higher than that of IMC at pH 1.2, suggesting that the more basic the release medium, the higher the rate of release of IMC; also the release rate of IMC from the micelles is considerably influenced by the pH of the medium. In the mechanism of release of IMC from the micelles, we assumed that the release rate of IMC from the micelles is controlled by the partition coefficient of IMC based on the pH of the medium and the hydrophobic interactions between IMC and the hydrophobic core of the micelles.

In the case of administration of drugs by the enteral route, absorption of drugs generally takes place along the whole length of the gastrointestinal (GI) tract. The large surface area of the intestinal villi, the presence of bile, and the rich blood supply all favor intestinal absorption. In the human body, the gastric juice is very acid (about pH 1), whereas the intestinal contents are nearly neutral (very slightly acid). The pH difference between plasma (pH 7.4) and the lumen of the GI tract plays a major role in determining whether a drug that is a weak electrolyte will be absorbed into plasma (*46*). Thus, we may assume, for practical purposes, that PEG-PBLA(IMC) micelles can be applied by the oral route to deliver IMC with decreased side effects in stomach as a result of protecting material; PEG-PBLA micelles hold promise for oral delivery of selective drugs with hydrophobicity or undesirable side effects such as irritation of GI mucosa from direct exposure of drug. In addition, *in vitro* stability of PEG-PBLA(IMC) micelles having different BLA units in phosphate buffered saline (PBS), with and without serum, revealed that the micelles in PBS, as well as in the presence of serum, remained very stable over 40 h (La, S. B.; Kataoka, K.; Okano, T.; Sakurai, Y., unpublished data). For the stability of PEG-PBLA(IMC) micelles, no significant difference can be detected, based on different BLA units of the micelles in PBS with and without serum. Further studies, such as biodistribution, are also currently under investigation.

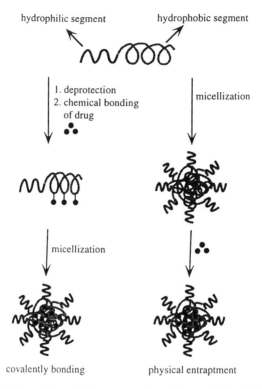

Figure 3. Chemical and physical incorporation of drugs within PEG-PAA micelles.

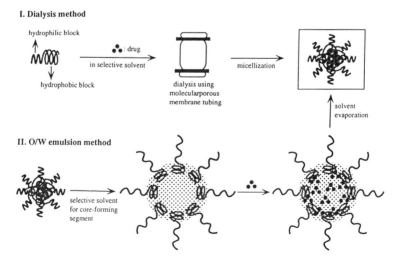

Figure 4. Drug loading of polymeric micelles by two different physical methods.

Polyion complex micelles

Recently, monodisperse, polyion-complex micelles having a spherical shape were developed in an aqueous medium (*47*). As the driving force to form micelles, the polyion-complex micelles were prepared through electrostatic interaction between a pair of oppositely-charged block copolymers containing PEG blocks. For the intention of using the micelles as a novel drug vehicle, biodegradable poly(L-lysine) and poly(aspartic acid) as the polycation and polyanion segments in the block copolymer, respectively, were selected. The particle size distribution by weight and number of the micelles was about 30 nm with a polydispersity (dw/dn) of 1.07 by DLS. The micelles showed a narrow size distribution without any secondary aggregates; polydispersity (m2/G2) of the micelles is less than 0.1. The results obtained by viscosity measurements and laser-Doppler electrophoresis of the micelles indicate that the polyion-complex micelles are formed stoichiometrically. Thus, polyion-complex micelles with promising distinctions can be applied as vehicles for delivery of charged drugs (e.g., peptides and nuclotides such as lysozyme (*48*) and DNA (*49*)).

Heterobifunctional PEGs and Reactive Micelles

As described above, polymeric micelle are very promising as non-toxic drug carriers. Functionalization of the PEG chain end provides a means for attaching anchoring systems for later chemical immobilization of pilot molecules on the micelle surface. Recently, we established a facile and quantitative synthetic method for the formation of heterobifunctional PEGs. By utilizing these techniques, polymeric micelles with functional groups on the surface can be prepared.

There are several reports on the synthesis of heterobifunctional PEG using homo-telechelic PEG as a starting material (*50*). The synthetic methods, however, are complicated because they require several reaction steps to derivatize the PEG terminus. In addition, the efficiency for derivatization is not high, meaning that the resulting PEG will be the mixture of starting homo-telechelic and the resulting heterotelechelic polymer.

Our strategy for heterotelechelic polymer synthesis starts from polymerization of ethylene oxide (EO) using initiators containing defined functionalities. So far, we have synthesized several types of heterotelechelic PEG (*51-54*).

A formyl group is useful for conjugation with proteins because of its stability in water and its rapid reactivity with primary amino groups. In addition, no charge variation takes place by the modification because the resulting Schiff base can be easily converted to a secondary amino group by reduction.

When potassium 3,3-diethoxypropyl alkoxide (PDA) is used as an initiator for EO polymerization, PEG with an acetal moiety at the α-terminus is obtained quantitatively (*53*). It should be noted that the ω-terminal of the obtained polymer is a hydroxyl group. It is generally known that acetal can be easily converted to aldehyde by weak acid treatment. For example, when the acetal-ended heteroPEG was treated with 0.01- 1 N HCl, the acetal terminal was converted to aldehyde completely (Figure 6). After acid treatment, the ^{13}C NMR signals derived from the acetal moiety completely diminished and the three signals derived from the aldehyde moiety appeared at 43.6, 64.6 and 200.9 ppm, which are assignable to -CH_2-$\underline{C}H_2$-CHO, -$\underline{C}H_2$-

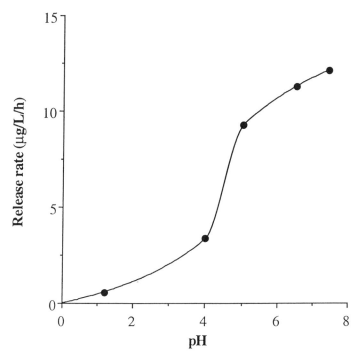

Figure 5. Release rate profile of IMC from PEG-PBLA(IMC) micelles in different pH solutions at 37 °C.

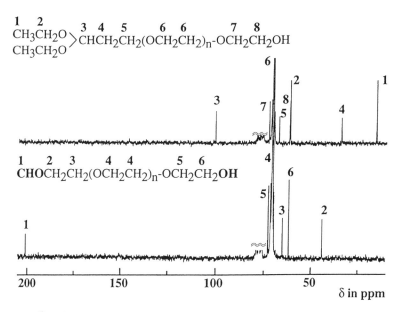

Figure 6. ^{13}C NMR spectra of acetal-ended PEG before (above) and after (below) acid treatment.

CH_2-CHO and -CH_2-CH_2-$\underline{C}$HO at the end of the polymer chain, respectively (Scheme 1).

This polymerization technique can be expanded to a sugar moiety as an initiator. As is well known, glucose has five hydroxyl groups. By the protection of four hydroxyl groups, a retained hydroxyl group can be utilized as an initiator for EO polymerization. In this technique, not only types of sugar but also the position of the hydroxyl group in the sugar can be selected for conjugation with PEG. The other terminal can be converted to another function such as -NH_2, -COOH or ally group (55).

In the same manner, primary-amino-ended hetero-PEG can be synthesized. Since acetonitrile shows high acidity due to the electron-withdrawing effect of the cyano group (pKa = 25 (56)), it is easy to metalate using an alkali metal alkyl such as potassium naphthalene (57). The cyanomethylpotassium thus prepared can be initiate the polymerization of EO to form PEG with a cyanomethyl group at the α-terminal (54). Transformation of the terminal cyano group to a primary amino group was carried out by reduction with lithium aluminum hydride. The reaction proceeds quantitatively to form primary-amino-ended PEG with a hydroxyl group at the other terminal (Scheme 2).

By expanding this synthetic technique, we prepared hydrophilic-hydrophobic block copolymers with functional group at the hydrophilic terminal. As stated above, PDA initiates EO polymerization to give a heterotelechelic PEG having an acetal moiety at one end and a potassium alkoxide at the other end. The potassium alkoxide at the PEG terminal can initiate block copolymerization of lactide (Scheme 3) (58).

The acetal end group can be converted into an aldehyde end group by acid treatment. After the acetal-polymeric micelle was prepared in water, the pH of the medium was adjusted to 2 to form a polymeric micelle with aldehyde groups on its surface. Under suitable conditions, the conversion to aldehyde attained up to 90%. It should be noted that no degradation of the PLA main chain is observed as a result of the acid treatment. This is one proof for the core-shell structure of the micelle. Protons hardly attack the ester bond in PLA unit of the core.

The size of the micelle, determined by DLS measurements, before and after the conversion of acetal group into aldehyde, does not change significantly (ca. 30 nm in diameter for PEG/PLA (5000/5000)). Also in the DLS measurements, polydispersity of the micelles is extremely low ($\mu/\Gamma^2 < 0.05$) and no angular dependency is observed. On the basis of these results, it is concluded the acetal-PEG/PLA shows completely spherical core-shell type polymeric micelle with aldehyde groups on the surface, with narrow polydispersity. This can be promising as a targeting drug vehicle(Figure 7).

Summary

In conclusion, this chapter focused on the feasibility of pharmaceutical application of nanoscopic polymer micelles based on PEG block copolymers for selective drug delivery. Advantages of polymeric drugs are passive targeting and active targeting (Kataoka, K. In Controlled Drug Delivery: The Next Generation, Park, K., Ed.; ACS, in press). For successful drug targeting, polymeric drugs also should retain special characteristics such as modulation of the cellular uptake of drugs as well as regulation of biodistribution and disposition.

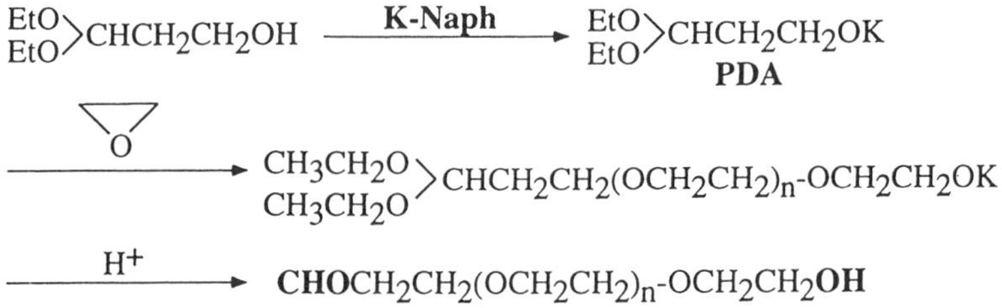

$$\underset{EtO}{\overset{EtO}{>}}CHCH_2CH_2OH \xrightarrow{\textbf{K-Naph}} \underset{EtO}{\overset{EtO}{>}}CHCH_2CH_2OK$$
$$\textbf{PDA}$$

$$\xrightarrow{\triangle_O} \underset{CH_3CH_2O}{\overset{CH_3CH_2O}{>}}CHCH_2CH_2(OCH_2CH_2)_n\text{-}OCH_2CH_2OK$$

$$\xrightarrow{H^+} \textbf{CHO}CH_2CH_2(OCH_2CH_2)_n\text{-}OCH_2CH_2\textbf{OH}$$

Scheme 1

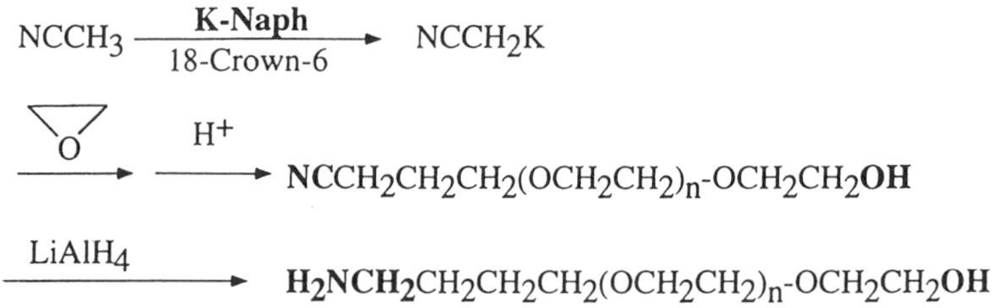

$$NCCH_3 \xrightarrow[\text{18-Crown-6}]{\textbf{K-Naph}} NCCH_2K$$

$$\xrightarrow{\triangle_O} \xrightarrow{H^+} NCCH_2CH_2CH_2(OCH_2CH_2)_n\text{-}OCH_2CH_2\textbf{OH}$$

$$\xrightarrow{LiAlH_4} \textbf{H}_2\textbf{NCH}_2CH_2CH_2CH_2(OCH_2CH_2)_n\text{-}OCH_2CH_2\textbf{OH}$$

Scheme 2

$$\textbf{PDA} \xrightarrow{\triangle_O} \underset{CH_3CH_2O}{\overset{CH_3CH_2O}{>}}CHCH_2CH_2O(CH_2CH_2O)_m\text{-}K$$

$$\xrightarrow{\text{Lactide}} \underset{CH_3CH_2O}{\overset{CH_3CH_2O}{>}}CHCH_2CH_2O(CH_2CH_2O)_m\text{-}(COCH(CH_3)O)_n\text{-}K$$

$$\xrightarrow{H^+} \textbf{CHO}CH_2CH_2O(CH_2CH_2O)_m\text{-}(COCH(CH_3)O)_n\text{-}H$$

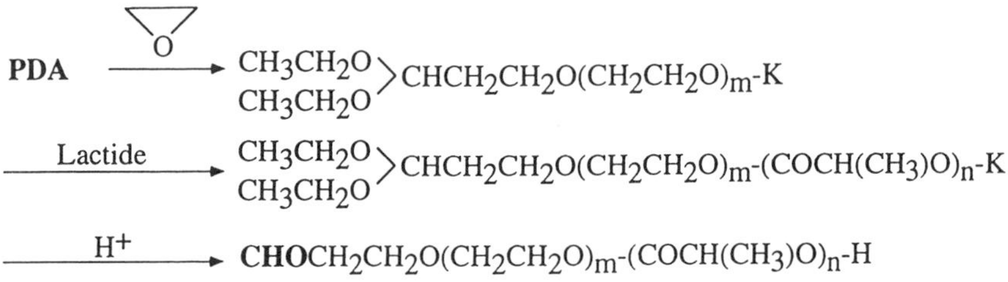

Scheme 3

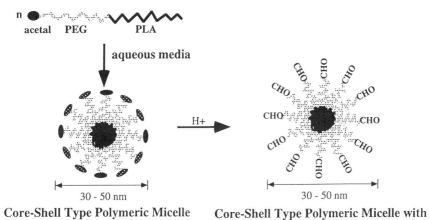

Figure 7. Schematic illustration of reactive polymeric micelle preparation

Furthermore, as pointed out in this chapter, PEG-PBLA micelles have promising features as carriers for drug delivery by means of physical entrapment. In the case of polyion-complex micelles and functional PEG-PLA micelles, systematic study is required to prove their suitability in drug delivery systems.

Acknowledgments

The authors gratefully acknowledge Professors Y. Sakurai, T. Okano, and T. Aoyagi, and Dr. M. Yokoyama, Tokyo Women's Medical College, and their colleagues at the ICBS for their kind cooperation and contribution. The authors would like to acknowledge the Ministry of Education, Science, and Culture, Japan for supporting a part of this work by Grant-in-Aid for Scientific Research.

Literature Cited

1. Xu, R.; Winnik, M. A.; Hallett, F. R.; Riess, G.; Croucher, M. D. *Macromolecules* **1991**, *24*, 87-93.
2. Kwon, G.; Naito, M.; Yokoyama, M.; Okano, T.; Sakurai, Y.; Kataoka, K. *Langmuir* **1993**, *9*, 945-949.
3. Ambler, L. E.; Brookman, L.; Brown, J.; Goddard, P.; Petrak, K. *J. Bioact. Compat. Polym.* **1992**, *7*, 223-241.
4. Matsushima, A.; Nishimura, H.; Ashihara, Y.; Yokota, Y.; Inada, Y. *Chem. Lett.* **1980**, 773-776.
5. Abuchowski, A.; van Es, T.; Palzuk, N. C.; Davis, F. F. *J. Biol. Chem.* **1977**, *252*, 3578-3581.
6. Savoca, K. V.; Abuchowski, A.; van Es, T.; Davis, F. F.; Palczuk, N. C. *Biochim. Biophys. Acta* **1979**, *578*, 47-53.
7. Herman, S.; Hooftman, G.; Schacht, E. *J. Bioact. Compat. Polym.* **1995**, *10*, 145-187.
8. Inada, Y.; Matsushima, A.; Kodera, Y.; Nishimura, H. *J. Bioact. Compat. Polym.* **1990**, *5*, 343-364.
9. Doillon, C. J.; Cote, M.-F.; Pietrucha, K.; Laroche, G.; Gaudreault, R. C. *J. Biomater. Sci. Polymer Edn.* **1994**, *6*, 715-728.
10. Vasanthan, N.; Shin, I. D.; Tonelli, A. E. *Macromolecules* **1996**, *29*, 263-267.
11. Weiner, B. Z.; Zilkha, A. Z. *J. Med. Chem.* **1973**, *5*, 573-574.
12. Cecchi, R.; Rusconi, L.; Tanzi, M. C.; Danusso, F.; Ferutti, P. *J. Med. Chem.* **1981**, *24*, 622.
13. Harris, J. M.; Yalpani, M.; van Alstine, J. M.; Struck, E. C.; Case, M. G.; Paley, M. S.; Books, D. E. *J. Polym. Sci., Polym. Chem. Ed.* **1984**, *22*, 341-352.
14. *Poly(ethylene glycol) Chemistry: Biotechnical and Biomedical Applications*, Harris, J. M., Ed.; Plenum Press: New York, NY, 1992.
15. Zalipsky, S. *Adv. Drug Delivery Rev.* **1995**, *16*, 157-182.
16. Francis, G. E.; Delgado, C.; Fisher, D.; Malik. F.; Agrawal, A. K. *J. Drug Targeting* **1996**, *3*, 321-340.

17. Kataoka, K.; Kwon, G. S.; Yokoyama, M.; Okano, T.; Sakurai, Y. *J. Controlled Release* **1993**, *24*, 119-132.
18. Matsumura, Y.; Maeda, H. *Cancer Res.* **1986**, *46*, 6387-6392.
19. Maeda, H.; Seymour, L. W.; Miyamoto, Y. *Bioconj. Chem.* **1992**, *3*, 351-362.
20. Maeda, H. *J. Controlled Release* **1992**, *19*, 315-324.
21. Bader, H.; Ringsdorf, H.; Schmidt, B. *Ang. Makromol. Chem.* **1984**, *123/124*, 457-485.
22. Pratten, M. K.; Lloyd, J. B.; Horpel, G.; Ringsdorf, H. *Makromol. Chem.* **1985**, *186*, 725-733.
23. Kataoka, K. *J. Macromol. Sci.-Pure Appl. Chem.* **1994**, *A31(11)*, 1759-1769.
24. Kwon, G. S.; Kataoka, K. *Adv. Drug Delivery Rev.* **1995**, *16*, 295-309.
25. Yokoyama, M.; Inoue, S.; Kataoka, K.; Yui, N.; Sakurai, Y. *Makromol. Chem. Rapid Commun.* **1987**, *8*, 431-435.
26. Yokoyama, M.; Miyauchi, M.; Yamada, N.; Okano, T.; Sakurai, Y.; Kataoka, K.; Inoue, S. *J. Controlled Release* **1990**, *11*, 269-278.
27. Yokoyama, M.; Kwon, G. S.; Okano, T.; Sakurai, Y.; Seto, T.; Kataoka, K. *Bioconj. Chem.* **1992**, *3*, 295-301.
28. Yokoyama, M.; Miyauchi, M.; Yamada, N.; Okano, T.; Sakurai, Y.; Kataoka, K.; Inoue, S. *Cancer Res.* **1990**, *50*, 1693-1700.
29. Yokoyama, M.; Sugiyama, T.; Okano, T.; Sakurai, Y.; Naito, M.; Kataoka, K. *Pharm. Res.* **1993**, *10*, 895-899.
30. Yokoyama, M.; Kwon, G. S.; Okano, T.; Sakurai, Y.; Naito, M.; Kataoka, K. *J. Controlled Release* **1994**, *28*, 59-65.
31. Yokoyama, M.; Okano, T.; Sakurai, Y.; Ekimoto, H.; Shibazaki, C.; Kataoka, K. *Cancer Res.* **1991**, *51*, 3229-3236.
32. Seto, T.; Fukushima, S.; Ekimoto, H.; Yokoyama, M.; Okano, T.; Sakurai, Y.; Kataoka, K. In *Advanced Biomaterials in Biomedical Engineering and Drug Delivery Systems*, Ogata, N.; Kim, S. W.; Feijen, J.; Okano, T., Eds.; Springer-Verlag: Tokyo, 1996; pp 327-328.
33. Kwon, G. S.; Yokoyama, M.; Okano, T.; Sakurai, Y.; Kataoka, K. *Pharm. Res.* **1993**, *10*, 970-974.
34. Kwon, G.; Suwa, S.; Yokoyama, M.; Okano, T.; Sakurai, Y.; Kataoka, K. *J. Controlled Release* **1994**, *29*, 17-23.
35. Yokoyama, M.; Okano, T.; Sakurai, Y.; Suwa, S.; Kataoka, K. *J. Controlled Release* **1996**, *39*, 351-356.
36. Kwon, G. S.; Naito, M.; Kataoka, K.; Yokoyama, M.; Sakurai, Y.; Okano, T. *Colloids Surf. B* **1994**, *2*, 429-434.
37. Kwon, G.; Naito, M.; Yokoyama, M.; Okano, T.; Sakurai, Y.; Kataoka, K. *Langmuir* **1993**, *9*, 945-949.
38. Matsumoto, T.; Kataoka, K.; Yokoyama, M.; Okano, T.; Sakurai, Y. *In Proceed. Intern. Symp. Control. Rel. Bioact. Mater.*; 23rd, Controlled Release Society Meeting; 1996, pp 637-638.
39. Kiserow, D.; Prochazka, K.; Ramireddy, C.; Tuzar, Z.; Munk, P.; Webber, S. E. *Macromolecules* **1992**, *25*, 461-469.
40. Dong, D. C.; Winnik, M. A. *Can. J. Chem.* **1984**, *62*, 2560-2565.

41. Cammas, S.; Harada, A.; Nagasaki, Y.; Kataoka, K. *Macromolecules* **1996**, *29*, 3227-3231.

42. Kwon, G. S.; Naito, M.; Yokoyama, M.; Okano, T.; Sakurai, Y.; Kataoka, K. *Pharm. Res.* **1995**, *12*, 192-195.

43. Cammas, S.; Kataoka, K. *Macromol. Chem. Phys.* **1995**, *196*, 1899-1905.

44. La, S. B.; Kataoka, K.; Okano, T.; Sakurai, Y. In *Advanced Biomaterials in Biomedical Engineering and Drug Delivery Systems*; Ogata, N.; Kim, S. W.; Feijen, J.; Okano, T., Eds.; Springer-Verlag: Tokyo, 1996; pp 321-322.

45. La, S. B.; Okano, T.; Kataoka, K. *J. Pharm. Sci.* **1996**, *85*, 85-90.

46. *Principles of Drug Action: The Basis of Pharmacology*; Goldstein, A.; Aronow L.; Kalma, S. M., Eds.; Harper & Row: New York, NY, 1968.

47. Harada, A.; Kataoka, K. *Macromolecules* **1995**, *28*, 5294-5299.

48. Harada, A.; Kataoka, K. In *Advances in Biomedical Engineering and Drug Delivery Systems*; Ogra, N.; Kim, S. W.; Feijen, J.; Okano, T.; Eds.; Springer-Verlag: Tokyo, 1996; pp 317-318.

49. Katayose, S.; Kataoka, K. In *Advances in Biomedical Engineering and Drug Delivery Systems*; Ogra, N.; Kim, S. W.; Feijen, J.; Okano, T.; Eds.; Springer-Verlag: Tokyo, 1996; pp 319-320.

50. Harris, J. .M; Yalpani, M. In *Partitioning in Aqueous Two-Phase Systems*; Walter, H.; Brooks, D. E.; Fisher, D; Eds.; Academic Press: New York, NY, 1985, pp 589-601.

51. Yokoyama, M.; Okano, T.; Sakurai, Y.: Kikuchi,. A.; Ohsako, N.; Nagasaki, Y.; Kataoka, K. *Bioconjugate Chem.* **1995**, 3, 275.

52. Kim, Y. J.; Nagasaki, Y.; Kataoka, K.; Kato, M.; Yokoyama, M.; Okano, T.; Sakurai, Y. *Polymer Bull.* **1994**, *33*, 1.

53. Nagasaki, Y.; Kutsuna, T.; Iijima, M.; Kato, M.; Kataoka, K. *Bioconjugate Chem.* **1995**, *6*, 231.

54. Nagasaki, Y.; Iijima, M.; Kato, M.; Kataoka, K. *Bioconjugate Chem..*, **1995**, 6, 702.

55. Nakamura, T.; Nagasaki, Y.; Kato, M.; Kataoka, K. In *Advanced Biomaterials in Biomedical Engineering and Drug Delivery Systems*; Ogata, N.; Kim, S. W.; Feijen, J.; Okano, T. Eds.; Springer, Tokyo, 1996; pp 323-324

56. March, J. *Advanced Organic Chemistry*, 4th Ed.; John Wiley & Sons: New York, NY; 1992

57. *The Chemistry of the Cyano Group*; Rappoport, Z. Ed.; John Wiley & Sons, New York, NY; 1970.

58. Scholz, C.; Iijima, M.; Nagasaki, Y.; Kataoka, K. *Macromolecules, Commun.* **1995**, *28*, 7295.

PROTEIN CONJUGATES

Chapter 8

The Use of Poly(ethylene glycol)-Enzymes in Nonaqueous Enzymology

P. A. Mabrouk

Department of Chemistry, Northeastern University, Boston, MA 02115

Polyethylene glycolated enzymes are increasingly being used for a variety of applications in non-aqueous media including organic synthesis, redox chemistry, and mechanistic biochemistry. PEG-enzymes function well in relatively hydrophobic organic solvents, in harsh mixed solvent mixtures, and at elevated temperatures. PEG-enzymes retain the unique reactive characteristics of enzyme powders while affording those benefits unique to homogeneous catalysts.

Most work on enzymes has been performed in aqueous solution. In the mid-eighties, Klibanov *(1-3)*, Inada *(4,5)*, and others forcefully demonstrated that, contrary to popular wisdom, enzymes are catalytically active in virtually anhydrous media including organic solvents *(1)*, supercritical fluids *(6-8)*, heterogeneous eutectic mixtures *(9)*, and gases*(10)*. In these unconventional media, enzymes were found to exhibit remarkable new reactive properties such as enhanced thermostability *(11)*, novel chemo- *(12)*, enantio- *(13,14)*, and regiospecificity *(15,16)* which offer a number of advantages with respect to applications in the field of biotechnology. For example porcine pancreatic lipase can withstand heating at 100°C for hours in dry alcohol and exhibit significantly greater catalytic rates compared to that possible in aqueous solution *(11)*.

Non-aqueous enzymology presents some unique advantages in biocatalysis. First, catalytic processes that are thermodynamically unfavorable in water can be reversed by maintaining a low concentration of water. For example, hydrolytic enzymes such as lipase, α-chymotrypsin, cholesterol esterase, and papain can be used to catalyze reverse hydrolysis/synthetic reactions such as ester synthesis and peptide bond formation in organic solvents *(2,3)*. Since most enzymes are insoluble in organic solvents, it is relatively easy to accomplish the clean, quantitative separation and recovery of the insoluble biocatalyst from the reaction mixture (downstream processing). In addition, hydrophobic as well as hydrophilic materials can be dissolved and utilized as substrates in the non-aqueous reaction medium.

Several different forms of enzymes including lyophilized enzyme powder *(2,3)*, cross-linked microcrystals *(16-19)*, and polymer modified enzymes *(4,5)* have been successfully investigated and used in non-aqueous enzymology.   The simplest approach involves the lyophilization of enzymes from aqueous solutions of optimal pH that can contain nonligand lyoprotectants including PEG *(20)* or KCl*(21)*.   Once the enzyme is freeze-dried, the powder is transferred to organic media containing the substrate of interest.   The advantages of this approach include its simplicity and the ability to readily recover the insoluble enzyme powder from the non-aqueous reaction mixture thus facilitating a clean preparation of the desired product with minimal processing.

A second approach involves the use of what are called cross-linked-enzyme crystals or CLEC's *(17-19)*.   This approach has recently been popularized and commercialized by Altus, Inc. First microcrystals of the specific enzyme of interest are grown in aqueous solution. When the microcrystals are of the appropriate dimensions, typically, ~0.1 mm, the enzyme is chemically cross-linked using gluteraldehyde. Finally, the cross-linked enzyme crystals are freeze-dried.   The advantages of this approach include the ability to cleanly recover the insoluble CLEC from the reaction mixture, the increased thermostability of the CLEC crystals (CLECs can be stored indefinitely at room temperature or used at elevated temperatures), and the retention of the novel reactive properties displayed by enzyme powders in organic media. Finally, CLECs are macroporous and have large internal channels which allow rapid diffusion of solvents, substrates, and products into and out from the enzyme crystal.*(19)* Unfortunately, CLECs are expensive -$400-1,000/g - at least at the present time.

A third approach involves the covalent binding of relatively high molecular weight poly(ethylene glycol) (PEG) via an appropriate linker to reactive groups on the surface of the protein.*(4,5)* It is the amphipathic nature of PEG that makes this particular polymer so useful in non-aqueous enzymology. The hydrophilicity of PEG makes it easy to prepare enzyme-PEG bioconjugates in aqueous solution under relatively mild reaction conditions. The hydrophobicity of PEG enables the modified enzymes to be dissolved and to function in water immiscible organic solvents such as benzene and chloroform.   Advantages of this approach include the ability to prepare homogeneous, optically transparent solutions of enzymes in organic solvents. Enzymes modified with PEG are very stable (> 3 months) and can be used and recovered repeatedly (at least three times) without loss of enzyme and enzymic activity *(22)*. Since the modified enzymes are dissolved in the organic medium, the catalysis is homogeneous rather than heterogeneous.   Thus, PEG-enzyme catalysis benefits from all of the advantages typically associated with a homogeneous process such as increased efficiency due to freedom from diffusional limitations.   It is the studies involving the use of pegylated enzyme bioconjugates in non-aqueous media that we wish to review in this chapter.

Strategies for Enzyme Modification

Surprisingly, in view of the wide range of linkages available for pegylation *(23,24)*, the majority of PEG-bioconjugates prepared for use in non-aqueous media (see Table I) have utilized the original cyanuric chloride coupling scheme *(25,26)*:

$$Cl-\underset{Cl}{\overset{OPEG}{\triangle}} + NH_2R \xrightarrow{\text{pH 9}} Cl-\underset{NHR}{\overset{OPE}{\triangle}}$$

ε-amino group from exposed lysine

To a certain extent this is likely due to the fact that the period of greatest research activity in this area coincided with the early years in pegylation chemistry. Nonetheless a number of very recent papers have also emphasized the use of this conjugation method *(27,28)*.

The chain length of the PEG used in the preparation of enzyme-PEG bioconjugates is an important consideration since it can exert a significant steric effect in non-aqueous catalytic conversions and influence bioconjugate solubility (*vide infra*). For example, Ohya *et al.* *(29)* found that esterase modified with a 1000 MW PEG produced significantly higher reactivity (x2) and a higher molecular weight distribution for the products in the polycondensation of ethyl glycolate in cyclohexanone than esterase modified with a 5000 MW PEG for the same degree of enzyme modification (~55%). In the same study Ohya *et al.* found that the lower the degree of modification, the higher the relative catalytic activity of the modified enzyme in organic media. This suggests that steric effects are an important consideration in preparing catalytically active PEG-enzymes. Thus, it is somewhat surprising to note that the majority of researchers investigating enzyme catalysis in organic media using pegylated preparations have chosen to use 5000 MW PEG (Table I).

Solubility of PEG-Enzymes

PEG-enzymes are readily soluble in water as well as a number of relatively hydrophobic organic solvents including benzene, toluene, dioxane, cyclohexane, 1,1,1-trichloroethane, chloroform, dichloromethane, dichlorobenzene, ethyl acetate, t-amyl alcohol, and n-decanol. Depending on the enzyme, the average molecular weight of the poly(ethylene glycol) used in the modification, and the degree of modification, millimolar non-aqueous solutions of pegylated enzymes can be prepared. Wirth *et al.* *(30)* reported for the modification of horseradish peroxidase that PEG with an average molecular weight less than 5000 did not produce soluble and active enzyme in the solvents studied (toluene, dioxane, and methylene chloride). The solubility of PEG-enzymes in organic solvents such as benzene has been shown to be significantly enhanced by increasing the extent of modification *(5)*.

While the majority of enzymes investigated to date such as lipase, peroxidase, α-chymotrypsin, etc., appear to tolerate pegylation very well, several enzymes including catalase *(31)* do not appear to tolerate extensive pegylation. Based on studies of pegylated horse cytochrome c, it appears likely that this behavior is due to a conformational change in some proteins resulting from the pegylation that has affected the heme active site.*(32)* A popular and useful strategy for increasing the solubility of pegylated enzymes that may not tolerate pegylation well or for the modification of enzymes that may not have many surface amino groups available for pegylation is based on the reagent 2,4-bis(o-methoxypolyethylene glycol)-6-chloro-s-triazine (PEG_2) *(33)*:

Cl—[triazine ring]—OPEG + NH_2R $\xrightarrow{\text{pH 9}}$ RHN—[triazine ring]—OPE

ε-amino group from exposed lysine

OPEG   OPEG

Activated PEG_2 facilitates the incorporation of two PEG chains per protein surface amino acid residue modified.

Factors Determining The Catalytic Activity of PEG-Enzymes in Organic Media

Although the experimental conditions determining the catalytic activity of pegylated enzymes in non-aqueous media have not been systematically and rigorously investigated, the experimental factors appear to be the same as those previously identified in the study of lyophilized enzyme powders suspended in non-aqueous media. These factors include: 1) the nature of the organic solvent; 2) the so-called "aqueous pH;" and 3) the solvent dryness.

The Nature of the Organic Solvent. The hydrophobic nature of the organic solvent appears to play an extremely important role in non-aqueous enzymology of pegylated enzymes *(34)*. Based on studies of enzyme powders in organic media, it is believed that the hydrophilic solvents inactivate enzymes by stripping them of their essential water layer *(35,36)*. Hydrophobic solvents, on the other hand, appear to partition the essential water away from the solvent and onto the enzyme.

Pegylated enzymes appear to function in the same manner: Pegylated enzymes do not dissolve in and are usually inactivated by relatively hydrophilic solvents. On the other hand, pegylated enzymes dissolve in and are catalytically most active in relatively hydrophobic solvents such as toluene *(37)*. This is by no means however the whole picture of the role played by the solvent in determining the efficiency of non-aqueous catalytic reactions. Ohya *et al.* *(29)*, studying the polycondensation of glycolic acid by esterase, has demonstrated that the solubility of the substrate in the reaction medium is an equally important factor in determining the conversion efficiency - the more soluble a substrate is in a water-immiscible solvent, the greater the enzymic efficiency is likely to be for that non-aqueous biocatalytic process.

Polar water-miscible solvents need not be dismissed out of hand when working with pegylated-enzymes. Souppe and Urritigoity *(38)* demonstrated that mixtures of water-miscible and water-immiscible organic solvents such as chloroform/DMSO could be used as an effective reaction medium for papain activity. In the same study, the authors also showed that rather harsh aqueous-water miscible solvent mixtures such as a 4:1 (v/v) 0.1 M phosphate buffer, pH 6.0 and ethanol mixture may also be

Table I. Pegylated Enzymes Investigated to Date for Use in Non-Aqueous Catalysis

Protein	Means of PEG Conjugation	PEG MW	Solvents Investigated	Reactions Studied	References
lipase	PEG₂	5000	benzene, toluene, chloroform, dioxane	ester synthesis	(41)
lipase	PEG₂	5000	benzene	ester synthesis	(22)
lipase	PEG₂	5000	n-decanol	asymmetric alcoholysis	(43)
lipase	PEG₂	5000	benzene	polymerization using ester bond formation	(42)
lipase	PEG₂	5000	benzene and 1,1,1-trichloroethane	ester synthesis	(71)
lipase, protease, esterase	PEG₂	5000, 1000	1,4-dioxane	polymerization using condensation and ester exchange	(29)
cholesterol esterase	PEG₂	5000	benzene	cholesterol linoleate synthesis	(72)
cholesterol esterase	PEG₂	5000	chloroform, benzene, toluene, cyclohexane	ester synthesis	(73)
cholesterol oxidase	anhydride	1500	chloroform, chloroform, 1,1,1-trichloroethane and dichloromethane	oxidation of cholesterol	(60)
lipase, catalase, peroxidase, and chymotrypsinogen	PEG₂	5000	benzene, chloroform, toluene, and dioxane		(74)
catalase	PEG₂	5000	benzene	unclear from text	(31)
thermolysin	phenylcarbonate	5000	benzene	peptide synthesis	(55)
hemoglobin	hydroxysuccinimide	3500	PEG	redox chemistry	(68,70)
myoglobin	hydroxysuccinimide	5000	PEG	redox chemistry	(67,69)

enzyme	PEG derivative	PEG size	solvent	application	ref.
peroxidase	mPEG	5000	ethyl acetate and 1,2-dichlorobenzene	redox chemistry	(27)
peroxidase	mPEG	5000	Benzene	spectroscopy	(28)
chymotrypsin and trypsin	mPEG	5000	benzene	peptide bond synthesis	(37)
chymotrypsin	three methods: s-triazine, carbamate, and amide	5000	t-amyl alcohol, benzene, and trichloroethylene	peptidase and esterase activity	(75)
chymotrypsin	PEG_2	5000	benzene	peptide synthesis	(51)
chymotrypsin	mPEG	5000	benzene	peptide bond synthesis	(49)
chymotrypsin	phenylcarbonate	5000, 2000	benzene, dimethyl formamide, and cyclohexane	transesterification	(47)
peroxidase	aldehyde	5000, 1900, 350	toluene, dioxane, and methylene chloride	oxidation of o-phenylene diamine	(30)
peroxidase and papain	thioamide	20000	chloroform and dimethyl sulfoxide	oxidation of p-anisidine and 9-methoxyellipticine	(38)
peroxidase	PEG_2	5000	benzene	oxidation of o-phenylene diamine	(58)
papain	succinimidyl succinate	4500	aqueous ethanol	peptide synthesis	(52)
papain	PEG_2	5000	benzene	oligopeptide synthesis	(39)
papain	PEG_2	5000	benzene	acid-amide bond formation	(76)

used competitively; a dipeptide Kyotorphin derivative was synthesized in 42% yield in 4:1 (v/v) 0.1 M phosphate buffer, pH 6.0 and ethanol mixture -twice the yield obtained by using an umodified papain powder.

The "Aqueous" pH. The so-called "aqueous" pH, the pH of the last aqueous solution to which the enzyme is exposed prior to being contacted by the non-aqueous solvent, is a second important factor in achieving high catalytic activity in non-aqueous media *(12,36)*. As in non-aqueous catalysis performed using enzyme powders, maximal catalytic activity for pegylated enzymes in organic media appears to be obtained when the "aqueous" pH is adjusted to correspond to that required for maximal enzymic activity in aqueous media. Unfortunately manipulation of the aqueous pH in order to maximize catalytic activity in non-aqueous media has been largely neglected by the majority of researchers using pegylated enzymes. In the only study to investigate the effect of aqueous pH on non-aqueous catalysis, Uemura *et al.* *(39)* observed no change in the catalytic rate when benzene was saturated with 50 mM buffer, the pH of which was varied between pH 6 and pH 8. Thus, it is impossible to reach any conclusions at this point regarding the significance, if any, of aqueous pH in the non-aqueous catalysis of pegylated enzymes.

The Solvent Dryness. A third factor that appears to be important in achieving non-aqueous enzymic activity for pegylated enzymes is the necessity for water. The catalytic activity of pegylated and non-modified enzymes alike in non-aqueous media appears to be extremely sensitive to changes in the amount of water bound to the enzyme, the so-called "essential" water, the thin layer of water on a protein's surface which is required by the enzyme in order to maintain its catalytically active conformation *(40)*. Unfortunately the majority of studies investigating the catalytic activity of pegylated enzymes in non-aqueous media make no mention of the total water content present in the reaction medium. Nonetheless, several studies *(37,39,41)* have reported that the addition of small amounts of water to the reaction medium appears to substantially increase catalytic rates for pegylated enzymes in non-aqueous media, as has been previously reported for enzyme powders in non-aqueous media. *(40)*

Catalytic Reactivity Characteristics of Specific PEG-Enzymes in Organic Media

To date a wide range of enzymes have been modified with poly(ethylene glycol) for use in non-aqueous media. While a large number of the studies reported to date have focused primarily on the preparation and exploitation of the catalytic activity of various pegylated enzymes, relatively few studies have examined the catalytic reactivity characteristics of pegylated enzymes in organic media. In the next several sections, we will examine the catalytic reactivity characteristics of several classes of pegylated enzymes in non-aqueous media.

PEG-Modified Lipase. One of the earliest enzymes to be pegylated and investigated for use in non-aqueous media is the lipoprotein lipase. Lipase is commercially available, inexpensive, has a broad substrate specificity, and requires no cofactor regeneration. The enzyme which consists of two polypeptide chains linked by a disulfide bridge and which has a molecular weight of 33,000 Da, catalyzes the hydrolysis of triglycerides and water-soluble esters in water. In organic solvents, the reverse reaction, ester synthesis by condensation or ester exchange, now becomes possible. Several groups have effectively demonstrated the utility of modified lipase to catalyze ester synthesis and aminolysis in non-aqueous solutions at room temperature *(29,39,41-43)*.

Lipase is robust and tolerates pegylation well: Modified lipase, prepared using 2,4-bis(o-methoxypolyethylene glycol)-6-chloro-s-triazine at pH 10 (borate buffer) 37°C using a 1 hr reaction time, has been shown to retain 80% of the esterolytic activity for triglycerides of the native enzyme in emulsified aqueous solution. Thus, it is clear that pegylation of this enzyme does not deleteriously affect its catalytic function in aqueous solution.

The modified enzyme has been shown to be a useful catalyst in benzene in ester synthesis, ester exchange, and aminolysis reactions *(41)*. Ester synthesis was accomplished at reasonable rates in benzene, toluene, and chloroform *(41)*. The authors obtained maximal activity using water saturated benzene (30 mM water present) suggesting that a small amount of water is required for the reaction. The modified enzyme has also been demonstrated to be useful in the preparation of linear polyester synthesis accomplished at room temperature *(42)* and has been investigated for use in the preparation of biodegradable, biocompatible polymers such as poly(glycolic acid) potentially useful as drug delivery carriers *(29)*. In non-aqueous media, the modified enzyme exhibits a preference for R-isomers and can be used to optically resolve a racemic alcohol producing an optical purity of 99% enantiomeric excess in a 7 hour reaction period *(39)*.

Several groups *(44-46)* have demonstrated that lipase powder can be used to effectively optically resolve lactones in organic media. This ability appears to be retained by PEG-lipase in non-aqueous media. Uemura *et al. (43)* used PEG-lipase to catalyze the asymmetric alcoholysis in neat mixtures of racemic d-decalactone and various straight-chain alcohols (2-12 carbon atoms) containing only trace amounts of water at temperatures up to 50°C. The R-lactone was preferentially converted to the R-hydroxy ester. In n-decanol at 50°C, a 79% yield of the R-hydroxy ester, representing a 78% enantiomeric excess, was obtained for a 6-hour reaction. Unfortunately, the authors did not use lipase powder so that a direct comparison between PEG-lipase and lipase powder cannot be made. However, the yield and percent enantiomeric excess from this experiment compare well with those obtained by other researchers using lipase powder and different substrates (hydroxyesters) in organic solvents (60-90% yield; 60-92% ee) *(44,45)*.

PEG-Modified α-Chymotrypsin. Several studies describing the preparation of pegylated chymotrypsin illustrate well the value in understanding the underlying biochemistry of an enzyme prior to choosing a pegylation strategy. Pina *et al. (47)* found that activity of PEG-modified α-chymotrypsin, prepared using reactive phenylcarbonates, in benzene to be strongly affected by the degree of modification with decreasing activity exhibited with increasing degrees of modification. The sensitivity of α-chymotrypsin to modification was likely due to the well-known importance of the terminal amino groups in the aqueous enzymatic catalysis of α-chymotrypsin *(48)*. Matsushima *et al.* *(49)* recognizing the significance of the terminal amino groups in the aqueous enzymatic catalysis of α-chymotrypsin, modified its zymogen prior to activation, and used the modified enzyme to form a peptide bond by aminolysis in benzene solution. Unfortunately, Matsushima provided no information on the relative activity of the pegylated enzyme compared with that of the native enzyme.

Inada *et al.* *(49)* recognizing the importance of the terminal amino groups in chymotrypsin, attempted to use a rather clever strategy to prepare pegylated chymotrypsin by reacting its zymogen form, chymotrypsinogen, with PEG_2 prior to activating the enzyme using trypsin. However, the modified enzyme prepared in this way surprisingly exhibited a substantially lower (57% activity) esterolytic activity compared to the native enzyme in aqueous solution than that exhibited by the modified α-chymotrypsin prepared by Pina *et al.* (~73% activity) when N-acetyl-L-tyrosine ethyl ester was used as a substrate.

Inada *et al.* *(49)* demonstrated that peptide bond formation can be accomplished quantitatively in benzene using modified α-chymotrypsin at relatively low temperature (37°C) by aminolytic reaction without the problem of ester hydrolysis of either the starting ester or the newly formed peptide product as typically occurs in either aqueous or biphasic aqueous organic systems *(50)*.

Gaertner and Puigserver *(37,51)* have investigated the kinetics of PEG-modified α-chymotrypsin and PEG-modified trypsin-catalyzed peptide-bond synthesis in benzene. The non-aqueous catalysis was found to follow Michaelis-Menten kinetics, as in water, for a ping-pong mechanism modified by a hydrolytic branch *(37)*. The authors found that the higher the water content, the higher the activity and the k_{cat} value. K_m, on the other hand, was not significantly affected. In addition, the authors found that both the substrate specificity and overall catalytic efficiency were strongly affected by the amount of water present in the reaction medium.

In the same study, the enzymatic activity was also observed to be strongly dependent on the nature of the organic solvent used as the reaction medium. The highest reaction rates were obtained in the most hydrophobic solvents used, specifically, benzene and toluene and a good correlation was obtained between the solvent polarity and enzyme activity when fragmental hydrophobic constants were used as the solvent characteristic. This finding suggests that the hydrophobicity of the solvent is an important parameter in non-aqueous catalysis using pegylated enzymes.

PEG-Modified Papain. The thiol protease papain was found to have little if any sensitivity to pegylation; When 39% of the surface amino groups on papain were modified, the pegylated enzyme exhibited 87% of the unmodified enzyme's specific activity *(39)*.

PEG-Papain, a cysteine proteinase, has been demonstrated by several groups *(38,39,52)* to be an effective catalyst for peptide synthesis in organic media. In anhydrous organic media, the enzyme has been used to form peptides between substrates with non-ionizable side chains *(38)*. More recently, in water miscible organic solvents such as ethanol the synthesis of peptides with ionizable side chains has been accomplished *(52)*.

Uemura *et al.* *(39)* observed a significant change in the substrate specificity for oligopeptide synthesis in benzene solution compared to buffered aqueous solution.   In aqueous solution, esters of L-glutamic acid were reactive while esters of L-aspartic acid were poor substrates.   In benzene, esters of L-aspartic acid became significantly more reactive than esters of L-glutamic acid.

PEG-Modified Thermolysin. Thermolysin, an extremely versatile metalloproteinase, is widely used in the synthesis of dipeptides and has been profitably exploited in the commercial synthesis of aspartame *(53,54)*.   Thus, it is somewhat surprising that only one study has appeared to date investigating the catalytic properties of pegylated-thermolysin.   Ferjancic *et al.* *(55)* investigated the substrate specificity of PEG-thermolysin in amino acid condensation reactions in benzene.   In water, the rate of cleavage by thermolysin of a peptide bond is significantly increased when a hydrophobic amino acid is the carboxyl group acceptor *(56,57)*.   High yields, up to 100%, were obtained when amides of hydrophobic amino acids such as phenylalanine, leucine, and valine were used as carboxyl acceptors.

PEG-Modified Horseradish Peroxidase (HRP). Souppe and Urrutigoity *(38)* examined the horseradish peroxidase-catalyzed oxidation of 9-methoxyellipticine in both aqueous and chloroform solutions.   The kinetics for the non-aqueous process were reported to be Michaelis as in aqueous solution.   The Michaelis constant K_m^{H2O2} was less favorable in chloroform than in water (6.25 mM vs. 1.30 mM).   The V_{max} value is only 17% of that for the native enzyme in water.   Takahashi *et al.* *(58)* observed the same behavior for the peroxidase-catalyzed oxidation of o-phenylenediamine in aqueous and benzene solutions.   The V_{max} value was 21% of that for the native enzyme in water.   The substantially lower value of V_{max} and unfavorable change in K_m^{H2O2} in the non-aqueous medium cannot be interpreted as due to changes induced in the enzyme structure due to pegylation since the specific activity in aqueous solution for the modified enzyme in both cases is comparable to that of the native enzyme.   This conclusion is also supported by recent spectroscopic study of pegylated HRP-PEG in benzene solution by Mabrouk *(28)* who showed that the active site structure in pegylated HRP in benzene appears to be the same as in aqueous solution.

Wirth *et al.* *(30)* examined the significance of polymer chain length in determining the solubility and activity of pegylated horseradish peroxidase prepared using different average molecular weight poly(ethylene glycol)s (MW 350, 1900, and 5000) in a variety of different organic solvents.   Modification of HRP, which has relatively few surface amino groups (six), using the low molecular weight polymers did not produce an organics-soluble enzyme.   The modified enzyme prepared using 5000-average molecular weight poly(ethylene glycol) on the other hand was soluble and active, using o-phenylenediamine as a substrate, in a relatively large number of different solvents including: toluene, methylene chloride, dioxane, and dimethylsulfoxide.   The oxidation of several water-insoluble substrates such as 1,2-dimethoxybenzene and 4-methoxybenzyl was attempted but no catalytic oxidation was was observed.   Another interesting finding was that hydrogen peroxide inhibited the modified enzyme in toluene when the hydrogen peroxide concentration exceeded 0.2

mM. This was attributed by the authors to replacement of water as the sixth ligand in the heme active site of peroxidase with peroxide.

Magnetite PEG-Enzyme Particles

While PEG-enzymes are clearly competitive with enzyme powders in accomplishing non-aqueous catalysis selectively and efficiently, due to their solubility in the medium they represent a problem for industrial catalysis in terms of downstream processing. Yoshimoto *et al.* *(22)* demonstrated that PEG-enzymes can be recovered by precipitation using hexanes or petroleum ether. Industrially this requires an additional, nontrivial processing step.   In this section we will examine a novel approach to the issue of the recoverability of PEG-enzymes from organic reaction mixtures.

Inada *et al.* *(4)* have conjugated PEG-enzyme bioconjugates to magnetite, Fe_3O_4, in order to facilitate recovery from a reaction mixture. Magnetite PEG-enzyme particles can be dispersed in organic media and then be completely recovered from the reaction medium by the application of magnetic force.   The recovered magnetic enzymes can be placed in fresh solvent without any loss of enzymic activity. To date a number of magnetic enzymes including lipase *(59)*, cholesterol oxidase *(60)*, and asparaginase *(61)* have been prepared in this way.

Two coupling strategies have been investigated *(59,61,62)*. In the first strategy (type I) *(62)* magnetite particles are produced in the presence of the PEG-bioconjugates at room temperature according to the following equation:

$FeCl_2 + FeCl_3 +$ PEG-lipase -> magnetite-PEG-lipase

The submicron magnetic enzyme-PEG particles produced in this way can be suspended in benzene and 1,1,1-trichloroethane but display relatively low catalytic activity. Although the exact nature of the binding interaction between the magnetite and enzyme is not known, it is clear that the pegylated enzyme is very tightly bound to the magnetite; extensive dialysis of the magnetic enzyme particles against water failed to release the PEG-lipase from the magnetite. The activity can be substantially improved if the magnetite particles are first prepared in the absence of the enzyme which is then covalently bound to the magnetite using a PEG-linker *(62)*. This strategy (type II), although producing conjugates that are significantly less soluble, allows control of the conjugate composition (magnetite/enzyme) which facilitates optimization of the non-aqueous catalytic efficiency for the magnetite-PEG-bioconjugate. Complete recovery of ferromagnetic-modified lipase, with full activity, from a reaction mixture can be effected within 3 min using a relatively low magnetic field of ~ 250 Oe.   This approach offers clear advantages in terms of simplifying the recovery and recycling processes of enzymes for industrial use.

Mechanistic Studies of Non-Aqueous Enzymology Utilizing PEG-Enzymes

Very recently a number of studies have begun to appear in which pegylated enzymes are being investigated in non-aqueous media in order to gain more fundamental information regarding the mechanism and kinetics of non-aqueous enzyme catalysis. In the next several sections, we will examine these studies.

Electrochemical Investigations of the Kinetics of PEG-Enzyme Catalysis in Organic Media

R. W. Murray *et al.* *(27)* recently examined the kinetics of organic-phase enzyme catalysis for the pegylated-peroxidase-catalyzed oxidation of several catechol and ferrocene derivatives in aqueous media and in two different organic solvents, ethyl acetate and 1,2-dichlorobenzene, spectrophotometrically and electrochemically. The rates of the reaction of PEG-modified peroxidase with the reductants in both organic solvents was significantly slower (~70x) than the corresponding reactions in aqueous solution. Significant differences were observed in the relative rates of oxidation of different substrates in aqueous compared to non-aqueous media. The relative hydrophobicity and steric bulk of the substrate appeared to correlate well with the relative rates observed for the oxidation of 4-methylcatechol, tetrachlorocatechol, and tetrabromocatechol.

The effect of aqueous pH on the reaction rate in non-aqueous media was examined using solvents saturated (total water content not determined) with aqueous tris buffers at different pH values (4.7 - 10.8). Under these conditions, the reaction rate in ethyl acetate and 1,2-dichlorobenzene appeared to be independent of the aqueous buffer pH.

The peroxidase-catalyzed oxidation reactions studied exhibited kinetic schemes more complex than that previously reported in aqueous solution. For example, no correlation was observed between the reaction rate and ferrocene redox potential in ethyl acetate even though previously similar studies by the same research group in aqueous solution found a Marcus-type relationship between these two parameters.

Mechanistic Study of PEG-Enzyme Catalysis in Organic Media

Another interesting application of the use of pegylated enzymes in non-aqueous media was recently reported by Mabrouk *(28)*. Horseradish peroxidase was pegylated and dissolved in benzene solution. Addition of hydrogen peroxide, extracted into benzene, to the peroxidase solution produced a new catalytically active form of the enzyme that appears to persist in benzene solution at room temperature for hours. The UV-vis spectrum is consistent with identification of the peroxidase species as the highly unstable, catalytically active oxyferryl peroxidase intermediate known as Compound II. In contrast to its well-known thermo- and photolability in aqueous solution *(63,64)*, the peroxidase- H_2O_2 intermediate was found to be surprisingly stable in benzene solution at room temperature even upon sustained laser irradiation (200 mW at 514.5 nm).

Electrochemistry of Pegylated-Redox Proteins in PEG

Polyethylene glycol readily dissolves small ions through ion-dipole interactions with the ether oxygen atoms and thus has recently been investigated for possible use as a polymer solvent in solid-state electrochemical applications such as solid state batteries and electrochromic displays (65,66). In this section, we will examine recent studies by Ohno et al. (67-69) who have investigated the redox activity of several pegylated heme redox proteins including myoglobin (67,69) and hemoglobin (69) dissolved in low molecular weight poly(ethylene glycol) oligomers at optically transparent indium tin oxide (ITO) electrode substrates. The highest reduction rate 7.6 x 10^{-4} s^{-1} was observed at 25°C for pegylated myoglobin having an average of ten 5,000-molecular weight PEG chains per myoglobin. More or fewer PEG chains per myoglobin resulted in significant decreases in the apparent reduction rate constant. Redox chemistry for the pegylated proteins studied appears to be quasi-reversible over a fairly wide temperature range (25 - 70°C).

In an effort to increase the unspectacular reduction rate, Ohno et al. (70) investigated electron transfer at ITO electrodes modified with hemoglobin-PEG. Hemoglobin-PEG films were cast on ITO electrodes which were subsequently soaked in PEG solutions of varying average molecular weight (200, 400, or 600) containing 0.2 M KCl. Although UV-vis spectroelectrochemical changes confirmed the redox reaction to be chemically reversible, high overpotentials and long time periods were required to reduce and then re-oxidize hemoglobin-PEG on the ITO electrodes; 90% of the hemoglobin-PEG was reduced at ~-1.0 V vs. Ag within 120 min in PEG_{200} containing 0.2 M KCl as the supporting electrolyte.

Conclusions

It is clear that non-aqueous PEG-enzyme catalysis is a relatively new and promising field. The study of pegylated enzymes in non-aqueous media thus far has neither been as systematic nor as thorough as the work done using enzyme powders. In the majority of studies performed to date the emphasis has been on the preparation of PEG-enzyme bioconjugates and the demonstration of their catalytic activity in non-aqueous media. Kinetic data and water content have frequently not been reported. This fact has made comparison between the activity of modified and unmodified enzymes in organic media impossible. Although relatively few kinetic studies have been carried out, it appears that the catalysis of PEG-modified enzymes in organic media follows conventional kinetic models. However, the values of k_{cat} and K_m may be very different from those of the enzyme in aqueous media.

Examination of the properties of pegylated enzymes with regard to the similarities and differences between these materials and enzyme powders or CLECs in aqueous and non-aqueous media has been largely ignored by researchers in the field. Thus it is clear that many fundamental questions remain to be answered: Is the enzyme specificity of pegylated enzymes in organic media generally the same as that of enzyme powers in organic media? Are pegylated enzymes more or less thermostable than enzyme powders? What are the differences if any between pegylated enzymes and enzyme powders in terms of their catalytic reactivity in organic media?

Future Prospects for PEG-Enzymes in Non-Aqueous Enzymology

Magnetic enzymes, the redox activity of pegylated enzymes and the demonstrated ability of pegylated enzymes to function in the absence of any solvents represent exciting prospects in terms of industrial use of these biomaterials. Another yet unexplored potential opportunity lies in the use of protein engineering in concert with pegylation and magnetite conjugation in order to improve the efficiency, stability, and recovery of enzyme catalysis in non-aqueous media. Finally, if other mechanistically significant enzyme intermediates can be generated and stabilized toward spectroscopic study in organic media, the study of PEG-enzymes in non-aqueous media may have general utility in mechanistic biochemistry. For example, it may finally be possible to gain much needed insight into the structure and redox character of these species and thus into the mechanism of heme catalysis for a number of very important enzyme systems including peroxidase and cytochrome P450.

Acknowledgments

The preparation of this review was supported by NSF CAREER award MCB 9600847.

Literature Cited

(1)   Klibanov, A. M. *CHEMTECH* **1986**, 354-359.
(2)   Klibanov, A. M. *Trends Biochem. Sci.* **1989**, *14*, 141-144.
(3)   Klibanov, A. M. *Acc. Chem. Res.* **1990**, *23*, 114-120.
(4)   Inada, Y.; Takahashi, K.; Yoshimoto, T.; Kodera, Y.; Matsushima, A.; Saito, Y. *Trends Biochem. Sci.* **1988**, *6*, 131-134.
(5)   Inada, Y.; Matsushima, A.; Kodera, Y.; Nishimura, H. *J. Bioactive Compat. Polymers* **1990**, *5*, 343-364.
(6)   Randolph, T. W.; Clark, D. S.; Blanch, H. W.; Prausnitz, J. M. *Science* **1988**, *239*, 387-390.
(7)   Randolph, T. W.; Blanch, H. W.; Prausnitz, J. M.; Wilke, C. R. *Biotechnol. Lett.* **1985**, *7*, 325-328.
(8)   Johnston, K. P.; Harrison, K. L.; Clarke, M. J.; Howdle, S. M.; Heitz, M. P.; Bright, F. V.; Carlier, C.; Randolph, T. W. *Science* **1996**, *271*, 624-626.
(9)   Gill, I.; Vulfson, E. *Trends Biotechnol.* **1994**, *12*.
(10)   Hammond, D. A.; Karel, M.; Klibanov, A. M. *Appl. Biochem. Biotechnol.* **1985**, *11*, 393-400.
(11)   Zaks, A.; Klibanov, A. M. *Science* **1984**, *224*, 1249-1251.
(12)   Zaks, A.; Klibanov, A. M. *Proc. Natl. Acad. Sci. USA* **1985**, *82*, 3192-3196.
(13)   Fitzpatrick, P. A.; Klibanov, A. M. *J. Am. Chem. Soc.* **1991**, *113*, 3166-3171.
(14)   Sakurai, T.; Margolin, A. L.; Russell, A. J.; Klibanov, A. M. *J. Am. Chem. Soc.* **1988**, *110*, 7236-7237.
(15)   Kazandjian, R. Z.; Klibanov, A. M. *J. Am. Chem. Soc.* **1985**, *107*, 5448-5450.
(16)   Rubio, E.; Fernandez-Mayorales, A.; Klibanov, A. M. *J. Am. Chem. Soc.* **1991**, *113*, 695-696.
(17)   St. Clair, N. L.; Navia, M. A. *J. Am. Chem. Soc.* **1992**, *114*, 7314-7316.
(18)   Lalonde, J. J.; Govardhan, C.; Khalaf, N.; Martinez, A. G.; Visuri, K.; Margolin, A. L. *J. Am. Chem. Soc.* **1995**, *117*, 68445-6852.
(19)   Persichetti, R. A.; St. Clair, N. L.; Griffith, J. P.; Navia, M. A.; Margolin, A. L. *J. Am. Chem. Soc.* **1995**, *117*, 2732-2737.
(20)   Dabulis, K.; Klibanov, A. M. *Biotechnol. Bioeng.* **1993**, *41*, 566-571.

(21)   Khmelnitsky, Y. L.; Welch, S. H.; Clark, D. S.; Dordick, J. S. *J. Am. Chem. Soc.* **1994**, *116*, 2647-2648.
(22)   Yoshimoto, T.; Takahashi, K.; Ajima, A.; Tamaura, Y.; Inada, Y. *Biotechnol. Lett.* **1984**, *6*, 337-340.
(23)   Harris, J. M. *Poly(Ethylene Glycol) Chemistry: Biotechnical and Biomedical Applications*; Harris, J. M., Ed.; Plenum Press: New York, 1992.
(24)   Zalipsky, S. *Bioconj. Chem.* **1995**, *6*, 150-165.
(25)   Abuchowski, A.; van Es, T.; Palczuk, N. C.; Davis, F. F. *J. Biol. Chem.* **1977**, *252*, 3578-3581.
(26)   Abuchowski, A.; McCoy, J. R.; Palczuk, N. C.; Van Es, T.; Davis, F. F. *J. Biol. Chem.* **1977**, *252*, 3582-3586.
(27)   Yang, L.; Murray, R. W. *Anal. Chem.* **1994**, *66*, 2710-2718.
(28)   Mabrouk, P. A. *J. Am. Chem. Soc.* **1995**, *117*, 2141-2146.
(29)   Ohya, Y.; Sugitou, T.; Ouchi, T. *Pure Appl. Chem.* **1995**, *A32*, 179-190.
(30)   Wirth, P.; Souppe, J.; Tritsch, D.; Biellmann, J.-F. *Bioorg. Chem.* **1991**, *19*, 133-142.
(31)   Takahashi, K.; Ajima, A.; Yoshimoto, T.; Inada, Y. *Biochem. Biophys. Res. Commun.* **1984**, *125*, 761-766.
(32)   Mabrouk, P. A. *Bioconj. Chem.* **1994**, *5*, 236-241.
(33)   Matsushima, A.; Nishimura, H.; Ashihara, Y.; Yokota, Y.; Inada, Y. *Chem. Lett.* **1980**, 773-776.
(34)   Zaks, A.; Klbanov, A. M. *J. Am. Chem. Soc.* **1986**, *108*, 2767-2768.
(35)   Gorman, L. A. S.; Dordick, J. S. *Biotechnol. Bioeng.* **1992**, *39*, 392-397.
(36)   Zaks, A.; Klibanov, A. M. *J. Biol. Chem.* **1988**, *263*, 3194-3201.
(37)   Gaertner, H.; Puigserver, A. *Eur. J. Biochem.* **1989**, *181*, 207-213.
(38)   Souppe, J.; Urrutigoity, M. *New J. Chem.* **1989**, *13*, 503-506.
(39)   Uemura, T.; Fujimori, M.; Le, H.-H.; Ikeda, S.; Aso, K. *Agric. Biol. Chem.* **1990**, *54*, 2277-2281.
(40)   Zaks, A.; Klibanov, A. M. *J. Biol. Chem.* **1988**, *263*, 8017-8021.
(41)   Inada, Y.; Nishimura, H.; Takahashi, K.; Yoshimoto, T.; Saha, A. R.; Saito, Y. *Biochem. Biophys. Res. Commun.* **1984**, *122*, 845-850.
(42)   Ajima, A.; Yoshimoto, T.; Takahashi, K.; Tamaura, Y.; Saito, Y.; Inada, Y. *Biotechnol. Lett.* **1985**, *7*, 303-306.
(43)   Uemura, T.; Furukawa, M.; Kodera, Y.; Hiroto, M.; Matsushima, A.; Kuno, H.; Matsushima, H.; Sakurai, K.; Inada, Y. *Biotechnol. Lett.* **1995**, *17*, 61-66.
(44)   Blanco, L.; Guibe-Jampel, E.; Rousseau, G. *Tetrahedron Lett.* **1988**, *29*, 1915-1918.
(45)   Makita, A.; Nihira, T.; Yamada, Y. *Tetrahedron Lett.* **1987**, *28*, 805-808.
(46)   Gutman, A. L.; Zuobi, K.; Bravdo, T. *Journal of Organic Chemistry* **1990**, *55*, 3546-3552.
(47)   Pina, C.; Clark, D.; Blanch, H. *Biotechnol. Tech.* **1989**, *3*, 333-338.
(48)   Oppenheimer, H. L.; Labousse, B.; Hess, G. P. *J. Biol. Chem.* **1966**, *241*, 2720-2730.
(49)   Matsushima, A.; Okada, M.; Inada, Y. *FEBS Lett.* **1984**, *178*, 275-277.

(50)   Fruton, J. S. In *Proteinase-Catalyzed Synthesis of Peptide Bonds*; Fruton, J. S., Ed.; John Wiley & Sons: New York, 1982; Vol. 53, pp 239-306.
(51)   Gaertner, H. F.; Puigserver, A. J. *Proteins* **1988**, *3*, 130-137.
(52)   Lee, H.-H.; Fukushi, H.; Oyama, K.; Aso, K. *Biotechnol. Lett.* **1993**, *15*, 833-836.
(53)   Matthews, B. W. *Acc. Chem. Res.* **1988**, *21*, 333-340.
(54)   Oyama, K.; Kihara, K.-i. *CHEMTECH* **1984**, *14*, 100-105.
(55)   Ferjancic, A.; Puigserver, A.; Gaertner, H. *Biotechnol. Lett.* **1988**, *10*, 101-106.
(56)   Oka, T.; Morihara, K. *J. Biochem.* **1980**, *88*, 807-813.
(57)   Wayne, S. I.; Fruton, J. S. *Proc. Natl. Acad. Sci. USA* **1983**, *80*, 3241-3244.
(58)   Takahashi, K.; Nishimura, H.; Yoshimoto, T.; Saito, Y.; Inada, Y. *Biochem. Biophys. Res. Commun.* **1984**, *121*, 261-265.
(59)   Takahashi, K.; Tamaura, Y.; Kodera, Y.; Mihama, T.; Saito, Y.; Inada, Y. *Biochem. Biophys. Res. Commun.* **1987**, *142*, 291-296.
(60)   Yoshimoto, T.; Ritani, A.; Ohwada, K.; Takahashi, K.; Kodera, Y.; Matsushima, A.; Saito, Y.; Inada, Y. *Biochem. Biophys. Res. Commun.* **1987**, *148*, 876-882.
(61)   Yoshimoto, T.; Mihama, T.; Takahashi, K.; Saito, Y.; Tamaura, Y.; Inada, Y. *Biochem. Biophys. Res. Commun.* **1987**, *145*, 908-914.
(62)   Tamaura, Y.; Takahashi, K.; Kodera, Y.; Saito, Y.; Inada, Y. *Biotechnol. Lett.* **1986**, *8*, 877-880.
(63)   Stillman, J. S.; Stillman, M. J.; Dunford, H. B. *Biochemistry* **1975**, *14*, 3183-3188.
(64)   Stillman, J. S.; Stillman, M. J.; Dunford, H. B. *Biochem. Biophys. Res. Commun.* **1975**, *63*, 32-35.
(65)   Wright, P. V. *Br. Polymer J.* **1975**, *7*, 319-327.
(66)   Ohno, H.; Wang, P. *Nippon Kagaku Kaishi* **1991**, 1588.
(67)   Ohno, H.; Tsukuda, T. *J. Electroanal. Chem.* **1992**, *341*, 137-149.
(68)   Ohno, H.; Yamaguchi, N. *Bioconjugate Chem.* **1994**, *5*, 379-381.
(69)   Ohno, H.; Tsukuda, T. *J. Electroanal. Chem.* **1994**, *367*, 189-194.
(70)   Ohno, H.; Yamaguchi, N.; Watanabe, M. *Polym. Adv. Technol.* **1992**, *4*, 133-138.
(71)   Kikkawa, S.; Takahashi, K.; Kataded, T.; Inada, Y. *Biochem. Intl.* **1989**, *19*, 1125-1131.
(72)   Mori, S.; Nakata, Y.; Endo, H. *Biotechnol. Appl. Biochem.* **1992**, *15*, 278-282.
(73)   Mori, S.; Nakata, Y.; Endo, H. *Biotechnol. Appl. Biochem.* **1992**, *16*, 101-105.
(74)   Takahashi, K.; Nishimura, H.; Yoshimoto, T.; Okada, M.; Ajima, A.; Matsushima, A.; Tamaura, Y.; Saito, Y.; Inada, Y. *Biotechnol. Lett.* **1984**, *6*, 765-770.
(75)   Babonneau, M.-T.; Jacquier, R.; Lazaro, R.; Viallefont, P. *Tetrahedron Lett.* **1989**, *30*, 2787-2790.
(76)   Lee, H.; Kodera, Y.; Ohwada, K.; Tsuzuki, T.; Matsushima, A.; Inada, Y. *Biotechnol. Lett.* **1988**, *10*, 403-407.

Chapter 9

Incorporation of Poly(ethylene glycol)–Proteins into Polymers

Janice L. Panza, Keith E. LeJeune, Srikanth Venkatasubramanian, and Alan J. Russell

Department of Chemical Engineering and Center for Biotechnology and Bioengineering, University of Pittsburgh, Pittsburgh, PA 15261

Combining polymer chemistry and enzyme catalysis can be achieved by using enzymes as co-monomers in polymerization reactions. This approach requires the protection of the enzyme from the polymerization environment, which is often organic rather than aqueous. In this chapter we describe how the modification of proteins with polyethylene glycol derivatives can prepare an enzyme for covalent incorporation into a polymer during the synthesis of that polymer. First, the polyethylene glycol serves to solubilize the enzyme in the polymerization environment. Second, the use of heterofunctional polyethylene glycols containing monomeric functionalities serves to involve the protein in the polymerization itself. Thus far, subtilisin (a key industrial enzyme used in biological detergents), thermolysin (the enzyme used for biocatalytic synthesis of aspartame) and chymotrypsin (one of the most studied enzymes commercially available) have been incorporated successfully into acrylic based materials.

Enzymes are proteins which catalyze a diverse range of reactions. Enzyme catalysis is widely studied because of the substrate selectivity and catalytic efficiency inherent to enzymatic processes. These attributes, along with regioselectivity and stereoselectivity have made enzymes attractive catalysts in synthetic chemistry (*1*), having a broad range of potential applications. In spite of the potential of biocatalysis, short catalytic lifetime and catalyst thermal sensitivity can limit process productivity. Catalyst recovery and re-use is often not straightforward due to the solubility of enzymes in predominately aqueous reaction media. Organic solvents have been examined as potential media for enzymatic processes in order to minimize these drawbacks; however, altered specificity and significant decreases in activity are often observed (*2*). The most important factor in the successful application of an enzyme is to immobilize the enzyme to an insoluble solid support.

Enzyme immobilization techniques are being continually developed and refined for applications in a wide variety of areas including the food, chemical synthesis, agricultural, pharmaceutical, and polymer industries (*1*). The benefits of enzyme immobilization include increased enzyme stability, decreased thermal sensitivity, and enhanced process versatility (*3*). For an immobilization technique to be practical, the immobilized enzyme should bind irreversibly (covalently) to the support matrix, retain a significant degree of native specific activity, and maintain structural stability in the presence of substrate, product, and reaction solvent. Enzyme immobilization has been performed on a wide variety of support materials, including alumina pellets (*4*), trityl agarose (*5*), and glass/silica beads (*6*). Organic polymers also make excellent enzyme support materials due to their structural flexibility and solvent resiliency. Numerous polymers, including nylons (*7,8*), acrylates (*9*), and several copolymer blends (*10,11*) have been utilized as effective supports.

While the attachment of proteins to polymers is commonplace, it is still limited by the need for an accessible polymer architecture. Consider, for example, how one could immobilize a protein onto an ultrafiltration membrane where the pore size of the membrane was smaller than that of the protein. In such an instance, the only way to introduce a protein into the membrane would be during the polymerization process itself. Protein immobilization during polymer synthesis is also attractive because it enhances the possibility for the formation of multiple bonds between the protein and the growing polymeric support. Such multi-point attachment can dramatically stabilize many proteins (*12,13*).

While there are many advantages associated with a single step immobilization and polymer synthesis, there is also a major limitation. Proteins are generally not soluble in the organic solvents which are necessary for effective polymer synthesis. Thus, if we are to take advantage of the opportunities which arise from multi-point attachment, we must address the question of how to solubilize proteins in organic media, and how to facilitate a reaction between the growing polymer chain in the solvent and the surface of the protein.

The low solubility of proteins in organic solvents can be overcome by covalently attaching a long chain amphiphilic molecule to the surface of the protein. Polyethylene glycol (PEG) has been shown to be an ideal molecule for such surface modification strategies (*14*). PEG-modified enzymes demonstrate dramatically enhanced enzyme solubility in organic solvents, close to native activity and specificity, and enhanced thermal stability (*14*). Confusion can arise about immunogenic effects because the addition of PEG tails to small hapten molecules increases immunogenicity by increasing molecular weight. In fact, there is a delicate balance between the degree of PEG modification, and the effect that the alteration in structure will have on the protein's properties.

Since a PEG-protein can be dissolved in an organic solvent while retaining its native structure, if the enzyme were also reactive (as a co-monomer) in a polymerization reaction, the enzyme would potentially become intrinsically coupled to the polymeric material. Undoubtedly, the number of cross-links would be related to the number of reactive sites on the soluble protein.

A "PEGylated" enzyme would not normally be reactive in a polymerization reaction. The most accessible and reactive groups on the protein surface will probably have reacted with the PEG itself, and the PEG will undoubtedly reduce access of many molecules to the surface of the protein. Indeed, it is this very ability of PEG to "protect" the surface of the protein that makes it so attractive an agent for increasing enzyme solubility in non-aqueous media. In order to react PEGylated proteins with monomers in organic solution a heterofunctional PEG must be employed. One end of the derivatizing PEG must be designed to couple to a protein, while the other should be a reactive group able to couple to a polymer. This approach is summarized in Figure 1, where the reactive PEG is capped with an NHS-ester (to react with the protein), and an allyl group (to incorporate the resulting PEGylated protein into an acrylic polymer). This example is particularly interesting because it demonstrates the additional opportunity to "self polymerize" the PEG-protein in the absence of additional monomers (*15*).

There are many approaches to modifying proteins with PEG, and most would be applicable to use with a heterofunctional PEG designed for application in Figure 1. Particularly popular approaches to PEG modification are outlined throughout this book, and include, cyanuric chloride activation of PEG hydroxyl groups (*16,17*), hydroxyl activation through use of phenylchloroformates, nitrophenylcarbonate (*18*), and succinimidylsuccinates (*19*). Some approaches have drawbacks, such as the inactivation of enzymes with thiol functionalities in close proximity to the active site by cyanuric chloride (*18*). For the most part, all these strategies react a PEG chain with the free amino group of lysine residues on accessible surfaces of the protein. Once again, the chemistry and methods for such reactions are detailed elsewhere in this text. In this chapter we will review the growing body of literature concerning the synthesis, activity, and stability of PEG-polymer-protein materials.

Protein Polymer Synthesis

Protein polymer synthesis can be performed by two methods. The first is a two step process where the enzyme is modified with PEG containing a polymerizable functional group at one terminus, followed by polymerization of the PEG-modified enzyme with other polymerizable monomers. This results in a polymer with a low percentage of incorporated protein (*20*). An example of this procedure is shown in Figure 2. Both subtilisin Carlsberg and thermolysin have been modified by this method (*9,20*). The PEG-protein conjugate is soluble in a variety of solvents (up to 5 mg/ml in chloroform, methylene chloride, 1,1,1-trichloroethane, carbon tetrachloride, toluene, benzene, dimethyl sulfoxide, tetrahydrofuran, dioxane, or acetonitrile), including the ones used for the polymer synthesis (chloroform, carbon tetrachloride). For modified subtilisin to remain in solution in tetrahydrofuran and acetonitrile, slight heating is required. The modified enzyme is not soluble in solvents such as hexane, acetone, ethyl ether and isoamyl alcohol.

The second method is a one step process where both the enzyme and the PEG molecule contain polymerizable functional groups which can be used to directly polymerize the protein and the PEG molecule together. This results in a hydrogel of PEG and protein (*21*). The enzyme α-chymotrypsin has been modified by this method (*21,22*).

PEG-Enzyme Reactions

Many studies have been performed on PEG-modification of proteins due its diverse biotechnological applications (*14*). In the case of PEG-protein-polymer synthesis, PEG is being used in order to increase the solubility of a protein in organic solvents (*23,24*). Other polymers, such as polystyrene, have also been used to increase the solubility of proteins in organic solvents (*25*). In general, PEG itself is first modified with a monomeric functional group at one end of the molecule to form an activated PEG. This functional group is a reactive terminal group that can react with a functional group on a protein. The reaction will most likely occur with an ε-amino group of a lysine residue on the protein (*26*).

One example of an activated PEG is PEG-aldehyde. Yang et. al. used PEG-aldehyde to modify the enzyme subtilisin Carlsberg (*20*). Subtilisin Carlsberg is a serine protease, best known for its activity in biological detergents. The annual market for sales of this enzyme in the US is approximately $300 million, and it has been extensively studied for many years. In this study, PEG-monomethacrylate was first converted to an aldehyde by a procedure adapted from Wirth (*27*). The PEG molecule was then converted to a difunctional PEG, with an aldehyde group at one end and a methacrylate group at the other. The PEG-aldehyde was then used to modify subtilisin by the reaction of the aldehyde with a lysine residue on subtilisin by reductive alkylation.

In a second study by Yang et al., both subtilisin and thermolysin were modified with PEG (*9*). Thermolysin is a metalloendopeptidase. PEG-A, an N-hydroxysuccinimide activated polyethylene glycol acrylate, which contained an acrylate group at one terminus and an active ester at the other terminus, was used in the modification of the enzymes. PEG-A reacted with the enzymes by nucleophilic substitution on the lysine residues of the enzymes.

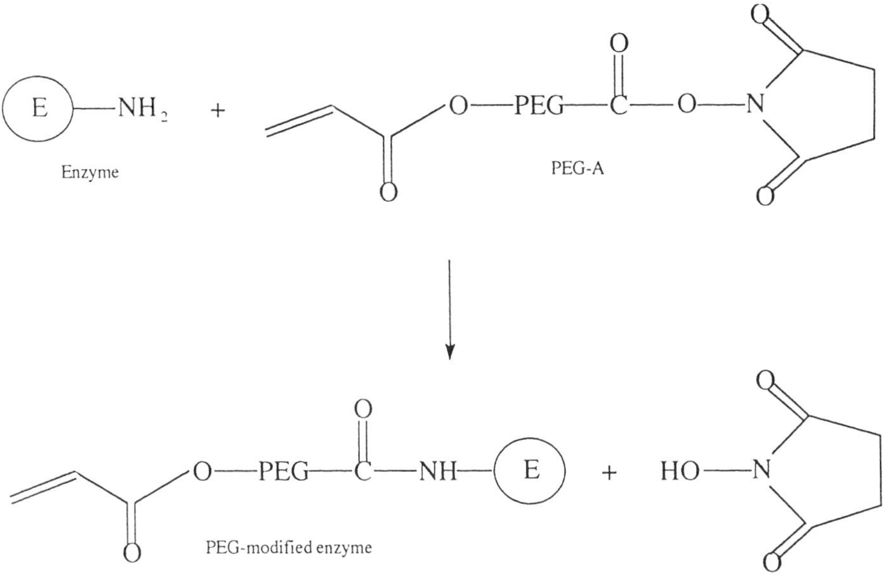

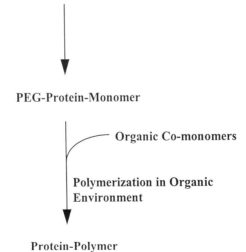

Figure 1. PEG-modification of a protein using PEG activated with an NHS-ester.

Protein + Activated Heterofunctional PEG Monomer

PEG-Protein-Monomer

Organic Co-monomers

Polymerization in Organic Environment

Protein-Polymer

Figure 2. The general method for incorporation of PEG-modified proteins into polymers during polymerization in organic solvents.

Of course, other functional groups can be used to modify the PEG molecule. For example, cyanuric chloride and p-nitrophenol chloroformate have been used to activate the hydroxyl group of the PEG molecule via nucleophilic substitution (18). Both trypsin (28) and subtilisin (26) have been modified with cyanuric chloride-activated PEG (CC-PEG) and p-nitrophenol carbonate-PEG (NPC-PEG).

Many different techniques exist that can be used to verify that a protein has been modified with PEG. Depending on the technique chosen, it can be determined whether, where, and how modification has occurred.

A simple test to ascertain modification of a protein with PEG is solubility data. As mentioned before, proteins have low solubility in organic solvents, attachment of PEG molecules increases the solubility of proteins in organic solvents. Therefore, if a PEG-modified enzyme is soluble in an organic solvent in which the unmodified protein was not, modification has been established. The solubilization of the PEG-modified protein can be used to purify the unmodified protein from the modified protein. Khan has speculated that the PEG-modified enzymes are not actually soluble in organic solvents, but they just form aggregates in the solvents that appear optically transparent (29). This does not change the fact that this property can be used to confirm PEG-modification.

High pressure liquid chromatography (HPLC) is another useful technique to substantiate PEG-modification. In addition, HPLC analysis can elucidate the degree of modification (what fraction of the protein molecules protein is modified with PEG). The retention times for native, singly-modified, and doubly-modified fractions of protein will vary. A comparison of the peaks for each fraction will indicate the degree of modification of the protein. HPLC can, however, give misleading results under some circumstances (33, 34, 35).

Both matrix assisted laser desorption time of flight (MALDI TOF-MS) and electrospray mass spectrometry (ES-MS) are also useful tools with which to determine the degree and yield of a modification strategy. Figure 3 shows the dramatic and detailed information that ES-MS and MALDI-MS in practice can provide. These methods cannot, however, distinguish where on the protein the modification has taken place, and as with all analysis tools they must be used judiciously and in combination with all the tools available for characterization (such as capillary electrophoresis) (36,37,38). The modification could take place at any lysine residue or the amino terminus of the protein (20). It is very unlikely that the modification occurs at the same position on each molecule. Because there is no convenient method to separate modified protein molecules based on where the modification is located, the properties of PEG-modified proteins are actually properties of a heterogeneous mixture.

Protein staining techniques such as the Bradford method, SDS-polyacrylamide gel electrophoresis, gel filtration chromatography, and amino acid analysis are also used in the analysis of the PEG-proteins. The Bradford method can determine the amount of protein in a sample (30). This assay involves the binding of a dye to the protein causing a shift in absorption which can be measured spectrophotometrically. SDS-polyacrylamide gel electrophoresis and gel filtration chromatography can be used to separate the modified fraction of protein based on size.

Activity and Stability

Activity and stability studies have been performed on many PEG-modified enzymes, prior to utilization in a given system or polymer. The primary purpose of these studies has been to investigate the stability of the enzyme against both temperature and pH before and after PEG modification, and to assess the role of the chemistry of modification on the eventual properties of the modified material. For example, activity studies were performed on PEG-modified subtilisin in aqueous solutions, organic solutions, and a mixture of the two (20,26). It was found that although PEG-modified enzymes showed less activity than the native subtilisin in the aqueous environments, the modified enzymes still retained substantial activity. In addition,

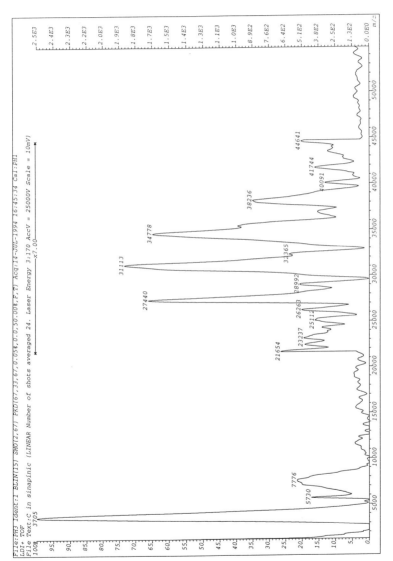

Figure 3. Matrix assisted laser desorption time of flight mass spectrometry (MALDI TOF-MS) of subtilisin modified with 3400 MW PEG using sinapinic acid as the dispersant in linear mode after averaging 24 shots of laser energy 3.170, with an accelerating voltage of 25000 V, and a full scale of 10 mV. The data were collected on a calibrated VG TofSpec E MALDI-TOF.

the PEG-modified subtilisin was active in organic solvents, thus the modified enzyme should remain active when solubilized in the organic solvents necessary to incorporate it into a polymer (20). The stability of some PEG-modified enzymes, particularly PEG-subtilisin, have also been shown to be markedly higher than the native forms of the enzymes with time and against temperature and pH (20,26,28). For subtilisin, one hypothesis which explains this increase in stability in buffer is a reduction in the rate of autolysis of the subtilisin via steric hindrance caused by the large PEG molecules (20,26). Alternatively, intrinsic changes in the stability of the enzyme may occur upon modification (26). A third hypothesis is that the stability of PEG-modified proteins may result from a highly hydrogen-bonded structure around the enzyme caused by the long PEG chains (28). Also, by attaching PEG molecules to an enzyme, the enzyme may become less flexible in aqueous solution which would make the enzyme less likely to unfold (31). The stability of PEG-modified enzymes is significant because the stability of the polymerized enzyme will most likely be more similar to that of the modified enzyme than the native enzyme (20).

Polymerization

Once a protein has been modified with PEG, it is now soluble in many organic solvents in which it was not soluble in its native state. In addition, an enzyme modified with PEG retains its activity in water, organic solvents, and some mixtures of the two. Usually an enzyme would have to be incorporated into a polymer through a polymerization in an aqueous solution due to the limited solubility of the protein in organic solvents. However, modification with PEG allows the choice of many new polymer systems since the enzyme is now soluble in organic solvents.

Synthesis

Enzymes have been incorporated into polymers by two methods. The first involves polymerizing the enzyme modified with PEG molecules containing polymerizable functional groups at the PEG terminus with other monomers and crosslinkers. An example of this is the polymerization of PEG-modified subtilisin with methyl methacrylate (9,20). The PEG molecules contained a terminal monomethacrylate that could undergo polymerization in the presence of monomers of methyl methacrylate, a crosslinker solution of trimethylolpropane trimethacrylate, and an initiator. Free radical polymerization of the mixture resulted in the incorporation of PEG-modified subtilisin in an acrylate polymer.

    An alternative method uses PEG and proteins which have been modified with polymerizable functional groups, enabling co-polymerization. Fulcrand et. al. used this method to cross-link PEG chains and α-chymotrypsin (21,22). First, both PEG and α-chymotrypsin were separately modified with acryloyl chloride to give acrylated derivatives of both PEG and α-chymotrypsin. These two reactants were allowed to co-polymerize along with bisacryloyl PEG as a crosslinker, the monomethylether of monoacrylated PEG as the matrix agent, and free radicals. The result was α-chymotrypsin combined with PEG in a polymeric hydrogel.

    Although both procedures led to an enzyme being incorporated in a polymer network, there is a distinct difference between the two polymerizations. Polymerization of subtilisin into the acrylate polymer took place in an organic solution of carbon tetrachloride and chloroform. The polymerization of the α-chymotrypsin into the PEG hydrogel took place in a aqueous solution of borate buffer.

    One of the major benefits of incorporating PEG-modified enzymes into polymers, as mentioned before, is the ability to have to reaction occur in organic solvent. This enables solvent engineering of polymer properties in order to design a polymer with appropriate properties that fit a given application. By keeping the protein content at less than 2 % by weight, the resulting polymer will have the same properties as the polymer without incorporated protein (*20*). For example, if diffusion limitations of a substrate are a concern, a polymer system that will yield a highly porous material can be used to incorporate the enzyme. In addition, a variety of shapes and morphologies of the polymers can be fabricated, including beads, fibers, or membranes (*20*).

There are three different ways that a protein can be immobilized in a polymer: covalent attachment, adsorption to the surface, or entrapment within pores (*9*). Since covalent attachment of the PEG-modified enzyme to polymer during synthesis is desired, the biopolymers are usually rinsed extensively after polymerization to remove adsorbed or trapped protein. The polymer may also be crushed or broken before rinsing. By assaying the rinsates for residual activity, the efficiency of the incorporation can be determined.

Characterization

The protein content in the polymer can be determined using the Bradford method as mentioned previously (*30*). Transmission electron microscopy or scanning electron microscopy can be used to confirm the size and shape of the pores within the polymer and morphology differences between polymers with incorporated proteins and polymers without proteins. Finally, BET analysis can be used to estimate the surface area of the polymers.

Activity and Stability of Biopolymers

Once PEG-modified enzymes have been incorporated in a polymer system, the biopolymer must be tested for activity and stability of the enzyme. It has been shown the PEG-modified enzymes retain activity and stability, but incorporation of the PEG-modified enzymes into polymers requires the enzyme to be exposed to harsh environments.

For the most part, PEG-modified enzymes retain considerable activity when incorporated into polymers. Yang compared the activity of native enzyme to the immobilized enzyme, and although reduced, the immobilized enzyme did exhibit approximately 12 % activity retention (activity reported as kcat/Km) (*9*). It is not surprising that the activity of the biopolymer was reduced in respect to the native enzyme because immobilization may decrease the accessibility of the substrate to the active site. In addition, the hydrophobicity of the polymer may attract hydrophobic substrates into hydrophobic areas of the polymer, away from the active sites of the immobilized enzymes (*9*). Fulcrand used the biopolymer hydrogel to access esterase activity in water, demonstrating that α-chymotrypsin retains significant activity in aqueous solutions (*21*).

Stability

In order to assess the stability of the biopolymer in the presence of substrate, Yang designed and constructed a continuous flow reactor (*9,20*). The flow cell apparatus consisted of substrate being first pumped into the reference cell of spectrophotometer; second, to a stirred reactor containing the biopolymer; third, to the sample cell of the spectrophotometer; and finally to a waste collection vessel. A schematic of the flow cell apparatus is shown in Figure 4. They tested the polymer-incorporated subtilisin in the system and found that the biopolymer remained active for over 100 days, at which time the experiment was ended. This established that enzymes covalently immobilized in a polymer have increased stability.

The enhanced stability of the enzyme is likely the result of a number of factors. By incorporating the enzyme into a polymer, the enzyme molecules are probably more rigid (*9*). That is, the enzyme is less likely to be denatured by unfolding since it is held in place by the covalent immobilization. Guisán has shown that stability of

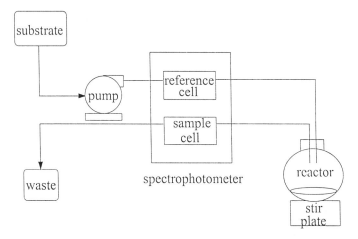

Figure 4. Flow cell apparatus for determination of the activity and stability of protein-polymers (Adapted from ref. 20.).

enzyme bound to supports is enhanced by multi-point enzyme-support attachments (*32*). In addition, improved stability may be due to decreased autolysis of the enzyme because of the steric hindrance of the PEG molecules and polymer network (*9*).

Organic Solutions

PEG-modified enzymes are soluble and display activity in organic solvents. Incorporating the PEG-modified enzymes into polymers may lead to new applications.

One of the advantages of PEG-modified enzyme incorporation into polymers is its activity in organic solutions. When polymerized with PEG in a hydrogel, α-chymotrypsin can be used to catalyze the synthesis of peptides in organic solvents (*21,22*). In addition, PEG-modified subtilisin has displayed activity in organic solvents (*9,20*). It should be noted that both enzymes required some water in the organic solvents in order for the enzymes to become active (*9,22*). The polymer may protect the enzyme from denaturation by the solvent due to the covalent attachment of the enzyme to the polymer, thereby allowing the enzyme to remain active.

In addition to remaining active in organic solvents, PEG-modified enzymes incorporated into polymers also show increased stability in organic solvents over their native forms. Even under very harsh conditions, polymer-incorporated subtilisin retained its activity after many cycles of exposure (*9*). Since the biopolymer was able to withstand the harsh conditions of the recycling experiments, the biopolymers will probably be able to function in most organic solvents.

Conclusions

PEG-modified proteins have been incorporated into polymers. One method first modifies an enzyme with a heterofunctional PEG molecule, that contains an activating group at one terminus to covalently attach to the enzyme and another functional group at the other terminus that is capable of polymerizing. The next step is to polymerize the PEG-modified protein with monomer and crosslinkers to form the polymer. A second method modifies both the enzyme and PEG molecules separately with polymerizable functional groups, then initiates a co-polymerization between them. Both methods result in biopolymers that are active and stable in aqueous and organic solutions. As a result of the greatly enhanced stability of the enzymes, the productivity of the biocatalysts is dramatically enhanced.

Literature Cited

1. Margolin, A. L. *Chemtech* **1991, March,** 160-167.
2. Dordick, J. S. *Applied Catalysis* **1991,** *1,* 1-51.
3. Martinek, K.; Klibanov, A. M.; Goldmacher, V. S.; Berezin, I. V. *Biochim. Biophys. Acta* **1977,** *485,* 1-12.
4. Vasudevan, P. T.; Thakun, D. S. *Appl. Biochem. Biotech.* **1994,** *49,* 173-189.
5. Caldwell, S. R.; Raushel, F. M. *Biotechnol. Bioeng.* **1991,** *37,* 103-109.
6. Vasudevan, P. T.; Weiland, R. H. *Biotechnol. Bioeng.* **1993,** *41,* 231-236.
7. Chellapandian, M.; Sastry, C. A. *Bioprocess Engineering* **1994,** *11,* 17-21.
8. Caldwell, S. R.; Raushel, F. M. *Appl. Biochem. Biotech.* **1991,** *31,* 59-73.
9. Yang, Z.; Mesiano, A. J.; Venkatasubramian, S.; Gross, S. H.; Harris, J. M.; Russell, A. J. *J. Am. Chem. Soc.* **1995,** *117,* 4843-4850.
10. Nguyen, A. L.; Luong, J. *Biotechnol. Bioeng.* **1989,** *34,* 1186-1190.
11. Hoshino, K.; Taniguchi, M.; Netsu, Y.; Fujii, M. *J. Chem. Engr. Japan* **1989,** *22,* 54-59.
12. LeJeune, K. E.; Russell, A. J. *Biotechnol. Bioeng.* **1996,** *51,* 450-457.
13. LeJeune, K. E.; Frazier, D. S.; Caranto, G. R.; Maxwell, D. M.; Amitai, G.; Russell, A. J.; Doctor, B. P. *Proc. Med. Def. Biosc. Rev.* **1996, June,** 1-8.

14. Harris, J. M. In: *Poly(ethylene glycol) Chemistry: Biotechnical and Biomedical Applications;* Harris, J. M., Ed.; Plenium Press: New York, 1992.
15. Andreopoulos, F. M.; Deible, C. R.; Stauffer, M. T.; Weber, S. G.; Wagner, W. R.; Beckman, E. J.; Russell, A. J. *J. Am. Chem. Soc.* **1996,** *118,* 6235-6240.
16. Abuchowski, A.; Van Es, T.; Palczuk, N. C.; Davis, F. F. *J. Biol. Chem.* **1977,** *252,* 3578.
17. Imoto, T.; Yamada, H. In *Protein Function: A Practical Approach;* Creighton, T. E., Ed.; IRL Press: Oxford, 1989, 247-277.
18. Veronese, F. M.; Boccu, R. L. E.; Benassi, C. A.; Schiavon, O. *Appl. Biochem. Biotech.* **1985,** *11,* 141-152.
19. Buckamann, A. F.; Morr M.; Johansson, G. *Makromol. Chem.* **1982,** *182,* 1379.
20. Yang, Z.; Williams, D.; Russell, A. J. *Biotechnol. Bioeng.* **1995,** *45,* 10-17.
21. Fulcrand, V.; Jacquier, R.; Lazaro, R.; Viallefont, P. *Tetrahedron* **1990,** *46,* 3909-3920.
22. Fulcrand, V.; Jacquier, R.; Lazaro, R.; Viallefont, P. *Int. J. Peptide Protein Res.* **1990,** *38,* 273-277.
23. Inada, Y.; Takahashi, K.; Yoshimoto, T.; Ajima, A.; Matsushima, A.; Saito, Y. *Trends Biotechnol.* **1986, 6,** 190-194.
24. Yoshinaga, K.; Ishida, H.; Hagawa, T; Ohkubo, K. In: *Poly(ethylene glycol) Chemistry: Biotechnical and Biomedical Applications;* Harris, J. M., Ed.; Plenium Press: New York, 1992, 103-114.
25. Ito, Y.; Fujii, H.; Imanashi, Y. *Biotechnol. Prog.* **1993,** *9,* pp 128-130.
26. Yang, Z.; Domach, M.; Auger, R.; Yang, F. X.; Russell, A. J. *Enzyme Microb. Technol.* **1996,** *18,* 82-89.
27. Wirth, P.; Souppe, J.; Tritsch, D.; Biellmann, J. F. *Biorg. Chem.* **1991,** *19,* 133-142.
28. Gaertner, H. F.; Puigserver, A. J. *Enzyme Microb. Technol.* **1992,** *14,* 150-155.
29. Khan, S. A.; Halling, P. J *Enzyme Microb. Technol.* **1992,** *14,* 96-100.
30. Bradford, M. M. *Anal. Biochem.* **1976,** *72,* 248-254.
31. Baillargeon, M. W.; Sonnet, P. E. *Ann. NY Acad. Sci.* **1988,** *542,* 244-249.
32. Guisán, J. M. *Enzyme Microb. Technol.* **1988,** *10,* 375-382.
33. McGoff, R.; Baziotis, A.C.; Maskiewicz, R. *Chem. Pharm. Bull.,* **1988,** *36,* 3079-3091.
34. Kurfurst, M.M., *Anal. Biochem.,* **1992,** *200,* 244-248.
35. Chowdhury, S.K.; Doleman, M.; Johnston, D.J., *J. Am. Mass. Spectrom.,* **1995,** *6,* 478-487.
36. Montaudo, G.; Montaudo, M.S.; Puglisi, C.; Samperi, F., *Rapid Commun. Mass Spectrom.,* **1995,** *9,* 453-460.
37. Cunico, R.L.; Gruhn, V.; Kresin, L.; Nitecki, D.E.; Wiktorowicz, J.E., *J. Chromatogr.,* **1991,** *559,* 467-477.
38. Bullock, J.; Chowdhury, S.; Johnston, D., *Anal. Chem.,* **1996,** *68,* 3258-3264.

Chapter 10

Biochemistry and Immunology of Poly(ethylene glycol)-Modified Adenosine Deaminase (PEG–ADA)

Michael S. Hershfield

Departments of Medicine and Biochemistry, Duke University Medical Center, Box 3049, Room 418, Sands Building,Durham, NC 27710

Poly(ethylene glycol)-modified bovine adenosine deaminase (PEG-ADA) was the first PEGylated protein to undergo clinical trial. It has been effective in correcting the toxic biochemical effects of ADA substrates, and in treating the fatal immune deficiency disease caused by ADA deficiency. The experience gained from monitoring PEG-ADA therapy over the past decade provides unique insight into the immunological response to chronic treatment with a PEGylated enzyme.

When enzyme replacement therapy was first attempted for lysosomal storage disorders, difficulties were quickly encountered with the rapid clearance and limited delivery to affected cells of parenterally administered enzyme. A neutralizing or allergic immune response to the therapeutic protein was also a potential barrier. Several solutions to these problems were explored during the 1970's, including chemically modifying or physically sequestering enzymes. Abuchowski, Davis and their collaborators showed that conjugating primary amino groups of enzymes with poly(ethylene glycol) of average molecular weight 5-10 kDa prolonged circulating life and reduced immunogenicity and antigenicity (1-3). Since these effects were achieved partly by blocking cellular uptake, this strategy was not ideal for treating lysosomal storage diseases. The first clinical test of a PEGylated enzyme involved a rare disorder of purine metabolism well-suited to this technology.

Adenosine deaminase (ADA) is a 41 kDa monomeric enzyme expressed at highest levels in the cytoplasm of lymphocytes. Inherited ADA deficiency profoundly impairs lymphoid development, causing the fatal syndrome Severe

Combined Immune Deficiency (SCID), which results from toxic
effects of the ADA substrates adenosine (Ado) and 2'-deoxy-
adenosine (dAdo) (4,5). Among these effects, dATP pool
expansion, due to excessive phosphorylation of dAdo, causes
immature lymphocytes to undergo apoptosis (programmed cell
death) (6). In addition, Ado inhibits and dAdo inactivates
the enzyme S-adenosylhomocysteine (SAH) hydrolase, which
may cause SAH-mediated inhibition of S-adenosylmethionine-
dependent transmethylation reactions (7,8). Unlike the
complex molecules that accumulate in lysosomal disorders,
nucleosides readily equilibrate across cell membranes.
Thus, eliminating circulating extracellular ADA substrates
could in theory prevent or reverse their toxic
intracellular effects, even if exogenous ADA were not taken
up by lymphoid cells.

The first form of replacement therapy for ADA
deficiency involved transfusing normal erythrocytes (9).
This was of limited clinical benefit, possibly because
erythrocytes possess relatively low ADA activity. In 1986,
we began testing PEG-modified bovine ADA (PEG-ADA,
ADAGEN™), manufactured by Enzon, Inc., as a potentially
safer and more effective treatment (10). The patients
involved lacked an HLA-identical (i.e. tissue-type matched)
donor for a bone marrow transplant, and had either failed
to engraft after transplantation, or were too ill to
undergo that procedure. Over the past decade we have
monitored most of the patients, now more than 60, who have
received PEG-ADA, both as a single therapy and as an
adjunct to the first human trials of somatic cell gene
therapy. In 1990, PEG-ADA was approved by the US Food and
Drug Administration as an Orphan Drug. We have reviewed the
clinical experience with PEG-ADA elsewhere (11,12). This
review will focus on biochemical efficacy and the
immunological response of patients to PEG-ADA.

Experimental

PEG-ADA (ADAGEN™, Enzon, Inc., Piscataway, NJ) was prepared
from purified bovine intestinal ADA using the succinimidyl-
succinate of monomethoxy-PEG, average molecular weight 5
kDa (SS-PEG (13)). PEG-ADA was supplied at 250 U/ml (one
unit = one μmol adenosine converted to inosine per minute,
measured at 25°). PEG-ADA was administered by intramuscular
injection once or twice weekly. As described in our
original report, we assessed plasma ADA activity at 37°,
and also followed the activity of SAHase in erythrocyte
hemolysates, and the level of total Ado and dAdo
nucleotides (AXP and dAXP, respectively) in acid extracts
of erythrocytes (10). The latter were measured by reversed-
phase HPLC after enzymatic digestion to the nucleosides
(the levels of free Ado and dAdo, measured in undigested
erythrocyte extracts, were negligible) (14). The
immunologic response to PEG-ADA was monitored by ELISA,
following IgG antibody that reacted with immobilized,

unmodified bovine ADA; we also characterized the specificity of anti-ADA antibodies and their effects on ADA activity, as described (15).

Results and Discussion

Metabolic Actions of PEG-ADA. When testing began in 1986, there was no animal model for ADA deficiency, and there were limited metabolic data from infants with SCID to use as a guide for replacement therapy. However, a potent ADA inhibitor, 2'-deoxycoformycin, was then in use for treating T lymphoblastoid leukemia, and its metabolic effects were similar to those due to inherited ADA deficiency: a marked increase in dAdo excretion and level of dAdo nucleotides (dAXP) in erythrocytes, and dAdo-mediated inactivation of SAHase (5,16,17). From the rates of dAdo excretion and red cell dAXP accumulation in deoxycoformycin-treated patients, we estimated daily dAdo production. This served as a guide for choosing an initial dosage of PEG-ADA, and for monitoring its biochemical efficacy, i.e., by following reduction in red cell dAXP and increase in red cell SAHase activity. Recovery of immune function could not be used to establish an effective dosage in the first patients since, at best, improvement was expected to be slow. For assessing PEG-ADA pharmacokinetics, we monitored plasma ADA activity, which is very low even in normal individuals (hence PEG-ADA, acting in plasma, rather than in cells, is an "ectopic" form of enzyme replacement).

Results in the first patients, older children who had been diagnosed several years earlier, indicated that PEG-ADA was absorbed within 24-48 hours after intramuscular injection and then was cleared from plasma with a half-life of 3 to 6 days (e.g., Figure 1A). In these patients, single weekly i.m. injections of 15 U/kg maintained pre-injection (trough) plasma ADA activity above the normal total blood (cellular) ADA level of 6 to 12 µmol/hour/ml (Figure 2). This resulted in correction of erythrocyte dAXP and SAHase levels (Figure 1B), even though there was no appreciable increase in ADA activity in erythrocytes or blood mononuclear cells (10). Later, when newly diagnosed, severely ill and malnourished infants were treated, we found that their clearance of PEG-ADA was more rapid. Twice-weekly injections of 30 U/kg (60 U/kg/week) were used to maintain 2 to 10 fold higher plasma ADA activity, in order to improve recovery of immune function (18). The dose was decreased after several months, depending on clinical status and the recovery of immune function.

At present weekly maintenance doses ranging from 15 to 60 U/kg, given as one or two i.m. injections, maintain erythrocyte dAXP at <10 nmol/ml in virtually all patients, compared with pretreatment dAXP levels averaging >600 nmol/ml (the normal level is <2) (5). Erythrocyte dAXP levels of <20 nmol/ml are apparently consistent with

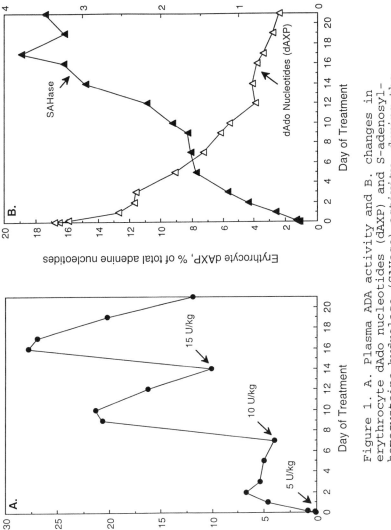

Figure 1. A. Plasma ADA activity and B. changes in erythrocyte dAdo nucleotides (dAXP) and S-adenosyl-homocysteine hydrolase (SAHase) activity during the first 3 weeks of treatment of a SCID patient with PEG-ADA. The dose and timing of intramuscular injections are shown in panel A.

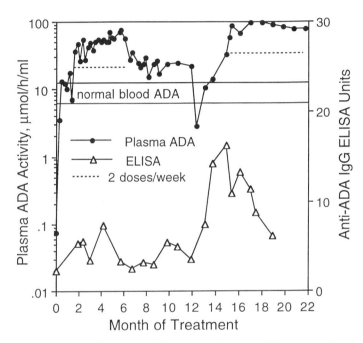

Figure 2. Transient decline in plasma ADA activity after 12 months (filled circles) with appearace of anti-ADA antibody (IgG) (open triangles) during treatment with PEG-ADA. The dotted lines indicate periods when the patient received twice weekly injections of PEG-ADA (each 30 U/kg). At other times he received 30 U/kg once a week.

adequate immunologic function, based on findings in several
individuals with "partial ADA deficiency" (identified in
screening programs), who have remained healthy into
adulthood without therapy (5,19,20).

Most SCID patients are seriously ill at the time of
diagnosis. Since any recovery of immune function requires
several weeks to a few months of PEG-ADA therapy, the
majority of deaths have occurred within the first few
months of treatment. In patients who survive this immediate
period, lymphocyte counts and immune function have
improved, though not to normal. Despite remaining
lymphopenic, most of these patients have done well
clinically, indicating a level of effective immune function
sufficient to sustain good health (5,11,12).

**Immunologic Response to PEG-ADA during Chronic
Therapy.** In addition to prolonging circulating life,
PEGylation is intended to minimize the immunologic response
to a therapeutic protein. ADA deficiency offered a means of
testing this in patients receiving an amount of bovine ADA
each week equivalent to the ADA present in 4.5 liters of
normal human erythrocytes. It should be understood that,
while intermittent enzyme replacement may be effective in
treating some lysosomal storage disorders, therapeutic
plasma ADA levels must be maintained continuously, since
toxic substrates can rapidly induce apoptotic death of
lymphocytes in patients lacking ADA (6,16,17).

During the first 5 years of investigation we
systematically monitored plasma ADA levels and assessed IgG
anti-ADA antibody levels by an ELISA in 17 (of the first
18) patients, all of whom had been treated for at least one
year (15). The ELISA became positive in 10 of these 17
patients, usually between 3 and 8 months of treatment,
coincident with return of immune function. Competition
studies showed that the antibodies reacted in solution
about equally well with PEG-modified and unmodified bovine
ADA, but they did not react with free PEG, or with a
different PEGylated protein. In almost all cases, the
antibodies reacted more strongly with bovine than with
human ADA, indicating specific recognition of bovine ADA
peptide epitopes (15).

In 8 of the 10 anti-ADA positive patients, there was
little effect on PEG-ADA pharmacokinetics, and no
correlation between ELISA titer and plasma ADA activity,
indicating that most of these antibodies neither block
catalytic activity nor enhance clearance of enzyme by the
reticuloendothelial system (15). In a few cases, the
appearance of anti-ADA antibody has coincided with a
transient fall in plasma ADA activity, which has reversed
spontaneously, or with increasing dosage (e.g Figure 2).

In two patients, plasma ADA activity decreased to pretreatment levels when the ELISA became positive at 4-5
months of treatment; this was associated with a
persistently enhanced rate of PEG-ADA clearance following

injection. Antibody (purified IgG) from both these patients, but not from the other ELISA-positive patients, directly inhibited the catalytic activity of both PEGylated and native bovine ADA (15). Scatchard analysis (based on the kinetics of enzyme inhibition (15)) suggested a single antibody species with a binding constant K_d of 0.5 nM. Depletion of plasma ADA activity caused a rise in erythrocyte dAXP to pre-treatment levels and a decline in lymphocyte counts. Doubling the dose of PEG-ADA by twice-weekly injection restored therapeutic plasma ADA levels in both patients, restoring metabolic and immunologic benefit. In one of these patients, treating with glucocorticoids and high dose intravenous immunoglobulin for four months suppressed the ADA-inhibitory antibody (but not anti-ADA ELISA reactivity) (15,21). Both of these patients have now received PEG-ADA for >7 years and have continued to do well clinically.

To date 63 patients have been treated with PEG-ADA, 45 for more than one year (40 have been treated for >2 y, average 5.7 years). We have continued to monitor plasma ADA levels and anti-ADA IgG ELISA in most US and several Canadian and European patients. We estimate that with chronic treatment, 80 to 85% of long-term patients develop a positive ELISA. As found in the original study (15), in most cases there has been no sustained effect on PEG-ADA clearance, or a decline of about two-fold in pre-injection (trough) plasma ADA activity. A more severe decline in ADA level has occurred in two additional patients, both of whom have discontinued treatment and have undergone successful bone marrow transplantation (neither had deteriorated clinically). In no patient treated to date has there been an allergic or hypersensitivity reaction to PEG-ADA, regardless of anti-ADA status.

A Genetic Approach For Enhancing The Effect Of Pegylation On Immunogenicity. The ability of PEGylation to reduce immunogenicity is related to the size and type of PEG, and chemistry used for modification, factors that to some extent can be controlled from protein to protein. However, suppression of immunogenicity and antigenicity may also be related in a protein-specific manner to the distribution of PEG-attachment sites (usually primary amino groups) relative to epitopes on the protein surface. Our studies of the antibody-mediated inhibition of PEG-ADA suggest such specificity (15). We found that ADA-inhibitory antisera directly inhibited both native and PEGylated bovine ADA, indicating a binding site close to the ADA active site (a deep pocket seen in the crystal structure of murine ADA (22)), and relatively free of bound PEG. Interestingly, these antisera inhibited native human ADA, but not PEGylated human ADA. Thus, human ADA may possess a site for PEGylation (a lysine residue) close to the active site, which is apparently not present near the bovine ADA active site (15).

The above results suggested that "clearing" or "neutralizing" antibodies to PEGylated enzymes may be directed to regions relatively free of PEG-attachment sites. We have tested this hypothesis in mice by asking whether introducing new lysine residues could enhance the ability of PEGylation to diminish the immunogenicity of a foreign protein. As a model, we used *E. coli* purine nucleoside phosphorylase (EPNP) (23). We used directed mutagenesis to substitute lysine for arginine codons in the gene for EPNP, hoping to maintain positive charge and minimize effects on structure and antigenicity (at the time, the crystal structure of EPNP was not known). We were able to introduce 3 such substitutions (at residues 38, 136, and 208) without affecting enzymatic activity of EPNP (23).

The purified triple-mutant EPNP retained full activity after modifying ~70% of accessible amino groups with excess 5-kDa SS-PEG. PEGylation increased the circulating life of both the wild type and mutant EPNP enzymes in mice from ~4 hours to >6 days. After 10 intraperitoneal injections at 7-14 day intervals, all mice treated with the unmodified enzymes, and 10 of 16 mice (60%) injected with PEGylated wild type EPNP, developed high levels of anti-PNP antibody and rapid clearance. In contrast, only 2 of 12 mice (17%) treated with the triple mutant PEG-EPNP developed rapid clearance; low levels of antibody in these mice did not correlate with circulating life (as in most patients treated with PEG-ADA). Antisera from animals treated with wild type PEG-EPNP did not react with the mutant PEG-enzyme (23). In theory, this strategy could be used to develop less immunogenic PEG-modified forms of some other proteins.

Summary

In the case of PEG-ADA (ADAGEN™), PEGylation has proved to be an effective strategy for prolonging enzyme circulating life and has enabled effective enzyme replacement therapy for a fatal metabolic disease. Although immunogenicity has not been eliminated, anti-ADA antibodies have usually had little effect on clearance or efficacy. These results suggest the importance of periodic monitoring of circulating levels of PEG-proteins to permit their safe and effective use in long-term therapy.

Acknowledgements

Supported by grants from Enzon, Inc., and the National Institutes of Health (DK 20902)

Abbreviations Used

ADA, adenosine deaminase; SCID, severe combined immunodeficiency disease; Ado, adenosine; dAdo, 2'-deoxyadenosine; dATP, 2'-deoxyadenosine triphosphate; dAXP, total 2'-deoxyadenosine nucleotides; SAHase, S-adenosyl-homocysteine hydrolase; PEG, poly(ethylene glycol).

Literature Cited

1. Abuchowski, A.; Van Es, T.; Palczuk, N.C.; Davis, F.F. *J Biol Chem* **1977**, *252*, 3578-3581.
2. Abuchowski, A.; McCoy, J.R.; Palczuk, N.C.; Van Es, T.; Davis, F.F. *J Biol Chem* **1977**, *252*, 3582-3586.
3. Davis, S.; Abuchowski, A.; Park, Y.K.; Davis, F.F. *Clin Exp Immunol* **1981**, *46*, 649-652.
4. Giblett, E.R.; Anderson, J.E.; Cohen, F.; Pollara, B.; Meuwissen, H.J. *Lancet* **1972**, *2*, 1067-1069.
5. Hershfield, M.S.; Mitchell, B.S. In *The Metabolic and Molecular Bases of Inherited Disease*; 7 th ed.; C.R. Scriver; A.L. Beaudet; W.S. Sly; D. Valle, Ed.; McGraw-Hill: New York, **1995**; pp 1725-1768.
6. Benveniste, P.; Cohen, A. *Proc Natl Acad Sci USA* **1995**, *92*, 8373-8377.
7. Kredich, N.M.; Martin, D.W., Jr *Cell* **1977**, *12*, 931-938.
8. Hershfield, M.S. *J Biol Chem* **1979**, *254*, 22-25.
9. Polmar, S.H. In *Enzyme Defects and Dysfunction, Ciba Foundation Symposium*; K. Elliot; J. Whelan, Ed.; Excerpta Medica: New York, **1979**; Vol. 68; pp 213-230.
10. Hershfield, M.S.; Buckley, R.H.; Greenberg, M.L.; Melton, A.L.; Schiff, R.; Hatem, C.; Kurtzberg, J.; Markert, M.L.; Kobayashi, R.H.; Kobayashi, A.L.; Abuchowski, A. *N Engl J Med* **1987**, *316*, 589-596.
11. Hershfield, M.S.; Chaffee, S.; Sorensen, R.U. *Pediatrics Research* **1993**, *33 (Suppl)*, S42-S48.
12. Hershfield, M.S. *Clin Immunol Immunopathol* **1995**, *76*, S228-S232.
13. Zalipsky, S.; Lee, C. In *Poly(Ethylene Glycol) Chemistry: Biotechnical and Biomedical Applications*; J.M. Harris, Ed.; Plenum: New York, **1992**; pp 347-370.
14. Hershfield, M.H.; Kredich, N.M. *Proc Natl Acad Sci USA* **1980**, 77, 4292-4296.
15. Chaffee, S.; Mary, A.; Stiehm, E.R.; Girault, D.; Fischer, A.; Hershfield, M.S. *J Clin Invest* **1992**, *89*, 1643-1651.
16. Hershfield, M.S.; Kredich, N.M.; Koller, C.A.; Mitchell, B.S.; Kurtzberg, J.; Kinney, T.R.; Falletta, J.M. *Cancer Res* **1983**, *43*, 3451.
17. Hershfield, M.S.; Kurtzberg, J.; Moore, J.O.; Whang Peng, J.; Haynes, B.F. *Proc Natl Acad Sci USA* **1984**, *81*, 253-257.
18. Weinberg, K.; Hershfield, M.S.; Bastian, J.; Kohn, D.; Sender, L.; Parkman, R.; Lenarsky, C. *J Clin Invest* **1993**, *92*, 596-602.

19. Hirschhorn, R. *Pediatr Res* **1993**, *33 (Suppl)*, S35-S41.
20. Ozsahin, H.; Arredondo-Vega, F.X.; Santisteban, I.;
    Fuhrer, H.; Tuchschmid, P.; Jochum, W.; Aguzzi, A.;
    Lederman, H.M.; Fleischman, A.; Winkelstein, J.A.;
    Seger, R.A.; Hershfield, M.S. *Blood* **1997**, *in Press*,
21. Chun, J.D.; Lee, N.; Kobayashi, R.H.; Chaffee, S.;
    Hershfield, M.S.; Stiehm, E.R. *Ann Allergy* **1993**, *70*,
    462-466.
22. Wilson, D.K.; Rudolph, F.B.; Quiocho, F.A. *Science*
    1991, *252*, 1278-1284.
23. Hershfield, M.S.; Chaffee, S.; Koro-Johnson, L.; Mary,
    A.; Smith, A.A.; Short, S.A. *Proc Natl Acad Sci USA*
    1991, *88*, 7185-7189.

Chapter 11

Conjugation of High-Molecular Weight Poly(ethylene glycol) to Cytokines: Granulocyte–Macrophage Colony-Stimulating Factors as Model Substrates

Merry R. Sherman[1], L. David Williams[1], Mark G. P. Saifer[1], John A. French[1], Larry W. Kwak[2], and Joost J. Oppenheim[2]

[1]Mountain View Pharmaceuticals, Inc., 871–L Industrial Park, Menlo Park, CA 94025
[2]National Cancer Institute, Frederick Cancer Research and Development Center, Frederick, MD 21702

The ability of the small receptor-binding protein, recombinant murine granulocyte-macrophage colony-stimulating factor (GM-CSF), to increase the abundance of certain blood cell types in mice was enhanced markedly by covalent attachment of a single long strand of PEG (30-40 kDa). Potency was not increased further by coupling a second strand. Such conjugates can be synthesized efficiently by reaction of protein amino groups with PEG propionaldehydes in the presence of $NaBH_3CN$ or with PEG p-nitrophenyl carbonates. Both methods have been used to prepare recombinant human GM-CSF conjugates of pre-determined composition, $e.g.$ PEG_1GM-CSF and PEG_2GM-CSF, in high yield. These compounds, or analogous derivatives of other cyto-kines, purified by ion-exchange and size-exclusion chromatography, may be suitable candidates for pharmaceutical development.

Many therapeutic proteins have been coupled to poly(ethylene glycol) (PEG)* in order to prolong their circulating life-times and increase their potencies *in vivo* (reviewed in *1-3*). The premise underlying the present research is that the activities of small proteins that function by interacting with receptors on cell membranes may not be enhanced optimally by modification with 5-kDa PEG under the usual conditions. These proteins have molecular weights (M_r) in the range of 10-30 kDa, based on their amino acid sequences, and are generally glycosylated *in vivo*. The nature, extent, and pattern of glycosylation of their recombinant homologues depend on the organism in which they are expressed. Cytokines, which function as intercellular signals among white blood cells and their precursors, represent a physiologically important group of such receptor-binding (glyco)proteins (reviewed in *4, 5*). The steric and chemical requirements for selective, high-affinity interactions of these small proteins with their receptors are not compatible with covalent modification by PEG (PEGylation) at many sites. Thus, the technique of attaching a large number of strands of 5-kDa or 6-kDa PEG, which has been applied successfully to dozens of larger proteins, particularly enzymes that act on small substrates, is unlikely to enhance, or even to preserve, the efficacy of most cytokines.

Effects of the Number and Size of Attached PEG Strands. In several successful efforts to enhance the activities of enzymes, of which L-asparaginase (6), adenosine deaminase (7), and Cu,Zn-superoxide dismutase (SOD; 8) are well-known examples, the increased plasma half-life resulting from the attachment of many strands of 5-kDa PEG more than compensated for the partial loss of catalytic activity observed *in vitro* (reviewed in 1-3). The use of fewer strands of larger PEG has been explored by only a few research groups (9-15). For example, studies of PEG conjugates of both bovine and recombinant human SOD showed that two to five strands of PEG with $M_r > 30$ kDa were more effective than seven or 15 strands of 5-kDa PEG in preserving the biological activity, increasing the plasma persistence, and reducing the antigenicity and immunogenicity of these enzymes (13).

Attempts to engineer more potent forms of cytokines and other small receptor-binding proteins by PEGylation have met with mixed success (1, 9-11, 14-20). Studies of PEG conjugates of interleukin-2 (IL-2) (11), granulocyte colony-stimulating factor (G-CSF) (14), tumor necrosis factor α (18), and human growth hormone (hGH) (19) have demonstrated that the coupling of several strands of PEG prolonged their plasma half-lives substantially. For example, Niven *et al.* (15) reported that coupling the amino terminus of recombinant human G-CSF (rhG-CSF) to one strand of 6-kDa or 12-kDa PEG extended its plasma half-life in rats from one hour to two or four hours, respectively. On the other hand, an inverse relationship has been observed between the number of attached PEG strands and the receptor-binding affinities of several small proteins, measured *in vitro* (18, 19). In contrast with the results obtained with several enzymes, the most extensively PEGylated preparations of cytokines and peptide hormones have not proven to be the most active *in vivo* (19, 20). For example, Satake-Ishikawa *et al.* (14) found that G-CSF conjugates containing an average of two to three strands of 10-kDa PEG displayed longer duration of activity and higher potency *in vivo* than conjugates containing an average of five strands of 5-kDa PEG, despite the larger size of the latter forms, inferred from their electrophoretic mobilities in polyacrylamide gels. In other words, bigger PEG-cytokine conjugates are not necessarily better!

Objective of This Study. The research reported here was undertaken to test the proposal that the potency of certain cytokines could be optimized by their covalent linkage to a small number of PEG strands of high molecular weight (at least 18 kDa). This size was selected as the minimum because, as shown below, the molecular radius of 18-kDa PEG exceeds that of serum albumin, which is effectively retained by the kidneys (21, 22). While the rates of clearance of proteins from the plasma by mammalian kidneys depend on their molecular weights, with a fairly sharp cut-off near 70 kDa (about the size of serum albumin), the filtration of PEGs into the urine has a shallower dependence on M_r (23). In addition, the clearance of PEGs by the liver has a molecular weight dependence resulting in minimal accumulation of PEGs with M_r in the range of 20-50 kDa (23). For these reasons, the present research has focused on conjugates prepared from monomethoxy, monohydroxy PEGs with M_r of 19-42 kDa.

We have explored three alternative chemical approaches for coupling one or a few long strands of PEG to several cytokines, starting with recombinant murine granulocyte-macrophage colony-stimulating factor (rmuGM-CSF). The discovery of GM-CSF was based on its profound stimulation of the proliferation of white blood cells of the monocytic lineage in cell culture (reviewed in 4). Because of the transient half-life of the unmodified cytokine *in vivo*, demonstration of its effects on peripheral blood cell counts in laboratory animals required repeated administration of high doses. This chapter includes results obtained using two conjugation chemistries to attach PEGs with M_r of 19-42 kDa to recombinant human GM-CSF (rhGM-CSF), and contrasts them with results obtained with 5-kDa PEG. Data are also presented to demonstrate that the conjugates formed by coupling a single strand of 36-kDa PEG to rmuGM-CSF promote dramatic increases in the abundance of several types of white blood cells in the peripheral circulation of mice.

Granulocyte-Macrophage Colony-Stimulating Factors as Model Substrates for PEG Conjugation. During the decades since the first descriptions of the actions *in vitro* of the cell lineage-selective colony-stimulating factors (CSFs) found in conditioned cell culture media (reviewed in *4*), knowledge regarding these proteins has increased exponentially. This has included the elucidation of their amino acid sequences, patterns of glycosylation, structures, receptors, signal transduction pathways, and physiological actions, particularly in cell culture and experimental animals (reviewed in *4, 24-27*). Despite the potential utility of these factors in the management of a wide variety of diseases, progress in implementing their therapeutic use has been impeded, in most cases, by the high cost of producing recombinant human CSFs, their rapid clearance following administration by various routes, and several serious side effects (*5, 28-32*).

Many academic laboratories and pharmaceutical companies have sought to overcome these problems by PEGylation of cytokines. The selection of rmuGM-CSF as the first model compound for our studies of cytokines coupled to high molecular weight PEG was based, in part, on the availability of a well-characterized laboratory model in which to test the potency of the conjugates *in vivo* (*33*). Another attractive feature of this cytokine is that, for a protein of its size, it has a relatively large number of potential sites of PEGylation: the amino-terminal alanine and 11 lysine residues (*34*). Our initial results with rmuGM-CSF included the demonstration that the covalent attachment of one or two strands of 36-kDa PEG via a urethane linkage increased its potency *in vivo* by more than an order of magnitude compared with the unmodified protein (*9, 10*). The resultant conjugates induced dose-dependent increases in the peripheral blood cell counts of all types of cells that had been shown previously to respond to GM-CSF *in vitro* (*9, 10*). Such effects on peripheral cell counts have not been reported in previous studies with conjugates of rhGM-CSF prepared by another PEGylation technique (*1, 16, 17*). Our successful preliminary studies with murine GM-CSF led us to extend this approach to recombinant human GM-CSF, which differs importantly in having only six lysine residues (*35*), and to recombinant murine stem cell factor.

Methods

Materials. Recombinant human GM-CSF (rhGM-CSF), expressed in yeast, was a gift from Immunex Corporation (Seattle, WA). Recombinant murine GM-CSF (rmuGM-CSF) and recombinant murine stem cell factor, both expressed in *Escherichia coli*, were from PeproTech (Rocky Hill, NJ). Monomethoxy PEGs of various molecular weights (19, 36, and 42 kDa) were from Polymer Laboratories, Ltd. (Church Stretton, Shropshire, UK). These preparations were characterized by very narrow polydispersities, *i.e.* the ratios of weight average to number average molecular weights (M_w/M_n) were less than 1.06. Their contamination with dihydroxy PEG (PEG diol) did not exceed 2%, which is the lower limit of quantitation by our current methods. Monopropionaldehyde derivatives of monomethoxy PEG (PEG aldehydes) were from Shearwater Polymers, Inc. (Huntsville, AL) or Polymer Laboratories. PEG standards for column calibration were from Polymer Laboratories and had the following values of M_r: 1.56, 4.82, 10.0, 18.3, 32.6, 50.1, 73.4, 120, 288, 448, 646, and 1,390 kDa.

The following proteins were used for calibration of size-exclusion chromatography columns (values for M_r and corresponding references are indicated): equine cytochrome c, 12,360 (*36*); equine myoglobin, 17,641 (*37*); bovine carbonic anhydrase, 29,000 (*38*); bovine SOD, 31,200 (*39*); bovine serum albumin (BSA), 66,430, and BSA dimer, 132,860 (*40*); equine liver alcohol dehydrogenase, 79,700 (*41*); human immunoglobulin G, 150,000 (*42*); equine ferritin, 476,000, and ferritin dimer, 952,000 (*43*), and bovine thyroglobulin, 620,000 (*44*). These proteins were from Sigma (St. Louis, MO) except for BSA, which was from Calbiochem-

Novabiochem (San Diego, CA), and SOD, which was from Oxis International, Inc. (Portland, OR). Sodium cyanoborohydride, p-nitrophenyl chloroformate, and most other reagents were from Aldrich Chemical Company (Milwaukee, WI).

Preparation and Coupling of PEG Derivatives. The attributes of various "activated" (*i.e.* electrophilic) derivatives of PEG, including three that we have coupled to GM-CSF, have been discussed in several recent reviews (*1-3, 45-47*). In the case of cytokines, the use of highly efficient coupling techniques is mandated by the high cost of the proteins as well as the cost of the purest available PEGs and PEG derivatives. Succinimidyl carbonate derivatives of monomethoxy PEG (SC-PEGs) were prepared and coupled to cytokines by modifications of published methods (*48, 49*). These modifications were necessitated by the small amounts of cytokines available for conjugation (*e.g.* <1 mg of stem cell factor). For the coupling of SC-PEG to rmuGM-CSF and stem cell factor, the input molar ratio of SC-PEG to protein was 2.5:1. Methods for cytokine PEGylation by reductive alkylation using PEG aldehydes and $NaBH_3CN$ were adapted from those of Friedman *et al.* (*50*). Related methods have been described by Chamow *et al.* (*51*) and have been applied to G-CSF (*15*) and to brain-derived neurotrophic factor (*52*), among other small receptor-binding proteins.

The use of p-nitrophenyl chloroformate to activate PEG is based on methods previously applied to the synthesis of affinity chromatography matrices (*53*). p-Nitrophenyl carbonate derivatives of PEG (pNPC-PEGs) have been used previously in the synthesis of PEG conjugates of various proteins including SOD (*54*), horseradish peroxidase (*55*), and a chimeric toxin composed of *Pseudomonas* exotoxin and transforming growth factor α (*56*). Various reaction conditions and input molar ratios of pNPC-PEG to GM-CSF (from 1.5:1 to 12:1) were tested in this study.

Fractionation and Characterization of PEGylated GM-CSF. Mixtures of proteins PEGylated to different extents can be fractionated chromatographically on the basis of size, hydrophobicity and/or ion exchange properties (*12, 57-59*). Each strand of high molecular weight PEG provides sufficient shielding of charges on the protein that the relative retention times on an ion exchange column of components in a PEGylated preparation of GM-CSF are: free PEG<PEG_2GM-CSF<PEG_1GM-CSF< GM-CSF. This sequential elution was used in the fractionation of PEG conjugates of GM-CSF on analytical or preparative Mono Q columns (Pharmacia, Piscataway, NJ). Other preparations were fractionated on DEAE Sepharose CL-6B columns (Pharmacia). Aliquots of PEGylation reaction mixtures or fractions from the Mono Q or Sepharose columns were characterized by size-exclusion high performance liquid chromatography (HPLC) on a TSK 5000 PW$_{XL}$ column or on TSK 4000 PW$_{XL}$ and TSK 2500 PW$_{XL}$ columns in series (TosoHaas, Montgomeryville, PA) at room temperature in 100 mM sodium phosphate buffer, pH 6.8, containing 100 mM NaCl. Proteins and PEGs were monitored by ultraviolet absorbance and/or refractive index detectors. The elution profiles were analyzed using the program PeakFit (Jandel Scientific, San Rafael, CA) and the results were interpreted with reference to standard curves for both protein and PEG molecular weights.

The fractions of PEG molecules linked to the amino-terminal alanine and to lysine residues in preparations of rhGM-CSF that were PEGylated by reductive alkylation were inferred from the results of amino acid analyses performed by Analytical Biotechnology Services (Boston, MA).

Evaluation of the Potency of PEGylated Murine GM-CSF in Mice. Methods used to assess the hematologic effects of PEGylated rmuGM-CSF in normal mice were adapted from Metcalf *et al.* (*33*). Briefly, groups of 10 female BALB/c mice were injected *i.p.* twice daily for six days with vehicle or with unmodified murine GM-CSF or PEG conjugates containing 100, 200 or 400 ng of GM-CSF protein. On the seventh and eighth days, retroorbital blood samples (*c.* 200 μl) were drawn and on the eighth day the mice were sacrificed and their spleens were weighed. Numbers of platelets and the following types of blood cells were determined by Maryland MetPath (Baltimore, MD): red cells; total white blood cells; total, normal and "activated" lymphocytes (the latter characterized by enlarged nuclei); polymorphonuclear (PMN) neutrophils and bands (immature neutrophils); eosinophils; basophils, and monocytes.

Results

Size-exclusion Chromatography of PEGs and Proteins. The size-exclusion HPLC columns were calibrated using both PEG and protein standards, and the data for log M_r as a function of retention time were fit to cubic equations. For 12 PEG standards with M_r of 1.6-1390 kDa, the fit was virtually perfect, with a correlation coefficient (r^2) of 1.000 (Figure 1). Analogous data for 11 protein standards with M_r of approximately 12-952 kDa fit less well to a cubic equation ($r^2 = 0.983$), since the molecular radii of proteins are influenced by factors other than M_r, particularly the shape and degree of hydration (*60*). The two standard curves were not parallel and diverged as the molecular weight increased. For example, the retention time of the pre-

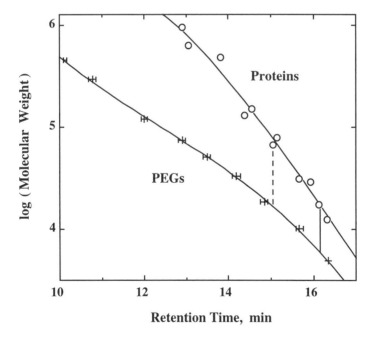

Retention Time, min

Figure 1. Calibration curves for PEG and protein standards on a size-exclusion HPLC column. Mean retention times ± s.d. are shown for nine PEG standards ($M_r = 4.8$-448 kDa). The 11 protein standards (○; $M_r = 12$-952 kDa) were cytochrome *c*, myoglobin, carbonic anhydrase, Cu, Zn-superoxide dismutase, bovine serum albumin (BSA), liver alcohol dehydrogenase, BSA dimer, immunoglobulin G, ferritin, thyroglobulin, and ferritin dimer. The retention time of BSA ($M_r = 66.4$ kDa) corresponds to an apparent M_r of 17 kDa on the PEG calibration curve (*dashed line*). The retention time of the peak in recombinant human GM-CSF corresponds to 17 kDa on the protein calibration curve and an apparent M_r of 6 kDa on the PEG calibration curve (*solid vertical line*).

dominant form of rhGM-CSF is nearly identical to that of myoglobin (M_r = 17,641; *37*), but corresponds to a molecular weight of only 6 kDa on the PEG calibration curve. In other words, these small proteins would coelute with PEGs that have molecular weights about one-third as large as theirs. For a very large protein, *e.g.* thyroglobulin, the discrepancy between its molecular weight and that of coeluting PEG species would be even more pronounced, corresponding to a ratio of approximately 9:1. The major virtue of this analytical method is that the range of molecular sizes it can resolve extends from those of the smallest unmodified cytokines to those of the multiply PEGylated conjugates, even when very large PEGs are coupled.

Kinetics of *p*-Nitrophenyl Carbonate-PEG Conjugation to GM-CSF. During 24 hours of incubation of rhGM-CSF with 19-kDa pNPC-PEG at 0 - 4 °C, the size of the predominant conjugates gradually increased as an increasing fraction of the rhGM-CSF was PEGylated (Figure 2). After two hours of incubation with an initial molar ratio of PEG to protein of 2.5:1, nearly half of the protein had been converted to mono-PEGylated GM-CSF (PEG$_1$GM-CSF). After 6.5 hours, PEG$_1$GM-CSF still constituted approximately half of the total protein, but more PEG$_2$GM-CSF had formed and the quantity of unmodified protein had decreased. After a further addition of pNPC-PEG to increase the total molar ratio of PEG to protein to 4:1, and incubation for a total of 24 hours, nearly all of the GM-CSF had been PEGylated and the predominant

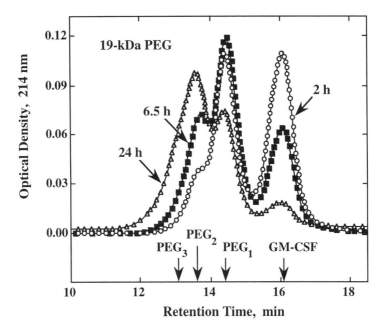

Figure 2. Sequential formation of rhGM-CSF conjugates containing one, two or three strands of PEG (*PEG$_1$-*, *PEG$_2$-* or *PEG$_3$GM-CSF*) by reaction with a *p*-nitrophenyl carbonate derivative of 19-kDa PEG (pNPC-PEG). Portions of a reaction mixture containing pNPC-PEG and rhGM-CSF at an initial molar ratio of 2.5:1 were analyzed by size-exclusion HPLC after incubation at 0-4 °C for 2 h (○) or 6.5 h (■). After the addition of pNPC-PEG to give a total molar ratio of 4:1, incubation was continued for a total of 24 h (△).

product was $PEG_2GM\text{-}CSF$ (Figure 2). Analysis of these elution patterns using the program PeakFit revealed that a small amount of $PEG_3GM\text{-}CSF$ had also formed in 24 hours under these conditions.

Effect of the Ratio of PEG Aldehyde to GM-CSF on Adduct Formation. Incubation of rhGM-CSF with aldehyde derivatives of PEG produced stable conjugates containing small numbers of strands of PEG only in the presence of the reducing agent, sodium cyanoborohydride ($NaBH_3CN$) (*10*). The number of strands of PEG coupled per molecule of rhGM-CSF, evaluated by size-exclusion HPLC, increased with the input molar ratio of PEG aldehyde to protein, which ranged from 1.5:1 to 6:1 in the studies shown in Figure 3. The observed retention times of the mono-PEGylated and di-PEGylated conjugates ($PEG_1GM\text{-}CSF$ and $PEG_2GM\text{-}CSF$) formed with 42-kDa PEG aldehyde were within two seconds of the retention times calculated from the standard curves shown in Figure 1.

During incubation of either 5-kDa or 42-kDa PEG aldehyde with rhGM-CSF and $NaBH_3CN$ for two weeks at $0\text{-}4\,^\circ C$ at a 6:1 molar ratio of PEG to protein, virtually all of the GM-CSF was PEGylated and the predominant conjugates contained two strands of PEG per molecule of protein. The relatively poor resolution among the conjugates formed with 5-kDa PEG precluded reliable quantitation of the extent of PEG modification, as shown in Figure 3 (top). In contrast, conjugates formed with 42-kDa PEG were clearly resolved from unmodified GM-CSF and partially resolved from each other by chrotography on the TSK 5000 PW_{XL} column (Figure 3, bottom). PeakFit analysis of these patterns indicated that approximately 25% and 50% of the GM-CSF molecules were present as $PEG_1GM\text{-}CSF$ and $PEG_2GM\text{-}CSF$, respectively. At lower input ratios of PEG to protein, *e.g.* 1.5:1 or 3:1, there was some residual unmodified protein and the predominant conjugates contained only one strand of PEG. With an input molar ratio of 1.5:1, approximately 80% of the PEGylated species consisted of $PEG_1GM\text{-}CSF$. Qualitatively similar results were obtained when rmuGM-CSF was coupled to 36-kDa SC-PEG (see Figure 4).

Hematologic Effects of PEGylated rmuGM-CSF in Mice. Recombinant murine GM-CSF was PEGylated by incubation for 2 hours at room temperature with 36-kDa SC-PEG at an input molar ratio of 2.5:1. Fractions obtained by chromatography on DEAE-Sepharose in isotonic buffer were analyzed by size-exclusion HPLC and combined into pools containing predominantly mono-PEGylated GM-CSF (PEG_1 Pool) or predominantly di-PEGylated GM-CSF (PEG_2 Pool), as shown in Figure 4.

Solutions of unmodified or PEGylated murine GM-CSF containing 200 ng of GM-CSF protein were injected into mice according to the protocol of Metcalf *et al.* (*33*). Under these conditions, unmodified GM-CSF caused no statistically significant increases in peripheral blood counts of any cell type, including PMN neutrophils, eosinophils or lymphocytes (hatched bars in Figure 5), or in the numbers of monocytes (*10*) or platelets (unpublished results). As observed previously (*33*), the unmodified cytokine had a dramatic effect on spleen weight, causing a 70% increase over the mean value in vehicle-treated mice (open bars in Figure 5). $PEG_1GM\text{-}CSF$ (light gray bars) and $PEG_2GM\text{-}CSF$ (dark gray bars) were similarly effective to each other in increasing the cell counts and spleen weight, and each was far more effective than unmodified GM-CSF. In mice treated with either $PEG_1GM\text{-}CSF$ or $PEG_2GM\text{-}CSF$, the increases in the numbers of circulating lymphocytes, PMN neutrophils, and eosinophils were approximately 3-fold, 7-fold, and 10-fold, respectively, relative to the corresponding cell counts in vehicle-treated mice. The numbers of circulating monocytes, platelets, and activated lymphocytes were also increased significantly after treatment with $PEG_1GM\text{-}CSF$ or $PEG_2GM\text{-}CSF$ (unpublished results). In addition, administration of $PEG_1GM\text{-}CSF$ or $PEG_2GM\text{-}CSF$ increased the spleen weight approximately 60% above the elevated weight observed after treatment with unmodified GM-CSF.

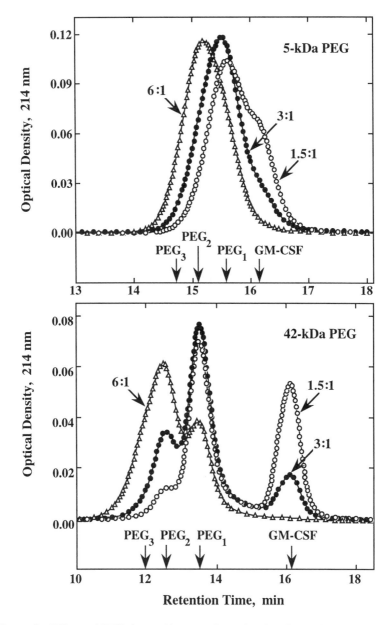

Figure 3. Effects of PEG size and input molar ratio of PEG to rhGM‐CSF on the
conjugates formed and resolution among them. Products of 2-week incubations of
rhGM‐CSF, NaBH$_3$CN, and either 5‐kDa PEG aldehyde (*top*) or 42‐kDa PEG
aldehyde (*bottom*) at PEG:protein molar ratios of 1.5:1 (O), 3:1 (●) or 6:1 (Δ)
were analyzed by size-exclusion HPLC. Vertical arrows indicate the mean retention
time of unmodified rhGM‐CSF and the calculated retention times of conjugates
containing one, two or three strands of PEG (*PEG$_1$*-, *PEG$_2$*- or *PEG$_3$GM‐CSF*).

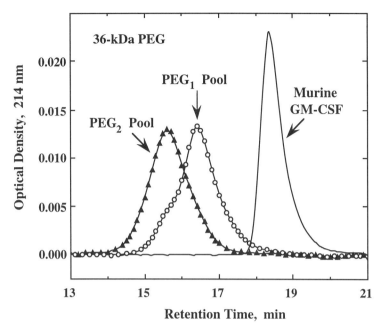

Figure 4. Size-exclusion HPLC analysis of the PEGylated rmuGM-CSF prepara-tions tested in mice (see Figure 5). Conjugates were prepared by incubation of GM-CSF with 36-kDa SC-PEG and fractionated by chromatography on DEAE-Sepharose. Optical density (OD) data for unmodified rmuGM-CSF (——) and for pools containing predominantly PEG$_1$GM-CSF (o) or PEG$_2$GM-CSF (▲) were normalized to give equal areas under the OD-time curves.

Discussion

Chromatographic Resolution and Characterization of PEG-Protein Con-jugates. High performance size-exclusion chromatography has been found to be a reliable method for characterizing and quantitating PEG-protein conjugates containing small numbers of PEG strands, as long as each strand of PEG is sufficiently large to alter the radius of the conjugate significantly (*12*). The results reported here confirm and extend those findings. In contrast, when the hydrodynamic radii of the PEG strands are smaller than those of the proteins to which they are coupled, as is generally the case when 5-kDa PEG is used, more laborious methods are required to quantitate the various conjugates (*16, 17*). As a corollary to these analytical results, the resolving power of preparative-scale size-exclusion chromatography is far greater for conjugates of proteins with high molecular weight PEGs than for conjugates with 5-kDa PEG.

The identification of cytokine species PEGylated to different extents was facil-itated by calibration of the columns with respect to M$_r$ of both PEGs and proteins and the use of the program PeakFit. The elution patterns from these columns are deter-mined primarily by the effective molecular radii of the components, but may be influenced by adsorption to the column matrix (*61*). Since the PEG standards are homologous molecules, data for the PEG standards fit a smooth curve more closely than data for this particular selection of proteins. Nevertheless, it is clear from the results in Figure 1 that the retention times of proteins in the molecular weight range studied (12-952 kDa) are similar to those of PEGs of much lower molecular weight.

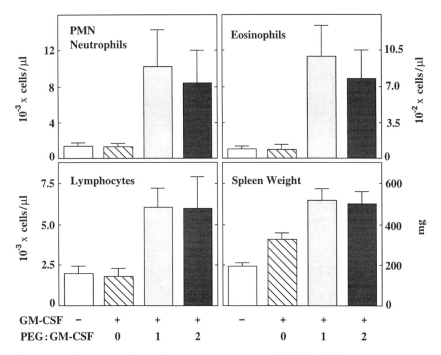

Figure 5. Dramatic effects of one or two strands of 36-kDa PEG on hematologic responses to rmuGM‑CSF in mice. On Days 1‑6, mice were injected twice daily with vehicle (*open bars*) or with 200 ng of GM‑CSF that was unmodified (*hatched*), coupled to one strand of PEG (*light gray*) or coupled to two strands of PEG (*dark gray*). Mean values ± s.d. ($n = 10$) are shown for the numbers of three types of white cells per microliter of blood sampled on Day 7 and for spleen weight on Day 8.

The pharmacokinetics and tissue distributions of PEGs and proteins are profoundly affected by their sizes in aqueous solution. The disproportionate increase in the effective molecular radii of PEGs of high molecular weight, reflected in the diverging curves in Figure 1, makes it possible to produce conjugates of cytokines with very large radii by coupling a small number of strands of PEG of moderately high molecular weight (*e.g.* 18 kDa). While it is known that molecular radius is only one of several factors that determine the rate of renal clearance (*21*), the results in Figure 1, as well as the data of Yamaoka *et al.* (*23*), suggest that the addition of one or two strands of 18-kDa PEG would increase the molecular radius of a cytokine such as GM‑CSF sufficiently that its clearance by the kidneys would be retarded substantially.

Comparison of Three Conjugation Chemistries. The rates of coupling of GM‑CSF and other proteins to the three types of activated PEG used in this study decrease in the order: succinimidyl carbonate > *p*-nitrophenyl carbonate >> aldehyde. When both the protein and the activated PEG are relatively stable under the coupling conditions, the slowest method permits the closest monitoring, and hence the most precise regulation of the degree of modification. For example, gradual coupling facilitates the optimization of the formation of either the mono-PEGylated product

(PEG$_1$GM-CSF) or the di-PEGylated product (PEG$_2$GM-CSF). This situation is illustrated by the results in the lower panel of Figure 3, obtained by incubating 42-kDa PEG aldehyde with rhGM-CSF for two weeks at 0-4°C at an input molar ratio of 1.5:1, 3:1 or 6:1. Based on amino acid analyses and additional data, including peptide maps (unpublished results), the first molecule of PEG attached under these conditions was linked nearly exclusively to the amino-terminal alanine residue. Assessment of the activities of these products *in vitro* and *in vivo* is in progress.

The data in Figure 2 illustrate the more rapid attachment of pNPC-PEG to GM-CSF. Under the coupling conditions used, the activated PEG is hydrolyzed to release *p*-nitrophenol (detectable by its yellow color) during several hours at 0-4°C. If the desired product is PEG$_1$GM-CSF, then a 6- or 7-hour incubation of pNPC-PEG with GM-CSF at an input molar ratio of 2.5:1 is optimal. On the other hand, the formation of PEG$_2$GM-CSF is favored by incubation for 24 hours at a molar ratio of 4:1.

Practical Advantages of Coupling Long Strands of PEG. In contrast to the coupling of cytokines to 5-kDa PEG (*1; Figure 3*, top), an advantage of coupling cytokines and other small proteins to high molecular weight PEG is the possibility of producing well-characterized conjugates of predetermined composition in high yield (see Figures 2, 3 bottom, and 4). From the standpoints of the cost-effectiveness of their synthesis, purification, and analysis, as well as their potential regulatory acceptability, mono-PEGylated or di-PEGylated conjugates that are active and stable *in vivo* are preferable to the heterodisperse mixtures obtained by coupling smaller PEGs to multiple sites on the protein.

Factors that Modulate the Clearance Rates of Proteins and PEGs. The attachment of PEG may retard several processes involved in the removal of proteins from the circulation, including receptor binding, proteolysis, immune complex formation, and filtration through the renal glomeruli. While anionic proteins that are at least as large as serum albumin are not generally subject to glomerular filtration by healthy kidneys (*21*), smaller proteins such as G-CSF (*62*), ribonuclease, various protein hormones, and Bence-Jones proteins (immunoglobulin fragments) are cleared rapidly from the circulation by the kidneys (*22*). Similarly, short strands of PEG are cleared more rapidly than longer strands. For example, 6-kDa PEG has a terminal half-life in mice of <10 minutes (*63*), while 20-kDa PEG has a terminal half-life of about 3 hours (*23*). It can be inferred from previous studies of neutral, anionic, and cationic forms of dextrans and derivatives of a model protein (horseradish peroxidase) (*64-66*), that the glomerular filtration rates of free PEGs and PEG-protein conjugates may be influenced by their charges and deformability, as well as by their molecular radii. The deformability of PEG is particularly relevant to the ability of high molecular weight PEGs, PEGylated forms of small proteins, and their proteolytic cleavage products to be cleared slowly by the kidneys, despite the fact that their apparent molecular radii, as determined by size-exclusion chromatography, may be much larger than that of serum albumin.

While the persistence of cytokines or peptide hormones in plasma is prolonged to some extent by the attachment of one strand (*15*) or a few strands of 5-kDa PEG (*67*), the half-lives of the complexes increase further as more PEG strands are coupled (*11, 18, 19*), up to at least four strands for PEG-IL-2 (*20*), five strands for PEG-G-CSF (*14*), and at least seven strands for PEG-hGH (*19*). The further increments in plasma persistence when additional strands of PEG are coupled to conjugates that are already larger than albumin may be explained, in part, by the deformability of PEG.

Optimization of the Composition of PEG-Cytokine Conjugates. As discussed above, the attachment of one strand of 18-kDa PEG to a cytokine or peptide hormone is expected to decrease its rate of clearance from the plasma markedly. In contrast, the coupling of one or two strands of 5-kDa PEG to small proteins that have molecular radii similar to that of 5-kDa PEG has been shown to produce conjugates that are smaller than serum albumin and are cleared from the plasma fairly rapidly (*11, 15, 19*). Since the attachment of PEG to many sites on these proteins may inhibit their ability to bind to receptors and to mediate the subsequent steps in signal transduction, it is difficult to achieve the ideal balance between the retention of intrinsic activity and the extension of plasma half-life by coupling many short strands of PEG.

Insights into the structure-function relationships of some cytokines may be provided by a comparison of the physiological activity of the amino-terminally PEGylated derivative, which is preferentially formed with PEG aldehyde at low pH, with that of a mixture of mono-PEGylated conjugates in which PEG is linked to either the amino terminal or one of the lysine residues. While the former is a unique species, it may be either more active or less active than a mixture of PEGylated isomers with the same average degree of modification. The relative activities of these products would depend on the proximity of the site(s) of PEG attachment to the domains of the cytokine that are critical for receptor binding and/or signal transduction.

Enhancement of Cytokine Potency by PEGylation. As shown in Figure 5, the covalent attachment of a single strand of 36-kDa PEG dramatically increased the activity of rmuGM-CSF in mice, and no further increase resulted from coupling a second strand. These observations and analogous results for several other hematologic parameters (unpublished results) imply that the attachment of each additional strand of PEG to a receptor-binding protein enhances its potency only if the resultant increase in half-life more than compensates for the decrease in its activity. This phenomenon has been observed with 5-kDa PEG conjugates of hGH (*19*) and IL-2 (*20*). Our results are consistent with the concept that coupling a few long strands of PEG to a cytokine interferes less with its function than coupling many short strands, making it possible to prolong its circulating life-time while preserving (more of) its intrinsic activity.

In addition to the results presented here, we have obtained encouraging preliminary data on high molecular weight PEG conjugates of recombinant murine stem cell factor (see *27*), including evidence of enhanced radioprotective activity in mice (unpublished results). These observations indicate that the methods described here for coupling PEG to GM-CSF are effective for some other cytokines and may be useful for other types of small receptor-binding proteins such as chemokines and peptide hormones. For each protein, the conditions of PEGylation need to be optimized with respect to the number and size of the attached strands, as well as the coupling chemistry.

An Alternative Explanation for the Extraordinary Potency of PEGylated GM-CSF. Reports that high concentrations of unbound PEG potentiate the activation of murine lymphocytes *in vitro* (*e.g. 68, 69*) may provide some additional insights into the remarkable increase in the potency of PEGylated GM-CSF compared with the unmodified cytokine. It is possible that the cytokine may function, in part, as a targeting device for the delivery of PEG to the membranes of receptor-bearing cells, thereby increasing the local concentration of PEG. This process might account, in part, for the discrepancy of many orders of magnitude between the circulating concentrations of PEG in the mice following injections of less than one microgram of PEG_1- or PEG_2GM-CSF and the concentrations of unbound PEG that were found to be maximally stimulatory in the *in vitro* experiments (at least 10 mg/ml).

Conclusions

The data obtained by size-exclusion chromatography of conjugates of recombinant human or murine GM-CSF with one to three strands of 19- to 42-kDa PEG (Figures 1-4), the demonstration of the enhanced potency *in vivo* of recombinant murine GM-CSF coupled to 36-kDa PEG (Figure 5), and dose-response data for the effects of PEGylated GM-CSF on several hematologic parameters published previously (*10*) lead to the following conclusions: 1) The covalent coupling of a single long strand of PEG to murine GM-CSF potentiates its ability to increase the numbers of several types of peripheral white blood cells in mice, either directly or indirectly, for example by stimulating the secretion of other cytokines (*70, 71*). 2) The potency of murine GM-CSF in mice is not increased further by coupling a second long strand of PEG. 3) The preceding results suggest that conjugates containing a small number of strands of high molecular weight PEG linked to human GM-CSF and other receptor-binding proteins, including other cytokines, chemokines, and hormones, may be suitable candidates for pharmaceutical development.

Abbreviations

The following abbreviations are used: BSA, bovine serum albumin; CSF, colony-stimulating factor; DEAE, diethylaminoethyl; G-CSF, granulocyte colony-stimulating factor; GM-CSF, granulocyte-macrophage colony-stimulating factor; hGH, human growth hormone; HPLC, high performance liquid chromatography; IL-2, interleukin-2; M_r, relative molecular weight; OD, optical density; PEG, poly(ethylene glycol); PEG_1-, PEG_2- or PEG_3GM-CSF, covalent complexes of GM-CSF with one, two or three molecules of PEG; PEGylation, covalent attachment of PEG; PMN, polymorphonuclear; pNPC-PEG, *p*-nitrophenyl carbonate derivative of PEG; rhG-CSF, recombinant human G-CSF; rhGM-CSF, recombinant human GM-CSF; rmuGM-CSF, recombinant murine GM-CSF; SC-PEG, succinimidyl carbonate derivative of PEG; SOD, Cu,Zn-superoxide dismutase.

Acknowledgments

Immunex Corporation generously provided the rhGM-CSF used in these studies. The collaboration of Drs. Oleg Osipovich and Kris Grzegorzewski in preliminary studies of PEGylated stem cell factor and the expert technical assistance of Mr. Orville Bowersox, all of the National Cancer Institute, are gratefully acknowledged. Mrs. Norma Wasserman and Drs. Allen Fay and Israel Schechter provided thoughtful reviews of the manuscript. Research on PEG activation chemistries was supported, in part, by National Institutes of Health Grant DK48529-02 to Mountain View Pharmaceuticals, Inc.

Literature Cited

1. Delgado, C.; Francis, G.E.; Fisher, D. *Crit. Rev. Ther. Drug Carrier Syst.* **1992**, *9*, 249-304.
2. Zalipsky, S.; Lee, C. In *Poly(Ethylene Glycol) Chemistry: Biotechnical and Biomedical Applications*; Harris, J.M., Ed.; Plenum Press: New York, 1992, pp 347-367.
3. Zalipsky, S. *Adv. Drug Delivery Rev.* **1995**, *16*, 157-182.
4. Burgess, A.W. In *Peptide Growth Factors and Their Receptors I;* Sporn, M.B.; Roberts, A.B., Eds.; Springer-Verlag: New York, 1991, pp 723-745.
5. Moore, M.A.S. *Annu. Rev. Immunol.* **1991**, *9*, 159-191.
6. Abuchowski, A.; Kazo, G.M.; Verhoest, C.R., Jr.; van Es, T.; Kafkewitz, D.; Nucci, M.L.; Viau, A.T.; Davis, F.F. *Cancer Biochem. Biophys.* **1984**, *7*, 175-186.
7. Hershfield, M.S.; Buckley, R.H.; Greenberg, M.L.; Melton, A.L.; Schiff, R.; Hatem, C.; Kurtzberg, J.; Markert, M.L.; Kobayashi, R.H.; Kobayashi, A.L.; et al. *N. Engl. J. Med.* **1987**, *316*, 589-596.
8. Fuertges, F.; Abuchowski, A. *J. Contr. Release* **1990**, *11*, 139-148.

9. Sherman, M.R.; Kwak, L.W.; Saifer, M.G.P.; Williams, L.D.; Oppenheim, J.J. *American Association for Cancer Research Conference on Cytokines and Cytokine Receptors, Bolton Landing, NY,* **1995,** Abstract B-17.
10. Saifer, M.G.P.; Williams, L.D.; Sherman, M.R.; French, J.A.; Kwak, L.W.; Oppenheim, J.J. *Polymer Preprints* **1997,** *38,* 576-577.
11. Knauf, M.J.; Bell, D.P.; Hirtzer, P.; Luo, Z.-P.; Young, J.D.; Katre, N.V. *J. Biol. Chem.* **1988,** *263,* 15064-15070.
12. Somack, R.; Saifer, M.G.P.; Williams, L.D. *Free Rad. Res. Commun.* **1991,** *12-13,* 553-562.
13. Saifer, M.G.P.; Somack, R.; Williams, L.D. *Adv. Exp. Med. Biol.* **1994,** *366,* 377-387.
14. Satake-Ishikawa, R.; Ishikawa, M.; Okada, Y.; Kakitani, M.; Kawagishi, M.; Matsuki, S.; Asano, K. *Cell Struct. Funct.* **1992,** *17,* 157-160.
15. Niven, R.W.; Whitcomb, K.L.; Shaner, L.; Ip, A.Y.; Kinstler, O.B. *Pharm. Res.* **1995,** *12,* 1343-1349.
16. Malik, F.; Delgado, C.; Knüsli, C.; Irvine, A.E.; Fisher, D.; Francis, G.E. *Exp. Hematol.* **1992,** *20,* 1028-1035.
17. Knüsli, C.; Delgado, C.; Malik, F.; Dómine, M.; Tejedor, M.C.; Irvine, A.E.; Fisher, D.; Francis, G.E. *Br. J. Haematol.* **1992,** *82,* 654-663.
18. Tsutsumi, Y.; Kihira, T.; Tsunoda, S.; Kanamori, T.; Nakagawa, S.; Mayumi, T. *Br. J. Cancer* **1995,** *71,* 963-968.
19. Clark, R.; Olson, K.; Fuh, G.; Marian, M.; Mortensen, D.; Teshima, G.; Chang, S.; Chu, H.; Mukku, V.; Canova-Davis, E., et al. *J. Biol. Chem.* **1996,** *271,* 21969-21977.
20. Katre, N.V.; Knauf, M.J.; Laird, W.J. *Proc. Natl. Acad. Sci. U.S.A.* **1987,** *84,* 1487-1491.
21. Venkatachalam, M.A.; Rennke, H.G. *Circ. Res.* **1978,** *43,* 337-347.
22. Maack, T.; Johnson, V.; Kau, S.T.; Figueiredo, J.; Sigulem, D. *Kidney Int.* **1979,** *16,* 251-270.
23. Yamaoka, T.; Tabata, Y.; Ikada, Y. *J. Pharm. Sci.* **1994,** *83,* 601-606.
24. Gasson, J.C. *Blood* **1991,** *77,* 1131-1145.
25. Lieschke, G.J.; Burgess, A.W. *N. Engl. J. Med.* **1992,** *327,* 28-35.
26. Lieschke, G.J.; Burgess, A.W. *N. Engl. J. Med.* **1992,** *327,* 99-106.
27. McNiece, I.K.; Briddell, R.A. *J. Leukoc. Biol.* **1995,** *58,* 14-22.
28. Scarffe, J.H. *Eur. J. Cancer* **1991,** *27,* 1493-1504.
29. Freund, M.; Kleine, H.-D. *Infection* **1992,** *20,* S84-S92.
30. Neidhart, J.A.; Mangalik, A.; Stidley, C.A.; Tebich, S.L.; Sarmiento, L.E.; Pfile, J.E.; Oette, D.H.; Oldham, F.B. *J. Clin. Oncol.* **1992,** *10,* 1460-1469.
31. Costello, R.T. *Acta Oncol.* **1993,** *32,* 403-408.
32. Jones, T.C. *Stem Cells* **1994,** *12 (Suppl. 1),* 229-239.
33. Metcalf, D.; Begley, C.G.; Williamson, D.J.; Nice, E.C.; DeLamarter, J.; Mermod, J.-J.; Thatcher, D.; Schmidt, A. *Exp. Hematol.* **1987,** *15,* 1-9.
34. Gough, N.M.; Gough, J.; Metcalf, D.; Kelso, A.; Grail, D.; Nicola, N.A.; Burgess, A.W.; Dunn, A.R. *Nature* **1984,** *309,* 763-767.
35. Kaushansky, K.; O'Hara, P.J.; Berkner, K.; Segal, G.M.; Hagen, F.S.; Adamson, J.W. *Proc. Natl. Acad. Sci. U.S.A.* **1986,** *83,* 3101-3105.
36. Brown, R.S.; Lennon, J.J. *Anal. Chem.* **1995,** *67,* 3990-3999.
37. Dautrevaux, M.; Boulanger, Y.; Han, K.; Biserte, G. *Eur. J. Biochem.* **1969,** *11,* 267-277.
38. Sciaky, M.; Limozin, N.; Filippi-Foveau, D.; Gulian, J.-M.; Laurent-Tabusse, G. *Biochimie* **1976,** *58,* 1071-1082.
39. Steinman, H.M.; Naik, V.R.; Abernethy, J.L.; Hill, R.L. *J. Biol. Chem.* **1974,** *249,* 7326-7338.
40. Hirayama, K.; Akashi, S.; Furuya, M.; Fukuhara, K. *Biochem. Biophys. Res. Commun.* **1990,** *173,* 639-646.

41. Jörnvall, H. *Eur. J. Biochem.* **1970**, *16*, 25-40.
42. Alexander, A.J.; Hughes, D.E. *Anal. Chem.* **1995**, *67*, 3626-3632.
43. Heusterspreute, M.; Crichton, R.R. *FEBS Lett.* **1981**, *129*, 322-327.
44. Lissitzky, S.; Mauchamp, J.; Reynaud, J.; Rolland, M. *FEBS Lett.* **1975**, *60*, 359-363.
45. Harris, J.M.; Dust, J.M.; McGill, R.A.; Harris, P.A.; Edgell, M.J.; Sedaghat-Herati, R.M.; Karr, L.J.; Donnelly, D.L. In *Water-Soluble Polymers. Synthesis, Solution Properties, and Applications*; Shalaby, S.W.; McCormick, C.L.; Butler, G.B., Eds.; American Chemical Society: Washington, DC, 1991, pp 418-429.
46. Zalipsky, S. *Bioconjug. Chem.* **1995**, *6*, 150-165.
47. Zalipsky, S. In *Biomedical Functions and Biotechnology of Natural and Artificial Polymers. Frontiers in Biomedicine and Biotechnology, Vol. 3*, Yalpani, M., Ed.; ATL Press: Mount Prospect, IL, 1996, pp 63-76.
48. Miron, T.; Wilchek, M. *Bioconjug. Chem.* **1993**, *4*, 568-569.
49. Zalipsky, S.; Seltzer, R.; Menon-Rudolph, S. *Biotechnol. Appl. Biochem.* **1992**, *15*, 100-114.
50. Friedman, M.; Williams, L.D.; Masri, M.S. *Int. J. Pept. Protein Res.* **1974**, *6*, 183-185.
51. Chamow, S.M.; Kogan, T.P.; Venuti, M.; Gadek, T.; Harris, R.J.; Peers, D.H.; Mordenti, J.; Shak, S.; Ashkenazi, A. *Bioconjug. Chem.* **1994**, *5*, 133-140.
52. Lopez, O.T.; Kinstler, O.; Cheung, E.; Welcher, A.A.; Yan, Q. *25th Annual Meeting of the Society for Neuroscience, San Diego, CA,* **1995**, Abstract 605.23.
53. Wilchek, M.; Miron, T. *Biochem. Int.* **1982**, *4*, 629-635.
54. Veronese, F.M.; Largajolli, R.; Boccù, E.; Benassi, C.A.; Schiavon, O. *Appl. Biochem. Biotechnol.* **1985**, *11*, 141-152.
55. Fortier, G.; Laliberté, M. *Biotechnol. Appl. Biochem.* **1993**, *17*, 115-130.
56. Wang, Q.-c.; Pai, L.H.; Debinski, W.; FitzGerald, D.J.; Pastan, I. *Cancer Res.* **1993**, *53*, 4588-4594.
57. Snider, J.; Neville, C.; Yuan, L.C.; Bullock, J. *J. Chromatogr.* **1992**, *599*, 141-155.
58. Jackson, C.-J.C.; Charlton, J.L.; Kuzminski, K.; Lang, G.M.; Sehon, A.H. *Anal. Biochem.* **1987**, *165*, 114-127.
59. McGoff, P.; Baziotis, A.C.; Maskiewicz, R. *Chem. Pharm. Bull.* **1988**, *36*, 3079-3091.
60. Sherman, M.R. *Methods Enzymol.* **1975**, *36*, 211-234.
61. Craven, J.R.; Tyrer, H.; Li, S.P.L.; Booth, C.; Jackson, D. *J. Chromatogr.* **1987**, *387*, 233-240.
62. Kuwabara, T.; Ishikawa, Y.; Kobayashi, H.; Kobayashi, S.; Sugiyama, Y. *Pharm. Res.* **1995**, *12*, 1466-1469.
63. Yamaoka, T.; Tabata, Y.; Ikada, Y. *J. Pharm. Sci.* **1995**, *84*, 349-354.
64. Rennke, H.G.; Patel, Y.; Venkatachalam, M.A. *Kidney Int.* **1978**, *13*, 278-288.
65. Rennke, H.G.; Venkatachalam, M.A. *J. Clin. Invest.* **1979**, *63*, 713-717.
66. Kanwar, Y.S. *Lab. Invest.* **1984**, *51*, 7-21.
67. Katre, N.V. *J. Immunol.* **1990**, *144*, 209-213.
68. Ben-Sasson, S.A.; Henkart, P.A. *J. Immunol.* **1977**, *119*, 227-231.
69. Ponzio, N.M. *Cell. Immunol.* **1980**, *49*, 266-282.
70. Lindemann, A.; Riedel, D.; Oster, W.; Meuer, S.C.; Blohm, D.; Mertelsmann, R.H.; Herrmann, F. *J. Immunol.* **1988**, *140*, 837-839.
71. Sisson, S.D.; Dinarello, C.A. *Blood* **1988**, *72*, 1368-1374.

Chapter 12

Preparation and Characterization of Poly(ethylene glycol)ylated Human Growth Hormone Antagonist

Kenneth Olson[1], Richard Gehant[1], Venkat Mukku[2], Kathy O'Connell[3], Brandon Tomlinson[4], Klara Totpal[2], and Marjorie Winkler[1]

Departments of [1]Recovery Sciences, [2]QC Biochemistry, and [3]Protein Chemistry and [4]Analytical Chemistry, Genentech, Inc., 460 Point San Bruno Boulevard, South San Francisco, CA 94080

In order to prepare long-lasting human growth hormone antagonist (GHA), the protein was conjugated with polyethylene glycol (PEG) to increase its molecular size. Following conjugation, the preparation was fractionated to select the optimal molecule for biological assessment. The molecules were characterized to assist in defining the molecules as drug substances for the purpose of receiving approval for clinical studies and eventually approval as therapeutic drugs. Tryptic peptide analysis was found to be an extremely useful tool to identify the PEGylation sites. Analysis revealed the pattern of PEGylation as well as the consistency of PEGylation from lot to lot. The optimal molecules were found to be between 40-50,000 kD in molecular mass. The results obtained by tryptic analysis were confirmed by MALDI-TOF mass spectrometry. In all cases, the PEGylated proteins were active with a sustained half-life.

Modification of proteins with polyethylene glycol (PEG) to improve their therapeutic properties is becoming an option for the preparation of protein pharmaceuticals which demonstrate a rapid clearance from the body or a neutralizing or toxic immunological response. For a recent review see reference 1. PEGylated human growth hormone (hGH) has been shown to have a sustained biological response in animal models (2). Human growth hormone antagonist (GHA) is a mutant of hGH which has been designed to block hGH-induced receptor dimerization. This is achieved by mutating glycine 120, preventing interaction with the receptor at one site, and enhancing receptor binding at the second receptor binding site through the use of several other site specific mutations (3,4). This potential pharmaceutical will also benefit from PEGylation by exhibiting a sustained antagonistic response (data not shown).

There are 9 potential amine PEGylation sites (3,4) in GHA including the α-amino group and the 8 ε-amino groups of lysyl residues. The propensity of these amines to be PEGylated depends upon their pKa values as well as their exposure on the surface of the molecule. Other than the α-amino group, it is diffucult to selectively PEGylate specific sites since a number of the exposed lysines have similar pKa values. Proteins with an effective molecular weight above approximately 70,000 daltons are not cleared by the glomerular filtration system of the kidney (5). In the case of hGH, MW 22,100 daltons, it was observed that a minimum of 4-5 PEG moieties with an average molecular weight of 5,000 (PEG5000) need to be incorporated into the protein. Excess PEGylation of a protein may decrease its activity. For example, it could block an active site on the protein preventing interaction with a receptor, or decrease the protein-receptor interaction by altering the diffusion of the molecule in solution. It was found that the selectivity process to obtain a specific molecular mass range for hGH could be achieved by cation-exchange chromatography (2). Fractionation of PEGylated proteins by cation-exchange chromatography is related to the number of lysines which are PEGylated. By incorporating a cation-exchange chromatographic step after PEGylation, molecules containing more or less than the desired number of polyethylene glycol groups can be selectively removed from the preparation.

During preclinical development of a PEGylated protein, it is not only necessary to make sure that the protein can be purified in a reproducible manner before PEGylation, but also to make sure that the PEGylation can be controlled to produce a product with the desired characteristics. This can be achieved by controlling the PEGylation reaction and by fractionation of the PEGylated product. Since pharmaceutical development is a lengthy and expensive process, it is critical to select the desired PEGylation product early in the development cycle.

One characterization tool that we have applied to PEGylated proteins is tryptic hydrolysis. Trypsin is a protease which cleaves a protein at the carboxyl side of lysyl or arginyl residue creating a number of peptides. These peptides can be fractionated by high pressure liquid chromatography (HPLC) creating a "tryptic map" of the protein (2). It was found that PEGylated lysine residues are resistant to trypsin hydrolysis. Analysis of the tryptic maps before and after PEGylation revealed which lysines were conjugated.

Experimental

PEG Derivatization. Purified GHA (molecular weight 22kD) was obtained from the Recovery Science Dept. at Genentech. The reaction of the succinimidyl ester of carboxymethylated methoxy-PEG or the succinimidyl derivative of PEG propionic acid (SPA, Shearwater Polymers, Inc, Huntsville, AL)(Figure1) with GHA was studied and optimized to produce a majority of the species with a molecular mass between 40,000-50,000 daltons. Approximately an equal molar amount of PEG5000 reagent for each potential reactive site was found to be optimal. A protein concentration of 10 mg/ml was used in 50mM sodium phosphate, pH 7.5. The components were mixed and incubated at room temperature for approximately one hr.

Column Chromatography. The reaction mixture was first fractionated on a Phenyl Toyopearl 650M (TosoHaas, Montgomeryville, PA) column to remove small amounts of high molecular weight cross-linked species. Sodium citrate was added to

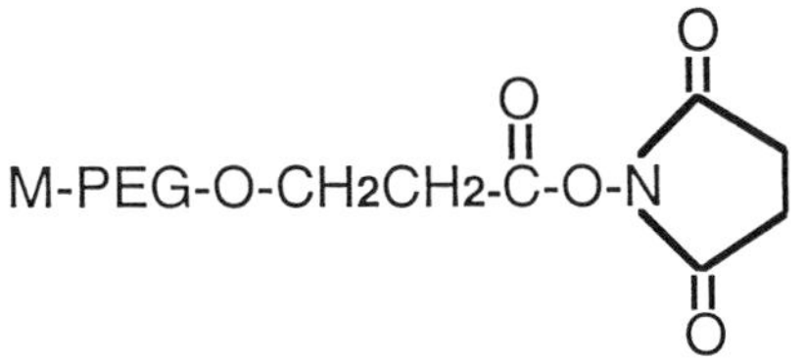

Succinimidyl Ester of the Propionic Acid of Methoxy-PEG

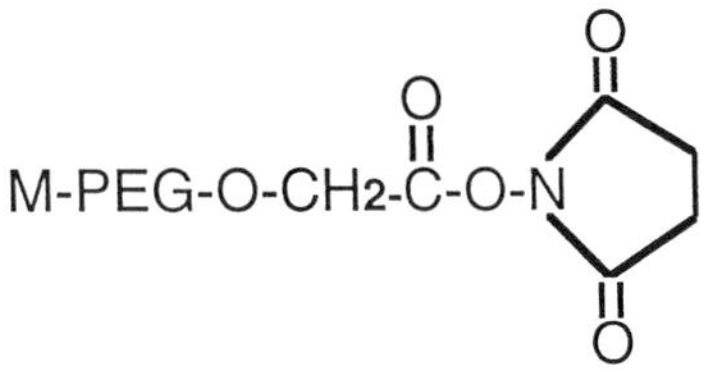

Succinimidyl ester of carboxymethylated Methoxy-PEG

Figure 1. Structure of PEG reagents used for derivatization.

the reaction mixture, to a final concentration of 0.35M, before application to the column. The preparation was loaded at a protein concentration of up to 3.7 grams per liter of resin onto the Phenyl Toyopearl column equilibrated in 0.35M sodium citrate/0.05M Tris, pH 7.6. The PEGylated protein was eluted with a 6-column volume (CV) linear gradient from 0.35M sodium citrate/0.05M Tris, pH 7.6 to 0.05M Tris buffer, pH 7.6. The main protein peak was pooled and desalted either by G-25 Sephadex (Pharmacia) or by diafiltration into 25mM sodium acetate, pH 4.0. The desalted preparation was applied (7 grams protein/liter of resin) to an S Sepharose Fast Flow column (Pharmacia, Piscataway, NJ) equilibrated in 25mM sodium acetate, pH 4.0, and eluted with a gradient generated using a 7-column volume linear gradient from 25mM sodium acetate, pH 4.0 to 25mM sodium acetate/250mM NaCl, pH 4.0. All chromatographic steps were performed at room temperature, and protein was detected by UV absorbance at 280nm. Fractions were collected and characterized.

SDS Gel Electrophoresis. Samples were analyzed by SDS polyacrylamide gel electrophoresis (SDS-PAGE)(*6*). Gradient polyacrylamide gels of 10-20% or 2-15% polyacrylamide were purchased from Integrated Separation Systems, Natick, MA. Samples were loaded onto a SDS polyacrylamide gel and run for approximately 50 minutes at 55mA. The protein bands were stained with Coomassie Blue.

MALDI-TOF Mass Spectrometry. Fractions or pools from the S Sepharose Fast Flow chromatography were analyzed by matrix-assisted laser desorption ionization time-of flight (MALDI-TOF) mass spectrometry (*9*). The analyte was mixed with alpha cyano-4-hydroxy-*trans*-cinnamic acid in 50% acetonitrile, 0.1% trifluoroacetic acid and dried on the sample holder. A Voyager Elite MALDI-TOF mass spectrometer (PerSeptive Biosystems, Framingham, MA) was operated in the linear mode. Fractions from the S Sepharose Fast Flow column elution which contained PEGylated protein in the 40kD-50kD molecular mass range were pooled.

Tryptic Map Analysis. The sites of PEGylation were determined by comparing the tryptic maps of the protein before and after PEGylation. PEGylated GHA was digested with bovine trypsin (Worthington, Freehold, NJ) 1:20 (w/w) trypsin:protein for 8 hr at 37°C in 1mM calcium chloride, 0.1M sodium acetate, and 10mM Tris at pH 8.3. The reaction was stopped with the addition of phosphoric acid to pH 2. The digested protein was loaded onto a C-18 Nucleosil column and eluted according to Reference 2. The peaks of the chromatogram (tryptic map) were integrated and normalized for comparative purposes using the peptides which did not contain lysyl residues. The reduction in the area of the peaks was used to determine the sites and degree of PEGylation.

Bioassays. The biological activity of PEGylated species was determined by a modified, non-radioactive version of the previously described cell-based biological assay (*3,8*). The cells are transfected with the gene for the hGH receptor and, therefore, grow upon interaction with hGH. The assay was run in the presence of 15ng/ml hGH to measure the inhibitory activity of growth hormone antagonists.

Results and Discussion

Preparation of PEGylated Proteins. PEGylation conditions were optimized to produce PEGylated GHA with a mass of primarily 40-50 kilodaltons. We were able to achieve similar results with either 5kD, 10kD, or 20kD PEG (data not shown). The fractionation of the PEGylated protein was achieved by cation-exchange chromatography (2). With increasing PEGylation of a protein, there is a reduction in the number of primary amino groups, resulting in decreased ionic interaction with the cation-exchange resin. Fractions were collected and analyzed by SDS-PAGE (Figure 2). Proteins of higher molecular weight, i.e., containing more PEGylated sites, elute first from the column consistent with their reduced basicity and decreased affinity for the cation-exchange column. This method can be scaled to prepare large quantities of PEGylated proteins with a specific molecular mass range.

Characterization by MALDI-TOF. MALDI-TOF mass spectrometry was used as an analytical tool to determine the actual molecular mass of fractionated PEGylated proteins. It was found that not only is gel filtration not very effective in the separation of various PEGylated species, but also their apparent molecular weight did not correlate with the molecular weight of protein standards. The effective molecular weight as determined by gel filtration is greatly influenced by the conformation and hydration of the PEGylated protein rather than the molecular mass. In 1988, Karas and Hillenkamp published a benchmark paper determining the molecular weight of proteins greater than 10kD using laser desorption ionization with nicotinic acid as a matrix (9). This technique has now become widely adopted and is used extensively for the mass analysis of proteins including PEGylated conjugates. The protein solution is mixed with an UV-absorbing matrix and deposited on the probe. The energy from a laser is adsorbed primarily by the matrix, vaporizing the matrix which carries the ionized protein polymer into the vapor phase generating a gas-phase polymer. The molecular mass is then determined by a time-of-flight mass analyzer. This method is semi-quantitative and has allowed the determination of the presence of small amounts of more highly PEGylated species, in addition to those with lower levels of conjugation. The SDS-PAGE results, which showed that the more highly-PEGylated proteins eluted first from the cation-exchange column, followed by fractions that had a lower level of PEGylation, were confirmed by MALDI-TOF mass spectrometry. Figure 3 shows the mass spectral result of a GHA preparation with primarily 4-5 PEG5000 molecules per protein molecule. In order to see any signal in mass spectroscopy, the protein needs to be ionized. The first signal in the 20,000 m/z range represents the doubly ionized species. The second signal (singly ionized) reveals the presence of 4 and 5 PEG5000 moieties conjugated to GHA and a small amount with 6 PEG5000. Now that PEGylated proteins can be fractionated and characterized, it is possible to correlate the mass of the molecule with the biological activity (2). Additional benefits of mass spectral analysis include verification of the reproducibility of the manufacturing process with regard to the molecular mass and determination of molecular mass limits for lot certification.

Characterization by Tryptic Map Analysis. Tryptic maps are an important tool in the determination of the site and extent of PEGylation. The peptides generated for

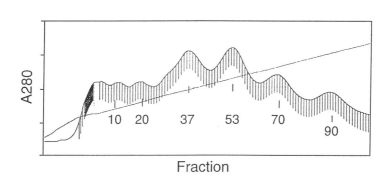

Fraction

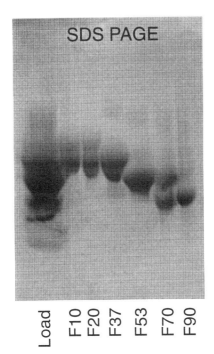

Figure 2. Chromatogram of S Sepharose Fast Flow fractionation of a GHA-PEG preparation with SDS-PAGE analysis of some selected fractions. Lane 1, 143 µg starting material; lanes 2-7, 20 µl of fractions 10, 20, 37, 53, 70, and 90, respectively.

GHA are shown in Table I. Figure 4 shows the tryptic map of GHA before and after PEGylation. Tryptic peptides of GHA were identified by amino acid sequence analysis and mass spectrometry (data not shown). Elimination or reduction in a tryptic peptide peak is indicative of PEGylation. For each complete PEGylation of a lysyl residue, two peptides disappear from their original positions in the tryptic map since cleavage after the PEGylated lysyl residue does not occur. Peptides containing PEGylated residues are more highly retained by the C18 Nucleosil column and elute as broad peaks at the end of the map. The results are summarized in Table II. Since two peptides are affected by PEGylation, a peptide which does not contain PEGylation sites adjacent to a peptide which can be PEGylated may be used to quantitate the extent of PEGylation. In the case of GHA only the N-terminal amine in F1 is 100% PEGylated. Lysines at positions 38, 145, 158 and 140 were PEGylated in approximately 80% of the protein molecules, while lysine in position 120 was PEGylated in 40% of the molecules. Lysine at position 158 was only PEGylated in 26% of the molecules. The tryptic peptides on either side of T15 are PEGylated so the extent of PEGylation on K145 can only be approximated. The average for this preparation was 4-5 PEG5000's/ GHA.

Table 1. Sequence and Tryptic Fragments of GHA

Peak Name	Fragment	Sequence
T1	1-8	FPTIPLSR
T1c	1-6	FPTIPL
T2	9-16	LFDNAMLR
T3	17-19	ADR
T4	20-38	LNQLAFDTYQEFEEAYIPK
T5	39-41	EQK
T6	42-64	YSFLQNPQTSLCFSESIPTPSNR
T7	65-70	EETQQK
T8	71-77	SNLELLR
T9	78-94	ISLLIQSWLEPVQFLR
T10	95-115	SVFANSLVYGASDSNVYDLLK
T10c1	95-99	SVFAN
T10c2	100-115	SLVYGASDSNVYDLLK
T10-11	95-120	SVFANSLVYGASDSNVYDLLKDLEEK
T11	116-120	DLEEK
T12	121-127	IQTLMGR
T13	128-134	LEDGSPR
T14	135-140	TGQIFK
T15	141-145	QTYSK
T16	146-158	FDTNSHNDDALLK
T6-17*	42-64 +	YSFLQNPQTSLCFSESIPTPSNR-
	159-172	NYGLLYCFNADMSR
T18	173-178	VSTFLR
T19-20*	179-	TVQCR-
	183+	
	184-191	SVEGSCGF

* Disulfide-bonded peptides

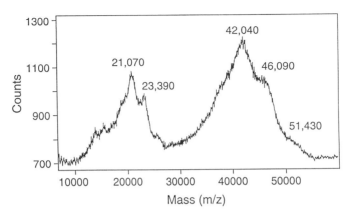

Figure 3. MALDI-TOF mass spectrum of GHA containing primarily 4 PEG5000 (42040 kD) and 5 PEG5000 (46090 kD) species.

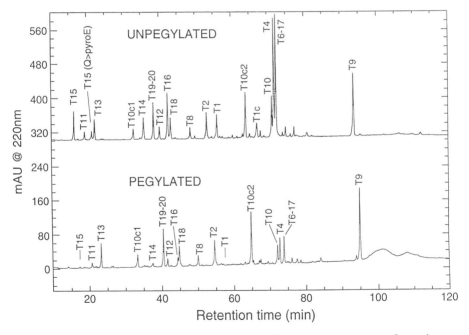

Figure 4. High performance reverse-phase liquid chromatogram of tryptic peptides from unmodified GHA and 4-5 PEG5000 GHA. The second T15 peptide reflects the cyclization of the N-terminal glutamine to pyroglutamic acid. The arrows identify the peptides which are reduced.

There are several buried lysines which are not conjugated even in the presence of excess PEG reagent. The reactivity of a lysine can be estimated by the extent that the peptide is reduced in the tryptic map. Those peptides that are completely missing in the map contain or are adjacent to the most reactive sites and those peptides that are reduced very little or not at all contain the least reactive sites. The order of PEGylation can also be determined by limited PEGylation followed by cation exchange chromatography and tryptic analysis. For example, the PEGylation site in the monoPEGylated protein can be analyzed to determine which residue was conjugated. This analysis can then be conducted on the two PEG molecule, three PEG molecule, etc. For GHA the most reactive site is the N-terminal phenylalanine due to its lower pKa value. Although tryptic analysis is not absolute due to incomplete tryptic digestion and the difficulties associated with the analysis of multiple PEGylation sites, it gives a good approximation of the level of PEGylation as well as the sites of conjugation. The results correlate well with mass spectrometry.

Table II. Tryptic analysis of PEGylated GHA

Tryptic Peptide Analyzed	Amino Acid Residue	% Reduction of PEGylated peptide
T1	F1	100
T4	K38	81
T14, T15, T16	K145	~80
T6-17	K158	79
T14	K140	78
T12	K120	39
T8	K70	26

In Vitro **Biological Activity.** A cell-based bioassay exists for hGH in which the full-length hGH receptor is stably transfected into a premyeloid cell line (FDC-P1 cells) which can then be induced to proliferate in the presence of hGH (*3,8*). Thymidine incorporation into the FDC-P1 cells was directly correlated with the ability of hGH to cause dimerization of the receptor (*10*). The results of PEGylated hGH in the cell-based assay were previously published (*2*). Potency decreased with the extent of PEGylation and a linear correlation was found to exist between the log of the increase in EC50 (the concentration of hormone required for 50% maximal stimulation of cell proliferation) and the number of PEG groups conjugated to hGH. The loss of activity with increased PEGylation appears to be substantial. This can be misleading, however, as it was found that the increase in the half-life of hGH more than compensates for the loss of binding to the receptor in the bioassay (*2*).

An adaptation of the hGH bioassay (*3,11*) was made to convert it into a useful assay for growth hormone antagonists. By maintaining a constant hGH concentration in the assay to effect a near maximal stimulation, the inhibition of activity can be measured in the presence of antagonists. The IC50 (concentration of antagonist where a 50% decrease in cell proliferation is seen) is a measure of the extent of inhibition of hGH receptor binding in the cell-based assay. Figure 5 shows the effect of GHA and PEGylated GHA on the bioassay. Inhibition occurs at a relatively low dose of GHA. The PEGylated molecule (MW 40-50kD) showed an 65-fold lower activity in the bioassay than the unpEGylated GHA. The traditional rat weight gain bioassay, designed to measure the growth effects of hGH, cannot be used for GHA since there is no constitutive exposure to growth hormone in these hypophysectomized animals. The cell-based bioassay is useful for measuring the bioactivity of hGH, the effect of PEG conjugation on receptor binding, and the reduction of activity of PEGylated GHA relative to GHA.

Conclusion

Considerable progress has been made in the PEGylation, fractionation, and analytical characterization of a product such as PEGylated recombinant GHA. Cation-exchange chromatography is a powerful tool to fractionate PEGylated proteins based on their extent of PEGylation (2). The desired PEG species can be produced through control of the pH and the stoichiometry of the components in the PEGylation reaction followed by cation-exchange chromatography. Tryptic map analysis and mass spectrometry have been applied to PEGylated proteins to assist in the characterization of the therapeutic protein produced. Tryptic map analysis provides a useful technique to determine the sites and extent of PEGylation in a conjugated protein which has been PEGylated through lysyl moieties and the N-terminal amine. It was found that certain sites are PEGylated more than other sites. The N-terminus was always 100% conjugated, while others were about 80%. Several lysines were conjugated in about 30% of the molecules. This tryptic map analysis is useful in that it provides an opportunity to measure the consistency of the PEGylation process from lot to lot. A well-characterized therapeutic protein should include data on the sites of PEGylation. This has not typically been done for complex PEGylated proteins.

It is not only useful but critical that biological activity be measured on PEGylated proteins prior to clinical studies in humans. The cell-based hGH bioassay allows a correlation of the extent of PEGylation to bioactivity. Studies in animals can demonstrate the combined effects of altered bioactivity and clearance. PEGylation must be controlled since over-PEGylation can decrease biological activity or create a molecule with a longer half-life than desired, while a protein with limited PEGylation may be cleared faster than desired. PEGylation had a similar effect on GHA as it did to hGH. The *in vitro* activity decreased as the extent of PEGylation increased. This is not expected to have an adverse effect on *in vivo* activity, however. In the case of hGH, the PEGylated protein had as much or more activity than the authentic protein (2).

Methods are described in this report to not only determine the molecular weight of PEGylated proteins but to also measure the relative abundance of various PEGylated species. In addition, MALDI-TOF mass spectrometry is a useful tool to determine the size of the conjugated proteins during the development of the PEGylation process. Other methods, such as gel permeation chromatography can provide an approximate molecular size, but the large hydrodynamic radius of PEG groups results in an apparent molecular size of the PEGylated protein which is

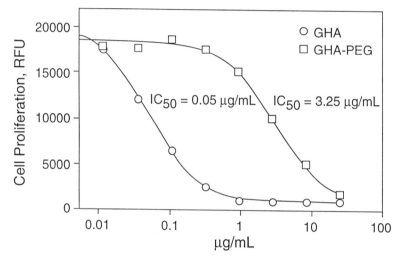

Figure 5. Cell proliferation assay of GHA and PEGylated GHA. The samples were incubated in the presence of 15 ng/ml of hGH. Cell growth was measured by fluorescence, measuring the uptake of alamar blue.   RFU = Relative Fluorescence Units

considerably higher than the actual mass. For instance, a PEGylated protein with a mass of approximately 50,000 daltons can have an apparent molecular mass of 300,000 daltons when compared with globular proteins on a gel permeation column. MALDI-TOF mass spectrometry provides an accurate molecular mass of a PEGylated protein. In addition, it can see the relative abundance of other PEGylated species in the preparation providing semi-qualitative analysis and assigning to them a molecular mass.

Acknowledgments
This work was sponsored by Sensus Corporation, Austin, Texas. Development of tryptic map analysis of PEGylated proteins was pioneered by Rosanne Chloupek, Glen Teshima, and Eleanor Canova-Davis.

Literature Cited

1. Zalipsky, S.; Lee, C. In *PEG Chemistry: Biotechnological and Biomedical Applications*; Harris, J.M., Ed.; Plenum Publishing Corp.: New York City, NY, **1992**, pp 347-370.
2. Clark, R.; Olson, K.; Fuh, G.; Marian, M.; Mortensen, D.; Teshima, G.; Chang, S.; Chu, H.; Mukku, V.; Canova-Davis, E.; Somers, T.; Cronin, M.; Winkler, M.; Wells, J. *Journal Biol. Chem.* **1996**, *271*, 21969-21977.
3. Fuh, G.; Cunningham, B.; Fukunaga, R.; Nagata, S.; Goeddel, D.; Wells, J. *Science,* **1992**, *256*, 1677-1680
4. Cunningham, B.; Wells, J. *Proc. Natl. Acad. Sci.* **1991**, *88*, 3407-3411.
5. Venkatachalam, M.; Rennke, H. *Circ. Res.* **1978**, *43*, 337-347.
6. Laemmli, U.K. *Nature.* **1970**, *227*, 680-685.
7. Watson, E.; Shah, B.; DePrince, R.; Hendren, R.W.; Nelson, R. B. *BioTechniques*. **1994**, *16*, 278-281.
8. Colose, P.; Wong, K.; Leong, S.; Wood, W.I. *J. Biol. Chem.* **1993**, *268*, 12617-12623.
9. Karas, M.; Hillenkamp, F. *Anal. Chem.* **1988**, *60*, 2299-2301.
10. Spencer, S.; Hammonds, R.G.; Henzel, W.J.; Rodriquez, H.; Waters, M.J.; Wood, W.I. *J. Biol Chem.* **1988**, *263*, 7862-7867.
11. Roswall, E.C.; Mukku, V.R.; Chen, A.B.; Hoff, E.H.; Chu, H.; McKay, P.A.; Olson, K.C.; Battersby, J.E.; Gehant, R.L.; Meunier, A.; Garnick, R.L. *Biologicals*, **1996**, *24*, 25-29

Chapter 13

New Synthetic Polymers for Enzyme and Liposome Modification

Francesco M. Veronese, Paolo Caliceti, and Oddone Schiavon

Department of Pharmaceutical Sciences, University of Padova,
35131 Padova, Italy

Although most enzyme and liposome modifications have been carried out using linear forms of poly(ethylene glycol), new polymers of natural or synthetic origin have also been employed and under study. A branched structure that, in many respects was found to be superior to the linear one, is similar to poly(ethylene glycol) in structure and single point attachment. New single point attachment polymers were recently synthesised and among these are poly(vinylpyrrolidone) and poly(acryloylmorpholine) oligomers which were successfully used for both enzyme and liposome modification, and poly(oxazoline) and poly(glycerols) which were used for long lasting liposome protection. Among the multifunctional reagents of note are dextran, divinyl ether-maleic acid anhydride and styrene-maleic acid anhydride that, although giving more complex reaction products with protein in comparison to the previous polymers, convey new and useful properties and *in vivo* behaviour on the conjugates.

Enzyme modifications for therapeutic applications or for bioconversions in organic solvents, have been carried out in recent years utilizing linear monomethoxy polyethylenglycol (PEG). This polymer shows unique and favorable properties such as the absence of toxicity, ample choice of activation procedures of the hydroxy terminating group for the covalent binding to a protein via an amino, guanidino, thiol or hydroxyl group, suppression of immunogenicity and antigenicity of proteins, enhanced residence time in blood of the conjugate and enhanced stability to proteolysis. These advantages of PEG in derivatizing proteins and enzymes for therapeutic applications were originally pointed out in a fundamental paper by Abuchowski et al. (1).

More recently PEG conjugates of enzymes have been shown to enhance enzyme solubility in organic solvents thus leading to new and unexpected applications in bioconversion processes and analysis (2). However one limit of PEG is that it cannot

182

be tailored to the varied needs for the efficient utilization of enzyme conjugates, since only the molecular weight or branching of the polymer can be changed, whereas other important and useful properties such as charge, backbone rigidity and hydrophilicity cannot be varied. These and also the need to overcome patent limitations have prompted the development of novel polymers for the purposes of enzyme conjugation. The properties of these polymers will be reviewed in this paper, paying greater attention to those providing single point attachment to proteins. Moreover the use of soluble polymers in enhancing the circulation time of liposomes will be briefly discussed.

Polymers Used in Enzyme-Polymer Conjugates. In recent years, a variety of natural or specially tailored synthetic polymers have been used for the preparation of enzyme conjugates. Polysaccharides and proteins have been used with some success, while numerous synthetic polymers have been widely applied. Divinylether-maleic acid anhydride (DIVEMA) styrene-maleic acid anhydride (SMA), poly(vinylalchool), poly(vinylpyrrolidone) (PVP), poly(acryloylmorpholine) (PAcM) are among these. Moreover the maleic acid-acrylic acid derivative of PEG, as well as branched PEG, have proved to be useful in a variety of situations.

One drawback of the enzyme conjugation reaction utilising polymers activated at different sites along the chain is that it produces inter- and intra-molecular cross links. On the one hand, this reaction also offers some advantages since conjugated proteins may be more resistant to denaturation due to the freezing of the three-dimensional protein structure by the multipoint attachment of the polymer, in very similar to enzyme immobilisation on insoluble supports. On the other hand, the intra-molecular cross-links leads to heterogeneity of the protein conjugates in terms of molecular weight and complex aggregates are also formed. This explains the success so far achieved with the use of PEG, branched PEG and other monofunctional polymers, which lead to more homogeneous products that are easily characterised.

Natural Polymers in Enzyme Conjugation

Proteins. Albumin was successfully used for the modification of several enzymes , for example them superoxide dismutase and asparaginase (3). Albumin, including human albumin, has the advantages of being easily available,. The immunogenicity problems of the protein conjugates with albumin are considerably reduced even if not completely. In fact, new epitopes at the protein surface may be formed during the activation and coupling reaction of albumin. Furthermore, the products are usually heterogeneous and of high molecular mass, since the addition of any albumin molecule to the enzyme increases the mass of the conjugate by 64 kDa (the albumin molecular weight). The enzyme protein-polymer conjugate is even more complex due to its heterogeneity derived from inter-molecular cross-links between the activated carboxyl groups of albumin and the protein amino groups.

Polysaccharides. Dextran is the polysaccharide that has been most closely studied among those used in enzyme modification. This is probably due to the fact that samples of fractionated dextran of low polydispersivity are available. Moreover, the chemistry of activation of this polysaccharide, as well as the results of using this polymer in binding low molecular weight drugs have been studied in depth(4). Dextran is certainly better than PEG in the preparation of low molecular weight drug conjugates since high ratios of drug to polymer can be achieved, due to the fact that any monomer in the dextran molecule may be the site of activation and coupling. The multivalence of dextran, that is generally seen as a limit in enzyme conjugation, allowed for, new biologically useful, functions to be added to the polymer molecule, as for example in the preparation of cationic or anionic dextran, as well as the glycosilated forms. These multiple forms of dextran have led to interesting comparative studies of enzyme conjugates, such as the analysis of four superoxide dismutase-dextran conjugates and PEG derivatives in delivery to kidney (5). The results of this study have demonstrated how the polymer structure is critically important in dictating biological effects, given that PEG and carboxymethyl-dextran for example decreased the enzyme glomerular filtration. On the other hand cationization enabled distribution to the kidney from the capillary side, while glycosilation with galactase and mannose reduced tubular reabsorption. In particular the mannosilated derivative was accumulated in the kidney via a mannose recognition mechanism.

As an additional example, mention should be made of the enhanced activity of pegylated superoxide dismutase in preventing heart ischemia in rats with respect to the corresponding dextran derivative The concentration of the PEG derivative in the heart tissue after systemic administration is of interest, especially as the level was higher than after similar administration of dextran (6).

An interesting dextran derivative for enzyme modification has recently been proposed (7). The oxidation of the polysaccharide yields a product where all the saccharide units are open and a single reactive group at the end. This product of a hydrophilic nature, appears particularly useful for single point attachment (7).

Lipids. Lipids are a class of non-polymeric compounds which now are successfully used to obtain enzyme conjugates with promising therapeutic effects (8). Several enzymes have been derivatized with properly activated lipids and the lipid-conjugates showed higher affinity for some tissues than displayed by PEG derivatives. This can be taken as an indication that the hydrophilic-hydrophobic character of the enzyme surface is critical in determining the biological properties of the enzyme conjugates.

Synthetic Polymers in Enzyme Conjugation

Synthetic polymers are expected to show unique advantages over natural ones, since they may be properly tailored to specific needs by modifying the hydrophobic-hydrophilic nature, monomer structure, molecular weight, shape, as well as type and amount of the reactive groups along the chain etc.. In practice, several limitations are posed by polymer preparation, purification and fractionation to obtain the adequate distribution of molecular weight. In this respect PEG stands out as a fortunate exception because of the simple and well described synthesis procedure, fractionation method, and terminal group activation yields a variety of derivatives. Furthermore, unlike other polymers (9), classical and most modern analytical methods, including NMR and mass spectrometry, can be utilized in the characterization of PEG.

Branched PEG. Two new branched PEGs have recently been prepared, the first being obtained by the reaction of trichlorotriazine with linear PEG at the level of two chlorides, while the third chloride remains free for the protein reaction (10) and the second one in which two linear PEG oligomers are bound to the lysine amino groups while the lysine α-carboxy group is activated as a succinimidyl ester in protein binding (11). This second branched PEG seems to be more selective for enzyme modification and may also present lower immunogenicity, which usually occurs when an aromatic compound is bound to proteins.

Branched PEG appears to be superior to the mostly used linear PEG. First of all, the protein conjugates have been found to be less immunogenic and antigenic, more stable to proteolysis and, furthermore, enzyme activity is usually better preserved better during the conjugation reaction (12). These properties seem to be related to the special spatial arrangement of the branched PEG that, like an "umbrella", protects a large protein surface from any interaction with antibodies and proteolytic enzymes. Moreover the enhanced hindrance of branched PEG with respect to the linear form does not allow penetration and binding of the activated polymer to enzyme active site cleft.

Copolymer PEG-Maleic Acid Anhydride. This polymer, called comb-shaped PEG, has mainly been studied with the aim of increasing the stability of the protein conjugates under denaturing conditions. Several proteins have been modified with polymers of 13,000 or 100,000 Da leading to a reduction in their immunogenicity and enhanced stability to urea and heat denaturation as well as when they are exposed to acidic conditions with respect to PEG modified proteins. The protective effect of bound polymers was found to be more significant when proteins were conjugated with a copolymer having a higher molecular weight (13). The reason for this appears to be related to the possibility of intramolecular cross-linking occurring with these polymers, even though protein aggregates can be formed due to intermolecular reaction (14).

Styrene-Maleic Acid Anhydride (SMA). The most important and successful study on this copolymer has been carried out with neocarcinostatin, a small antitumor protein. The aromatic character of SMA gives the modified protein a compact and globular structure in water due to the presence of the phenyl groups, while the hydrolysed maleic acid groups are exposed to water. The unusual structure of the polymer is probably the reason for the favorable properties of the neocarcinostatin conjugate. The presence of only two accessible amino groups in neocarzinostatin leads to homogeneous derivatives, a quite important property in obtaining approval for commercialisation (15).

Divinylether-maleic acid anhydride (DIVEMA). This polymer, also known as "piran copolymer", possesses "per se" therapeutic potential, in particular as an anti viral agent (16). Special efforts have been devoted to obtaining products with a low polydispersivity. The polymer was utilised for the preparation of polymeric prodrugs, such as the polymer adducts of doxorubicine and superoxide dismutase. In the SOD conjugate, in order to reduce the enzyme inactivation during the modification, 2,3-dimethylmaleic anhydride was employed to protection of the lysines near the active site which was reversible reaction. SOD conjugation has shown by circular dichroism spectra, was not accompanied by any change in the secondary structure content of the native enzyme, but increase heat stability was evident. The pharmacokinetic profile showed an increased residence time in blood although it was lower than the PEG conjugate. Furthermore, unlike the native or PEG derivatized SOD, the DIVEMA enzyme, in its apo-form, does not completely recover Cu^{++}, a metal ion which is essential in enzyme activity. An interesting favorable property of DIVEMA-SOD conjugate with respect to other polymer derivatives is its increased stability to oxidation, which derives from the chelating capacity of DIVEMA (19-17).

Polymers Formed Directly on the Enzyme Surface. An original method recently reported for the synthesis of vinyl polymers is based on the growth of the polymer chain directly onto the amino acid residue of a protein. The procedure starts with the covalent binding of an aliphatic initiator to the protein amino groups, 4,4'-azobis(4-cyanovaleric acid), and the desired protein-polymer conjugate is obtained by irradiation of the derivatized protein with the specific monomer. Depending upon the monomer used, conjugates responded to external signals, as pH or oxidation, were obtained (20). This method appears to be useful in adding a variety of new functions to the protein in one single step. However, sophisticated applications as the use in therapy, seem to be limited, because the side products of polymerization may remain bound to the protein and, moreover, chains of different length may grow on the same or different protein molecules resulting in heterogeneous products.

New End-functionalized Polymers. To avoid heterogeneity in conjugation chemistry, new end-functionalized polymers have been synthesized. The idea was to prepare new polymers with a similar molecular weight (in the range of 2000-8000 Da), with an hydroxy terminal group with the view taking advantage of the results of previous studies on the PEG chemistry on the activation and conjugation reactions. The aim, in developing these new polymers was that, these would have new and favorable *in vivo* behaviour as a result of the different structure and different physico-chemical properties of the chain.

Poly(acryloylmorpholine) (PAcM). The interesting physico-chemical and biological properties of high molecular weight poly(acryloylmorpholine) has been known for long time in polymer. In particular the lack of toxicity and amphiphilicity prompted its biological applications as a blood expander, in the preparation of membranes for blood filtration, as a support for chromatography as well as for liquid-phase peptide synthesis. Nevertheless low molecular weight oligomers have not been described and have only recently been prepared by the chain transfer technique, using functionalized thiols as chain-transfer agents. During polymerization, 2-mercaptoethanol, 2-mercaptoacetic acid and mercaptoethylamine were used to introduce as an end terminal group, primary alcoholic, carboxylic and primary amino function, respectively. In order to avoid any hydrogen transfer addition reaction to the acryloylmorpholine double bond, the polymerization was always effected under slightly acidic conditions. Oligomer fractions ranging from 1000 to 7500 Da were obtained with a polydispersivity of 1.1. Among the three terminating groups the most convenient for conjugation with protein was found to be the carboxylic one since the active succinimidyl ester could be easily obtained (21).

Few studies of enzyme modification with PAcM have been reported, one of these makes a comparison of PEG, branched PEG and PAcM conjugated tyrosinase using oligomers of the same size and similar degree of modification (22). The study pointed out the higher resistance of the PAcM-conjugates to the oxidation products of the enzyme reaction, with respect to the conjugates obtained with other polymers. On the other hand no significant differences with the PEG conjugates were observed in Km, Vmax, thermostability an *in vivo* residence time. Studies performed on ribonuclease, asparaginase and superoxide dismutase have shown that the PAcM conjugates were still active after modification and were always highly resistant to proteolysis and had longer *in vivo* residence time. The immunogenicity and antigenicity of the PAcM conjugates have also been investigated and the results are reported below.

The hydroxy terminated-PAcM has been studied in two quite different applications as an alternative to PEG: as a support in liquid phase oligonucleotide synthesis (23) and in a lipase modification for bioconversion in organic solvents (24).

The advantages and disadvantages of PAcM over PEG, in the liquid phase oligonucleotide synthesis, have been defined during the preparation of an octonucleotide. Using PAcM as a support matrix made the detrytilation step easier in comparison to PEG, that, instead, allows for easily following the reaction by 1H NMR because of its lower absorption. The purification of the products by ether precipitation was also more complete when PEG was used as a support(23).

On the other hand PAcM lipase conjugates prepared for bioconversion in organic solvents, were less soluble in chlorinated solvents (between 1/4 to ½ depending on the solvent) than the PEG derivatives, but were two to four times more activity catalitically (24).

Poly(vinylpyrrolidone) (PVP). The lack of toxicity and the favorable immunological properties of PVP, long known in literature, has led many investigators to use it in enzyme conjugation. One of the procedures for the preparation of PVP conjugates involves the opening of a few pyrrolidone rings by alkaline treatment followed by acetilation of the unmasked amino groups, while the carboxylic groups are activated for protein binding (25). A more recent method involves the copolymerization of vinylpyrrolidone with the acetal form of acrylaldehyde. After polymerization and fractionation, the aldehyde groups are unmasked allowing for amino group binding, to give Shiff bases. The conjugate is finally stabilized by reduction to secondary-amines. The difficulty in obtaining oligomers with a desired molecular weight and low polydispersivity with, furthermore, the presence of many reactive groups randomly distributed along the chain yields heterogeneous conjugates poses several limits to both methods.

Unfortunately, the controlled polymerization using thiol derivatives as chain transfer agents, successful in poly(acryloylmorpholine) oligomer preparation, have only partially solved the problem of obtaining homogenous PVP products, although satisfactory results have recently been obtained using 2-propanol or isopropoxyethanol as chain transfer agents to introduce an alcoholic group as a terminal chain (26). The best procedure in activating these new PVP oligomers has been the one that takes advantage of 4-nitrophenylchloroformate, a reagent that has already been successfully used in PEG activation (27).

The suitability of these new PVP oligomers for enzyme modification has been demonstrated using ribonuclease, uricase and superoxide-dismutase (28). As in the case of other polymers, the protein conjugates presented increased stability to heat, proteolytic digestion and increased residence time in rats.

Of note, was the finding that PVP-SOD conjugate presented a higher elution volume in gel filtration in comparison compared to the enzyme modified to the same extent with PEG. This behaviour, interpreted on the basis of hydrodinaminc volume, suggested that, while PEG chains are extended towards the water surrounding the protein, the PVP ones may interact more closely with the protein surface resulting in reduced overall volume of the conjugate (28).

Comparative Study of Imunogenic, Pharmacokinetic and Biodistribution Behaviour of Linear PEG, Branched PEG, PAcM and PVP Conjugates. A useful investigation on relevant biological properties of enzyme conjugates with different polymers was carried out using uricase as model. Uricase is an enzyme of therapeutic interest in the treatment of hyperuricemia connected to metabolic disorders and in excess uric acid removal that often occurs during chemotherapy (29). In order to compare the properties of the conjugates, a similar number of chains of the four polymers were bound to uricase and polymers with similar molecular weight were also used.

The results, as reported in table I, demonstrated that the immunogenicity in mice is significantly different depending upon the polymer used in the conjugation. Taking the antibody titles raised by unmodified uricase treated animals as a reference, PVP conjugates showed only limited reduction in immunogenicity.

Table I. Immunogenicity of Unmodified and Polymer Conjugate Uricase

	% reduction of uricase immunogenicity after polymer conjugation *		
Immunization time	2nd week	5th week	12th week
Native uricase	-	-	-
Uricase-PEG	100	98	98
Uricase-PEG2	100	100	100
Uricase-PVP	0	80	80
Uricase-PAcM	100	99	99

*The conjugates used for the immunological studies were prepared with the four different polymers to reach the same extent of modification. The immunological studies were carried out by weekly subcutaneous immunization of Balb/c male mice with native or conjugated uricase. IgG+IgM were estimated by ELISA (30) using native enzyme for well coating and the reduction in immunogenicity was calculated as reported in literature (31).

Linear PEG and PAcM greatly reduced immunogenicity of uricase while the effect of branched PEG was impressive. The results were similar for both IgG and IgM. A detailed investigation on the composition of antibodies in mice was carried out using in the ELISA well coating native protein, protein conjugates of the four polymers and finally the four polymers bound to a different protein. It has been possible to demonstrate that PAcM and PVP polymers, when bound to the protein, display higher immunogenicity than PEG.

The four different uricase conjugates labelled with tritium were also administered to mice in order to investigate the distribution pattern in the body for two days after bolus administration. It was found that the polymer structure dictates both the concentration and residence time in tissues. In fact, while PEG conjugates were distributed in organs without any specific tropism, PVP and PAcM conjugates were, on the other hand, accumulated in tissues. A higher concentration of the PAcM conjugate was found in liver, heart, spleen and kidney while the PVP conjugate was found in the liver, lung and spleen.

Finally the pharmacokinetic behaviour of uricase derivatives in blood was investigated and some relevant data are reported in table II. To note that for all the conjugates the molecular volume considerably overcomes the renal ultrafiltration limit. Differences in residence time should therefore reside on different stability to degradation or different body dispoition. Both $t1/2\beta$ Vd and MRT indicate the longest residence and highest distribution volume for PVP and branched PEG conjugates, lower values were found for the linear PEG and PAcM derivatives.

Table II. **Pharmacokinetic Parameters of Native Uricase and Polymer Conjugated Species**

	$t1/2\beta$ (min)[a]	Vd (mL)[b]	MRT (min)[c]
Native uricase	352	2.5	429
Uricase-PEG	1604	2.08	2218
Uricase-PEG2	2176	2.12	2979
Uricase-PVP	2371	2.4	3030
Uricase-PAcM	1146	3.13	1542

[a]$t1/2\beta$: half life of the elimination phase; [b]Vd: distribution volume; [c]MRT: mean residence time. The pharmacokinetic experiments were performed on Balb/c male mice and the enzyme content in blood estimated by enzymatic activity assay.

These results do not allow for the drawing conclusions about the general behaviour of PVP and PAcM oligomers. Further experiments with other protein models are required, but they are however sufficient to demonstrate that these two polymers can change the properties of protein conjugates with respect to the other.

The Effect of Different Polymers in Liposome Protection *In Vivo*. A great many of studies have recently been dedicated to increasing the blood liposome residence time and reduce uptake by their reticulo endotelial system (RES). Modification of the phospholipidic bilayer composition and charge to liposome surface were the first useful parameters found although surface modification with polymers was much more effective. Liposome protection by polymers is similar to what has already been found in other nanoparticular drug carriers such as nanospheres, nanocapsules, micelles and enzymes (32).

The mechanisms involved in the increased permanence of liposomes or in blood enzymes after polymer conjugation are different in some respects. The polymers act mainly to inhibit the interaction of liposomes with destructive substances such as lipoproteins or opsonins, while the increased size obtained by polymer binding is not important as it is for proteins or polypeptydes, in that when these have a molecular weight of less than 50-60000 Da they are lost by renal ultrafiltration. On the other hand, liposome size is always well above this value and can not therefore escape. The hydrophilicity of the polymer which creates protecting water shield is of great importance for long circulating liposomes, while a hydrophobic polymer destabilizes the liposome double layer. It has also been suggested that flexibility plays a critical role, and this statement (33) is supported by the finding that dextran, which is a hydrophilic but rigid polymer, does not increase circulation time or offer protection from opsonins (34).

Polyethylene glycol, that meets to these requisites, is thus very effective in liposome protection and similarly to what has recently been found in proteins, branched PEG protects liposomes from being removed from plasma even better than linear PEG and also greatly reduces liver uptake by the liver (35). Favorable behaviour both in plasma and liver was found in liposomes with bound PVP or PAcM, a not a totally unexpected finding since both polymers possess good hydrophilicity accompanied by moderate hydrophobicity and flexibility, all properties that have made PEG so successful.

The behaviour of five liposome forms, namely plain liposomes or containing 3% of phosphatidyl ethanolamine derivatized with different a polymer was compared in rats. The plasma half life, after bolus injection with an unmodified liposome, or with bound linear PEG, branched PEG, PAcM, or PVP was 10, 80, 140, 90, and 40 min respectively (36). The low protection given by PVP may be related to its behaviour in water where it appeared less solvated than PEG.

It should be observed that no correlation was not found with protein protection for the same four polymers.

Other polymers were successfully used in liposome protection, among these the polyglycerols that, for liposome preparation, were bound to dipalmitoyl phosphatidic acid by an original enzymatic reaction. They showed great efficiency in protecting liposomes from RES uptake and furthermore a corrrelation between the length of the polyglycerol chain and liposome protection was found. This polymer appeared even more hydrophilic than PEG although a comparison of the efficiency between the two polymers, was difficult to assess (36).

Two other amphiphilic polymers, poly(2-methyl-2-oxazoline) and poly(2-ethyl-2-oxazoline) were bound to disteroyl-phosphatidylethanolamine for the preparation of long lasting liposome. The favorable properties of these polymers have been known for a long time, but unfortunately they have not, so far, been properly exploited. When used at the same molecular weight as PEG to derivatize liposomes, they gave comparable results in that to these vescicles had long circulation time and low RES uptake (37).

Conclusion

A variety of polymers of natural or synthetic origin have been proposed in recent years for enzyme or liposome modifications. Nevertheless, the results obtained in a number of studies do not allow for suggesting general strategies for their use, since few comparative studies utilizing different polymers have, so far, been carried out. It seems that no single polymer can be considered really superior to another since the negative properties of one application may be useful in a different one. However, using polymers with the greatest possible homogeneity and with a good capacity for binding is highly recommended, since the complexity of proteins, and liposomes also, suggests the use of the most simple polymeric counterpart so that experimental results can be clearly interpreted and so determine where they may be most usefully applied.

Literature Cited

1. Abuchowski, A.; Van Es, T; Palcank, N.C.; Davis, F.F. *J. Biol. Chem.* **1977**, 252, 3578.
2. Inada, Y; Matsushima, A.; Kodera, Y.; Nishimura, H. *J. Bioact. Compt. Polym.* **1990**, 5, 343.
3. Wong, K.; Cleland, L.G.; Polnansky, M.J. *Agent and Action*, **1980**, 10, 231.
4. Molteni, L. *Methods in Enzymol.* 1985, Vol. 112, 285.
5. Takakura, Y.; Mihara, K.; Hashida, M. *J. Controlled Rel.* **1994**, 28, 111.
6. Maksimenko, A.V.; Petrov, A.D.; Caliceti P.; Konovalova, G.G.; Grigorjeva E.L.; Schiavon, O.; Tishenko, E.G.; Lankin, V.Z.; Veronese, F.M. *Drug Delivery* 1995, 2, 39.
7. Papisov, M.; Garrido, L.; Pass, K.; Wright, C.; Weissleder, R.; Bradi, T.Y. *Proc. Int. Symp. Control. Rel. Bioact. Mater.* **1996**, 23, 107.
8. Igaroshi, R.; Hoshino, M; Takenaga, M.; Kawai, S.; Morirawa, Y.; Yashuda, A.; Otani, M.; Misushima, Y. *J. Pharmacol. Exper. Ther.* **1992**, 262, 1214.
9. Montando, G.; Montando, M.S.; Puglisi, C.; Samperi, F. *Rapid Commun.Mass Spectrom.* **1995**, 9, 453.
10. Matsushima, A.; Nishimura, H.; Ashihara, Y.; Yokota, J.; Inada, Y. *Chem. Lett.* **1980**, 8 bis, 773.
11. Monfardini, C.; Schiavon, O.; Caliceti, P.; Morpurgo, M.; Harris J.M.; Veronese, F. *Bioconjugate Chem.* **1994**, 6, 62.
12. Veronese, F.; Caliceti, P.; Schiavon, O.; J. Bioact.Comp. Pol., *submitted.*
13. Kodera, Y.; Sekine, T.; Yasukohchi, T.; Kirin, Y.; Hirato, M.; Matsushima, A.; Inada, Y., *Biocojugate. Chem.* **1994**, 5, 283.
14. Klibanov, A.M. *Anal. Biochem.* **1979**, 93, 1.
15. Maeda, H.; Takashita, J.; Kanamarun, R.; Int. J. Pept. Protein Res. **1987**, 14, 81.
16. Regelson, W., In *Anionic Polymeric Drug*, L.G. Donaruma, R.M. Ottembrite, O. Vogl, Eds, Wiley, New York, 1980 pp. 303-325.

17. Hirano, T.; Todaroki, T.; Kato, S.; Yamamoto, H.; Caliceti, P.; Veronese, F.M.; Maeda, H.; Ohashi, S. *J. Controlled Rel.* **1994**, *28*, 203.

18. Caliceti, P.; Schiavon, O.; Hirano, T.; Ohashi, S.; Veronese, F.M., *J. Controlled Rel.* **1996**, *39*, 27.

19. Yuichiro, K.; Takaaki, A.; Keizo, S.; Hiroshi, M.; Takashi, I. *J Bioact. Comp. Pol.* **1996**, *11*, 169.

20. Ito, Y.; Kotawa, M.; Chung, D.J.; Imanishi, Y. *Bioconjugate Chem.* **1993**, *4*, 358.

21. Ranucci, E.; Spagnoli, G.; Sartore, L.; Ferruti, P.; Caliceti, P.; Schiavon, O.; Veronese, F.M. *Macromol. Chem. Phys.* **1994**, *195*, 3469.

22. Morpurgo, M.; Schavon, O.; Caliceti, P.; Veronese, F.M. *Appl. Biochem. Biotechnol.* **1996**, *56*, 59.

23. Bonora, C.; Baldi A.; Schiavon, O.; Ferruti, P.; Veronese, F.M. *Tetraedron Letters* **1996**, *37*, 4761.

24. Bovara, R.; Ottolina, G.; Carrea, G.; Ferruti, P.; Veronese, F.M. *Biotechnol. Letters* **1994**, *16*, 1069.

25. VanSpect, B.W.H.; Seinfeld, H.; Brendol, W. Hope-Seyler's Z. *Physiol. Chem.* **1973**, *374*, 1964.

26. Sartore, L.; Ranucci, E; Ferruti, P.; Caliceti, P.; Schiavon, O.; Veronese F.M. *J. Bioact. Comp. Pol.* **1994**, *9*, 411.

27. Ferruti P.; Ranucci, E.; Sartore, L.; Bignotti, F.; Mrchisio, M.A.; Biancardi, P., Veronese, F.M. *Biomaterials* **1994**, *15*, 1235.

28. Caliceti, P.; Schiavon, O.; Morpurgo, M.; Veronese, F.M.; Sartore, L.; Ranucci, E.; Ferruti, P. *J. Bioact. Comp. Pol.* **1995**, *10*, 103.

29. Masera, G.; Jankovic, M.; Zurlo, M.G.; Locasciulli, A.; Rossi, M.R.; Oderro, C.; Biostat, M.R. *J. Pediatrics* **1982**, *100*, 152.

30. Veronese, F.M.; Monfardini, C.; Caliceti, P.; Schiavon, O.; Scrawen, M.D.; Beer D. *J. Control. Rel.*, **1996**, *40*, 199.

31. Lang, G.M.; Kierek-Jaszczuk, D.; Rector, E.S.; Milton, A.D.; Emmrich, F.; Sehon A.H. *Immunol. Lett.* **1992**, *32*, 247

32. Allen, T.M. *Adv. Drug Del. Rev.* **1994**, *13*, 285.

33. Torchilin, V.P.; Papisov, M.J. *Liposome Res.* **1994**, *4*, 725.

34. Blume, G.; Cevc, G. *Biochem. Biophys. Acta* **1993**, *1146*, 157.

35.Torchilin, V.P.; Trubetskoy, V.S.; Whiteman, K.R.; Caliceti, P.; Ferruti, P.; Veronese, F.M. *J. Pharm. Sciences* **1995**, *84*, 1049.

36. Moruyama, K; O. K. Kuiumi, S; Ishida, O; Yamauchi, M; Kikuchi, H.; Iwatsuru, M. *Int. J. Pharm.***1994**, *111*, 103.

37. Woodle, M.C.; Engbers, C.M., Zalipsky, S. *Bioconjugate Chem.* **1994**, *5*, 493.

Chapter 14

Covalent and Noncovalent Adducts of Proteins with Water-Soluble Poly(alkylene oxides)

Irina N. Topchieva

Department of Chemistry, Lomonosov State University, Lenin Hills, Moscow 119899, Russia

Protein conjugates with PEG and amphiphilic block copolymers of ethylene oxide and propylene oxide (Proxanols) were synthesized. Four types of conjugates ranging in the placement of hydrophobic block and type of polymer chains distribution were obtained. In parallel methods of thermoinduced and high-pressure induced complexation were developed for the synthesis of non-covalent adducts between proteins and Proxanols. Covalent and non-covalent adducts based on α-chymotrypsin (CHT) retain high enzymatic activity. They were characterized by higher thermostability with regard to the native enzyme. Membranotropic properties of conjugates were demonstrated through the study of their translocation across the cell membrane of T-lymphocytes and the investigation of catalytic properties in hydrated reversed micelles. Conformational models of polymer-protein conjugates were suggested. It was assumed that conjugates form in aqueous solutions compact structures resembling intramolecular micelles.

Covalent attachment of poly(ethylene glycol) (PEG) to proteins is a progressive way of imparting new useful properties to a protein molecule, such as reduced immunogenicity, protracted retention in circulation, and, in some cases, by the enhanced stability (1-4). These features are due to the effect of "steric stabilization" of protein globule caused by the attachment of polymer chains. One of the approaches to the construction of new polymer-protein conjugates is the introducing of additional fragments (blocks) in a polymer using copolymerization process. The forming polymers are Proxanols - diblock copoly(ethylene oxide)-poly(propylene oxide) copolymers (Pro) (5). They are amphiphilic compounds, in which poly(ethylene oxide) (PEO) and poly(propylene oxide) (POP) are hydrophilic and hydrophobic parts of compounds, respectively. The coupling of functionalized derivatives of these copolymers to surface functional groups of the protein led to the formation of conjugates of proteins with Proxanols (Pro) (6).

Compared with the PEO-protein conjugates, the conjugates based on Proxanols are characterized by a greater structural diversity, manifested, firstly, by the possibility of attaching copolymers of different numbers, compositions, and arrangements of the blocks and, secondly, by the possibility of different arrangements of the polymer chains on the surface of the protein globule. The amphiphilic character of Proxanol-protein conjugates provides new functions of modified proteins such as capability of translocation across biological membrane (*7, 8*), maintenance of effective solubilization and transport of insoluble biologically active compounds (*9, 10*), as well as the ability to interact with surfactants of different chemical nature. In this review I report new approaches to the synthesis of polymer-protein adducts based on the formation both covalent and non-covalent bonding between the polymer and protein molecules.

Synthesis and Properties of the Conjugates

Conjugates of proteins with poly(alkylene oxide)s were synthesized by the interaction of protein ε-amino groups with monoaldehyde derivatives of poly(alkylene oxide)s resulting in the formation of labile iminium groups, which were subsequently reduced by sodium cyanoborohydride. As a result, polymer chains became grafted to the protein molecule through stable secondary amine bonds. The total charge of the protein was not affected by the process of modification. The grafting of polymers to CHT was performed in the presence of reversible inhibitor of the active site of the enzyme, N-benzoyl or N-acetyl-L-tyrosine.

The following monoaldehyde derivatives of poly(alkylene oxide)s were used for protein modification:

1. Monomethoxy-PEG monoaldehyde (PEG-CHO)

$$CH_3-(OCH_2CH_2-)_{m+n}OCH_2CHO \qquad (1)$$

2. RPE block copolymer monoaldehyde (RPE-CHO)

$$C_4H_9-(OCHCH_2)_m-(OCH_2CH_2)_n-OCH_2CHO \qquad (2)$$
$$\quad\quad\quad\; |$$
$$\quad\quad\quad CH_3$$
$$\quad\quad POP \qquad\quad PEO$$

3. REP block copolymer monoaldehyde (REP-CHO)

$$C_4H_9-(OCH_2CH_2)_n-(OCHCH_2)_m-OCH_2CHO \qquad (3)$$
$$\quad\quad\quad\quad\quad\quad\quad\quad |$$
$$\quad\quad\quad\quad\quad\quad\quad\quad CH_3$$
$$\quad\quad\; PEO \qquad\quad POP$$

where m = 14, n = 25.

Evidently the only difference between block copolymers 2 and 3 is in the position of PEO and POP blocks relative to the aldehyde group. Using these polymer reagents two main types of conjugates with hydrophobic blocks at the periphery or inside conjugate structure were synthesized.

Molecular models of the conjugates are presented in Figure 1. It is seen that conjugates have a star-shaped structure where protein occupies the central place with polymer chains growing from it as branches. The length of polymer chains is of the same order or even exceeds the diameter of protein molecule.

The following approach was developed to evaluate the type of the distribution of polymer chains at the protein surface. We studied the interaction between polymers grafted onto the protein and bulky cyclic reagent β-cyclodextrin (CD), that forms insoluble inclusion-type complexes with POP-blocks of Pro (*11*). The main criterium of statistic character of distribution is the capacity of polymers interacting with CD. In the other case when polymer chains form clustered structures their complexation with CD becomes highly improbable. Really it was shown that conjugates based on RPE synthesized in aqueous buffer media do not form complexes with CD, whereas conjugates obtained in water-alcohol media do form insoluble complexes (*12*). Hence, in the first case polymer chains form clustered structures at the surface of the protein (type IIa), and in the second - polymers are randomly distributed on the protein surface (type IIb).

Enzymatic Properties of Conjugates. Kinetic characteristics of the hydrolysis of BTEE, catalyzed by CHT-conjugates are given in Table 1. These data show that conjugates retain high enzymatic activity toward both investigated substrates - N-benzoyl-L-tyrosine ethyl ester (BTEE) and azocasein. It should be noted that the attempts to modify CHT in the absence of the inhibitor led to the considerable loss of enzymatic activity (*13*). There are insignificant changes in the values of Michaelis constant, K_m and catalytic constant, k_{cat}, for hydrolysis of BTEE by conjugates with the degree of modification 3-6, comparing to the unmodified enzyme. Analogous results were obtained for hydrolysis of high molecular weight substrate, azocasein (data not shown). Noticeable decrease of the enzymatic activity was observed only for conjugate with high degree of modification ($n = 11$). The same tendency was described for PEG-CHT conjugates with the degree of modification 10-14 (*14*). It is seen from Table 1 that the loss of activity may be accounted for either by a decrease in content of active protein in conjugates, or an increase in K_m (obviously due to steric hindrance).

Thermal Stability of Conjugates. The characteristic feature of conjugates on the basis of Pro consists in the possibility to regulate thermal stability of modified protein by the variation of a number of parameters, such as the degree of modification, the number and placing of blocks in copolymres, hydrophilic-lipophilic balance and the type of polymer distribution on the surface of protein globule. We studied the thermal inactivation of all types of conjugates on the basis of CHT and compared it with thermal stabilities of PEG-CHT conjugate and of native CHT (*13*). Time dependences of thermoinactivation of CHT and its conjugates with polymers are presented in Figure 2. The

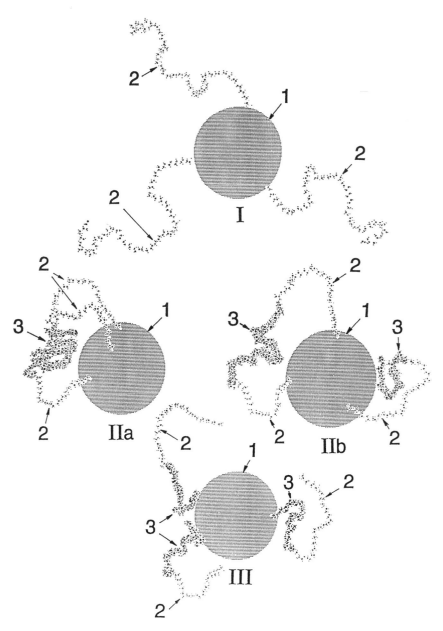

Figure 1. Molecular models of conjugates based on CHT: **(I)** with PEG (M = 1900); **(IIa)** with RPE having a clustered distribution of polymer chains; **(IIb)** with PRE having an uniform distribution of polymer chains; **(III)** with REP. The POP blocks of Proxanols are shown shaded; the radius of the protein globule is 2.5 nm. 1 - protein, 2 - PEG, 3 - POP-block. (Adapted from ref. 6).

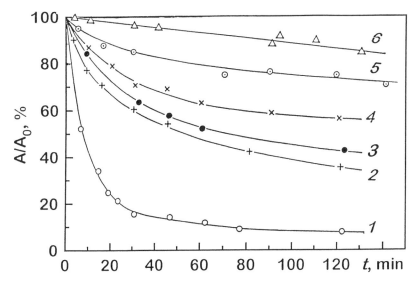

Figure 2. Kinetic curves for the thermal inactivation of CHT and its conjugates: (*1*) CHT; (*2*) (REP)$_3$-CHT; (*3*) (RPE)$_3$-CHT-a; (*4*) (RPE)$_6$-CHT-a; (*5*) (RPE)$_5$-CHT-b; (*6*) (PEG)$_9$-CHT; 0.2 M Tris-HCl buffer, pH 8.0, 45°C; the enzyme concentration in each specimen was 0.1 μM. (Adapted from ref. 6).

thermostability of conjugates of CHT decreases in the following order: (PEG)$_9$-CHT > (RPE)$_5$-CHT-b > (RPE)$_6$-CHT-a > (RPE)$_3$-CHT-a > (REP)$_3$CHT-a > CHT. It should be noted that all Pro-CHT conjugates with the exception of (RPE)$_5$-CHT-b are of the IIa type.

Table I. Enzymatic Properties of CHT Conjugates with PEG and Proxanols

Protein, conjugate	Enzymatic hydrolysis of caseine	Enzymatic hydrolysis of BTEE			Protein containing active site, %
	Activity, %	Activity, %	k_{cat}, s^{-1}	$K_m \cdot 10^5$, M	
ChT	100	100	31.2±0.5	6.27±0.94	80
(RPE)$_5$-ChT-a	102	97	42.4±2.8	13.0±2.9	61
(RPE)$_3$-ChT-b	103	115	-	-	80
(RPE)$_6$-ChT-b	96	98	32.5±1.2	6.18±0.82	67
(RPE)$_{11}$-ChT-b	50	57	37.0±0.4	21.7±3.5	40
(PEG)$_6$-ChT	97	98	40.4±0.8	9.3±1.5	80
(REP)$_3$-ChT	98	100	-	-	70

Adapted from ref. 6.

It is seen that the rate of thermal inactivation of the enzyme decreases significantly as a result of the attachment of three polymer chains. The stabilizing effect becomes even more pronounced when the degree of CHT modification is increased. It is remarkable that conjugate (RPE)$_5$-CHT-b with random mode of distribution of polymer chains possesses higher thermostability than conjugate (RPE)$_6$-CHT-a with clustered distribution. At the same time the difference in the placement of hydrophobic block in conjugates with equal number of copolymer chains leads to the change in their thermostability: conjugates of type IIa and more stable than conjugates of reversed type (III).

The data on the thermal stability of the conjugates, obtained by the kinetic method have been supplemented by the results of the thermal denaturation of the conjugates with differential scanning calorimetry (DSC). A significant alteration of the shape of the heat absorption peak is observed.

Apart from the first maximum at 47.6-49.6°C corresponding to the unmodified CHT, the thermograms of the conjugates are characterized by a second maximum or shoulder at higher temperatures (54.5-55.5°C). This fact demonstrates that the conjugates are thermally more stable that the unmodified protein. Certain experiments show that a second maximum on the heat absorption curves of the conjugates is apparently due to the melting of polymer-protein complexes, forming at the heating.

Interaction of Conjugates with Diphilic Compounds. One of the unique properties of protein conjugates with Pro is their ability to interact with both synthetic (surfactants) and natural (lipids) amphiphilic compounds (*13, 15*). The functional aspect of these interactions was demonstrated in the study of thermoinactivation. Thermal stability of conjugates may be enhanced by the addition of free Pro or PEG into the incubation system. One can assume that an increase in the stability of Pro-CHT conjugates compared to the unmodified protein is due to the ability of polymer chains to screen hydrophobic domains at the surface of protein (*16, 17*). An additional increase in the stability of CHT-Pro conjugate in the presence of free polymers can be a result of interaction between free and bound polymers leading to the formation of a thick coat surrounding protein globule.

Addition of egg yolk lecithin and CD produces an opposite effect on the stability of protein in conjugate (*18*). All additives themselves do not affect the behavior of the native enzyme. Probably these additives are stronger competitors for hydrophobic bond formation with Proxanols than the corresponding domains of the protein globule. As a result the steric arrangement of polymer chains is changed with the hydrophobic domains of protein being exposed again, thus destabilization of the enzyme structure takes place.

Kinetic Properties of Conjugates in Hydrated Reversed Micelles. Another aspect of using surface active compounds for studying conjugate properties is the application of hydrated reversed micelles (HRM) in octane. These systems are micelles with polar heads interior comprising polar heads and hydrophobic tails as surface layer. The most popular systems are based on sodium salt of bis(2-ethylhexyl)-sulfosuccinic acid (AOT) (*19*). Solubilization of water resulting in the formation of HRM is characterized by narrow size and form distribution. Enzymes entrapped in a water pool of HRM are able to preserve their catalytic activity. Studying of catalysis of various enzymes in HRM may give useful information about the enzymes, namely about their dimensions and membranotropic properties. Two sets of experiments with conjugates in HRM were performed (*20*). The correlation between the catalytic activity (deacylation constant k_3) of native CHT and its conjugates with poly(alkylene oxides), entrapped in HRM, and the value of the $[H_2O]/[AOT]$ ratio, corresponding to the maximum on the curves k_3 versus $[H_2O]/[AOT]$ ratio, promotes estimating M and dimensions of the enzyme (*21*). Thus the thickness of the polymer layer in all of the conjugates was calculated as 0.65 nm. From the comparison of this value with the length of an isolated polymer chain, estimated as ~3 nm (*22*), it was concluded that polymer chains in conjugates form a dense layer near the surface of the protein globule. Membranotropic properties of enzymes entrapped in HRM can be demonstrated through the study of the catalytic activity at various concentrations of surfactant. The enzymes which are able to interact with biological membranes are characterized by the catalytic activity - concentration of AOT (at constant $[H_2O]/[AOT]$) relationship of the catalytic activity when AOT concentration is varied. The dependences of k_3 value on the concentration of AOT at $[H_2O]/[AOT]$, corresponding to the maximum of a bell-shaped curve, show that the value of k_3 for native CHT as well as PEG-CHT conjugate remains constant on variation of [AOT]. However, in the case of conjugates of types II and III, the value of k_3 decreases markedly with increasing of AOT concentration. These results suggest that the conjugates studied interact with micellar matrix, that may be considered as a demonstration of their membranotropic properties.

Conformational Models of Conjugates on the Basis of Proteins and Proxanols. Using reverse-phase hydrophobic chromatography for evaluation of the relative hydrophobicity of Pro-protein conjugates based on Bauman-Birk proteinase inhibitor (BBI) and CHT it was shown that the retention volume of the conjugate is less than that of the native protein, thus indicating its lesser affinity to the hydrophobic sorbent (23). This may be accounted for by the fact that surface of the protein globule of the conjugate is covered mainly with the hydrophilic blocks of the polymer due to the intramolecular interactions of the hydrophobic blocks of Pro and the hydrophobic regions of the protein molecule.

These combined data, received on the basis of physico-chemical and kinetic properties of the conjugates, indicates that the excess of hydrophobicity may be compensated by the formation of intramolecular structures similar to micelles in which poly(propylene oxide) blocks "adhere" to the surface of protein. It should be particularly emphasized that these structures possess certain conformational lability. This property is revealed in the presence of hydrophobic or amphiphilic "partners", such as lipids (15), Pro (13), CD (12), and is realized in the formation of mixed associates or complexes.

Membranotropic Properties of the Conjugates. A fundamentally new property of the conjugates of proteins with Pro is their ability to interact with biological membranes (7, 8). Since the initial proteins are not membranoactive, one can postulate that this property is due to the presence of the membranotropic polymer chains of Pro.

In order to investigate the translocation of the conjugates across a biological membrane, use was made of the approach involving the study of the influence of these compounds on the rate of consumption of oxygen by liver mitochondria and the thymus lymphocytes. It is known that diphilic substances, capable of being translocated through a cell membrane, can reach the NADH-dehydrogenase complex, located in the mitochondria responsible for cell respiration, and inhibit the NADH-dehydrogenase activity. This is manifested by a decrease in the rate of consumption of oxygen. When such experiments were carried out with the conjugates, it was found that some of them are capable of suppressing cell respiration of T-lymphocytes. This indicates their translocation through the cell. We investigated the dependence of the translocation of the conjugates on the nature of the attached polymer chains and of their distribution relative to the protein globule. It was shown that only type **II** conjugates in which the hydrophobic blocks is on the periphery of the molecule, are able to suppress respiration of the lymphocytes. On the other hand, neither the PEG-protein conjugate, nor the conjugate with the hydrophilic block at the periphery of the molecule (type **III**), and the unmodified CHT affect respiration in the lymphocytes. The same results were obtained for the conjugates based on BSA and cytochrome *c*. This means that the capacity for translocation is determined by the polymer chains and does not depend on the nature of the protein. The conjugates of the three proteins investigated showed a similar effect also on the respiration in isolated mitochondria. It was shown also that only the conjugates with a clustered distribution of the chains cause a significant inhibition of respiration. At the same time conjugates of IIb and III types, which display membranotropic properties in HRM, did not affect the respiration of lymphocytes. Probably, these conjugates are able to interact with biological membranes, for example, by intercalation in the lipid bilayer, but do not pass through them. Some steps involving the mechanism of the translocation may be as follows: the insertion of the conjugate in the membrane via the interaction of diphilic polymer chains with lipids, structural changes in protein globule facilitated by membrane environments and the following import of partly denatured intermediate .This suggestion would be consistent with the hypothesis of O.B.Ptitsyn about the role of membrane environment in the transmembrane translocation of proteins (*24*).

Non-Covalent Complexes between Proteins and Poly(alkylene oxides)

Until recently it was assumed that PEG does not form complexes with proteins. Whereas certain functions of PEG contradict to this idea. For instance as a cosolvent PEG could enhance the refolding of some proteins and inhibit the aggregation of intermediates without changing the characteristics of the proteins (*25, 26*). In this respect it behaves similarly to chaperones. It was naturally to suggest that PEG under certain conditions is able to form non-covalent complexes with proteins.

We demonstrated in relation to CHT that stable Pro-protein complexes can be obtained when their solutions are heated to temperatures corresponding to the onset of the thermal denaturation of the protein (*27, 28*). The DSC methods appeared to be very sensitive to complex formation. It was shown that PEG does not affect the fusion of the protein globule. In contrast to this, in the presence of Pro, a shift of the fusion peak toward higher temperatures and an increase of the temperature of fusion were observed. Hence it follows that during heating the CHT and Pro are bound to one another in some way, giving rise to a new product, in which the protein component acquires a somewhat higher thermal stability. The thermally induced binding of CHT and Pro is irreversible under the experimental conditions.

The number of polymer chains incorporated into the complex increases from 4 to 11 when the temperature rises from 44 to 60°C. At higher temperatures the complexes are enriched with polymers and have lower enzymatic activity. The same tendency towards an increase in the number of polymer chains in the complex is observed also with increase in the molar excess of the polymeric component at a fixed upper temperature limit.

It should be noted that the protein fraction isolated from the mixture of CHT and PEG after heating at 60°C consists of pure protein. This means that there are no interactions between these two components under the experimental conditions.

The most remarkable feature of the investigated complexes is their ability to retain enzyme properties (28). Kinetic constants of hydrolysis of BTEE in the presence of complexes are presented in Table 2. It is seen that the increasing the upper temperature limit of heating of the mixture of CHT and Pro (or time of incubation at fixed upper temperature) decreased the content of the active sites in the complex. The values of K_m in this series of experiments regularly increase, indicating a decrease in affinity between complex and substrate. However, the increase in k_{cat} in this series indicates an increase in catalytic efficiency of the active sites of CHT in the complexes. Analogous effects of superactivity were observed earlier in studies of enzymatic hydrolysis of N-acetyl-L-tyrosine ethyl ester in the presence of CHT in HRM (29-31).

Table II. Enzymatic Properties of Complexes Based on α-Chymotrypsin and Proxanol

Complex	Preparation conditions		Protein containing active site, %	$K_m \cdot 10^5$, M^a	k_{cat}, s^{-1}
	temperature of heating, °C	CHT:RPE molar ratio			
(RPE)$_5$-CHT	44	1:20	82	7.1±0.8	25±3
(RPE)$_5$-CHT	44	1:20	83	9.5±0.7	27±1
(RPE)$_6$-CHT	48	1:20	74	13.2±0.8	30±2
(RPE)$_7$-CHT	48	1:20	57	16.7±0.8	32±2
(RPE)$_{10}$-CHT	52	1:20	32	13.9±0.6	48±4
(RPE)$_{13}$-CHT	52	1:20	14	18±2	166±19
Native CHT			92	5.6±0.4	24±2

[a] The kinetic constants were obtained by studying the hydrolysis of BTEE (pH 8.0; 25°C). The errors were determined with a legal probability of 0.95.

It was shown that non-covalent complexes as well as the conjugates are characterized with the increased thermal stability of the protein (28).

The study of enzymatic properties of temperature-induced complexes showed that complex formation is accompanied by side reactions resulting in inactivation of the enzyme. Thus the problem to elaborate new methods of synthesis of non-covalent adducts under conditions retaining their enzymatic activity has arose. Therefore high pressures methods were used (*32*). It is well documented that the denaturation of proteins takes place at high pressures (~400 MPa) (*33*). In the field of moderate non-denaturating pressures the reversible conformational changes stemming from the intrinsic compressibility of proteins and changes in their solvation state take place. It is known too that the binding of proteins with specific ligands may be accompanied with the decrease of the volume (*34*). In this case the increase of pressure would favor the shift of equilibrium towards the complexation.

Taking into account these facts and cooperative character of interaction between the macromolecular components (protein and Pro) we showed that the increase of pressure in the system under study favored the complexation processes. Really, polymer-protein adducts are formed in the system containing a mixture of protein and Proxanol under elevated pressures above 1 MPa. An average number of polymer chains in complexes obtained in the range of pressure from 1.1 to 400 MPa does not reveal systematic dependence from the value of applied pressure and corresponds to 7 in contrast to complexes forming at heating. The most important property of non-covalent adducts forming at high pressures is the retaining of catalytic activity. Kinetic analysis of enzymatic properties of these complexes shows that the content of active protein, the values of K_m and k_{cat} for complexes do not differ from the corresponding values for native CHT over the whole range of pressures used.

The comparison between the structure of temperature induced and pressure induced complexes was done on the basis of their thermal stability (Table 3). Kinetic data are treated assuming that they may be described by the sum of two exponential terms:

$$A/A_0 = A_1 \exp(-k_1 t) + A_2 \exp(-k_2 t)$$

where A/A_0 is the relative enzyme activity; A_1 and A_2 are the relative contribution of fast and slow stages correspondingly ($A_1 + A_2 = 1$); k_1 and k_2 are the constants of the rate of thermoinactivation for each stage. The first stage of the process is so fast that the value k_1 in all the cases except for the complex (Pro)$_4$-ChT, can not be determined. The correlation of values of k_2, characterizing the slow stage of the process shows that the change of CHT for its complexes with Pro produces marked stabilization of the enzyme, with k_2 decreases as the temperature or pressure increases. Based on these regularities and on the numerical values of k_2 we can say that complexes of both types are characterized by approximately similar thermal stability and therefore identical type of steric arrangement of polymer chains at the periphery of protein globule.

Table III. Parameters of Thermal Inactivation of α-Chymotrypsin and Its Complexes with PEG and Proxanol

Complex	Preparation conditions*	A_1	k_1, s^{-1}	A_2	k_2, s^{-1}
CHT	-	0.39 ±0.02	-	0.61 ±0.02	0.16 ±0.001
(REP)$_7$-CHT	1.1 Mpa	0.44 ±0.02	-	0.56 ±0.02	0.0087 ±0.006
(REP)$_4$-CHT	13.0 Mpa	0.53 ±0.02	0.09 ±0.01	0.47 ±0.02	0.0043 ±0.005
PEG	16.0 Mpa	0.08 ±0.01	-	0.92 ±0.01	0.0112 ±0.0004
(REP)$_6$-CHT	48°C	0.21 ±0.02	-	0.79 ±0.02	0.0055 ±0.0004
(REP)$_{10}$	52°C	0.21 ±0.02	-	0.79 ±0.02	0.0042 ±0.0003

*The molar ratio of CHT : polymer is 1 : 20

At the same time catalytic properties of temperature and pressure induced complexes differs essentially that demonstrates structural alterations between them. Probably thermoinduced complexation results in the deeper penetration of PPO blocks into protein globule than in the case of pressure induced complexation. Probably in the last case the desolvation of protein and Pro being a driving force for interaction between macromolecules does not result in the change of the of protein globule structure.

The remarkable feature of pressure induced method is the possibility of receiving polymer-protein complexes based on both Pro and PEG. The complexes between CHT and PEG (M = 1900) have approximately the same composition as for complexes based on Pro, characterized by the same kinetic properties and possess higher thermal stability as compared to native CHT (Table 3). Thus, we propose a new soft method of manufacturing of catalytically active non-covalent adducts between CHT and poly(alkylene oxides), Pro and PEG, based on the action of high pressure.

Conclusion

In these review two general approaches to the synthesis of polymer-protein adducts are demonstrated . The first is the covalent modification of proteins with polymeric reagents, monofunctional derivatives of poly(ethylene oxide) - poly(propylene oxide) copolymers. This method made it possible to realize a fundamental property of conjugates - membranotropicity and ability for the translocation across the biological membranes. These properties are not distinctive for PEG-protein conjugates. The second approach is the supramolecular assembly of proteins with poly(alkylene oxides) favoured by the action of elevated temperatures or pressures. Now it is prematurely to tell about the advantages or disadvantages of these approaches. The practice will be the main criterium. Nevertheless we shall now highlight the distinctive features of non-covalent adducts that were revealed or may be predicted on the basis of our results:

1. The simplicity of synthetic procedure, in particular, the employment of commercial polymers instead of functionalized derivatives.
2. A full retention of enzymatic properties relative to low molecular substrates.
3. The ability of adducts for dissociation to starting materials that may be used for controlled release of enzyme in experiments *in vivo*.

Thus, both types of polymer-protein adducts may be regarded as the basis for designing new materials in biotechnology and medicine.

List of Abbreviations: AOT - bis(2-ethylhexyl)-sulfosuccinic acid; BBI - Bauman-Birk proteinase inhibitor; BTEE - N-benzoyl-L-tyrosine ethyl ester; CD - β-cyclodextrin; CHT - α-chymotrypsin; DSC - differential scanning calorimetry; HRM - hydrated reversed micelles; PEG - poly(ethylene glycol); PEO - poly(ethylene oxide); POP - poly(propylene oxide); Pro - Proxanol; REP - diblock poly(ethylene oxide)-poly(propylene oxide) copolymer; RPE - diblock poly(propylene oxide)-poly(ethylene oxide) copolymer.

Acknowledgments. The study was supported by Russian Fundamental Research Fund (Project code 96-03-33519a/300); International Scientific Fund (Grant MPE000) and INTAS (Grant 93-2223).

Literature Cited

1. Abuchowski, A.; Mc Coy, I. R.; Palczuk, N. C.; Davis, F. F. *J.Biol. Chem.* **1977**, *252*, 3582.
2. Abuchowski, A.; Van Es, T.; Palczuk, N. C.; Davis, F. F. *J. Biol. Chem.* **1977**, *252*, 3578 .
3. *Poly (Ethylene Glycol) Chemistry: Biotechnical and Biomedical Applications;* Harris J. M., Ed; Plenum Press, New York, 1992.
4. Veronese, F. M.; Caliceti, P.; Pastorino, A.; Schiavon, O.; Sartore, L. *J. Control. Release.* **1989**, *10*, 145.
5. Lundsted, L. G.; Schmolka, I. R. In *Block and Graft Copolymerization*; Ceresa R.J., Ed; John Wiley, New York, 1976, pp. 174-205.
6. Topchieva, I. N.; Efremova, N. V.; Khvorov, N. V.; Magretova, N. N. *Bioconj. Chem.* **1995**, *6*, 380.
7. Kirillova, G. P.; Mochova, E. N.; Dedukhova, V. L.; Tarakanova, A. N.; Ivanova, V. P.; Efremova, N. V.; Topchieva, I. N. *Biotechnol. Appl. Biochem.*, **1993**, *18*, 329.
8. Topchieva, I. N.; Mochova, E. N.; Kirillova, G. P.; Efremova, N. V. *Biochemistry (Moscow)*, **1994**, *59*, 11.
9. Topchieva, I. N.; Momot, I. G.; Kurganov, B. I.; Chebotareva, N. A. *Dokl. Acad. Nauk SSSR*, **1991**, *316*, 242 (in Russian).
10. Topchieva, I. N.; Efremova N. V. *Biotechnol. Gen. Eng. Rev.* **1994**, *12*, 1008.

11. Topchieva, I. N.; Blumenfel'd, A. L.; Klyamkin, A. A.; Polyakov, V. A.; Kabanov, V. A. *Polymer Science,* **1994,** *36(2)*, 221.
12. Efremova, N. V.; Topchieva, I. N. *Biokhimiya,* **1993,** *58,* 1071 (in Russian).
13. Efremova, N. V.; Mozhaev, V. V.; Topchieva, I. N. *Biokhimiya,* **1992,** *57,* 342 (in Russian).
14. Chiu, H-C; Zalipsky, S.; Kopeckova, P.; Kopecek, J. *Bioconj. Chem.,* **1993,** *4,* 290.
15. Topchieva, I. N.; Osipova, S. V.; Banazkaya, M. I.; Val'kova, I. A. *Dokl. Acad. Nauk SSSR,* **1989,** *308,* 910 (in Russian).
16. Mozhaev, V. V.; Martinek, K. *Enzyme Microb.Technol.,* **1984,** *6,* 50.
17. Mozhaev, V. V.; Martinek, K.; Berezin, I. V. *CRC Crit.Rev.Biochem.* **1988,** *23,* 235.
18. Topchieva, I. N.; Efremova, N. V.; Kurganov, B. I. *Rus. Chem. Rev.,* **1995,** *64(3)* , 277.
19. Martinek, K.; Levashov, A. V.; Klyachko N. L.; Beresin, I.V. *Dokl. Acad. Nauk SSSR,* **1977,** *236,* 920.
20. Efremova, N. V.; Kurganov, B. I.; Topchieva, I. N. *Biochemistry (Moscow),* **1996,** *61,* 121.
21. Eicke, H-F.; Rehak, J. *Helv.Chem.Acta,* **1976,** *59,* 2883.
22. Bailey, F. E., Jr.; Koleske, J. V. In *Nonionic Surfactants,* Schick M.J., Ed; Marcel Dekker Inc., New York 1966. pp`.794-822
23. Larionova, N. I.; Gladysheva, I. P.; Topchieva, I. N.; Kazanskaya, N. F. *Biokhimiya,* **1993,** *58,* 1658.
24. Bychkova, V. E.; Ptitsyn, O. B. *Chemtracts-Biochem. Molec. Biol.,* **1993,** *4,* 133.
25. Arakawa, T.; Timasheff, S. N. *Biochemistry,* **1985,** *24,* 6756.
26. Otamiri, M.; Adlercreutz, P.; Mattiasson B. *Biotechnol. Bioengin.,* **1994,** *44,* 73.
27. Topchieva, I. N.; Efremova, N. V.; Snitko, Ya. E.; Khvorov, N. V. *Dokl. Ross. Akad. Nauk,* **1994,** *339,* 498.
28. Topchieva, I. N.; Snitko, Ya. E.; Efremova, N. V.; Sorokina, E. M., *Biochemistry (Moscow),* **1995,** *60,* 131.
29. Ishikawa, H.; Noda, K.; Oka, T. *J. Ferment. Bioeng.,* **1990,** *70,* 381.
30. Ishikawa, H.; Noda, K.; Oka, T. *An. Enzyme Eng.* **1990,** *613,* 529.
31. Martinek, K.; Levashov, A. V.; Klyachko, N. L.; Khmel'nitskii, Yu. L.; Beresin, I. V. *Biol. Membrany* **1985,** *2,* 669 (in Russian).
32. Topchieva, I. N.; Sorokina, E. M.; Kurganov, B. I.; Zhulin, V. M.;, Makarova, Z. G. *Biochemistry (Moscow),* **1966,** *61,* 746.
33. Cross, M.; Jaenicke, R. *Eur. J. Biochem.,***1994,** 221, 617.
34. Jaenicke, R.; Gregori, E.; Laepple, M. *Biophys. Struct. Meth.* **1979,** *6,* 57.

Chapter 15

Isolation of Positional Isomers of Mono-poly(ethylene glycol)ylated Interferon/α-2a and the Determination of Their Biochemical and Biological Characteristics

Seth P. Monkarsh, Cheryl Spence, Jill E. Porter, Alicia Palleroni, Carlo Nalin, Perry Rosen, and Pascal Bailon[1]

Departments of Biopharmaceuticals and Oncology, Hoffmann-La Roche, Inc., Nutley, NJ 07110

MonoPEGylated interferon α-2a (monoPEG-IFN) is produced by conjugating interferon α-2a to a chemically reactive form of polyethylene glycol-5000 in 1:1 ratio, via a urea linkage. Potentially, all 11 lysine residues and the N-terminal amino group in interferon α-2a are available for PEGylation. Thus monoPEG-IFN is a mixture of individual species, each having a unique site of PEGylation. A method involving sulfopropyl-high performance liquid chromatography (SP-HPLC) which exploits the minute local charge differences between the individual monoPEG-IFN α-2a species (positional isomers) is used to separate them. Eleven positional isomers were isolated by this method, each having a specific lysine as their PEGylation site indicating all eleven lysines of interferon α-2a were PEGylated. However, N-terminal amino group was not derivatized. All isolated species exhibit antiviral and antiproliferative activities.

In recent years pharmaceutically active proteins have been conjugated with polyethylene glycol (PEG) to improve circulating half-life, bioavailability, solubility, stability, as well as to reduce immunogenicity and toxicity. For example interferon α-2a which is an anti-viral and anti-tumor agent has been conjugated to PEG resulting in improved pharmacokinetic properties and retention of biological activity (1).

Typically, conjugation of interferon α-2a (IFN) to chemically activated PEG results in a mixture of mono, di, tri and oligo PEGylated protein conjugates. These individual species could possibly be separated using a combination of conventional methods such as sodium dodecyl sulfate-polyacrylamide gel electrophoresis (SDS-PAGE), isoelectric focusing (IEF), high performance size exclusion chromatography (HPSEC), high performance ion exchange chromatography (HPIEC), reversed-phase high performance liquid chromatography (RP-HPLC) and high performance hydrophobic interaction chromatography (HPHIC), among others (2-7). Each of these individual species

[1]Corresponding author.

are comprised of positional isomers which has PEG attached to a different lysine of interferon α-2a. Here we report the isolation and characterization of the positional isomers of monoPEG-IFN. Main focus of this chapter is on separation methods for monoPEG-IFN isomers. The biochemical and biological characteristics of the separated isomers will be reviewed using data from another publication (Monkarsh et al., J. Anal Biochem., in press). A method (8), originally developed for the separation of deamidated and amidated forms of human interleukin-1α, as well as for isoforms of human interleukin-1β with and without N-terminal methionine, has been modified and adapted for the separation of monoPEG-IFN positional isomers. The theoretical number of positional isomers in a PEGylated molecule can be calculated from the following equation:

For N PEGylation sites, taken K at a time;

$$\frac{N!}{(N-k)! \bullet k!} = P;$$ wherein k is the number of sites actually

occupied by PEG and P is the possible number of positional isomers.

For monoPEG-IFN, k=1 and the possible number of positional isomers, P = 12. Similarly for diPEG interferon α-2a, k=2 and P= 66; for triPEG interferon α-2a, k = 3 and P = 220, and soon.

Materials and Methods

Preparation of mono-PEGylated interferon α-2a (PEG-IFN). MonoPEG-IFN is produced by conjugating interferon α-2a to a chemically reactive form of polyethylene glycol 5000 (PEG-5000, average M.W. 5300 daltons) in 1:1 molar ratio via a urea linkage (Figure 1). The aforementioned PEG reagent was obtained from Roche Kilo Lab 9). The PEGylation reaction was performed by mixing interferon (~6mg/ml) and the PEG reagent at a protein:reagent molar ratio of 1:3 in 5 mM sodium acetate, pH 10 buffer. After 1hour of mixing at room temperature, the reaction was stopped by adding one-twentieth volume of 1M glycine, pH 7.5. The pH of the reaction mixture was adjusted to 5-6 by the addition of one-twentieth volume of 1M sodium acetate, pH 4.

The reaction mixture was diluted 4-fold with 40mM ammonium acetate pH 4.5 and applied onto a Whatman CM-52 cellulose column (2mg protein/mL resin), previously equilibrated with the same buffer. The unabsorbed materials were washed away and the adsorbed PEG-IFN was eluted using a 0-0.5M linear sodium chloride gradient in the equilibration buffer. The column effluents were monitored by UV absorbance at 280nm. Fractions containing monoPEG-IFN was identified by SDS-PAGE.

Alternatively, the PEG-IFN was made using a solid-phase method (10). In the solid-phase method IFN was bound to a strong anion exchange resin and the PEGylation reagent in the appropriate buffer and concentration was recirculated through the column for 30 minutes. The excess reagent and reaction byproducts were washed away from the column and the PEGylated IFN was eluted with increasing salt concentrations. Elution conditions were chosen such that the unmodified IFN remained on the column. After elution was complete, the column was reequilibrated with the buffer. Then an amount of IFN equal to the amount PEGylated (eluted) was applied onto the column and the whole PEGylation process was repeated. This continuous PEGylation process could be repeated as many times as needed. Fractions containing the monoPEG-IFN from both the solution and solid-phase methods was pooled, concentrated by ultrafiltration and stored in 40 mM ammonium acetate, pH 5.0 buffer at a concentration of 9 mg/ml.

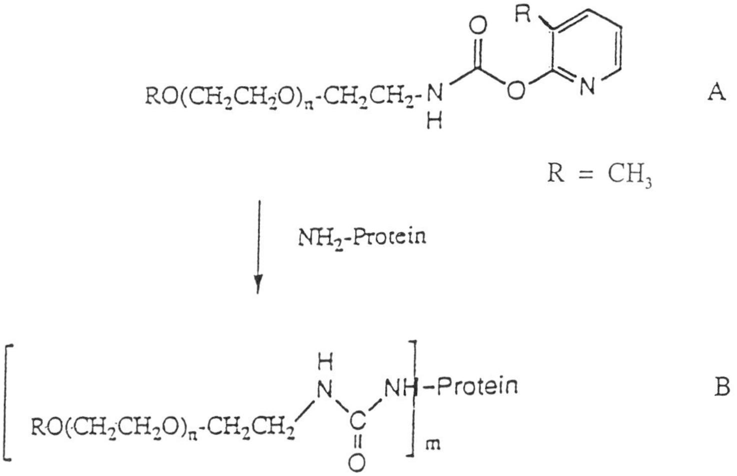

Figure 1. Reaction scheme of the formation of urea linkage between protein and the PEG reagent. A. PEG reagent and B. urea-linked PEG-protein conjugate.

Separation of PEG-IFN Positional Isomers

Analytical Scale. A Waters 626 biocompatible, analytical high-performance liquid chromatography system (Millipore, Milford, MA) was used. This HPLC system consisted of Model 626 LC pump, Model 600S powerline control module, model 486 variable-wavelength absorbance detector, a model 717 Plus autosampler and data was collected and processed using the Millenium software package version 2.00. A commercially available stainless-steel HPLC column containing TSK-Gel SP-5PW type, hydrophilic SP-strong cation exchange support (10 µm, 7.5 mm I.D. x 7.5 cm L.; TOSO HAAS Mahwah, NJ) was equilibrated with 3.7 mM acetic acid, pH 4.3 (pH adjusted very carefully with 10N sodium hydroxide; buffer A) at a flow rate of 1 ml/min. The stock monoPEG-IFN was diluted 10-fold with buffer A, pH adjusted to 4.3 with 1 N hydrochloric acid (Fisher Scientific) and 70 µg protein was applied to the column via autosampler. The column was washed with buffer A for 20 minutes, followed by a linear ascending pH gradient (pH 4.3-6.4) was performed from 0-100% of 10 mM dibasic potassium phosphate, pH adjusted to 6.4 with 6N hydrochloric acid (Pierce, Rockford, IL); buffer B). The pH gradient was completed in 220 minutes. The column effluents were monitored by UV absorbance at 280 nm. The column was regenerated by washing with 0.2 M dibasic potassium phosphate containing 0.5 M sodium chloride pH 8.0 for 15 minutes at 2.0 ml/min. The column was re-equilibrated with buffer A for 40 minutes at 2 ml/min to attain the starting pH conditions for the next run. All operations were carried out at ambient temperature (23-25°C) and all buffers were sparged continuously with ultra-high-purity-helium (99.999% pure; JWS Technologies, Piscataway, NJ).

Preparative Scale. A Waters Delta Prep 3000 preparative HPLC system equipped with a preparative TSK-gel SP-5PW HPLC column (13µm, 21.5 mm I.D. x 15cm L; TOSO HAAS) was used under the same buffer and gradient conditions as in the analytical scale except that the flow rate was 6 ml/min. Data were collected and analyzed using the Dynamax PC HPLC data system (Rainin Instruments, Woburn, MA). The PEG-IFN stock (9 mg/ml) was diluted 10-fold with buffer A and pH adjusted to 4.3, as before. Ten milliliters of the diluted PEG-IFN stock equivalent to 9 mg was applied onto the column. As each of the monoPEG-IFN species were eluted individual pool fractions were collected manually and rechromatographed on the same column to determine any changes in elution profile. Protein in each of the pooled samples was individually concentrated in a Centriplus-10 concentrator (Amicon, Beverly, MA, USA).

Protein Determination. Protein concentrations were determined by the amino acid composition analyses performed on a post-column amino acid analyzer with reaction with fluorescamine *(11).*

SDS-PAGE Analysis. Protein from the individual pools was subjected to SDS-PAGE analysis using non-reducing conditions according to the methods of Laemmli *(12) .*

Antiviral Assay. The antiviral activity of both IFN α-2a and the isolated PEG-IFN species were determined by the reduction of the cytopathic effect of vesicular stomatitis virus on Madin-Darby bovine kidney (MDBK) cells *(13)*

Antiproliferative Assay. Human Daudi cells (Burkitt's lymphoma) were maintained as stationary suspension cultures in RPMI 1640 media supplemented by 10% fetal bovine serum and 2mM glutamine (Grand Island Biologicals, Grand

Island, NY). The cells were screened and found to be free of mycoplasma. The assay was performed as previously described *(14)*. Antiproliferative activity of monoPEG-IFN and its separated positional isomers was determined by its inhibition of cell proliferation in human Daudi cells (a Burkitt's lymphoma) as measured by the decrease in incorporation of 3-H-thymidine into cellular DNA. The results were expressed as % inhibition calculated from the following formula:
% Inhibition = [(A - B)/A] x 100, where;

    A = cpm in control culture (cells incubated in medium alone)

    B = cpm in experimental culture

The 50% inhibition or IC_{50} value was caculated from % inhibition of of cell growth based upon 15.6-1000 pg/ml protein concentrations.

Results and Discussion

Application of the analytical SP-HPLC method to both the solution-phase and solid-phase derived monoPEG-IFN resulted in separation of monoPEG-IFN into at least 9 species for each application (Figure 2). These peaks elute between pH 5.6 and pH 6.3 and appear to have similar relative peak heights and retention time for both monoPEG-IFN preparations. Unmodified interferon α-2a elutes at pH 6.4, around 245 minutes (data not shown). In general, we have noted minor variations in the profiles of the 3 major peaks mainly due to column-to-column differences. However, the results show that the monoPEG-IFN from the solid-phase and solution-phase methods display similar fingerprint patterns indicating that the same species are present in the two preparations.

    The preparative scale elution profiles obtained with 9 mg monoPEG-IFN are illustrated in Figure 3. Both analytical and preparative elution profiles have similar finger prints except that the major peaks 5, 6 and 7 are broader and poorly resolved in the preparative mode. In addition, the preparative-scale separation of the monoPEG-IFN resulted in eleven species for each application. Elution profiles of the samples did not change upon rechromatography (data not shown). Amino acid composition analyses performed on the rechromatographed samples indicate agreement between the amino acid compositions of unmodified and modified IFN samples. SDS-PAGE analysis indicated that each preparative peak pool contained monoPEG-IFN, free of aggregates or degradation products (data not shown).

    Recovery of monoPEG-IFN from the preparative SP-HPLC was fairly high. When 9mg PEG-IFN was applied to the column, the total protein recovery was 73-78%, as determined by amino acid composition analysis. An aliquot of the pooled samples (all peaks collected as a single pool) was rechromatographed by analytical SP-HPLC and the resulting fingerprint was unchanged demonstrating that there is no selective loss of any of the species.

    MonoPEG-IFN is interferon α-2a monoPEgylated with polyethylene glycol (PEG-5000), in a 1:1 molar ratio, via a urea linkage. This conjugation results in the modification of primary amino groups of IFN α-2a. When the separated PEG-IFN species were subjected to peptide mapping, sequencing and mass spectrometric analyses, they demonstrated that the individual species are positional isomers which differ only in their site of PEGylation (Mokarsh, S. P. et al., Anal. Biochem., in press)). There are 12 primary amino groups (N-terminus plus 11 lysine side chains) in interferon α-2a, all of which are potentially available for PEGylation. Eleven out of the 12 possible positional isomers were identified, all of which involved lysines as their PEGylation sites (Table 1). However, no positional isomer involving the N-terminus was detected with the SP-HPLC method (data not shown). Since the N-terminal amino acid of interferon α-2a is a cysteine which is involved in a disulfide bond, it is presumed that the primary amino group of the N-terminus may not be readily available for PEGylation and may also be due to the reaction pH.

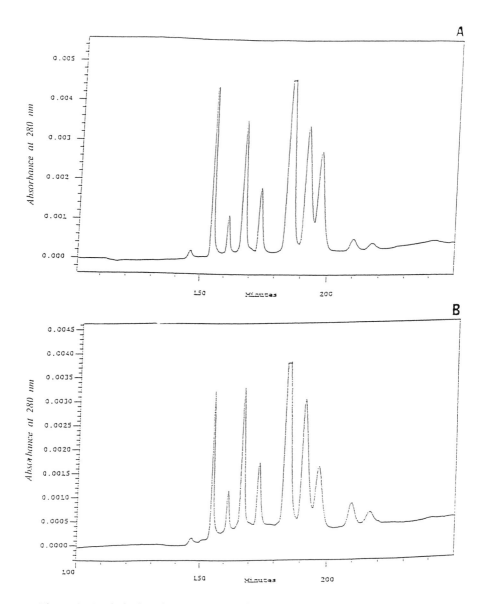

Figure 2. Analytical scale SP-HPLC profiles of PEG-IFN produced by the solid-phase (A) and solution-phase (B) methods. Seventy µg PEG-IFN were used in each analysis.

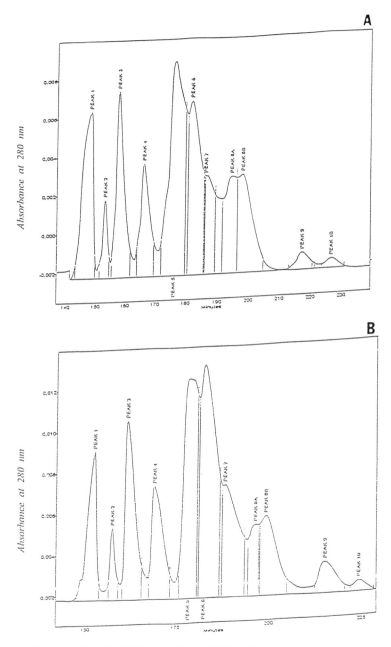

Figure 3. Preparative SP-HPLC profiles of PEG-IFN produced by the solid-phase (A) and solution-phase (B) methods. Ten sample pools were collected as indicated by the vertical bars on the peaks in the profile. The peaks were numbered 1-7, 8A, 8B, 9 and 10.

Table I. Antiviral and Antiproliferative Activities of MonoPEG-IFN Species
produced by the solution-phase method.

Samples	Peak No.	Bioactivities		**PEG Sites
		Antiviral	*Antiproliferative	
		10^7 x IU/mg	IC_{50} (pg /mL)	
IFN	NA	21.0	33	NA
MonoPEG-IFN	NA	7.2	180	NA
Species-1	1	6.7	227	K^{31}
Species-2	2	8.4	213	K^{133}
Species-3	3	6.4	113	K^{13}
Species-4	4	4.3	273	K^{23}
Species-5	5	4.0	200	K^{131}
Species-6	6	3.0	267 $K^{121, 164, 131}$	
Species-7	7	4.5	340	$K^{121, 164}$
Species-8	8A	5.9	170	K^{70}
Species-9	8B	6.3	293	K^{83}
Species-10	9	6.7	250	K^{49}
Species-11	10	1.2	361	K^{112}

NA = Not Applicable
*The % inhibition of cell growth based upon the concentrations of PEG-IFN (15.6-1000 pg/mL) was utilized to calculate 50% inhibition values (IC_{50}).
**Data taken from Monkarsh et al., Anal. Biochem. (in press) and K is the symbol for lysine.

MonoPEG-IFN is a mixture of positional isomers, each being conjugated with a PEG-5000 by a urea linker at a different site. The sites (mainly lysines) and the extent to which each of the lysines become PEGylated are dependent upon individual pKa and steric accessibility, as well as the chemical process itself. The pKa of each of the 11 lysine residues is expected to be slightly different due to the location on the polypeptide chain and the local microenvironment. Therefore, all lysines of interferon a-2a are not equally active. Modification of the basic lysines not only lowers the local pI, but also changes the overall pI of the protein. Each positional isomers may have a slightly different pI and a unique position of attachment of a single PEG molecule. The end-result is heterogeneity in charge, site of PEGylation and distribution of positional isomers in the PEG-IFN preparation. The SP-HPLC method takes advantage of minute local charge differences to separate and isolate the PEG-IFN positional isomers into 10 peaks or fractions. The results listed in Table 1 (last column) indicate that each peak except 6 and 7, corresponds to a single positional isomer with a unique PEGylation site, whereas peaks 6 and 7 are mixtures containing 2-3 positional isomers.

MonoPEG attachment at various sites in interferon α-2a may produce physicochemical differences (e.g., changes in conformation, electrostatic binding properties, hydrophobicity, local lysine pKa, pI etc.) in the individual monoPEG-IFN conjugate species. These physicochemical differences could influence the binding affinity to the IFN-receptor and consequently affect their bioactivities. However, each of the 11 monoPEG-IFN positional isomers exhibited antiviral and antiproliferative activities. The sum of activities of individual species appears to be slightly lower than that of the parent monoPEG-IFN mixture which was produced by the solution-phase method (see Table I). It is also possible that some loss of activity might have occured during additional handling of the samples, as well as due to non-ideal storage (buffer) conditions. However, the differences in activity observed in both assays fall within the range of variations usually associated with these types of bioassay. The species derived from the solid-phase method also exhibited similar biological activities (data not shown). All monoPEG-IFN species exhibited bioactivity, some more than others, indicative of the effect of site of PEGylation on IFN/receptor interactions and the resulting biological activity.

The results listed in Table I indicate a substantial loss in *in vitro* bioactivities upon PEGylation of interferon α-2a. In contrast, the unfractionated monoPEG-IFN exhibited significant enhancements in circulating half-life and *in vivo* biological activity (Palleroni, A, Hoffmann-La Roche, Nutley, NJ, unpublished data). The positional isomers were not tested individually for *in vivo* biological activities.

In summary, PEG-IFN is a mixture of positional isomers in which each PEG moiety is attached at a unique site on interferon via a urea linkage. A method involving SP-HPLC which exploits minute local charge differences between the positional isomers is used to separate them. All isolated PEG-IFN positional isomers exhibit antiviral and antproliferative activities.

Acknowledgments

We want to thank Dr. Wen-Jian Fung for her critical review and Ms. Audrey Faronea for preparing the manuscript.

Literature Cited

1.   Gilbert, C.W. 1995, International patent publication number WO95/13090
2.   Abuchowski, A.; van Es, T.; Palczuk, N.C.; Davis, F.F. *J. Biol. Chem.* **1977,** 252, 3578.

3.   Sharp, K.A.; Yalpani, M.; Howard, S.J.; Brooks, D.E. *Anal. Biochem.*
      1986, 154, 110.
4.   McGoff, P.; Baziotis, A.C.; Maskiewicz, R. *Chem. Pharm. Bull.* **1988,** 36,
      3079.
5.   Snider, J.; Neville, C.; Yuan, L.-C.; Bullock, J. *J. Chrom.* **1992,** 599, 141.
6.   Vestling, M.M.; Murphy, C.M.; Fenselau, C.; Dedinas, J.; Doleman, M.S.;
      Harrsch, P.B.; Kutney, R.; Ladd, D.L.; Olsen, M.A. In: *Techniques in
      Protein Chemistry III;* Angeletti, R.H., Ed.; Academic Press: NY, NY,
      1992, pp. 477-485.
7.   Vestling, M.M.; Murphy, C.M.; Keller, D.A.; Fenselau, C.; Dedinas, J.;
      Ladd, D.L.; Olsen, M.A. *Drug Metabolism and Disposition* **1993**, 21, 911.
8.   Monkarsh, S.P.; Russoman, E.A.; Roy, S.K. *J. Chromatog.* **1993,**. 631,
      277.
9.   Karasiewcz, R.; Nalin, C.; Rosen, P. 1995, U.S. Patent No. 5,382,657.
10.   Porter, J.; Cavilhas, C.; Prinzo, K.; Schaffer, C.A.; Monkarsh, S.; Bailon,
      P. 1996, "Continuous solid-phase PEGylation of proteins", Recovery of
      Biological Products VIII Symposium, October 20-25, Tuscon, AZ.
11.   Pan, Y.-C.; Stein, S. In *Methods of Protein Characterization*, Shively,J.
      Ed.; Humana Press: Clifton, NJ, 1986, pp 105-119.
12.   Laemmli, U.K. *Nature* **1970**, 227,.680.
13.   Rubinstein, S.; Familletti, P.; Pestka, S. *J. Virology* **1988,** 37, 755.
14.   Borden, E.C.; Hogan, T.F.; Voelkel, J.G. *Cancer Research* **1982,** 42,
      4948.

Conjugates of Oligomeric and Low-Molecular Weight Substrates

Chapter 16

Site-Specific Poly(ethylene glycol)ylation of Peptides

Arthur M. Felix

Department of Chemistry and Physics, William Paterson College,
Wayne, NJ 07470

In contrast to proteins, peptides are ideal targets for PEGylation since they may be PEGylated using modern orthogonal protection strategies. Synthetic conditions have been developed for the site-specific solid phase PEGylation (Fmoc/tBu strategy) at the NH_2-terminal, internal (at the side-chain of Lys/Orn or Glu/Asp) and at the COOH-terminal positions. Conditions for site-specific solution phase PEGylation have also been developed in which a Cys residue is introduced at any position into the peptide and subsequently PEGylated using an activated dithiopyridyl-PEG reagent. Using these procedures, a number of biologically active mono-PEGylated peptides were prepared in which the PEG moieties were varied in molecular weight (i.e. degree of PEGylation; 750, 2000, 5000 and 10,000). Site-specific PEGylation of peptides enables the evaluation of the structure-activity relationships of biologically important peptides (i.e. the effect of site of PEGylation and degree of PEGylation on biological activity). These studies provide an important groundwork for the design of PEGylated peptides which retain the desirable properties imposed by the incorporation of PEG (enhanced solubility, increased resistance to proteolytic degradation, decreased antigenicity and immunogenicity) and have improved *in vivo* profiles.

Bioactive peptides are known to play key pharmacological roles and have found widespread application in a variety of fields. Active biomedical research using peptides includes their application for autoimmune, inflammatory, cardiovascular, hormonal and neurological disorders and for wound healing, infectious diseases and cancer therapy. Peptides have found utility in the diagnosis and treatment of such diverse fields as diabetes, HIV, Alzheimer's Disease, asthma and growth deficiencies. Although many peptides have found application for a variety of clinical indications, their use has been limited due to their rapid metabolism from the action of extracellular proteases. Chemists have therefore developed ingenious methods for the modification of peptides (e.g. peptidomimetic design, cyclization etc.) with the goal of maintaining or improving their receptor selectivity while stabilizing them to enzymatic degradation. Although these efforts have yielded important progress, the problems of poor absorption of peptides, rapid proteolytic degradation and potential immunogenicity have severely hampered the application of peptides as potential therapeutics.

In the last decade considerable progress has been made on the application of poly(ethylene glycol) (PEG) as a covalent linking agent to polypeptides and proteins. The resultant PEG-protein conjugates have been compared to the parent protein and have been shown to have enhanced resistance to proteolytic degradation (increased plasma half life). In addition, PEG-proteins have been reported to have improved water solubility, increased resistance to proteolytic degradation, reduced antigenicity and decreased immunogenicity (*1-6*). Unfortunately, these important advantages of PEG-proteins have been offset by their decrease in biological activity *in vitro*. These observations may be rationalized as a result of

the indiscriminate incorporation of PEG into the protein by the standard methods of protein PEGylation (attachment to the ε-NH$_2$ of lysine). Unfortunately, the specific position of PEGylation in proteins usually cannot be controlled using the conditions of protein PEGylation thereby resulting in the incorporation of a multitude of PEG units per mol of protein.

In contrast to proteins, peptides are ideal targets for PEGylation since they may be specifically PEGylated using modern orthogonal protection strategies. The enormous progress that has been made in peptide synthetic methodology has provided an important basis for the site-specific PEGylation of peptides. This manuscript describes the chemical procedures that our laboratories have developed for site-specific PEGylation of peptides at the NH$_2$-terminal, internal (side-chain of Lys/Orn or Glu/Asp) and COOH-terminal positions (Figure 1). In addition, we have prepared several biologically important PEGylated peptides using this methodology, evaluated structure-activity relationships, and compared their biological and physical properties to those of the parent peptides.

Solid Phase Synthesis of PEGylated Peptides: NH$_2$-Terminal PEGylation

The derivatization of the terminal hydroxyl function of monomethoxypoly-(ethyleneglycol), *via* a urethane linkage covalently linked to norleucine (*7,8*) enables the analytical determination of the number of PEG units/mol of protein by amino acid analysis. We recently modified this procedure in our laboratory using derivatives derived from monomethoxypoly(ethyleneglycol), **1** (Figure 2). Synthetic methods were developed for the preparation of PEG$_n$-CH$_2$-CO-Nle-OH, **2**, in 3 steps from monomethoxypoly-(ethyleneglycol) (*9*). We observed that PEG$_n$-CH$_2$-CO-Nle-OH could be used as an acylating reagent for NH$_2$-terminal PEGylation in solid phase synthesis by standard procedures using BOP-activation (*10*). PEGylation typically proceeds in (1:9 CH$_2$Cl$_2$-DMF or 1:9 CH$_2$Cl$_2$-NMP) using 2 equivalents of BOP followed by 4 equivalents of DIEA for 24 hours. Completion of PEGylation is monitored in the usual way using the Kaiser ninhydrin test (*11*). The PEGylation is repeated if the Kaiser test is positive. Initially, we compared the purity of the PEGylated peptide using TFA cleavage to that from the low-high HF cleavage. In a model study, we observed that TFA cleavage was preferred since it resulted in the preparation of PEGylated peptides that revealed a single peak by analytical HPLC. In contrast, using the HF procedure we observed impurities by HPLC which corresponded to small amounts of low molecular weight fragments resulting from degradation of poly(ethyleneglycol) on exposure to HF. From these studies it was concluded that the Fmoc/tBu solid phase strategy for the synthesis of PEGylated peptides was preferred since this is compatible with TFA cleavage in the final stage.

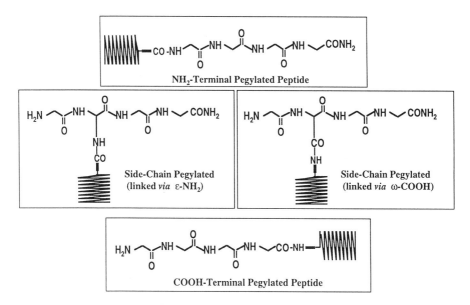

Figure 1. Peptide pegylation sites.

$$CH_3\text{-}O\text{-}CH_2\text{-}CH_2\text{-}O\text{-}(CH_2\text{-}CH_2\text{-}O)_n\text{-}CH_2\text{-}CH_2\text{-}OH$$

PEG$_n$ 1

$$CH_3\text{-}O\text{-}CH_2\text{-}CH_2\text{-}O\text{-}(CH_2\text{-}CH_2\text{-}O)_n\text{-}CH_2\text{-}CH_2\text{-}O\text{-}CH_2\text{-}CO\text{-}NH\text{-}Nle\text{-}COOH}$$

N^{α}-PEG$_n$-CH$_2$-CO-Nle-OH   2

Fmoc-Lys-OH   Fmoc-Orn-OH

PEG$_n$-CH$_2$-CO-Nle⌐   2a   PEG$_n$-CH$_2$-CO ⌐   2b

$$CH_3\text{-}O\text{-}CH_2\text{-}CH_2\text{-}O\text{-}(CH_2\text{-}CH_2\text{-}O)_n\text{-}CH_2\text{-}CH_2\text{-}NHCO\text{-}Nle\text{-}NH_2}$$

H-Nle-NH-CH$_2$-CH$_2$-PEG$_n$ 3

Fmoc-Asp-OH

3a   ∟Nle-NH-CH$_2$-CH$_2$-PEG$_n$

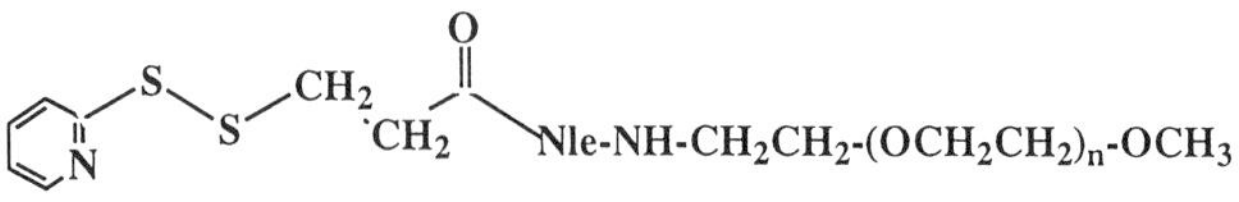

Pyr-S-S-CH$_2$-CH$_2$-CO-Nle-NH-CH$_2$-CH$_2$-PEG$_n$ 4

Figure 2. Monomethoxypoly(ethylene glycol) derivatives.

Accordingly, resins that are cleaved with TFA [e.g. Wang-resin (*12*), Rink-resin (*13*), PAL-resin (*14*) and Rink amide MBHA-resin (*15*)] are compatible with the solid phase synthesis of PEGylated peptides. The procedure for site-specific solid phase NH_2-terminal PEGylation is outlined in Figure 3 starting with Rink amide MBHA-resin.

Solid Phase Synthesis of PEGylated Peptides: Side-Chain PEGylation

The preparation of side-chain-PEGylated amino acids has been reported and used for the liquid-phase synthesis of hydrophobic peptides of medium size (*16*). We have extended these observations and found that side-chain solid phase PEGylation of Lys or Asp can be readily achieved by one of two pathways as outlined in Figure 4; direct side-chain PEGylation or post side-chain PEGylation (*17*). Starting with Rink amide MBHA-resin, we assemble the C-terminal portion of the peptide (Peptide$_a$) by Fmoc/tBu solid phase peptide synthesis. This intermediate is subjected to direct side-chain PEGylation utilizing an N^α-Fmoc-protected derivative of Lys or Asp in which poly(ethyleneglycol) is already linked to the side-chain of Lys or Asp. This derivative was prepared by BOP-catalyzed solid phase coupling with Fmoc-Lys(PEG$_n$-CH$_2$-CO-Nle)-OH, **2a** [prepared by the condensation of Fmoc-Lys(Nle)-OH with PEG$_n$-CH$_2$-COOSu (*16*)]. Direct side-chain PEGylation of Asp was accomplished in a similar manner by BOP-catalyzed coupling with Fmoc-Asp(Nle-NH-CH$_2$-CH$_2$-PEG$_n$)-OH, **3a** [prepared by the condensation of Fmoc-Asp-OtBu with H-Nle-NH-CH$_2$-CH$_2$-PEG$_n$, **3**], followed by TFA deprotection. Alternatively, the intermediate may be subjected to post side-chain PEGylation using an N^α-Fmoc-protected derivative of Lys or Asp with an allylic side-chain protecting group. Following the complete assemblage of the peptide-resin, the allylic group was selectively removed by palladium-catalyzed deprotection using PdCl$_2$[P(C$_6$H$_5$)$_3$]$_2$. Solid phase PEGylation of the side-chain of Lys or Asp was then carried out in the final stage by BOP-catalyzed acylation with PEG$_n$-CH$_2$-CO-Nle-OH, **2**, or H-Nle-NH-CH$_2$-CH$_2$-PEG$_n$, **3**, respectively. Deprotection of the tBu-protected side-chains and resin cleavage was accomplished by standard procedures using TFA. The two pathways for side-chain PEGylation are outlined in Figure 4 for the site-specific Lys-side-chain PEGylation starting with Rink amide MBHA-resin -resin.

Solid Phase Synthesis of PEGylated Peptides: COOH-Terminal PEGylation

Fmoc-Orn(PEG$_n$-CH$_2$-CO)-OH, **2b** [prepared by the condensation of Fmoc-Orn-OH with PEG$_n$-CH$_2$-COOSu], was attached directly to Rink amide MBHA-resin. The use of ornithine as the COOH-terminal residue, with PEG attached at the side-chain position, also enables the analytical determination of the number of PEG units/mol of peptide by amino acid analysis.   Following this attachment stage, solid-phase assemblage of the PEGylated peptide-resin was carried out using the standard Fmoc/tBu strategy (*17*). The presence of the PEG moiety at the COOH-terminus did not alter the course of solid phase synthesis. Deprotection of the tBu-protected side-chains and resin cleavage was accomplished by standard procedures using TFA. The procedure for site-specific solid phase COOH-terminal PEGylation is outlined in Figure 5 starting with Rink amide MBHA-resin -resin.

Multiply Site Specific PEGylated Peptides

Using the solid phase procedures outlined in Figures 3, 4 and 5 we have prepared NH_2-terminal,   side-chain   and   COOH-terminal   PEGylated   peptides   using

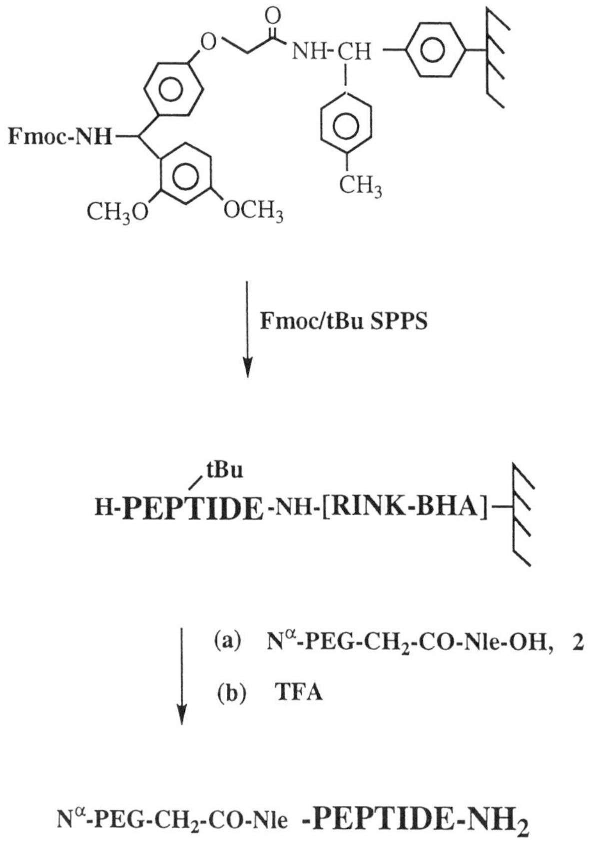

Figure 3. Site-specific solid phase NH$_2$-terminal pegylation.

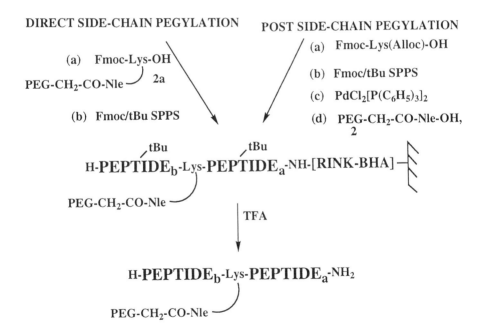

Figure 4. Site-specific solid phase Lys-side-chain pegylation.

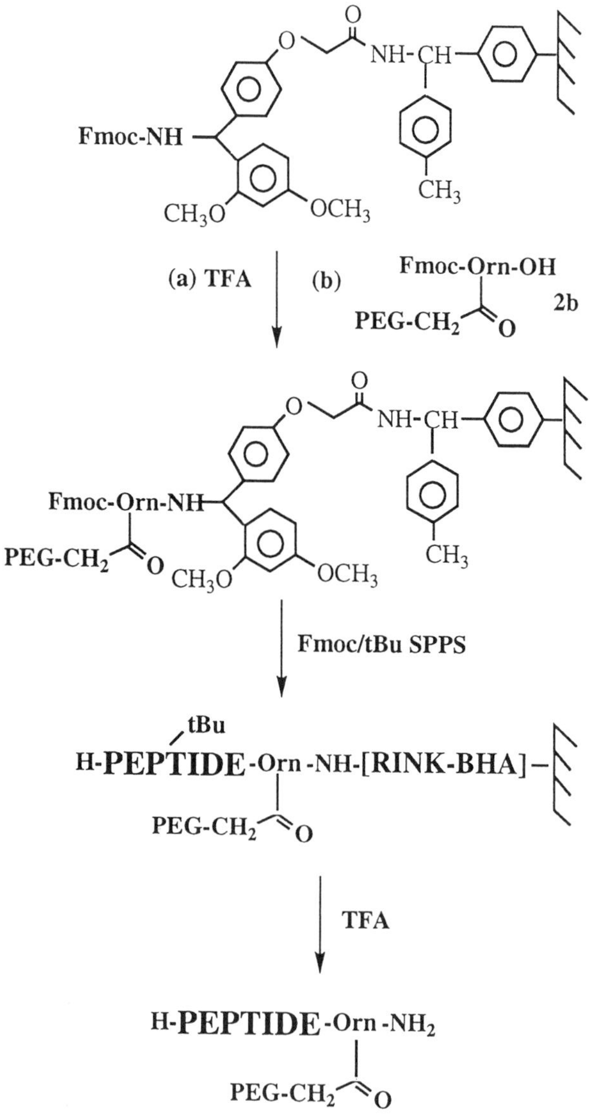

Figure 5. Site-specific solid phase COOH-terminal pegylation.

poly(ethyleneglycol) of varying molecular weight (PEG$_{750}$, PEG$_{2000}$, PEG$_{5000}$ and PEG$_{10000}$). Attempts to extend these solid phase site-specific PEGylation methods to diPEGylated peptides were also successful. However, the second stage of PEGylation proceeded with some difficulty when PEGylation was attempted with high molecular weight poly(ethyleneglycol) derivatives (i.e. PEG$_{5000}$ and PEG$_{10000}$). We observed that multiple solid phase PEGylation proceeds more efficiently with poly(ethyleneglycol) derivatives of lower molecular weight (i.e. PEG$_{750}$ and PEG$_{2000}$). In addition, the second stage of PEGylation at the NH$_2$-terminus is less efficient when the NH$_2$-terminal residue is sterically hindered (e.g. Ile) (*18*).

Solution Synthesis of PEGylated Peptides: COOH-Terminal PEGylation

The methodology for site-specific COOH-terminal PEGylation was extended to include a novel solution phase PEGylation methodology. This procedure features the incorporation of a cysteine residue at the COOH-terminus. Earlier biotinylation studies had shown that it is preferred to incorporate the cysteine residue at a site distant from the bioactive core of the peptide and a spacer arm consisting of -Gly-Gly-Cys-NH$_2$ was incorporated into the peptide (*19*). We have extended these earlier observations and found that the -Gly-Gly-Cys-NH$_2$ extended peptides are ideally suited for solution phase PEGylation. The conversion of H-Nle-NH-CH$_2$-CH$_2$-PEG, **3**, to the dithiopyrridyl-activated PEG, **4**, was carried out by reaction with N-succinimidyl 3-(2-pyridyldithio)propionate (SPDP) (*20*). Figure 6 outlines the general scheme for site-specific solution phase COOH-terminal PEGylation through disulfide formation at COOH-terminal cysteine. This process is not confined to COOH-terminal PEGylation. Using this procedure, a residue of Cys may be introduced at any position into the peptide and subsequently site-specifically PEGylated using dithiopyrridyl-activated PEG, **4**.

Purification of PEGylated Peptides

The PEGylated peptides are best purified by a 2-stage process using dialysis and preparative HPLC. Dialysis was carried out for 48 hours using dialysis membrane tubing with a nominal cutoff of 1000, 3500 or 7000 (used respectively for PEG$_{750}$-peptides, PEG$_{2000}$-peptides or PEG$_{5000}$-peptides). This process removes all salts and low molecular impurities including any non-PEGylated peptides. The final purification step, reversed-phase preparative HPLC was carried out in the same manner as for the free peptide on Vydac C-18 columns using a gradient of H$_2$O-CH$_3$CN (containing 0.1% TFA). The PEGylated peptides usually elute from the column after the free peptide and can therefore be separated from trace amounts of free peptide that were not removed in the dialysis step.

Characterization of PEGylated Peptides

Since the elution properties of PEGylated peptides are similar to those of the corresponding free peptides, it is essential to thoroughly characterize the final product and to determine whether the PEGylated peptide is contaminated with free peptide. Surprisingly, analytical HPLC of PEGylated peptides results in homogeneous, sharp peaks. Moreover, PEG-peptides elute ~1-5 minutes after the corresponding free peptides. Application of "spiking" studies using analytical HPLC enables the detection of small amounts of free peptide in the presence of PEG-peptide. Although poly(ethyleneglycol) does not absorb UV light, HPLC of the PEG-peptides are tracked at 214 nm where the peptide chromophore enables their detection. Figure 7

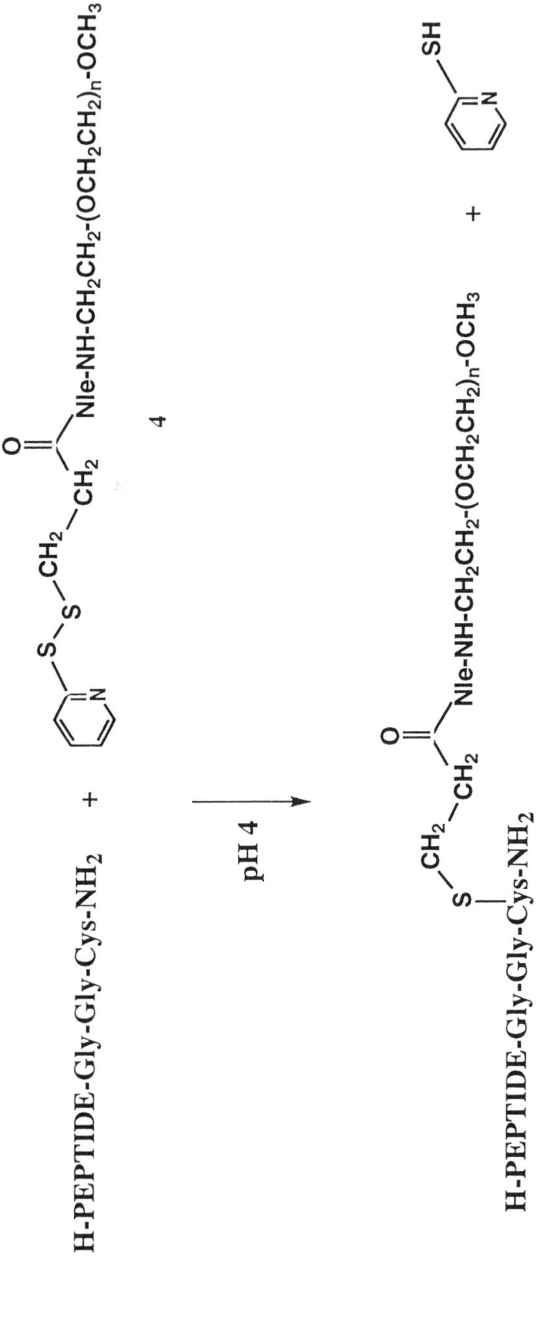

Figure 6. General scheme for site-specific solution phase COOH-terminal pegylation through disulfide formation at cysteine.

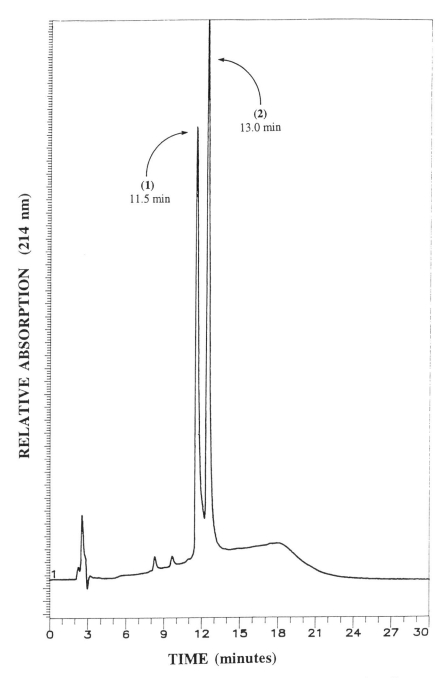

Figure 7. Analytical HPLC of a mixture of (1) [His^1,Val^2,Gln^8,Ala^{15},Leu^{27}]-GRF(1-32)- Gly-Cys-NH$_2$ (5μg) with (2) [His^1,Val^2,Gln^8,Ala^{15},Leu^{27}]-GRF(1-32)-Gly-Cys(S-CH$_2$-CH$_2$-CO-Nle-NH-PEG$_{5000}$)-NH$_2$ (10μg). Column: Lichrosorb RP-8 (5μ); Eluant: (A) 0.1M NaClO$_4$ (pH 2.5)-(B) CH$_3$CN; Gradient: 40-60% (B) in 20 min; Detection: 214 nm; Flow rate: 1 ml/min.

demonstrates that baseline resolution can be achieved for a mixture of peptide and PEGylated peptide. In this admixture chromatogram, the peptide, [His[1],Val[2],Gln[8],Ala[15],Leu[27]]-GRF(1-32)-Gly-Cys-NH$_2$, elutes at 11.5 min and the PEGylated analog, [His[1],Val[2],Gln[8],Ala[15],Leu[27]]-GRF(1-32)-Gly-Cys(S-CH$_2$-CH$_2$-CO-Nle-NH-PEG$_{5000}$)-NH$_2$, elutes at 13 min. Similar resolution have been observed for a number of peptides and their corresponding PEGylated peptides.

As mentioned earlier, amino acid analysis is used to determine the extent of PEGylation (i.e. the number of PEG moieties in the PEGylated peptide). Since norleucine and ornithine elute at different retention times than any of the natural amino acids, when these residues are covalently linked to PEG their detection by amino acid analysis can be used to determine the number of PEG moieties in the final PEG-peptide product. Therefore, determination of Nle (from PEG-CH$_2$-CO-Nle-OH, **2**) was used to confirm the presence of PEG in NH$_2$-terminal PEGylated peptides. Similarly, determination of Nle (from either Fmoc-Lys(PEG-CH$_2$-CO-Nle)-OH, **2a**, or Fmoc-Asp(Nle-NH-CH$_2$-CH$_2$-PEG$_n$)-OH, **3a**) was used to confirm the presence of PEG in side-chain PEGylated peptides. Finally, the determination of Orn (from Fmoc-Orn(PEG-CH$_2$-CO)-OH, **2b**) was used for confirmation of PEG in COOH-terminal PEGylated peptides.

Since internal and COOH-terminal PEGylated peptides undergo Edman degradation, sequence analysis of these site-specific PEGylated peptides can be used to provide indirect evidence of the position of PEGylation in the PEGylated peptide. The cycle of Edman degradation corresponding to the release of the PEGylated amino acid is undetected since the resultant thiohydantoin is PEGylated and does not co-elute with the known thiohydantoins. This absence of a thiohydantoin derivative may be used to confirm the position of the PEGylated residue in the sequence.

^{1}H-NMR spectroscopy is also extremely useful for characterization of the site-specific PEGylated peptides. ^{1}H-NMR spectra of PEGylated peptides reveals a major off-scale peak at δ 3.5-3.8 from -(CH$_2$CH$_2$O)$_n$- which integrates for the expected number of protons (e.g. 452 H for PEG$_{5000}$). In addition, the ^{1}H-NMR spectra of PEGylated peptides also reveal a singlet at δ 3.39 which is used as a standard and integrates for the expected number of protons (3H).

Laser desorption mass spectrometry of PEGylated peptides results in a bell-shaped spectrum centered at the expected position for the PEGylated peptide with a family of adjacent peaks at 44±1 Da corresponding to the varying molecular weights of poly(ethyleneglycol). The laser desorption mass spectrum of poly(ethyleneglycol), PEG$_{5000}$, shown in Figure 8 reveals the bell-shaped curve with a maxima at 5027.4 Da and a family of peaks at 44±1 Da starting with adjacent peaks at 4983.5 and 5072.4. Similar spectra have been observed for PEGylated amino acids and PEGylated peptides of poly(ethyleneglycol) of varying molecular weights.

Site-Specific PEGylation of Growth Hormone-Releasing Factor

Human growth hormone-releasing factor, hGRF(1-44)-NH$_2$, is an amidated peptide containing 44-amino acids. It is produced and released by the hypothalamus and transported *via* a portal network system to the pituitary gland where it triggers the release of human growth hormone. The physiological profile of endogenous human growth hormone-releasing factor has been recently reviewed (*21*). Studies have demonstrated the potential use of human growth hormone-releasing factor for the

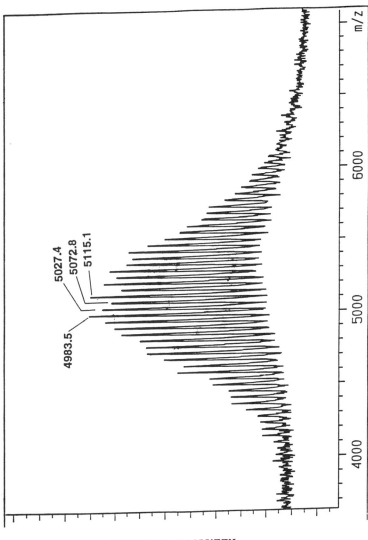

Figure 8. Laser desorption mass spectrum of poly(ethyleneglycol), PEG5000. The spectrum was obtained on a Bruker ReflexTime of Flight Mass Spectrometer in the reflector mode at 30 kV.

treatment of dwarfism resulting from failure of the hypothalamus to produce sufficient quantities of the releasing factor. Structure-activity studies on human growth hormone-releasing factor have shown that an analog, [Ala[15]]-hGRF(1-29)-NH_2, is 3-4 times more potent *in vitro* than the parent peptide, hGRF(1-44)-NH_2 (*22*). With the goal of producing potent, long acting analogs of human growth hormone releasing-factor, a series of monoPEGylated derivatives of [Ala[15]]-hGRF(1-29)-NH_2 were prepared. Site-specific PEGylation was carried out by the procedures outlined earlier at the NH_2-terminus (position 1), side-chain positions (Asp[8], Lys[12], Lys[21] and Asp[25]) and at the COOH-terminus (position 30). Each PEGylated peptide was purified by dialysis and preparative HPLC and characterized by laser desorption mass spectrometry, analytical HPLC, amino acid analysis and ^{1}H-NMR spectrometry.

In vitro Biological Activity of PEGylated Growth Hormone-Releasing Factor

Each [Ala[15]]-hGRF(1-29)-NH_2 analog was studied *in vitro* to determine the effect of (a) site of PEGylation and (b) degree of PEGylation (i.e. molecular weight of PEG moiety) on the growth hormone releasing potency (*23*). The results of these studies are summarized in Table I and demonstrate that both the site of PEGylation and, in some cases, the degree of PEGylation are critical to biological activity. In each case the biological activity of the PEGylated peptide was compared to that of the corresponding acetylated or amidated peptide. This ensures an objective evaluation of the effect of PEGylation since the acetylated or amidated derivatives have different potencies than the parent peptide, [Ala[15]]-hGRF(1-29)-NH_2.

Acetylation at the NH_2-terminus (compound **7**) resulted in reduced potency compared to [Ala[15]]-hGRF(1-29)-NH_2, **6**. Compound **6** showed no further change when the NH_2-terminus was PEGylated, regardless of the degree of PEGylation (compounds **7a**, **7b**). Amidation at position 8 (compound **8**) did not significantly alter the potency of **6**. However, PEGylation at this site resulted in a drastic decrease of potency, which decreased further with increased molecular weight of PEG moiety (compounds **8a**, **8b**). Acetylation at position 12 (compound **9**) resulted in a decrease of potency compared to **6**, which decreased further with increasing molecular weight of PEG moiety (compounds **9a**, **9b**). On the other hand, acetylation at position 21 (compound **10**) or amidation at position 25 (compound **11**) had no significant effect on biological activity *vs* **6**. In addition, an increase in the molecular weight of the PEG moiety at position 21 (compounds **10a**, **10b**) or at position 25 (compounds **11a**, **11b**) also did not result in any change in relative potency. Finally, acetylation at the COOH-terminus (compound **12**) provided the most potent analog, which retained its high potency regardless of the degree of PEGylation (compounds **12a**, **12b**).

These observations demonstrate the importance of both the site and degree of PEGylation on the *in vitro* biological activity in the growth hormone releasing factor peptide. It is anticipated that similar results will be observed for other biologically important peptides. These studies reveal that it is not possible to predict the effect of PEGylation on the biological activity of the peptide. Indiscriminate PEGylation of a peptide should be avoided since biological activity may be drastically reduced. Each peptide should be thoroughly evaluated to determine the optimal site of PEGylation.

Table I. Biological Activity (*in vitro*) of Pegylated [Ala^{15}]-GRF(1-29)-NH$_2$ Analogs

Compound	hGRF Analog	Pegylation Site	Relative Potency[a]
5	hGRF(1-44)-NH$_2$	---	1.00
6	[Ala^{15}]-hGRF(1-29)-NH$_2$	---	3.81
7	Ac-[Ala^{15}]-hGRF(1-29)-NH$_2$	---	1.02
7a	N$^\alpha$-PEG$_{2000}$-CH$_2$-CO-Nle-[Ala^{15}]-hGRF(1-29)-NH$_2$	1	1.04
7b	N$^\alpha$-PEG$_{5000}$-CH$_2$-CO-Nle-[Ala^{15}]-hGRF(1-29)-NH$_2$	1	1.28
8	[Asp(NHEt)^{8}Ala^{15}]-hGRF(1-29)-NH$_2$	---	4.68
8a	[Asp(Nle-NH-CH$_2$CH$_2$-PEG$_{2000}$)-Ala^{15}]-hGRF(1-29)-NH$_2$	8	1.59
8b	[Asp(Nle-NH-CH$_2$CH$_2$-PEG$_{5000}$)-Ala^{15}]-hGRF(1-29)-NH$_2$	8	0.17
9	[Lys(Ac)^{12}Ala^{15}]-hGRF(1-29)-NH$_2$	---	0.77
9a	[Lys(PEG$_{2000}$-CH$_2$-CO-Nle)^{12}-Ala^{15}]-hGRF(1-29)-NH$_2$	12	0.34
9b	[Lys(PEG$_{5000}$-CH$_2$-CO-Nle)^{12}-Ala^{15}]-hGRF(1-29)-NH$_2$	12	0.03
10	[Ala^{15},Lys(Ac)^{21}]-hGRF(1-29)-NH$_2$	---	2.86
10a	[Ala^{15},Lys(PEG$_{2000}$-CH$_2$-CO-Nle)^{21}]-hGRF(1-29)-NH$_2$	21	3.21
10b	[Ala^{15},Lys(PEG$_{5000}$-CH$_2$-CO-Nle)^{21}]-hGRF(1-29)-NH$_2$	21	3.24
11	[Ala^{15},Asp(NHEt)^{25}]-hGRF(1-29)-NH$_2$	---	3.20
11a	[Ala^{15},Asp(Nle-NH-CH$_2$CH$_2$-PEG$_{2000}$)^{25}]-hGRF(1-29)-NH$_2$	25	3.09
11b	[Ala^{15},Asp(Nle-NH-CH$_2$CH$_2$-PEG$_{5000}$)^{25}]-hGRF(1-29)-NH$_2$	25	3.27
12	[Ala^{15},Orn(Ac)^{30}]-hGRF(1-30)-NH$_2$	---	4.36
12a	[Ala^{15},Orn(PEG$_{2000}$-CH$_2$-CO)^{30}]-hGRF(1-30)-NH$_2$	30	4.63
12b	[Ala^{15},Orn(PEG$_{5000}$-CH$_2$-CO)^{30}]-hGRF(1-30)-NH$_2$	30	4.83

a Potency relative to hGRF(1-44)-NH$_2$ as determined by ability to stimulate growth hormone release by rat pituitary cells *in vitro* (concentrations tested: 3.1-400 pM) (reproduced with permission from reference 23)

In vivo Biological Activity of PEGylated Growth Hormone-Releasing Factor

The potent analog of hGRF, [Ala^{15}]-hGRF(1-29)-NH$_2$, undergoes rapid NH$_2$-terminal proteolytic degradation (24) initiated by dipeptidylpeptidase-IV (25). Other more potent analogs have been developed in which the NH$_2$-terminus is protected against dipeptidylpeptidase-IV degradation. These include [desNH$_2$Tyr^1,D-Ala^2,Ala^{15}]-hGRF(1-29)-NH$_2$, 13, and [His^1,Val^2,Gln^8,Ala^{15},Leu^{27}]-hGRF(1-32)-OH, 14, which are potent in vivo since they have enhanced plasma stability (26, 27). Homologs of these potent hGRF analogs were prepared with -Gly-Gly-Cys-NH$_2$ at the COOH terminus and shown to have similar in vivo biological activity. These peptides were then subjected to site-specific solution phase PEGylation through disulfide formation at the COOH-terminal cysteine using dithiopyrridyl-activated PEG, 4, as outlined in Figure 6. This procedure enabled us to prepare a series of PEGylated hGRF analogs with varying degree of PEGylation. The PEG$_{5000}$ analogs (compounds 13a and 14a) were evaluated in vivo as summarized in Table II (20).

These in vivo studies (Table II) revealed that [desNH$_2$Tyr^1,D-Ala^2,Ala^{15}]-hGRF(1-29)-NH$_2$, 13, was ~10 times more potent than hGRF(1-44)-NH$_2$, 5, in both mice and pigs in vivo. COOH-Terminal PEGylation of this peptide (compound 13a) resulted in a further increase of in vivo potency by factors of 3.4 and 5.2 times in the mouse and pig in vivo models, respectively. Similar observations were made for [His^1,Val^2,Gln^8,Ala^{15},Leu^{27}]-hGRF(1-32)-OH, 14, which was ~7-10 times more potent than 5. When this analog was PEGylated at the COOH-terminus (compound 14a) the in vivo potency increased by factors of 1.7 and 5.2 times, respectively, in the mouse and pig in vivo models. Therefore COOH-terminal PEGylation of 13 and 14 resulted in an increase of in vivo potency by 50-fold (pigs) and 20-fold (mice) when compared to the parent peptide, hGRF(1-44)-NH$_2$, 5. Since in vitro plasma stability is not altered by COOH-terminal PEGylation, the increased potency of these PEGylated peptides may be attributed to longer half lives resulting from decreased clearance from the blood and tissues (20).

Conformational studies of PEGylated peptides

Conformational studies were undertaken to better understand the observation that the PEGylation of [desNH$_2$Tyr^1,D-Ala^2,Ala^{15}]-hGRF(1-29)-NH$_2$, 13, to [desNH$_2$-Tyr^1,D-Ala^2,Ala^{15}]-hGRF(1-29)-Gly-Gly-Cys(S-CH$_2$-CH$_2$-CO-Nle-NH-PEG$_{5000}$)-NH$_2$, 13a, dramatically increases the in vivo potency. Figure 9 shows the circular dichroism of 13 and 13a and reveals that the spectra are superimposable in the helix-supporting solvent, 75% methanol-H$_2$O. The CD spectra were also shown to be superimposable in the helix-breaking solvent, H$_2$O. Furthermore, the degree of PEGylation had no effect on the CD spectra. PEGylation of 13 with PEG$_{750}$, PEG$_{2000}$, or PEG$_{10000}$, resulted in the corresponding PEGylated peptides which gave CD spectra that were superimposable with the PEG$_{5000}$-peptide, 13a in both 75% methanol-H$_2$O and in pure H$_2$O. Similar observations were made for the CD spectra of [His^1,Val^2,Gln^8,Ala^{15},Leu^{27}]-hGRF(1-32)-OH, 14, and that of [His^1,Val^2,Gln^8,Ala^{15},Leu^{27}]-hGRF(1-32)GlyCys(S-CH$_2$-CH$_2$-CO-NleNHPEG$_n$)-NH$_2$, 14a, which were also shown to be superimposable in both 75% methanol-H$_2$O and in pure H$_2$O. The CD spectra were also superimposable regardless of the degree of PEGylation; i.e. PEG$_{750}$, PEG$_{2000}$, or PEG$_{5000}$ and PEG$_{10000}$ gave identical CD spectra. From these observations we conclude that COOH-terminal PEGylation of hGRF peptides does not alter the conformation of the parent peptide. Therefore the

Table II. Biological Activity (*in vivo*) of COOH-Terminal Pegylated hGRF Analogs

Compound	hGRF Analog	Relative Potency Mice[a]	Pigs[b]
5	hGRF(1-44)-NH$_2$	1.00	1.00
13	[desNH$_2$Tyr^1,D-Ala^2,Ala^{15}]-hGRF(1-29)-NH$_2$	9.81	10.5
13a	[desNH$_2$Tyr^1,D-Ala^2,Ala^{15}]-hGRF(1-29)-Gly-Gly-Cys(NH$_2$)-S-Nle-PEG$_{5000}$	33.6	54.9
14	[His^1,Val^2,Gln^8,Ala^{15},Leu^{27}]-hGRF(1-32)-OH	6.97	9.48
14a	[His^1,Val^2,Gln^8,Ala^{15},Leu^{27}]-hGRF(1-32)-Gly-Cys(NH$_2$)-S-Nle-PEG$_{5000}$	12.0	47.1

[a] Injected into C57BL/6 mice (retro-orbital, 0-30 µg/kg i.v.). Mice were bled at 0, 5, 10, 15, 30, 60 min post-injection (n=5)

[b] Injected into pigs (0-30 µg/kg s.c.). 1st hr post-dosing, blood was collected at 10 min intervals. 2nd hour post-dosing blood was collected at 15 min intervals. 3rd hr post-dosing blood was collected at 30 min intervals. Total collection time: 6.5 hr.

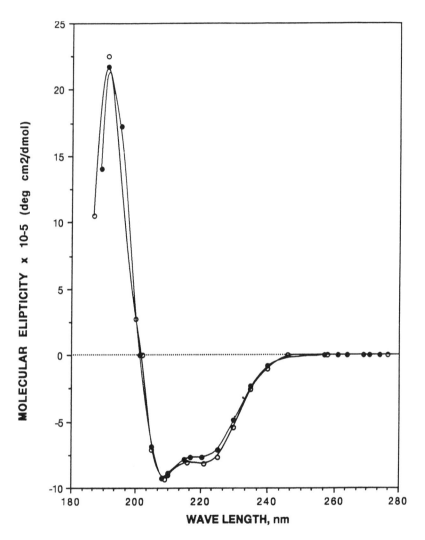

Figure 9. Circular dichroism of non-pegylated and pegylated peptides:
o [desNH$_2$Tyr^1,D-Ala^2,Ala^{15}]-hGRF(1-29)-NH$_2$ and ● [desNH$_2$Tyr^1,
D-Ala^2,Ala^{15}]- hGRF(1-29)-Gly-Gly-Cys(S-CH$_2$-CH$_2$-CO-Nle-NH-
PEG$_{5000}$)-NH$_2$ in 75% methanol-H$_2$O.

increase in the*in vivo* biological activity of these peptides is unrelated to conformation.

Site-specific PEGylation of β-amyloids

Alzheimer's Disease is a progressive dementia that is characterized by the presence of proteinaceous deposits causing lesions in the brain. These deposits are composed mainly of a hydrophobic family of peptides (*28*) known as the β-amyloids which are derived from a larger amyloid β-protein. Both the hydrophobic COOH-terminus (*29*) and the hydrophobic region between residues 17-20 (*30*) have been reported to be responsible for the amyloid properties of amyloid β-protein. Recent studies have shown that a structural transition may be required for the conversion of β-amyloid peptide into amyloid plaque(*31*).

βAP(1-40)-NH$_2$, **15**, forms stable amyloid fibers and its deposition is reported to proceed by a nucleation-dependent mechanism (*32*). We have prepared COOH-terminal PEGylated derivatives of βAP(1-40)-NH$_2$ and evaluated the rate of aggregation using turbidity measurements modeled after those previously reported (*32*). Figure 10 shows the rate of aggregation of βAP(1-40)-NH$_2$, **15**, and βAP(1-40)-Orn(PEG$_{5000}$-CH$_2$-CO)-NH$_2$, **15a**. It was observed that the aggregation of βAP(1-40)-NH$_2$ began almost immediately, but the PEGylated analog, **15a**, required a substantial period of time before aggregation was even detectable. These studies demonstrate the advantages of PEGylation of β-amyloid peptides since this will enable us to carry out more detailed studies, including conformational studies, on the structural transitions that are proposed during the formation of amyloid plaque.

Conclusions

Conditions for the site-specific PEGylation of peptides have been developed. The solid phase PEGylation procedures that are described for NH$_2$-terminal and internal PEGylation are convenient since they can be incorporated into a standard solid phase peptide synthesis by simply removing an aliquot of peptide-resin at the appropriate cycle. The solution phase PEGylation that is described enables one to carry out post-synthesis site-specific PEGylation at a Cys residue at any position in the peptide. Both the solid phase and the solution phase PEGylation procedures are compatible with poly(ethyleneglycol) derivatives of varying molecular weight (i.e. PEG$_{750}$, PEG$_{2000}$, PEG$_{5000}$ and PEG$_{10000}$). The incorporation of two or more PEG units by solid phase synthesis (multiple PEGylation) can best be carried out with poly(ethyleneglycol) derivatives of lower molecular weight (i.e. PEG$_{750}$, PEG$_{2000}$). PEGylated peptides should be purified by a 2-stage process using dialysis and preparative HPLC. PEGylated peptides should be held to the same analytical standards as are currently required for free peptides. Homogeneity can be demonstrated by analytical HPLC and complete characterization includes amino acid analysis, laser desorption mass spectrometry and ^{1}H-NMR spectroscopy. in special cases sequence analysis can also be carried out.

Our studies have demonstrated PEGylation by a site-specific procedure is of paramount importance. This was clearly demonstrated in the structure-activity studies that were carried out on the human growth hormone-releasing factor system. PEGylation at the NH$_2$- terminus and at positions 21 and 25 had no significant effect on *in vitro* biological activity compared to the non-PEGylated peptide. However, PEGylation at positions 8 and 12 drastically reduced the *in vitro* potency of the peptide. Finally, the observation that PEGylation at the COOH-terminus resulted in

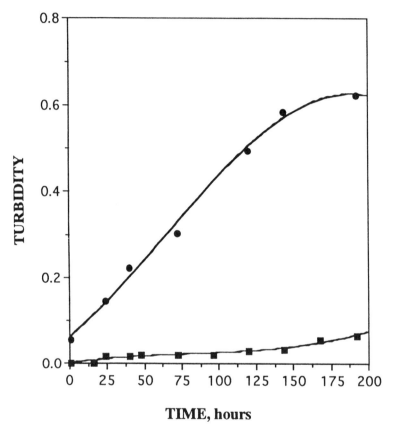

Figure 10. Time course formation of amyloid fibrils: Rate of aggregation
of non-pegylated and pegylated β-amyloid peptides: ● βAP(1-40)-NH$_2$
and ■ βAP(1-40)-Orn(PEG$_{5000}$-CH$_2$-CO)-NH$_2$. Concentration: 20μM.
Solutions were prepared in DMSO (50μL) and aqueous buffer (950μL) and
turbidity measured at 400 nm as previously described (reference 33).

an increase of the *in vivo* potency provides strong evidence of the importance of site-specificity.

Abbreviations used: Standard abbreviations are used for the amino acids and protecting groups as recommended by the IUPAC-IUB Commission on Biochemical Nomenclature: *J. Biol Chem.* **1985**, 260, p. 14. Other abbreviations are as follows: BOP, benzotriazol-1-yloxytris(dimethylamino)phosphonium hexafluorophosphate; βAP, β-amyloid peptide; CD, circular dichroism; DIEA, diisopropylethylamine; Fmoc, 9-flourenylmethyloxycarbonyl; hGRF, human growth hormone-releasing factor; HIV, human immunodeficient virus; HPLC, high performance liquid chromatography; PEG, monomethoxypoly-(ethyleneglycol); SPDP, N-succinimidyl 3-(2-pyridyldithio)propionate; tBu, tert.butyl; TFA, trifluoroacetic acid.

Acknowledgments.

The author thanks Dr. Robert M. Campbell and his associates for carrying out the biological studies on the PEGylated GRF analogs and Dr. John Maggio and his associates for carrying out the biological studies on the β-amyloid PEGylated peptides. The author also thanks Dr. Y.-A. Lu and Dr. E. Heimer and his associates for carrying out the synthesis of the PEGylated peptides. The author also acknowledges Dr. Y.-C. Pan and Dr. H. Michel for the laser desorption ionization mass spectrometric measurements reported in this manuscript.

References.

1. Abuchowski, A. and Davis, F. F. in *Enymes as Drugs;* Hosenberg, J. and Roberts, J., Eds; J. Wiley: New York, NY, **1981**, p 367.

2. Abuchowski, A., Kafkewitz, A. D. and Davis, F. F. *Prep. Biochem.* **1979**, 9, pp. 205-211.

3. Rajagopaiain, S., Gonias, L. and Pizzo, S. V. *J. Clin. Invest.* **1985**, 75, pp. 413-419.

4. Katre, N. V., Knaug, M. J. and Laird, W. J. *Proc. Natl. Acad. Sci. USA* **84**, pp. 1487-1491.

5. Ho, D. H., Brown, N. S., Yen, A., Holmes, Y. R., Keating, M., Abuchowski, A., Newman, R. A. and Krakoff, I. H. *Drug Metab. Disp.* **1986**, 14, pp. 349-352.

6. Nucci, C. L., Shorr, R. and Abuchowski, A. *Adv. Drug Delivery Res.* **1991**, 6, pp. 133-151.

7. Sartore, L., Caliceti, P., Schiavon, O. and Veronese, F. M. *Appl. Biochem. Biotech.* **1991**, 27, pp. 45-54.

8. Sartore, L., Caliceti, P., Schiavon, O., Monfardini, C. and Veronese, F. M. *Appl. Biochem. Biotech.* **1991**, 31, pp. 213-222.

9. Lu, Y.-A. and Felix, A.M. *Pept. Res.* **1993**, 6, pp. 140-146.

10. Fournier, A., Wang, C.-T. and Felix, A. M. *Int. J. Pept. Prot. Res.* **1988**, 31, pp. 86-97.

11. Kaiser, E., Colescott, R. L., Bossinger, C. D. and Cook, P. I. *Anal. Biochem.* **1970**, 34, pp. 595-599.

12. Wang, S.-S. *J. Am. Chem. Soc.* **1973**, 95, pp. 1328-1333.

13. Rink, H. *Tetrahedron Lett.* **1987**, 28, pp. 3787-3790.

14. Albericio, F., Kneib-Cordonier, N., Gera, L., Hammer, R. P., Hudson, D. and Barany, G. in *Peptides: Chemistry and Biology; Proceedings of the Tenth American Peptide Symposium;* Escom: Leiden, The Netherlands, **1988**, pp. 159-161.

15. Albericio, F., Kneir-Cordonier, N., Biancalana, S., Gera, L., Masada, R. I., Hudson, D. and Barany, G. *J. Org. Chem.* **1990**, 55, pp. 3730-3743.

16. Mutter, M., Oppliger, H. and Zier, A. *Makromol. Chem., Rapid Commun.* **1992**, 13, pp. 151-157.

17. Lu, Y.-A. and Felix, A.M. *Int. J. Pept. Prot. Res.* **1994**, 43, pp. 127-138.

18. Lu, Y.-A. and Felix, A. M. *Reactive Polymers* **1994**, 22, pp. 221-229.

19. Campbell, R. M., Lee, Y., Mowles, T. F., McIntyre, K. W., Ahmad, M., Felix, A. M. and Heimer, E. P. *Peptides* **1992**, 13, pp. 787-793.

20. Campbell, R. M., Ahmad, M., Heimer, E. P., Lambros, T., Lee, Y., Miller, R., Stricker, P. and Felix, A. M. *Int. J. Pept. Prot. Res.* (in press).

21. Campbell, R. M. and Scanes, C. G. *Growth Regulation* **1992**, 2, pp. 175-191.

22. Felix, A. M., Heimer, E. P., Mowles, T. F., Eisenbeis, H., Leung, P., Lambros, T., Ahmad, M., Wang, C.-T. and Brazeau, P. *Proc. 19th European Peptide Symposium;* Walter de Gruyter, Berlin, **1987**, pp. 481-484.

23. Felix, A. M., Lu, Y.-A. and Campbell, R.M. *Int. J. Pept. Prot. Res.* **1995**, 46, pp. 253-264.

24. Campbell, R. M., Bongers, J. and Felix, A. M. *Biopolymers* **1995**, 37, pp. 67-88.

25. Frohman, L. A., Downs, T. R., Williams, T. C., Heimer, E. P., Pan, Y.-C. and Felix, A. M. *J. Clin. Invest.* **1986**, 78, pp. 906-913.

26. Heimer, E. P. Bongers, J., Ahmad, M., Lambros, T., Campbell, R. M. and Felix, A. M. in *Peptides: Chemistry and Biology; Proceedings of the Twelfth American Peptide Symposium;* Escom: Leiden, The Netherlands, **1992**, pp. 80-81.

27. Campbell, R. M., Stricker, P., Miller, R., Bongers, J., Heimer, E. P. and Felix, A. M. in *Peptides: Chemistry Structure and Biology; Proceedings of the Thirtennth American Peptide Symposium;* Escom: Leiden, The Netherlands, **1994**, pp. 378-380.

28. Glenner, G. G. and Wong, C.-W. *Biochem. Biophys. Res. Commun.* **1984**, 122, pp. 1131-1135.

29. Jarrett, J. T., Berger, E. P. and Lansbury, P. T., Jr. *Biochemistry* **1993**, 32, pp. 4693-4697.

30. Hilbich, C., Kister-Woike, B., Reed, J., Masters, C. L. and Beyreuther, K. *J. Mol. Biol.* **1992**, 220, pp. 460-473.

31. Lee, J. P., Stimson, E. R., Ghilardi, J. R., Mantyh, P. W., Lu, Y.-A., Felix, A. M., Llanos, W., Behbin, A., Cummings, M., Van Criekinge, M., Timms, W. and Maggio, J. E. *Biochemistry* **1995**, 34, pp. 5191-5200.

32. Jarrett, J. T. and Lansbury, P. T., Jr. *Biochemistry* **1992**, 31, pp. 12345-12352.

Chapter 17

Poly(ethylene glycol)-Containing Supports for Solid-Phase Synthesis of Peptides and Combinatorial Organic Libraries

George Barany[1], Fernando Albericio[1,2,3], Steven A. Kates[3], and Maria Kempe[4]

[1]Department of Chemistry, University of Minnesota, Minneapolis, MN 55455
[2]Department of Organic Chemistry, University of Barcelona,
E–08028 Barcelona, Spain
[3]PerSeptive Biosystems, Inc., 500 Old Connecticut Path, Framingham, MA 01701
[4]Department of Pure and Applied Biochemistry, University of Lund,
S–221 00 Lund, Sweden

The choice of a polymeric support is a key factor for the success of solid-phase methods for syntheses of organic compounds and biomolecules such as peptides and oligonucleotides. Classical Merrifield solid-phase peptide synthesis, performed on low cross-linked hydrophobic polystyrene (PS) beads, sometimes suffers from sequence-dependent coupling difficulties. The concept of incorporating PEG into supports for solid-phase synthesis represents a successful approach to alleviating such problems. This chapter reviews the preparation of families of poly(ethylene glycol)-polystyrene (PEG-PS) graft as well as (highly) *C*ross-*L*inked *E*thoxylate *A*crylate *R*esin (CLEAR) supports developed in our laboratories, and demonstrates their applications to the syntheses of a wide range of targets in connection with numerous research objectives.

Solid-phase synthesis (SPS) is a powerful tool for the preparation of a wide range of molecules (1-5). The methodologies were first applied to the assemblies of biomolecules such as peptides and oligonucleotides (2-8), and the past few years have witnessed a great proliferation of work extending the principles to the synthesis of organic compounds, particularly in conjunction with combinatorial libraries (9-22). The efficiency of solid-phase methods depends on a number of parameters, of which the choice of polymeric support is a key factor for achieving success. For solid-phase peptide synthesis (SPPS), the majority of work through the 1980's has been carried out with the low divinylbenzene-cross-linked polystyrene (PS) beads that were introduced to the area in Merrifield's original studies (1, 2). Sequence-dependent coupling difficulties were ascribed by Sheppard and others to the hydrophobic nature of PS, and spurred the development of more hydrophilic supports (5). While it is not certain that this reasoning to move away from PS was entirely valid, there can be no doubt that a wide range of materials, many of them now commercially available, have since been established as useful for SPPS. These include polyamide supports (regular or encapsulated in rigid materials such as kieselguhr or highly cross-linked polystyrene) (23-26), polystyrene-Kel F (PS-Kel F) grafts (27-29), polyethylene-polystyrene (PE-PS) films (30), membranes (31-33), cotton and other carbohydrates (34, 35), and chemically modified polyolefins ("ASPECT") (36). Solid-phase synthesis of oligonucleotides (7) has been carried out mainly on controlled pore glass (CPG). Much solid-phase synthesis of organic compounds is carried out on various functionalized PS resins, and some of the other materials mentioned in this paragraph have been used as well.

Arguably, the most significant set of practical advances in materials for solid-phase synthesis has come from the recognition that properties of supports are often improved by the incorporation of multiple hydrophilic ethylene oxide units (Table 1). The present chapter focuses on families of poly(ethylene glycol)-polystyrene (PEG-PS) graft and Cross-Linked Ethoxylate Acrylate Resin (CLEAR) supports developed and established from our own laboratories. Grafting of PEG onto PS, for other applications, was carried out first in the 1970's and early-1980's by Inman (37), Regen (38), Warshawsky and Patchornik (39), Sherrington (40), and Mutter (41, 42). Our initial studies and contemporaneous pioneering independent work of Bayer with Rapp (50-54) were reported in the mid-1980's; the commercial viability and advantages of PEG-containing supports [including Meldal's PEGA (56,57)] were realized in the 1990's. Scanning electron micrographs of four such supports are shown in Figure 1, and should be contrasted with respect to the size, spherical nature, and uniformity of beads. Herein, we review methods for the preparation of PEG-PS and CLEAR supports, and provide examples for their applications to synthesis.

Goals for Solid-Phase Supports

Criteria for what constitutes an optimal set of features for a suitable SPS support are controversial, and much of the conventional wisdom in the field is based on empirical observations. While the term "solid-phase" might imply a static resin support, it turns out that this is not at all the case with those materials which serve best for SPPS (64-66). Reactions commonly occur on mobile, well-solvated, and reagent-accessible polymer strands throughout the *interiors* of the supports; relatively few of the sites occur in surface regions (67). The conventional wisdom in the field has therefore been that supports should have the minimal level of cross-linking consistent with stability, in order to allow the material to swell. PEG-PS fits this model, since the parent hydrophobic PS is low-crosslinked and the hydrophilic pendant PEG chains enhance the swelling to cover a broad range of solvents (Table 2, middle columns). Nevertheless, we have shown that CLEAR supports, which are similar to PEG-PS in terms of the presence of ethylene oxide units but differ insofar as CLEAR's are highly cross-linked (≥ 95% by weight of cross-linker), also demonstrate substantial swelling in many solvents (Table 2, right columns). The mechanical and physico-chemical properties of PEG-PS and CLEAR translate into excellent performance characteristics for stepwise SPPS, both continuous-flow and batchwise, in all cases in conjunction with Fmoc/tBu chemistry and in some cases with the Boc/Bzl strategy. Information is also accumulating for solid-phase oligonucleotide synthesis and solid-phase organic synthesis (SPOS) with these materials.

PEG-PS Supports

The defining idea behind grafting PEG onto PS was to combine in the same support a hydrophobic core of PS with hydrophilic PEG chains. A further concept was that PEG might act as a "spacer" separating the starting point of solid-phase synthesis from the PS core (Figure 2). Our general approach for the preparation of PEG-PS supports relies on the covalent attachment via amide linkages of PEG of defined molecular weight onto suitable amino-functionalized microporous PS. Bayer and Rapp have championed an alternative way to prepare a version of PEG-PS, which they refer to as POE-PS (known commercially as TentaGel or NovaSyn), by means of

Table 1. Poly(ethylene glycol)-containing synthetic polymers used as supports in solid-phase synthesis [a]

Support	Description	Reference(s)
PEG-PS	poly(ethylene glycol) grafted covalently onto 1% cross-linked microporous poly(styrene–co–divinylbenzene)	43-49
POE-PS (TentaGel or NovaSyn)	poly(ethylene glycol) polymerized onto poly(styrene–co–divinylbenzene)	50-54
PEG/PS (ArgoGel)	poly(ethylene glycol) polymerized onto a malonate-derived 1,3-dihydroxy-2-methylpropane "branching" unit bound at the 2 position to benzyl group from poly(styrene–co–divinylbenzene)	55
PEGA	poly(N,N-dimethylacrylamide–co–bisacrylamido poly(ethylene glycol)–co–monoacrylamido poly(ethylene glycol))	56, 57
TEGDA-PS	poly(styrene–co–tetra(ethyleneglycol) diacrylate)	58
PEO-PEPS	3,6,9-trioxadecanoic acid coupled to polyethylene-polystyrene (PE-PS) films	59
CLEAR	poly(trimethylolpropane ethoxylate (14/3 EO/OH) triacrylate–co–allylamine)	60-63

[a] Abbreviations and trademarks for PEG-containing polymers are literally those of the referenced inventors and/or companies commercializing the appropriate supports, and are listed without any judgement by the authors of this review on the similarities and differences among the materials.

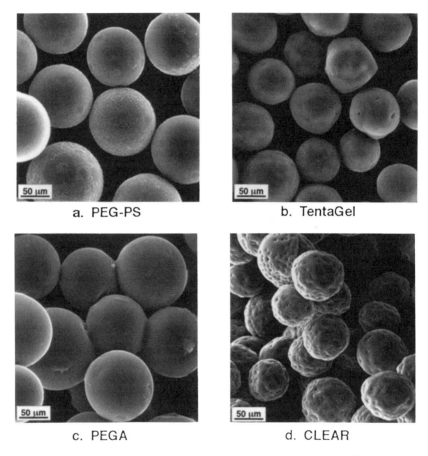

a. PEG-PS

b. TentaGel

c. PEGA

d. CLEAR

Figure 1. Scanning electron micrographs [accelerating voltage: 10 kV] showing the shape and texture of (a) PEG-PS, (b) TentaGel, (c) PEGA, and (d) CLEAR-IV.

Table 2. Swelling of PS, PEG-PS, and CLEAR Supports [a]

Solvent	Bed volume (mL) of 1 g polymer				
	PS[b]	PEG-PS low-load[c]	PEG-PS high-load[d]	CLEAR-I[e]	CLEAR-IV[f]
Hexane	nd	nd	nd	3	3
Toluene	nd	nd	nd	5	5
tBuOMe	nd	nd	nd	3.5	3
EtOAc	2	4	4	5	4.5
THF	5	4	4.5	6.5	5
TFA	nd	nd	nd	12	6.5
CH$_2$Cl$_2$	5.5	5.5	6	10	5.5
MeOH	5	4.5	5	7	5
DMF	5.5	5	5	8	5.5
CH$_3$CN	2	4	3.5	6.5	5
H$_2$O	2	3	3.5	8	4

[a] Solvents listed in order of increasing dielectric constant
[b] MBHA-PS: 1.0 mmol/g
[c] Low load PEG-PS: 0.15 mmol/g
[d] High load PEG-PS: 0.55 mmol/g
[e] CLEAR I,: 0.26 mmol/g
[f] CLEAR-IV: 0.17 mmol/g
nd = not determined

Figure 2. General structure for PEG-PS supports ("spacer" model). PS = low cross-linked polystyrene, PEG = poly(ethylene glycol), typically of average molecular weight 2000 (~ 44 repeating units); X = starting point for peptide (or other) synthesis.

anionic copolymerization of ethylene oxide onto a PS-bound initiator. While we are not in a position to contrast the relative merits of these two ways of accessing PEG-PS, it should be noted that our methodology allows the final weight ratio of PEG:PS, as well as the final loading, to be controlled by both the starting loading of amino groups on PS and the molecular weight of PEG. Final loadings (0.1 to 0.6 mmol/g) can be readily tailored: lower loadings appear to be better for large macromolecular targets and higher loadings are preferred for combinatorial SPOS.

Our original "first generation" approach (43, 47) started from commercially available homobifunctional PEG (diol). This was converted in six chemical steps and one key chromatographic step to the *pure* heterobifunctional N^ω-Boc-protected PEG-ω-amino acid (Figure 3). The first chemical step, reaction with thionyl chloride in the presence of pyridine, was *intentionally* prevented from reaching completion, so as to *partially* derivatize PEG. The resultant statistical mixture of dichloride, monochloride, and unreacted diol reacted further with ethyl isocyanatoacetate to convert PEG hydroxyls to the corresponding urethanes, sodium azide in DMF to convert PEG chlorides to azides, and hydroxide to saponify the pendant esters. At this stage of the overall process, the desired derivative of *defined* structure 4 [boxed in Figure 3] was isolated efficiently by ion-exchange chromatography with a stepwise ammonium bicarbonate gradient. Catalytic hydrogenolysis then smoothly converted the azide to the amine, which was finally protected as its Boc derivative. It should be stressed that all chemical transformations other than the first step proceed *quantitatively*, hence avoiding potential problems should extraneous functional groups be carried over to the final product. Polymeric intermediates and the final derivative were purified based on the physical properties of PEG, yet they were amenable to accurate characterization by analytical and spectroscopic techniques commonly applied to low molecular weight organic compounds. The next step in preparing PEG-PS suitable for evaluation in peptide synthesis was to carry out oxidation-reduction coupling of the heterobifunctional protected PEG 5 onto aminomethyl-PS, followed by deprotection/coupling cycles to introduce an "internal reference" amino acid (IRAA), usually Nle, and a handle such as 5-(4-Fmoc-aminomethyl-3,5-dimethoxyphenoxy)valeric acid (Fmoc-PAL-OH) (68, 69). Typically loadings were 0.25-0.30 mmol/g (loadings for starting PS was 0.6 mmol/g), which implies a ratio of PEG:PS of 0.55:0.45.

Although the original route (Figure 3) to PEG-PS is certainly unambiguous, we sought simpler alternatives that might avoid the time-consuming process — particularly the chromatography step — to obtain the required PEG intermediates. Particularly attractive was to couple onto *p*-methylbenzhydrylamine (MBHA) polystyrene resins, already containing an IRAA, *homobifunctional* PEG-diacids [Figure 4, ref. 46]. The PEG-diacids (9) were obtained easily by reaction of the corresponding inexpensive PEG-diamines (8) with succinic anhydride. This approach accepts a modest level of accompanying cross-linking (e.g., PS with 0.6 mmol/g substitution typically gave grafts with PEG:PS = 0.45:0.55, from which it can be calculated that ~ 2/3 of the PEG remains available to act as a spacer). The pendant carboxyl groups of PEG-PS derived from the PEG-acids were next converted to amino groups (final loading typically 0.15-0.25 mmol/g) by a coupling reaction with excess ethylenediamine (EDA).

Through a fortuitous set of observations described elsewhere (45, 70) we were led to consider the possibility that *environmental* effects (Figure 5) contribute to the efficacious properties of PEG-PS, and to devise an unambiguous formulation to test this idea, as follows: N^α-Fmoc, N^δ-Boc-ornithine was coupled onto aminomethyl-PS (0.95 mmol/g), the Fmoc-group was removed, and a *monofunctional* methoxy-PEG-acid was added. The resultant graft comprised PEG:PS = 0.60:0.40, with a loading of 0.31 mmol/g; Boc removal exposed the ornithine side-chain for handle incorporation and further peptide chain assembly (44, 45, 70). Such supports were successful when applied to challenging target sequences, allowing us to conclude that the PEG environment was much more critical than any spacer effects.

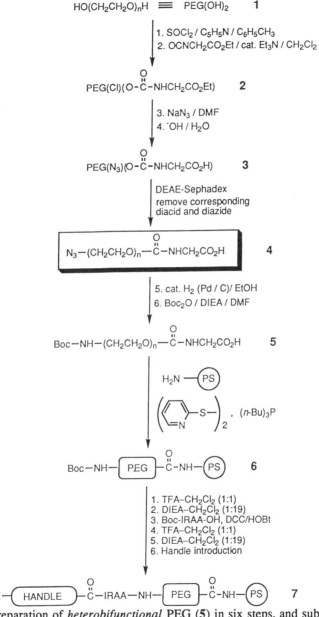

Figure 3. Preparation of *heterobifunctional* PEG (**5**) in six steps, and subsequent use to generate "first generation" PEG-PS supports. Compounds **2** and **3** are statistical mixtures, whereas the boxed compound **4** is the compound with the shown defined structure as isolated after ion-exchange chromatography. The Figure shows the grafting of **5** onto amino-functionalized supports, introduction of an "internal reference" amino acid (IRAA), and introduction of a carboxyl group-containing handle. X = starting point for peptide (or other) synthesis. The indicated chemistry was carried out both for n = 45 (PEG-2000) and n = 90 (PEG-4000). Adapted from Zalipsky *et al.* (ref. 47).

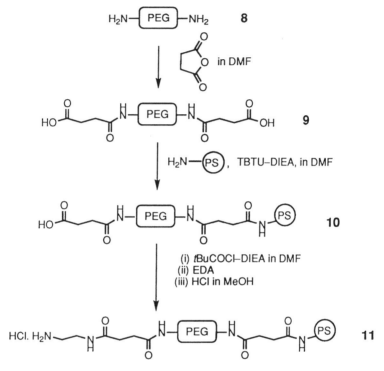

Figure 4. Preparation of "second generation" PEG-PS supports from *homobifunctional* PEG-diacids. Adapted from Barany *et al.* (ref. 46).

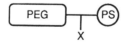

Figure 5. General structure for PEG-PS supports ("environmental" model). PS = low cross-linked polystyrene, PEG = poly(ethylene glycol), typically of average molecular weight 2000 (~44 repeating units); X = starting point for peptide (or other) synthesis.

A "high load" PEG-PS (Figure 6, ref. 49) was prepared following a combination of both the "spacer" and "environmental" strategies. N^α-Fmoc, N^δ-Boc-ornithine was coupled onto MBHA-PS, the Fmoc group was removed, and homobifunctional PEG-diacids were added as already described. The N^δ-Boc group was then removed, and PAL handle (1 equiv) was coupled for optimal results. Remaining free amino groups were capped with Ac_2O, to achieve typical final loadings of 0.35-0.6 mmol/g.

PEG-PS supports were first commercialized on the basis of MBHA-PS as the parent resin. As a consequence, the linkage between PEG and PS was labile to HF, and such PEG-PS's were compatible only with the Fmoc/*t*Bu strategy for SPPS. However, by starting with aminomethyl-PS, HF-stable supports compatible with Boc/Bzl chemistry are obtained readily (48). For such cases, handles such as Boc-aminobenzhydryloxyacetic acid (BHA-linker) (71) are used in place of PAL.

CLEAR Supports

We described recently (60-62) a novel approach to PEG-containing supports, the CLEAR family (Table 1), which is very much complementary to the chemistry of PEG-PS discussed above. Rather than derivatizing and combining preformed polymers as is the case with PEG-PS, we opted to create supports *de novo* by copolymerizing appropriate monomers and cross-linkers that might lead to biocompatible structures with good mechanical stabilities. For a number of reasons, our work focused on branched PEG-containing cross-linkers such as trimethylolpropane ethoxylate (14/3 EO/OH) triacrylate [structure 12 in Figure 7]. Each branch of 12 contains a polymerizable vinyl endgroup as well as a chain with on average four to five ethylene oxide units. A high molar ratio of 12 was copolymerized with amino-functionalized monomers, such as allylamine (13) or 2-aminoethyl methacrylate (14), which were chosen in anticipation of the later need of starting points for the solid-phase synthesis. The fact that amino groups could be introduced *directly* rather than by transformation of another functional group or by deprotection of a protected amino monomer, was an unexpected yet advantageous discovery in the CLEAR family. Incorporation of amines into synthetic polymers has been reported to be difficult due to (i) addition of the amine to activated vinylic double bonds (72), and (ii) $O \rightarrow N$ acyl migration resulting in hydroxylated acrylamides when starting with amino acrylates (73). In addition, various non-functionalized monomers and cross-linkers (15-17) have been copolymerized successfully with 12. Five different CLEARs will be discussed in the following: CLEAR-I and CLEAR-V (made from 12 and 13); CLEAR-II (from 12, 14, and 15); CLEAR-III (from 12, 14, and 16); and CLEAR-IV (from 12, 13, and 17). All of these polymers have hydrophilic PEG-like character, even though individual oligo(ethylene oxide) chains are quite short compared to chains in PEG-PS (Figures 8 and 9).

CLEAR particles of irregular shape are prepared by bulk polymerization followed by grinding and sieving, whereas the more preferred spherical beads are obtained by a suspension polymerization procedure. The loadings of the resins are affected by the amount of the amino-functionalized monomer used during the polymerization. Typical loadings are in the range of 0.1-0.3 mmol/g. As with PEG-PS, the free amino groups are usually acylated with an IRAA and extended with PAL or another handle.

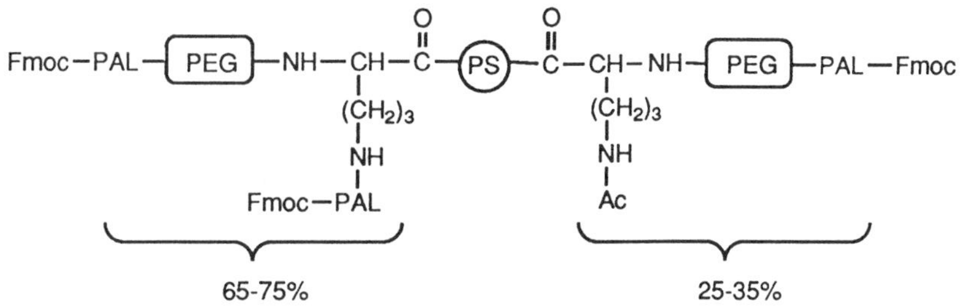

Figure 6. General structure for "high load" PEG-PS supports. PS = low cross-linked polystyrene, PEG = poly(ethylene glycol), typically of average molecular weight 2000 (~44 repeating units); Fmoc-PAL = tris(alkoxy)benzylamide handle. Adapted from McGuinness *et al.* (ref. 49).

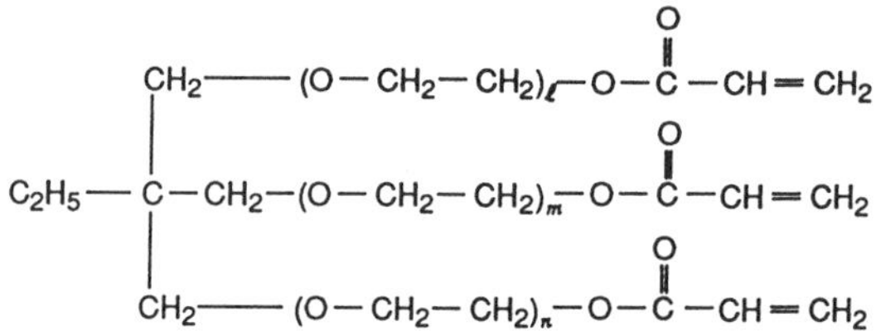

12 $\ell + m + n \sim 14$

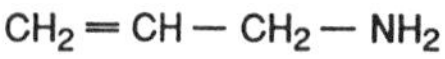

13

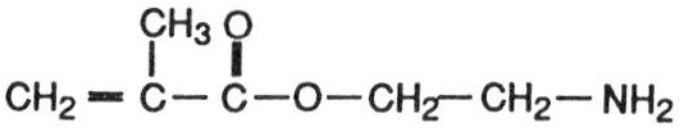

14

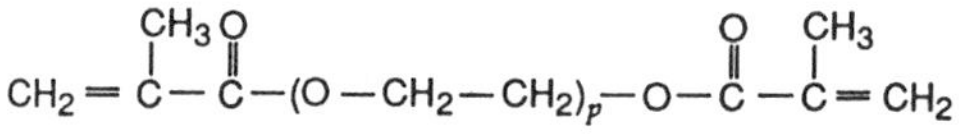

15 $p \sim 9$

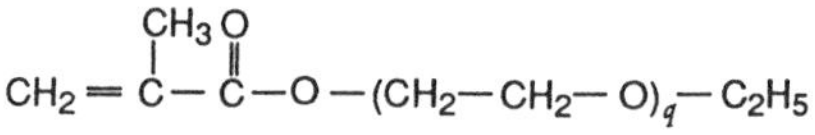

16 $q \sim 3\text{-}5$

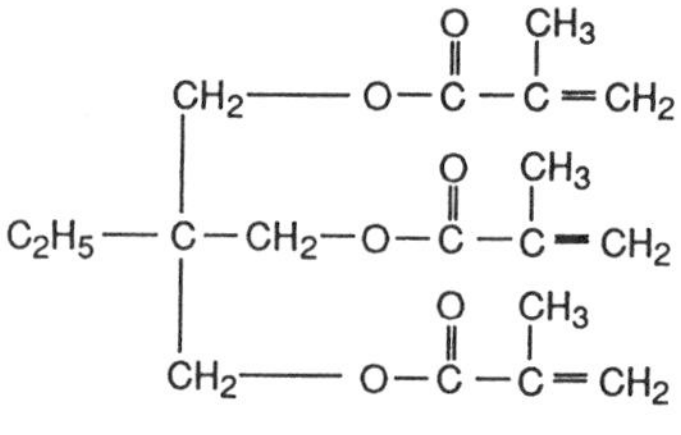

17

Figure 7. Cross-linkers and monomeric building blocks of CLEAR supports.

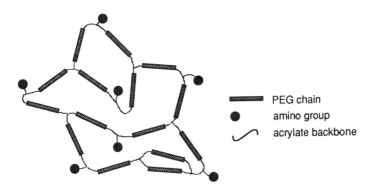

Figure 8. Schematic representation of CLEAR supports. The supports are prepared by copolymerizing a PEG-containing branched cross-linker (**12** in Figure 7) with an amino-functionalized monomer. See text for details.

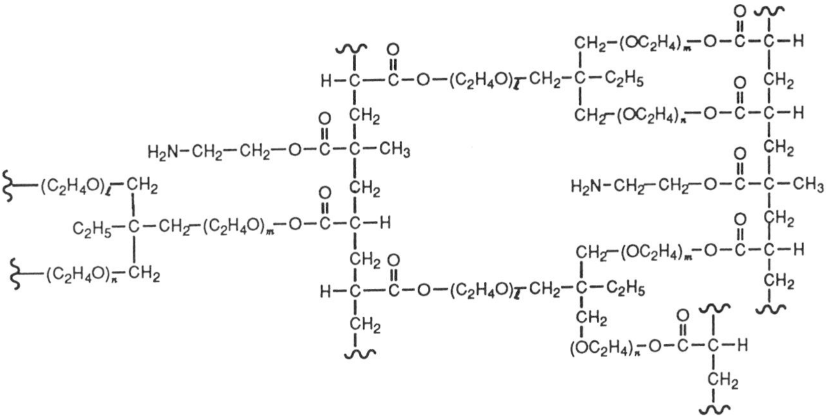

Figure 9. Structure of the polymeric network of CLEAR supports.

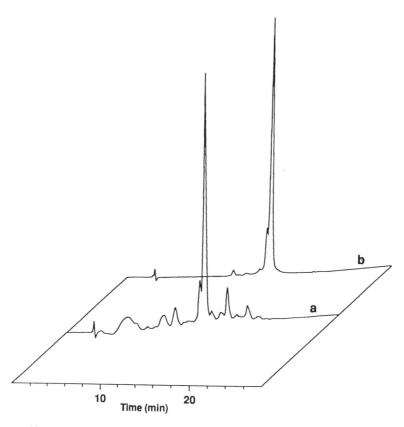

Figure 10. Analytical HPLC chromatogram of (a) crude human gastrin-I, prepared on a PerSeptive Biosystems 9050 continuous-flow peptide synthesizer, starting with CLEAR-II (0.26 mmol/g) and using Fmoc chemistry with a BOP/HOBt/NMM protocol (1-h couplings), and (b) human gastrin-I after purification by preparative HPLC. Elutions at 1 mL/min with a linear gradient over 28 min from 3:1 to 2:3 of 0.1% aqueous TFA and 0.1% TFA in CH_3CN (Vydac C-18, 4.6 x 250 mm). Peptides were detected at 220 nm. Adapted from Kempe and Barany (ref. 62).

Solid-Phase Peptide Synthesis Examples

The usefulness of PEG-PS and CLEAR for the syntheses of challenging peptides, including hydrophobic peptides and peptides that contain post-translational modifications, have been demonstrated in a variety of ways. Comparative syntheses carried out in parallel with control studies using a variety of other solid supports have allowed us to draw conclusions about the relative superiority of PEG-PS and/or CLEAR. It is noteworthy that PEG-PS and CLEAR each have excellent mechanical stability that allows their use in conjunction with both batchwise and continuous-flow reactors. Whereas numerous commercially available supports for SPPS are unusable, or usable only with major difficulties, in the continuous-flow mode, PEG-PS and CLEAR in conjunction with optimized stepwise Fmoc chemistry give excellent results, i.e., negligible back-pressures during assembly, and good yields and purities of the initial cleaved product mixtures. Successful illustrations of these points are provided with human gastrin-I (17 residues; see Figure 10), ovine corticotropin releasing factor

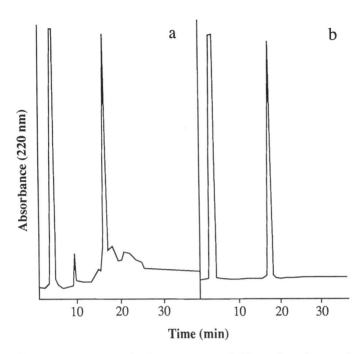

Figure 11. Analytical HPLC chromatogram of (a) crude ovine corticotropin releasing factor (CRF), prepared on a 9050 PerSeptive continuous-flow peptide synthesizer, using PEG-PS, Fmoc chemistry, and DIPCDI/HOBt-mediated couplings (2-h), and (b) CRF after purification by preparative HPLC. Elutions at 1 mL/min with a linear gradient over 30 min from 1:9 to 7:3 of 0.1% aqueous TFA and 0.1% TFA in CH_3CN (Vydac C-18, 4.6 x 250 mm). These data were shown first in the poster associated with Barany *et al.* (ref. 44).

(CRF) (43 residues, see Figure 11), and bovine pancreatic trypsin inhibitor (BPTI) (58 residues, see refs. 70 and 74).

The acyl carrier protein (ACP) (65-74) amide decapeptide sequence, H-Val-Gln-Ala-Ala-Ile-Asp-Tyr-Ile-Asn-Gly-NH_2, has long been recognized as a challenging target, dating to the initial report by Hancock and Marshall (75). While efficient assemblies of this structure on a range of supports by a variety of protocols are known, the fact that several of the requisite peptide bond forming steps are quite slow legitimizes comparative studies in which relatively short *single* couplings are used. Two sets of parallel batchwise syntheses were carried out on PAL-PEG-PS *vs.* PS using a Fmoc/*t*Bu strategy and DIPCDI-mediated couplings (Figure 12). When the conventional solvent mixture CH_2Cl_2–DMF (1:2) was used, both qualitative ninhydrin analyses of the peptide-resins after coupling steps and HPLC analyses of crude products showed clearly that syntheses on PEG-PS gave better rates and initial homogeneities than on PS. Next, the same experiments were repeated except that all coupling and washing steps were carried out in acetonitrile (CH_3CN). In this more demanding case, the purity of the crude ACP peptide obtained on PEG-PS in CH_3CN was comparable to that observed on the same support with CH_2Cl_2–DMF, while the synthesis carried out on PS did not lead to any of the target product. Essentially the

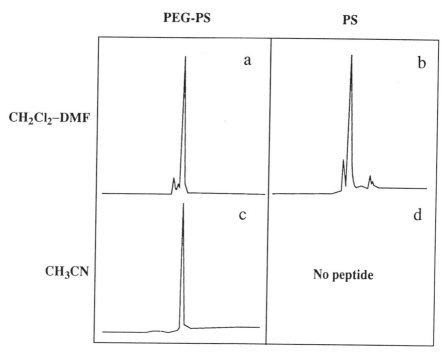

Figure 12. Analytical HPLC chromatograms of crude ACP(65-74) amide, prepared by batchwise manual synthesis, Fmoc chemistry, DIPCDI-mediated couplings (1-h), on PEG-PS (left) and PS (right). Elutions at 1.2 mL/min with a linear gradient over 15 min from 9:1 to 1:1 of 0.01 N aqueous HCl and CH_3CN (Vydac C-18, 4.6 x 250 mm). Peptides were detected at 215 nm. (a) Synthesis carried out on PAL-PEG-PS support *in parallel* with (b) synthesis using CH_2Cl_2–DMF (1:2) as the solvent for all coupling steps and washes. (c) Synthesis carried out on PAL-PEG-PS support, using CH_3CN as the solvent for all coupling steps and washes. Adapted from Zalipsky *et al.* (ref. 47).

same experimental design with CLEAR supports showed that excellent peptide purities of ACP (65-74) were attained when either neat DMF or neat CH_3CN were used as the reaction as well as wash media (Figure 13). The main conclusion from these studies is that the expanded range of solvents that swell PEG-PS and CLEAR can be translated to additional options for peptide synthesis. Most other support materials are incompatible with CH_3CN as the solvent for stepwise incorporation of amino acids into peptides.

The exceedingly "difficult" (68) retro-sequence 74-65 of ACP, i.e., H-Gly-Asn-Ile-Tyr-Asp-Ile-Ala-Ala-Gln-Val-NH$_2$, was prepared in side-by-side batchwise experiments (Figure 14) using CLEAR and PEG-PS along with a number of commercially available supports, including two other PEG-containing supports (TentaGel and PEGA). Although each synthesis could likely be optimized further, the data reveal that performance characteristics on CLEAR and PEG-PS were at least comparable, and in some cases better, to any found previously in our hands or in the literature.

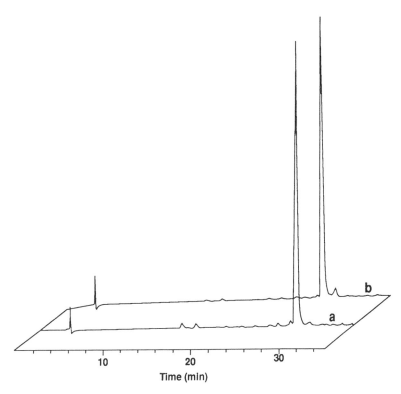

Time (min)

Figure 13. Analytical HPLC chromatogram of crude ACP (65-74) amide, prepared by batchwise manual synthesis, Fmoc chemistry, on CLEAR-I (0.29 mmol/g). Elutions at 1 mL/min with a linear gradient over 30 min from 19:1 to 3:1 of 0.1% aqueous TFA and 0.1% TFA in CH_3CN (Vydac C-18, 4.6 x 250 mm). Peptides were detected at 220 nm. (a) Couplings (2-h) were mediated by DIPCDI/HOAt, and couplings and washings were performed with DMF as solvent. (b) Couplings (2-h) were mediated by DIPCDI, and couplings and washings were performed with CH_3CN as solvent. Adapted from Kempe and Barany (ref. 62).

    Preparations of oligo(alanyl)valine sequences on either polystyrene or polydimethylacrylamide supports are known to pose significant difficulties due to secondary structure formation and aggregation during chain assembly (76-79). The superiority of PEG-PS for the hydrophobic peptide deca(alanyl)valinamide was reflected in the purity of final products obtained in two parallel syntheses: PEG-PS *vs.* PS and PEG-PS *vs.* polydimethylacrylamide encapsulated in kieselguhr (Pepsyn K) (Figure 15). Further evidence of this (data not shown) stems from the fact that syntheses on PEG-PS gave normal Fmoc deprotection peaks upon UV monitoring, whereas in the syntheses on PS, these peaks were broad and abnormally shaped, as had been reported earlier (compare our results in ref. 79 to those of Wade *et al.* in ref. 77).

    Peptides containing Met or Trp are more difficult to synthesize due to the tendency of these residues to undergo oxidation and back-alkylation at several stages of standard SPPS procedures. Such side reactions can be more severe when these

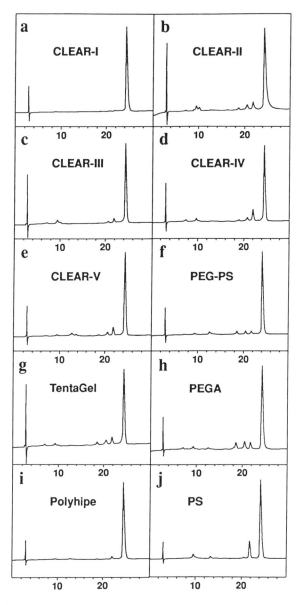

Figure 14. Analytical HPLC chromatograms of crude retro-ACP (74-65) amide, prepared by batchwise manual synthesis, Fmoc chemistry, on (a) CLEAR-I; (b) CLEAR-II; (c) CLEAR-III; (d) CLEAR-IV; (e) CLEAR-V; (f) PEG-PS; (g) TentaGel; (h) PEGA; (i) Polyhipe; and (j) PS. Couplings (1-h) mediated by DIPCDI/HOAt in DMF. Elutions at 1 mL/min with a linear gradient over 30 min from 17:3 to 3:1 of 0.1% aqueous TFA and 0.1% TFA in CH$_3$CN (Vydac C-18, 4.6 x 250 mm). Peptides were detected at 220 nm (ordinates); time scale is in min (abscissa). Adapted from Kempe and Barany (ref. 62).

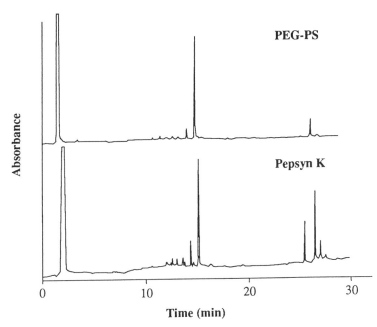

Figure 15. Analytical HPLC chromatograms of crude H-(Ala)$_{10}$-Val-NH$_2$, prepared on a PerSeptive Biosystems 9050 continuous-flow peptide synthesizer, starting with PEG-PS (upper) and Pepsyn K (lower). PyBOP/HOBt/DIEA-mediated couplings (30 min) were in DMF. Elutions at 1 mL/min with a linear gradient over 30 min from 1:9 to 7:3 of 0.1% aqueous TFA and 0.1% TFA in CH$_3$CN (Vydac C-18, 4.6 x 250 mm). Peptides were detected at 220 nm. Adapted from Kates *et al.* (ref. 79).

residues are used in conjunction with allyl protecting groups. The model peptide H-Ala-X-Gln-Lys-Thr-Asp(OAl)-Thr-Pro-NH$_2$ (X = Ala, Met, or Trp) was assembled on PEG-PS and PS supports with the PAL handle (80). On-resin removal of allyl groups was achieved by treatment with Pd(PPh$_3$)$_4$ and morpholine in DMSO–THF–0.5 M HCl (2:2:1) under Ar for 2 h, followed by washings with THF, DMF, CH$_2$Cl$_2$, DIEA–CH$_2$Cl$_2$ (1:19), 0.02 M sodium diethyldithiocarbamate in DMF, DMF, and CH$_2$Cl$_2$ to remove metal ions and other side products. HPLC chromatographic analysis indicated that the purities of peptides were greater when PEG-PS supports were used with respect to PS (Figure 16). The special suitability of PEG-PS in conjunction with allyl chemistry has been exploited for the synthesis of head-to-tail cyclic peptides (81), bicyclic peptides that included side-chain–to–side-chain lactams (82), sulfated peptides (83), N-glycopeptides (84), and multiple antigenic peptides (80).

Commercially available PEG-containing supports do not survive the HF cleavage conditions of Boc/Bzl chemistry. Earlier in this review, an HF-stable variant of PEG-PS was described (48). This support, and standard PS as a control, were loaded with a BHA linker (71) and used in the synthesis of a rather challenging 19-residue fragment [including three Met(O) residues] corresponding to the sequence 29–47 of uteroglobin. Chain assembly was with an ABI 430A instrument programmed with a small scale rapid cycle (SSRC) standard protocol (0.1 mmol scale, 5 equiv of *in situ* symmetrical anhydride in CH$_2$Cl$_2$–DMF), and products after

PEG-PS   PS

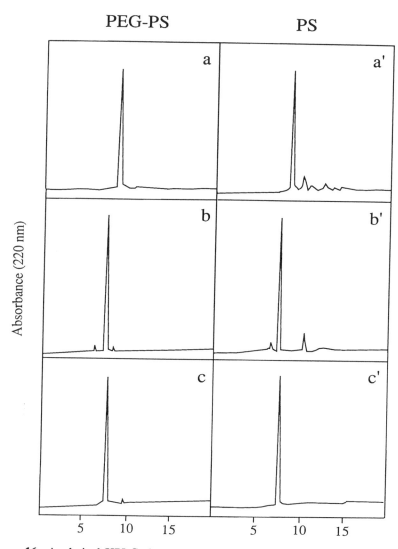

Figure 16. Analytical HPLC chromatograms of H-Ala-X-Gln-Lys-Thr-Asp-Thr-Pro-NH$_2$, illustrating differences in removal of allyl protection from β-carboxyl of Asp. Left side, on Fmoc-PAL-PEG-PS (0.27 mmol/g), right side on Fmoc-PAL-PS (0.30 mmol/g). Panels a, X = Trp, b, X = Met, c, X = Ala. PyBOP/HOBt/DIEA-mediated couplings; allyl removal as outlined in text; acidolytic cleavage with reagent R. Elutions at 1 mL/min with a linear gradient over 30 min from 1:0 to 3:2 of 0.1% aqueous TFA and 0.1% TFA in CH$_3$CN (Vydac C-18, 4.6 x 250 mm). Adapted from Kates *et al.* (ref. 80).

cleavage with   HF–*p*-cresol (9:1), 1 h at 0 °C were evaluated by HPLC (Figure 17). Synthesis clearly was more efficient on PEG-PS than on PS (48). In a further example of the usefulness of these HF-stable supports, a 17-mer "peptide nucleic analogue" (PNA) sequence was assembled in conjunction with Boc protection for primary amino groups, Z on exocyclic amino groups, and the BHA linker (48).

Combinatorial Organic Libraries Synthesis Examples

The usefulness of PEG-PS extends far beyond solid-phase peptide synthesis. The robust nature and combination of hydrophobic and hydrophilic character in these materials promise compatibility with a substantial subset of the transformations of organic chemistry. In particular, our earlier results on the optimal features of PEG-PS to facilitate removal of allyl protecting groups in the presence of Pd(0) has augured well for adaptation of stable mild palladium-mediated carbon-carbon bond-forming reactions such as those associated with the names of Suzuki, Heck, and Stille. To illustrate this point (Figure 18, ref. 85), we demonstrated that resin-bound aryl iodides (**19**) undergo efficient Suzuki coupling with 3-aminoaryl boronic acids. The resultant biphenyl **20** was extended further by reductive amination, and acid cleavage of the PAL anchor gave (3-alkylamine)arylbenzamides (**22**).

In another application, a library of potential angiotensin converting enzyme (ACE) inhibitors **26** derived from *N*-alkylated dipeptides was prepared (Figure 19, ref. 86). PAL-PEG-PS was acylated with Fmoc-Pro-OH, and after removal of the Fmoc group, a mixture containing the Fmoc derivatives of 19 genetically-encoded amino acids (excluding cysteine) was incorporated. The resin-bound dipeptide library **25** was deblocked, followed by reductive alkylation using ethyl-2-oxo-4-phenylbutyrate in the presence of NaBH$_3$CN.

Conclusions

PEG-PS and CLEAR supports swell in a broad range of hydrophobic and hydrophilic solvents, and have excellent physical and mechanical properties for both batchwise and continuous-flow systems. The general value of both PEG-PS and CLEAR (loadings 0.1-0.3 mmol/g) supports has been demonstrated by syntheses of numerous lengthy and especially difficult peptide sequences by Fmoc chemistry. Furthermore, PEG-PS supports have been proven in complicated chemistry such as multi-dimensional orthogonal schemes to prepare homodetic (lactam) and heterodetic (disulfide) mono and bicyclic peptides. Versions of PEG-PS supports can be prepared with a "high load" (0.35-0.6 mmol/g) and/or with the acid-stability required for use along with Boc chemistry. All of the supports reviewed herein have demonstrated promise for the convenient preparation of small organic molecules. A unique feature of PEG-containing supports is that they are compatible with on-resin screening assays conducted in aqueous milieus. Further developments and applications in these burgeoning areas can be predicted with a high degree of confidence.

Abbreviations

ACP	acyl carrier protein
BHA	Boc-aminobenzhydryloxyacetic acid linker
Boc	*tert*-butyloxycarbonyl
BOP	benzotriazolyl *N*-oxytris(dimethylamino)phosphonium hexafluorophosphate
Bzl	benzyl
CLEAR	cross-linked ethoxylate acrylate resin
DCC	*N,N'*-dicyclohexylcarbodiimide
DEAE-Sephadex	diethylaminoethyl-Sephadex

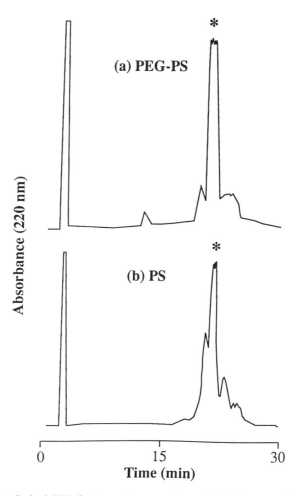

Figure 17. Analytical HPLC chromatograms of crude 29-47 uteroglobin fragment prepared by batch syntheses, Boc chemistry, DCC/HOAt (30 min) couplings, on an ABI-430A peptide synthesizer, starting with (a) PEG-PS and (b) PS. Elutions at 1 mL/min with a linear gradient over 30 min from 1:9 to 3:2 of 0.1% aqueous TFA and 0.1% TFA in CH3CN (Vydac C-18, 4.6 x 250 mm). Due to the three methionine sulfoxide residues, each of which can exist as either of two diastereomers, the final product must have intrinsic microheterogeneity as indicated by the asterisk.

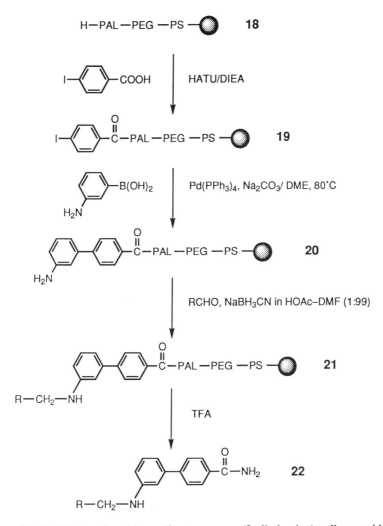

Figure 18. Solid-phase Suzuki coupling to prepare (3-alkylamine)arylbenzamides.

DIEA	*N,N*-diisopropylethylamine
DIPCDI	*N,N'*-diisopropylcarbodiimide
DMF	*N,N*-dimethylformamide
DMSO	dimethyl sulfoxide
EDA	ethylenediamine
Fmoc	9-fluorenylmethyloxycarbonyl
HATU	*N*-[(dimethylamino)-1*H*-1,2,3-triazolo[4,5-b]pyridin-1-ylmethylene]-*N*-methylmethanaminium hexafluorophosphate *N*-oxide
HOAt	1-hydroxy-7-azabenzotriazole
HOBt	1-hydroxybenzotriazole
HPLC	high performance liquid chromatography

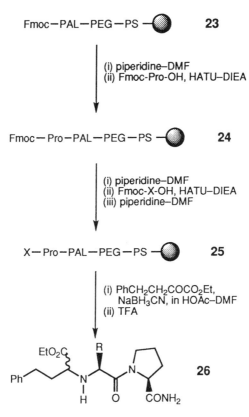

Figure 19. Solid-phase preparation of a library of ACE inhibitors. X = any one of the 19 genetically encoded L-amino acids (excluding cysteine).

IRAA	"internal reference" amino acid
MBHA	p-methylbenzhydrylamine (resin)
Nle	norleucine
NMM	N-methylmorpholine
PAL	tris(alkoxy)benzylamide linker, 5-(4-aminomethyl-3,5-dimethoxyphenoxy)valeric acid
PEG	poly(ethylene glycol)
PEGA	poly(ethylene glycol) acrylamide
PEG-PS	poly(ethylene glycol)-polystyrene graft
POE	poly(oxyethylene)
PS	polystyrene
PyBOP	benzotriazole-yl N-oxy-tris-pyrrolidinophosphonium hexafluorophosphate
SPOS	solid-phase organic synthesis
SPPS	solid-phase peptide synthesis
SPS	solid-phase synthesis
tBu	tert-butyl
TBTU	N-[(1H-benzotriazol-1-yl)(dimethylamino)methylene]-N-methylmethanaminium tetrafluoroborate N-oxide

TEGDA	tetraethyleneglycol diacrylate
TFA	trifluoroacetic acid
THF	tetrahydrofuran
Z	benzyloxycarbonyl

Acknowledgments

We thank Dr. Lin Chen for invaluable assistance in completing this manuscript. Seminal contributions to the development of PEG-PS were made by Drs. Samuel Zalipsky, Jane L. Chang, and Nuria A. Solé. The SEM photomicrographs (Figure 1) were taken at the Microscopy Facility, Center for Interfacial Engineering, University of Minnesota.with the assistance of Dr. Stuart McKernan. M.K. held a Hans Werthén Fellowship from The Royal Swedish Academy of Engineering Sciences. Research at Minnesota was supported by NIH (GM 42722 and 51628) and at Barcelona by CICYT (PB95-1131 and PB96-1490).

References

1. Merrifield, R.B. *J. Am. Chem. Soc.* **1963**, *85*, 2149-2154.

2. Review: Barany, G.; Merrifield, R.B. In *The Peptides: Analysis, Synthesis, Biology,* Gross, E.; Meienhofer, J., Eds.; Academic Press, Inc.: New York, NY, 1979; Vol.2, pp 1-284.

3. Review: Merrifield, R.B. *Angew. Chem., Int. Ed. Engl.* **1985**, *24*, 799-810.

4. Gait, M.J., Ed. *Oligonucleotide Synthesis: A Practical Approach;* IRL Press: Oxford, 1985.

5. Atherton, E.; Sheppard, R.C. *Solid Phase Peptide Synthesis: A Practical Approach;* IRL Press: Oxford, 1989.

6. Review: Fields, G.B.; Tian, Z.; Barany, G. In *Synthetic Peptides. A User's Guide*; Grant, G.A. Ed.; W.H. Freeman and Co.: New York, NY, 1992; pp 77-183.

7. Review: Beaucage, S.L.; Iyer, R.P. *Tetrahedron* **1992**, *48*, 2223-2311.

8. Review: Merrifield, B. In *Peptides: Synthesis, Structures, and Applications;* Gutte, B., Ed.; Academic Press, Inc.: San Diego, CA, 1995; pp 93-169.

9. Review: Crowley, J.I.; Rapoport, H. *Acc. Chem. Res.* **1976**, *9*, 135-144.

10. Review: Leznoff, C.C. *Acc. Chem. Res.* **1978**, *11*, 327-333.

11. Review: Fréchet, J.M.J. *Tetrahedron* **1981**, *37*, 663-683.

12. Review: Gordon, E.M.; Barrett, R.W.; Dower, W.J.; Fodor, S.P.A.; Gallop, M.A. *J. Med. Chem.* **1994**, *37*, 1385-1401.

13. Review: Terret, N.K.; Gardner, M.; Gordon, D.W.; Kobylecki, R.J.; Steele, J. *Tetrahedron* **1995**, *51*, 8135-8173.

14. Review: DeWitt, S.H.; Czarnik, A.W. *Acc. Chem. Res.* **1996**, *29*, 114-122.

15. Review: Armstrong, R.W.; Combs, A.P.; Tempest, P.A.; Brown, S.D.; Keating, T.A. *Acc. Chem. Res.* **1996**, *29*, 123-131.

16. Review: Ellman, J.A. *Acc. Chem. Res.* **1996**, *29*, 132-143.

17. Review: Gordon, E.M.; Gallop, M.A.; Patel, D.V. *Acc. Chem. Res.* **1996**, *29*, 144-154.

18. Review: Still, W.C. *Acc. Chem. Res.* **1996**, *29*, 155-163.

19. Review: Thompson, L.A.; Ellman, J.A. *Chem. Rev.* **1996**, *96*, 555-600.

20. Review: Früchtel, J.S.; Jung, G. *Angew. Chem., Int. Ed. Engl.* **1996**, *35*, 17-42.

21. Review: Hermkens, P.H.H.; Ottenheijm, H.C.J.; Rees, D. *Tetrahedron* **1996**, *52*, 4527-4554

22. Review: Barany, G.; Kempe, M. *In* Combinatorial Chemistry: A Short Course; DeWitt, S.H.; Czarnik, A.W., Eds.; American Chemical Society Books, Washington, DC, in press (1997), and other chapters in this monograph.

23.   Atherton, E.; Clive, D.L.J.; Sheppard, R.C. *J. Am. Chem. Soc.* **1975**, *97*, 6584-6585.
24.   Atherton, E.; Brown, E.; Sheppard, R.C. *J. Chem. Soc., Chem. Commun.* **1981**, 1151-1152.
25.   Arshady, R.; Atherton, E.; Clive, D.L.J.; Sheppard, R.C. *J. Chem. Soc., Perkin Trans. 1* **1981**, 529-537.
26.   Small, P.W.; Sherrington, D.C. *J. Chem. Soc., Chem. Commun.* **1989**, 1589-1591.
27.   Tregear, G.W. In *Chemistry and Biology of Peptides*, Meienhofer, J., Ed.; Ann Arbor Sci. Publ.: Ann Arbor, MI, 1972; pp 175-178.
28.   Kent, S.B.H.; Merrifield, R.B. *Isr. J. Chem.* **1978**, *17*, 243-247.
29.   Albericio, F.; Ruiz-Gayo, M.; Pedroso, E.; Giralt, E. *React. Polym.* **1989**, *10*, 259-268.
30.   Berg, R.H.; Almdal, K.; Batsberg Pedersen, W.; Holm, A.; Tam, J.P.; Merrifield, R.B. *J. Am. Chem. Soc.* **1989**, *111*, 8024-8026.
31.   Daniels, S.B.; Bernatowitcz, M.S.; Coull, J.M.; Köster, H. *Tetrahedron Lett.* **1989**, *30*, 4345-4348.
32.   Frank, R. *Tetrahedron* **1992**, *48*, 9217-9232.
33.   Frank, R. *Bioorg. Med. Chem. Lett.* **1993**, *3*, 425-430.
34.   Eichler, J.; Bienert, M.; Stierandova, A.; Lebl, M. *Peptide Res.* **1991**, *4*, 296-307.
35.   Englebretsen, D.R.; Harding, D.R.K. *Int. J. Peptide Protein Res.* **1994**, *43*, 546-554.
36.   Hudson, D.; Cook, R.M. In *Peptides: Chemistry, Structure & Biology, Proceedings of the Fourteenth American Peptide Symposium*, Kaumaya, P.T.P., Hodges, R.S., Eds.; Mayflower Scientific Ltd.: Kingswinford, England, 1996; pp 39-41.
37.   Inman, J.K.; Du Bois, G.C.; Appella, E. In *Solid Phase Methods in Protein Sequence Analysis (Proceedings of the Second International Conference)*, A. Previero; M.A. Coletti-Previero, Eds.; North-Holland Publishing Co.: Amsterdam, The Netherlands, 1977; pp 81-94.
38.   Regen, S.L.; Dulak, L. *J. Am. Chem. Soc.* **1977**, *99*, 623-625.
39.   Warshawsky, A.; Kalir, R.; Deshe, A.; Berkovitz, H.; Patchornik, A. *J. Am. Chem. Soc.* **1979**, *101*, 4249-4258.
40.   Heffernan, J.G.; MacKenzie, W.M.; Sherrington, D.C. *J. Chem. Soc., Perkin Trans.* **1981**, *2*, 514-517.
41.   Becker, H.; Lucas, H.-W.; Maul, J.; Pillai, V.N.R.; Anzinger, H.; Mutter, M. *Makromol. Chem., Rapid Commun.* **1982**, *3*, 217-223.
42.   Hellerman, H.; Lucas, H.-W.; Maul, J.; Pillai, V.N.R.; Mutter, M. *Makromol. Chem.* **1983**, *184*, 2603-2617.
43.   Zalipsky, S.; Albericio, F.; Barany, G. In *Peptides: Structure and Function, Proceedings of the Ninth American Peptide Symposium*; Deber, C.M., Hruby, V.J., Kopple, K.D., Eds.; Pierce: Rockford, IL, 1985; pp 257-260.
44.   Barany, G.; Albericio, F.; Biancalana, S.; Bontems, S.L.; Chang, J.L.; Eritja, R.; Ferrer, M.; Fields, C.G.; Fields, G.B.; Lyttle, M.H.; Solé, N.A.; Tian, Z.; Van Abel, R.J.; Wright, P.B.; Zalipsky, S.; Hudson, D. In *Peptides—Chemistry and Biology: Proceedings of the Twelfth American Peptide Symposium*; Smith, J.A.; Rivier, J.E. Eds.; ESCOM Science Publishers, Leiden, The Netherlands, 1992, pp 603-604.
45.   Barany, G.; Solé, N.A.; Van Abel, R.J.; Albericio, F.; Selsted, M.E. In *Innovations and Perspectives in Solid Phase Synthesis: Peptides, Polypeptides and Oligonucleotides, 1992*, Epton, R., Ed.; Intercept Ltd.: Andover, UK, 1992; pp 29-38.
46.   Barany, G.; Albericio, F.; Solé, N.A.; Griffin,, G.W.; Kates, S.A.; Hudson, D. In *Peptides 1992: Proceedings of the Twenty-Second European Peptide*

Symposium; Schneider C.H; Eberle, A.N., Eds.; ESCOM Science Publishers: Leiden, The Netherlands, 1993; pp 267-268.
47.   Zalipsky, S.; Chang, J.L.; Albericio, F.; Barany, G. *React. Polym.* **1994**, *22*, 243-258, and references cited therein.
48.   Albericio, F.; Bacardit, J.; Barany, G.; Coull, J.M.; Egholm, M.; Giralt, E.; Griffin, G.W.; Kates, S.A.; Nicolás, E.; Solé, N.A. In *Peptides 1994. Proceedings of the Twenty-Third European Peptide Symposium*, H.L.S. Maia, Ed.; ESCOM Science Publishers: Leiden, The Netherlands, 1995; pp 271-272.
49.   McGuinness, B.F.; Kates, S.A.; Griffin, G.W.; Herman, L.W.; Solé, N.A.; Vágner, J.; Albericio, F.; Barany, G. In *Peptides: Chemistry, Structure and Biology. Proceedings of the Fourteenth American Peptide Symposium*, P.T.P. Kaumaya; R.S. Hodges, Eds.; Mayflower Scientific Ltd.: Kingswinford, England, 1996; pp 125-126.
50.   Bayer, E.; Hemmasi, B.; Albert, K.; Rapp, W.; Dengler, M. In *Peptides: Structure and Function, Proceedings of the Eighth American Peptide Symposium*, Hruby, V.J., Rich, D.H., Eds.; Pierce: Rockford, IL, 1983; pp 87-90.
51.   Bayer, E.; Dengler, M.; Hemmasi, B. *Int. J. Peptide Prot. Res.* **1985**, *25*, 178-186.
52.   Bayer, E. *Angew. Chem., Int. Ed. Engl.* **1991**, *30*, 113-129.
53.   Bayer, E.; Rapp, W. In *Poly(ethylene Glycol) Chemistry: Biotechnical and Biomedical Applications;* Harris, J.M., Ed.; Plenum Press: New York, NY, 1992; pp 325-345.
54.   Review: Rapp, W. In *Combinatorial Peptide and Nonpeptide Libraries: A Handook,* Jung, G., Ed.; VCH: Weinheim, Germany, 1996; pp 425-464.
55.   For further information, see the following web site of Argonaut, Technologies (San Carlos, CA): http://www.argotech.com/f-agdata.htm
56.   Meldal, M. *Tetrahedron Lett.* **1992**, *33*, 3077-3080.
57.   Auzanneau, F.-I.; Meldal, M.; Bock, K. *J. Peptide Science* **1995**, *1*, 31-44.
58.   Renil, M.; Nagaraj, R.; Pillai, V.N.R. *Tetrahedron* **1994**, *50*, 6681-6688.
59.   Winther, L.; Almdal, K.; Batsberg Pedersen, W.; Kops, J.; Berg, R.H. In *Peptides: Chemistry, Structure and Biology, Proceedings of the Thirteenth American Peptide Symposium;* Hodges, R.S.; Smith, J.A., Eds.; Escom Science Publishers: Leiden, 1994; pp 872-873.
60.   Kempe, M.; Barany, G. In *Peptides: Chemistry, Structure & Biology, Proceedings of the Fourteenth American Peptide Symposium*, Kaumaya, P.T.P., Hodges, R.S., Eds.; Mayflower Scientific Ltd.: Kingswinford, England, 1996; pp 865-866.
61.   Kempe, M.; Barany, G. In *Innovations and Perspectives in Solid Phase Synthesis and Combinatorial Chemical Libraries, 1995,* Epton, R., Ed.; Mayflower Worldwide Ltd.: Birmingham, England, 1996; pp 191-194.
62.   Kempe, M.; Barany, G. *J. Am. Chem. Soc.* **1996**, *118*, 7083-7093.
63.   Kempe, M.; Keifer, P.A.; Barany, G. In *Peptides 1996, Proceedings of the Twenty-Fourth European Peptide Symposium*, R. Ramage, Ed.; Mayflower Scientific Ltd.: Kingswinford, England, 1996; in press.
64.   Sarin, V.K.; Kent, S.B.H; Merrifield, R.B. *J. Am. Chem. Soc.* **1980**, *102*, 5463-5470.
65.   Merrifield, B. *Br. Polym. J.* **1984**, *16*, 173-178.
66.   Fields, G.B.; Fields, C.G. *J. Am. Chem. Soc.* **1991**, *113*, 4202-4207.
67.   Vágner, J.; Barany, G.; Lam, K.S.; Krchňák, V.; Sepetov, N.F.; Ostrem, J.A.; Strop, P.; Lebl, M.. *Proc. Natl. Acad. Sci. USA* **1996**, *93*, 8194-8199, and references cited therein.
68.   Albericio, F.; Kneib-Cordonier, N.; Biancalana, S.; Gera, L.; Masada, R.I.; Hudson, D.; Barany, G. *J. Org. Chem.* **1990**, *55*, 3730-3743, and references cited there.
69.   Review: Songster, M.F.; Barany, G. In *Meth. Enzymol.,* G.B. Fields, Ed.; Academic Press, Orlando, 1997; in press.

70.   Ferrer, M.; Woodward, C.; Barany, G. *Int. J. Peptide Protein Res.* **1992**, *40*, 194-207.
71.   Gaehde, S.A.; Matsueda, G.R. *Int. J. Peptide Protein Res.* **1981**, *18*, 451-458.
72.   Ghatge, N.D.; Shinde, B.M.; Jagadale, S.M. *J. Polym. Sci., Polym. Chem. Ed.* **1984**, *22*, 985-994.
73.   Smith, D.A.; Cunningham, R.H.; Coulter, B. *J. Polym. Sci. Part A-1* **1970**, *8*, 783-784.
74.   Barany, G.; Gross, C.M.; Ferrer, M.; Barbar, E.; Pan, H.; Woodward, C. In *Techniques in Protein Chemistry VII*, Marshak, D., Ed.; Academic Press: San Diego, CA, 1996; pp 503-514.
75.   Hancock, W.S.; Prescott, D.J.; Vagelos, P.R.; Marshall, G.R. *J. Org. Chem.* **1973**, *38*, 774-781.
76.   Merrifield, R.B.; Singer, J.; Chait, B.T. *Anal. Biochem.* **1988**, *174*, 399-414.
77.   Wade, J.D., Bedford, J.; Sheppard, R.C.; Tregear, G.W. *Peptide Research* **1991**, *4*, 194-199.
78.   Larsen, B.D.; Holm, A. *Int. J. Peptide Protein Res.* **1994**, *43*, 1-9.
79.   Kates, S.A.; Solé, N.A.; Beyermann, M.; Barany, G.; Albericio, F. *Peptide Research* **1996**, *9*, 106-113, and references cited therein.
80.   Kates, S.A.; Daniels, S.B.; Albericio, F. *Anal. Biochem.* **1993**, *212*, 303-310.
81.   Kates, S.A.; Solé, N.A.; Johnson, C.R.; Hudson, D.; Barany, G.; Albericio, F. *Tetrahedron Lett.* **1993**, *34*, 1549-1552.
82.   Solé, N.A.; Kates, S.A.; Albericio, F.; Barany, G. In *Innovation and Perspectives in Solid-Phase Synthesis: Peptides, Proteins and Nucleic Acids, Biological and Biomedical Applications*, Epton, R., Ed.; Mayflower Worldwide Ltd: Birmingham, England, 1994; pp 105-110.
83.   Han, Y.; Bontems, S.L.; Hegyes, P.; Munson, M.C.; Minor, C.A.; Kates, S.A.; Albericio, F.; Barany, G. *J. Org. Chem.* **1996**, *61*, 6326-6339.
84.   Kates, S.A.; de la Torre, B.G.; Eritja, R.; Albericio, F. *Tetrahedron Lett.* **1994**, *35*, 1033-1034.
85.   Blackburn, C.; Herman, L.W.; Solé, N.A.; Triolo, S..; Kates, S.A. Poster presented at Solid-Phase Synthesis: Developing Small Molecule Libraries, February 1-2, 1996, Coronado, CA.
86.   Blackburn, C.; Pingali, A.; Kehoe, T.; Herman, L.W.; Wang, H.; Kates, S.A. *Bioorg. Med. Chem. Lett.* **1997**, in press.

Chapter 18

Oligonucleotide–Poly(ethylene glycol) Conjugates: Synthesis, Properties, and Applications

Andres Jäschke

Institut für Biochemie, Freie Universität Berlin, Thielallee 63, D–14195 Berlin, Germany

The use of DNA and RNA oligonucleotides for therapeutic and diagnostic purposes has created great interest in methods of enhancing their stability and delivery. One promising strategy is covalent conjugation of oligonucleotides to polymers. This chapter discusses polyethylene glycol and related polymers suitable for conjugation, methods of conjugation, the properties of DNA-PEG and RNA-PEG conjugates and the possibilities for designing nucleic acid based drugs using these conjugates. In addition, an overview on the use of short oligoethylene glycol based linkers as synthetic loop replacements and for the development of molecular tools is presented.

While the conjugation of proteins with polyethylene glycol has been studied for many years, oligonucleotide-PEG conjugates have received attention only recently. Interest has been arising along different lines. The development of novel therapeutic principles using oligonucleotides as drugs and the corresponding commercial interest have initiated vigorous research in all fields related to nucleic acids. In contrast to conventional therapeutic strategies targeting a specific function at the protein level, nucleic acid based drugs can interfere with the flow of genetic information at any level; antisense oligonucleotides are targeting single-stranded RNA molecules (1), antigene strategies are directed against double-stranded DNA by formation of triple helices (2), ribozymes are designed for the sequence-specific catalytic cleavage of RNA-molecules (3), and aptamers bind to and specifically inhibit proteins, peptides or small molecules (4). The practical application of these strategies in living organisms or biological fluids necessarily requires chemical modification, as unmodified RNA is degraded instantly by nucleolytic enzymes. For use as drugs, biodistribution and pharmacokinetics must be controlled. One of the strategic lines being followed is conjugation, and - as in the protein and peptide field (5) - PEG seems to be a promising candidate for practical application.

Another scientific discipline has also demonstrated considerable interest in PEG conjugation. Short oligoethylene glycols have been used in a number of studies on the structure and function of nucleic acids. They have been used to substitute for single-stranded loops in hairpin-like structures, and they have assisted in the investigation of duplex and triplex structures. Moreover, oligoethylene glycols are used quite often as mixed-polarity spacers in complex conjugates.

It is a combination of specific properties that makes PEG interesting for a number of applications. PEG is soluble in aqueous solutions as well as in most organic solvents, due to the sole occurrence of single bonds it is highly flexible, and PEG shows low tendency to the formation of ordered structures. PEG is known to interact with cellular membranes in a complex manner, and it cannot specifically recognize nucleic acids. PEG is of low toxicity and immunogenicity, it has been used to improve the pharmacokinetic properties of drugs, and it has a rather simple chemistry (5,6).

In this review, the synthesis of DNA and RNA conjugates with polyethylene glycol and related polymers as well as their use and properties will be summarized.

Synthesis of Oligonucleotide-PEG Conjugates

Like any other conjugate group, polyethylene glycol may be coupled to oligonucleotides either through sites present naturally in nucleic acids or through some other reactive group introduced specifically for the purpose. Generally, conjugate groups may be introduced either during chemical or enzymatic synthesis or postsynthetically.

While this review is restricted to PEG as a conjugate group, a more comprehensive treatment of oligonucleotide conjugation is provided by Goodchild and Agrawal (7,8).

Incorporation of PEG during Chemical Synthesis of Oligonucleotides. From the synthetic point of view, incorporation of conjugate groups during the assembly of an oligonucleotide rather than afterward is the most rigorous approach. It gives greatest control over the number and location of the modifications, side reactions are minimized by the protecting groups on the nucleotides, and advantage is taken of the benefits of solid-phase synthesis for workup and purification (7).

Both DNA and RNA oligonucleotides are commonly synthesized by solid phase synthesis using the phosphoramidite approach (9,10). Synthesis proceeds in 3'→5' direction, starting with the 3'-nucleoside already attached to a solid support, and reactive phosphoramidites are added sequencially. To incorporate chemically modified or unnatural building blocks into oligonucleotides, their chemistry must be compatible with the phosphoramidite method. Poly- and oligoethylene glycols are much simpler synthetic targets for derivatization than nucleosides, since the only two reactive groups are the terminal hydroxyls, avoiding the use of complex protection schemes.

Most publications on the incorporation of oligo- and polyethylene glycols into synthetic oligonucleotides use phosphoramidites **1**. The first published procedure used hexaethylene glycol for internal incorporation into a DNA hairpin structure (*11*). First, one hydroxyl group was protected by reaction with 4,4'-dimethoxytrityl (DMT) chloride, the mono-DMT-product was chromatographically isolated, and then the other hydroxyl group was reacted with 2-cyanoethyl N,N,N',N'-tetraisopropyl phosphordi-amidite. After workup and chromatographic purification, the phosphoramidite was reported to couple with 98-99 % yield.

Phosphoramidites of this type were synthesized and used from triethylene glycol upwards to a degree of polymerization of 35 (*12-25*). Shorter oligoethylene glycols (up to n=8) were synthesized from monodisperse oligoethylene glycols, the longer oligomers were synthesized from polydisperse mixtures (PEG 400, PEG 1000). The DMT protecting group was used in all cases with only slight modifications of the synthetic procedure. Three different protocols for phosphitylation have been used, all of which are standard procedures in nucleotide chemistry (*11,14,22*).

For the incorporation of PEG at 3'-terminal positions, PEG-derivatized solid supports **2** were synthesized (*14-16,25*). The PEG is linked to the controlled pore glass (CPG) support via a succinic ester bond that is cleaved during the final alkaline cleavage/deprotection step. Mono-DMT-PEG was treated with succinic anhydride to give succinate which was activated with *p*-nitrophenol and the activated ester was reacted with aminopropylated CPG. Synthesis on these supports proceeded essentially as with standard supports. The use of Tentagel supports (PEG-polystyrene copolymers) for the preparation of 3'-PEGylated oligonucleotides using a similar derivatization scheme was described recently (*26*).

Numerous DNA- and RNA-PEG conjugates have been synthesized using these phosphoramidites and supports. The products are easily purified using standard procedures (reversed phase HPLC, gel electrophoresis). No de-PEGylation during deprotection has been reported. Besides simply structured molecules, complex conjugates having one or more oligoethylene glycols incorporated internally have been synthesized (*12,20,21*). Other non-nucleoside phosphoramidites (chromophores, labels, cross-linkers, reactive groups) can be incorporated into an oligonucleotide after coupling the PEG phosphoramidite, too. Zhang *et al.* incorporated N-trifluoroacetyl-6-aminohexyl-2-cyanoethyl-N,N'-diisopropyl phosphoramidite after the hexaethylene glycol phosphoramidite (*19*). The final products were oligonucleotides having an oligoethylene glycol spacer at the 5'-end, terminating with a primary aliphatic amino group. These conjugates were immobilized on carboxylated nylon membranes employing carbodiimide coupling. Maskos and Southern synthesized oligonucleotides immobilized to microscope slides and solid sphere glass beads via ethylene glycol, pentaethylene glycol and hexaethylene glycol linkers, respectively (*27*). The supports were first derivatized with 3-glycidoxypropyltrimethoxysilane and the epoxy-ring was then brought to reaction with the diol. The terminating primary hydroxyl group served as the starting point for automated or manual oligonucleotide synthesis. Oligonucleotides synthesized on the support remained tethered to the support after ammonia treatment and were used for sequence-specific hybridization reactions.

Oligoethylene glycol linkers are contained in several commercially available phosphoramidites and glass supports for oligonucleotide modification, like biotin-triethylene glycol or cholesterol-triethylene glycol. Triethylene glycol serves as a mixed polarity spacer between two functional elements and prevents steric hindrance.

The automated synthesis of 3'- and 5'-PEGylated oligonucleotides by H-phosphonate chemistry has been described (*28*). Preparation of the support was done by reac-

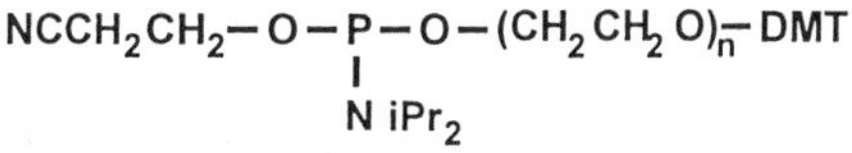

$$NCCH_2CH_2-O-P-O-(CH_2CH_2O)_n-DMT$$
$$\underset{N\ iPr_2}{\overset{|}{}}$$

1

$$CPG-CH_2CH_2CH_2\overset{}{N}H-\overset{\overset{O}{\|}}{C}CH_2CH_2\overset{\overset{O}{\|}}{C}-O-(CH_2CH_2O)_n-DMT$$

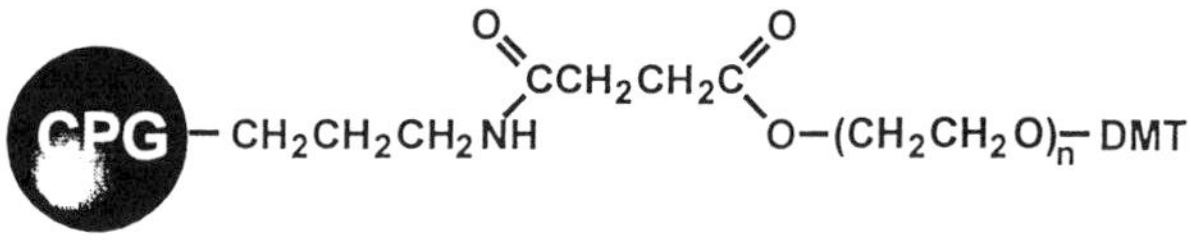

2

tion of the pentafluorophenylester of DMT-PEG-phthalate with CPG, and the DMT-PEG-H-phosphonate was prepared by standard procedures. PEGs with a molecular weight of up to 6000 were coupled to oligonucleotides using the normal protocols for synthesis and workup.

A different class of oligonucleotide-PEG conjugates has been reported recently (*29,30*). For the preparation of antisense oligonucleotides, phosphoramidites having an O-methylated oligoethylene glycol coupled to the 2'-hydroxyl group via an ether bond were synthesized and incorporated. Synthesis of the 2'-modified nucleosides was performed in 11 steps starting from ribose. The modified nucleosides were 5'-O-DMT-protected and then reacted with 2-cyanoethyl N,N,N',N'-tetraisopropyl phosphordiamidite. The simplest structure prepared contained 2'-O-methoxyethyl, and analogs derived from O-methyl-di- and -triethylene glycol have also been described. The resulting phosphoramidites were used in solid phase synthesis and gave oligonucleotides that had the respective O-methyl-oligoethylene glycol attached to the C-2' of the ribose moieties. Pure all-2'-O-derivatized oligonucleotides were synthesized as well as chimeric products, where part of the oligonucleotide was unmodified phosphodiester (*29*) or phosphorothioate (*30*).

Incorporation of PEG during Enzymatic Synthesis of Oligonucleotides. There are different ways to incorporate unnatural residues during enzymatic polynucleotide synthesis. If site-specific attachment is not required, a variety of DNA and RNA polymerases can be used for statistic or permodification employing modified nucleotides. Positions where several polymerases tolerate even bulky modifications are the C5 of pyrimidines and the N7 and C8 of purines, and some of the more commonly used modified nucleoside triphosphates are commercially available. Livak *et al.* reported the incorporation by *Taq* and *Tli* DNA polymerases of biotinylated nucleotides with hexaethylene glycol derived linker arms attached to the C5 of the pyrimidine bases and to C7 of deazaadenosine. The mobility shift caused by the incorporation allowed the identification of a single-base polymorphism in a 500 nucleotide DNA segment (*31*).

Site-specific incorporation of a modified building block requires different approaches. Labels can be selectively incorporated at the 5'-position of RNA transcripts by using chemically modified initiator nucleotides (*32*), and work from the authors laboratory shows that initiator nucleotides with PEG 600 attached to the 5' phosphate of guanosine monophosphate are effectively incorporated into transcripts generated by T7 RNA polymerase. Initiator nucleotides containing 6 to 18 ethylene glycol units were all incorporated with similar efficiencies   (Seelig & Jäschke, submitted). Primer-dependent polymerases that extend the 3'-end of a given primer hybridized to a template can be easily used to produce 5'-conjugates. Several polymerases efficiently elongate primers with basically anything attached at or near their 5'-ends, from biotin to polyamine to membranes or chromatography supports (*33*).

Another general method of polynucleotide synthesis is joining two smaller oligonucleotides by enzymatic ligation. The template-independent T4 RNA ligase has often been used to introduce a single, modified mono-, di- or trinucleotide at the 3'-end of RNA or DNA oligonucleotides (*34*). T4 DNA ligase in contrast joins two oligo-

or polynucleotides in a template-dependent manner. Therefore, it can be used to join an enzymatically prepared transcript with a chemically synthesized oligonucleotide bearing a modification (35). We have used both approaches for the incorporation of up to three consecutive hexaethylene glycol spacers into randomized libraries of RNA transcripts. The conjugates are easily identified by their reduced electrophoretic mobility or by specific reporter groups attached to the end of the PEG linker (Jäschke et al., to be published).

Postsynthetic Coupling of PEG to Oligonucleotides. Conjugate groups are often introduced after the chemical or enzymatic synthesis of oligo- or polynucleotides. Most frequently, a specific linker group is incorporated into the oligonucleotide during chemical synthesis which is used for conjugation purposes after deprotection and purification. Primary alkylamines, hydrazines, thiols, and carboxyl groups are most commonly used. The site-specific incorporation at naturally occurring sites in oligonucleotides is rather limited; the 3'-terminal ribose in RNA can be oxidized using periodate and then coupled to amino- or hydrazine-derivatized conjugate groups (36). A terminal 3'- or 5'-phosphate groups can be derivatized using carbodiimides or cyanogen bromide (37).

Kawaguchi et al. incorporated a hexylamine linker at the 5'-end of a 15mer oligonucleotide during automated synthesis. The primary amino group was reacted with 2,4-bis-(O-methoxypolyethylene glycol)-6-chloro-S-triazine, yielding oligo-nucleotides coupled to two chains of PEG 10000 at their 5'-terminus (38). Without specifying synthetic details, Manoharan et al. described a similar strategy employing "PEG activated esters" of average molecular weight 550, 2000, and 5000, for attaching them to aminoderivatized antisense oligonucleotides (39).

Pieken and Carter used activated ester chemistry for derivatizing oligonucleotides with high-molecular weight polyethylene glycol. In a typical reaction, an oligonucleo-tide bearing an aminolinker attached to the C5 of thymidine was reacted with 2 to 3 equivalents of the N-hydroxysuccinimide ester of monomethoxy-PEG 20.000-propionate to give PEG-derivatized oligonucleotide in > 90% yield. The exocyclic amino groups on the nucleobases were found not to react with the activated PEG under these conditions (W.A. Pieken, personal communication).

Jones et al. reacted a diamino-substituted PEG 3350 with 3,5-bis-(iodoacetamido)-benzoic acid N-hydroxysuccinimidyl ester (40). This derivative was reacted with a chemically synthesized thiol-containing 20mer oligonucleotide to provide an oligonucleotide-PEG conjugate of precisely four oligonucleotides on each PEG carrier. These conjugates were annealed with a partialy complementary 50mer conjugated to keyhole limpet hemocyanin.

Properties and Applications of Oligonucleotides Terminally Coupled to PEG

Chromatographic and Electrophoretic Properties. Reversed phase HPLC was often used for the purification and analysis of oligonucleotide-PEG conjugate mixtures (14-16,28). PEG-coupling caused an increase in hydrophobicity leading to an increased retention time. Single degrees of polymerization were eluted sequentially, giving a characteristic elution pattern (Figure 1). The different retention of unmodified oligonucleotides caused by their different lengths and compositions was found with the PEG conjugates, too (*15*). Generally, the resolution of single degrees of polymerization decreased with the size of both the oligonucleotide and the attached PEG. The retention times for 3'- and 5'-conjugates of the same nucleotide sequence were almost identical, while internally PEG-coupled molecules eluted somewhat earlier. Coupling of each one chain of PEG to the 3'- and 5'-end of an oligonucleotide caused a further increase in retention time and a permutation of the molecular size distribution. These conjugates, however, were eluted earlier than the single-sided conjugates with the same degree of polymerization. RNA-PEG conjugates eluted significantly earlier than the respective DNA conjugates. In anion exchange chromatography, retention of PEG-derivatized oligonucleotides is shifted towards earlier retention times.

Due to the lower charge/mass ratio, electrophoretic mobility decreases as a function of the PEG size. For terminally PEG-coupled pentanucleotides, the resolution of single degrees of polymerization is possible up to PEG 1000 on native polyacrylamide gels (*14-16,28*). For 18mers, there is also a significant decrease in electrophoretic mobility, with the 3'-PEG 4000 conjugates moving about half as fast as the unmodified oligonucleotides. Bayer *et al.* described two sets of conditions for the characterization of oligonucleotide-PEG conjugates by capillary electrophoresis. In the first case, gaussian distributions corresponding to the different molecular weights were obtained with the polydisperse products. A slight change in the buffer conditions resulted in a separation completely dominated by the oligonucleotide part of the molecule, and failure sequences could easily by distinguished (*26*).

Multiple PEG units were added to oligonucleotides as mobility modifiers in oligonucleotide ligation assays. One ligation probe for each target carries a fluorescent tag, while the other carries a defined non-nucleotide mobility modifier. Each ligation product has a unique electrophoretic mobility determined by the both the ligated oligonucleotide and the assigned modifier (*41*). These oligonucleotide ligation assays were used for allelic discrimination in highly polymorphic genes (*41,42*).

Mass Spectrometry. Both MALDI and ESI mass spectrometry were successfully used for the characterization of oligonucleotide-PEG conjugates. Complete reaction mixtures, single peaks or fractions with a narrow size distribution were analyzed by MALDI-mass spectrometry (*15,16*). Up to molecular weights of 7000, mass accuracies of ± 1 Dalton were obtained, and neighboring peaks in the HPLC chromatograms were directly shown to differ by 44 Dalton, i.e., by the size of one CH_2CH_2O unit. Analysis of conjugate mixtures by electrospray mass spectrometry also revealed typical distributions (*26*).

Substrate Properties and Nuclease Stability. 3'-PEG-conjugates were efficiently phosphorylated at their 5'-terminus by T4 polynucleotide kinase, while 5'-conjugates - as expected - gave no product. Internally PEG-coupled oligonucleotides were phosphorylated with a somewhat reduced yield; the yield increased with increasing distance between PEG and phosphorylation site (*14-16*). Oligonucleotide-PEG conjugates were accepted as substrates by T4 DNA ligase and by T4 RNA ligase both on the 3' and 5' side, as long as the direct vicinity of the ligation site remained unmodified (Jäschke et al., to be published). Exonuclease stability, however, was drastically increased, by the factor 10 for phosphodiesterase I, and no cleavage products at all were detected using phosphodiesterase II (15). A slight increase in stability towards the single-strand specific endonuclease S1 by 5'-PEG conjugation has been reported (factor 1.6, compared to the unmodified oligonucleotide), and similar values were found for the stability in human and mouse serum (38).

Thermal Stability of Oligonucleotide Duplexes. In contrast to internal coupling, terminal attachment of PEG has only marginal effects on the thermostability of oligonucleotide duplexes. The terminal attachment of PEGs with molecular weights of >1000 increases the melting temperature T_m by 2-3°C (15,38).

Pharmacological Properties. Kawaguchi et al. reported plasma concentration-time profiles for oligonucleotide-PEG conjugates after intravenous administration in mice. While unmodified oligonucleotides rapidly disappeared with a half-life of less than 0.2 min, retention of the 5'-conjugates was prolonged about 5 times to 1 min (38).

Manoharan et al. have prepared oligonucleotides conjugated to cholic acid, cholesterol, polyamines and PEG to study their effects on enhancing absorption of antisense agents. These conjugates were targeted against human or murine intercellular adhesion molecule mRNA, and bioavailablity was determined by measuring the inhibition of gene expression in target cells. Unexpectedly, the PEG conjugates exhibited less activity than the unconjugated parent compound. Very little cellular localization of 5'-fluorescein labeled PEG conjugates was observed (39).

Scientists at NeXstar Pharmaceuticals using the SELEX combinatorial oligonucleotide technology, have pursued the development of oligonucleotide aptamers for use as therapeutic agents (4). In vivo studies have shown that both size and stability affect the clearance of oligonucleotides. Even metabolically stable 10 kD oligonucleotides are cleared relatively rapidly by renal filtration. To significantly reduce renal clearance, an overall molecular weight of $\geq$ 50 kD is required. It was shown that metabolically stable oligos composed of either 2'-O-methyl nucleosides, or 2'-O-methyl-purines and 2'-fluoro-pyrimidines demonstrate very slow clearance rates when conjugated to PEG 20.000-40.000. The clearance of an all 2'-O-methyl oligonucleotide- PEG 40.000 was 0.08 ml/kg*min in rats with a $t_{1/2}$ of 10.6 hours. The clearance and half-life of an identical oligonucleotide without PEG were 13 ml/kg*min and 0.6 hours, respectively (S.C. Gill, M. Willis, and R. Fielding, unpublished data).

Linking of Two or More Oligonucleotides by PEG

Nucleic acid loops are an ubiquitous part of RNA and DNA secondary structure. Most biologically significant RNAs, like tRNAs, ribosomal RNAs, mRNAs and ribozymes, contain extended loop domains. There has been increasing interest in synthetic DNAs and RNAs containing hairpin loops. Studies in modification of these structures have led to the development of non-nucleotide linking groups. Oligoethylene glycols have been used in several combinations which are illustrated in Figure 2 and described in more detail below.

Intramolecular Duplexes. Durand *et al.* reported the use of hexaethylene glycol as a replacement for a tetranucleotide loop in a palindromic DNA sequence and the investigation of its conformation by CD spectroscopy (*11*). The hexaethylene glycol loop was found more stabilizing than the natural T_4 loop, yielding a 3°C higher T_m. Compared to the non-linked structure, a 35°C increase in T_m was found.

Altmann *et al.* reported the synthesis of three singly cross-linked duplexes d(GTGGAATTC)-linker-d(GAATTCCAC) (*22*). Linker I was an assembly of a propylene, a phosphate, and a second propylene group and was thought to mimic the backbone of two nucleotides. Linkers II and III consisted of five and six ethylene glycol units, respectively. The melting temperatures of the cross-linked duplexes are 65 °C for I and 73 °C for II and III, as compared with 36 °C for the corresponding non-linked duplex. NMR spectroscopy revealed that the nucleotides flanking the propylene-phosphate-propylene-linker in duplex I did not form a Watson-Crick base pair, whereas in duplexes II and III the entire DNA stem was in a B-type double helix conformation.

Rumney and Kool investigated length and structural effects of oligoethylene glycol linkers in duplex DNAs by incorporating three to eight ethylene glycol units into a hairpin structure, thus spanning 13 to 31 Å in distance (*12*). They compared melting temperatures and free energies with a hairpin structure containing a natural tetranucleotide loop. Their results show that the stability of folded helical structures strongly depends on linker length and structure. The optimum linker is heptaethylene glycol giving complexes considerably more stability than those bridged by tetra-nucleotide loops. The optimum linker is ~12 Å longer than simple straight-line distances derived from models would indicate (*12*).

Intramolecular Triplexes. Durand *et al.* described the investigation of triple-helix formation by the oligonucleotide $(dA)_{12}$-x-$(dT)_{12}$-x-$(dT)_{12}$, where x was a hexaethylene glycol unit (*43*). Thermal denaturation analysis showed that this single-stranded oligonucleotide is able to fold back on itself twice to give a triple helix at low temperature. Upon an increase in temperature, two cooperative transitions were observed: first the formation of a double-stranded structure with a dangling x-$(dT)_{12}$, then formation of a single-stranded coil structure. Due to the intramolecular character of the transition, the triplex was much more stable than that formed by the reference mixture $(dA)_{12} + 2(dT)_{12}$. CD spectrometry demonstrated that the conformation of the linked triplex structure was essentially identical to that of the reference mixture.

Dagneaux *et al.* described an intramolecular triple helix with dT*dA-dT base triplets in which the pyrimidine third strand was oriented antiparallel to the dA strand, formed by folding back twice the conjugate $(dT)_{10}$-x-$(dA)_{10}$-x-$(dT)_{10}$ (x = triethylene glycol) (*44*). Third-strand base pairing to the target strand, sugar conformation, and thermal denaturation of the triplex have been studied by Fourier transform infrared spectroscopy. The results confirmed that when the third-strand orientation is reversed from parallel to antiparallel with respect to the target strand, the third-strand hydrogen-bonding scheme is changed from Hoogsteen to reverse Hoogsteen.

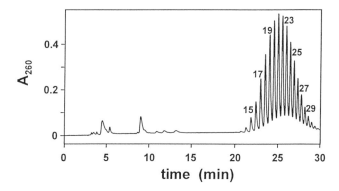

Figure 1. Reversed phase HPLC chromatogram of conjugate TTCGA-PEG 1000. The numbers at the peaks indicate the degree of polymerization of coupled PEG. Column: Nucleosil 300/5 C4, flow rate: 1 ml/min, gradient: 0.8 to 32 % acetonitrile in 100 mM triethylammonium acetate, pH 7.0 (Adapted from ref. 15).

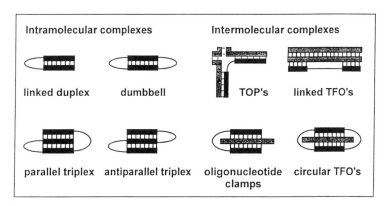

Figure 2. Schematic representation of different oligonucleotide-PEG conjugates. (TOP's: tethered oligonucleotide probes, TFO's: triplex forming oligonucleotides). ██████ oligonucleotide sequence ────── PEG linker ▨▨▨▨ target sequence.

Intermolecular Triplexes. Giovannangeli *et al.* reported the use of a hexaethylene glycol linker in a folded oligopyrimidine triplex-forming oligonucleotide (*45*). When hybridized with an all-purine target sequence, this oligonucleotide clamp formed a triplex being more stable than either the corresponding trimolecular complex or the double helix formed upon binding of the oligopyrimidine complement to the same oligopurine target. Attaching a psoralen derivative to the 5' end of the clamp allowed to photoinduce a covalent linkage to the target sequence, thereby making the oligonucleotide clamp irreversible. These crosslinking reactions introduced strong stop signals during DNA replication, and even in the absence of photocrosslinking, the psoralen-oligonucleotide clamp was able to arrest DNA replication.

Kandimalla and Agrawal reported similar constructs which they call "foldback triplex-forming oligos (FTFOs)" (*46*). Single-stranded oligonucleotides containing both Watson-Crick and Hoogsteen hydrogen bonding domains joined by either a 5-nucleotide loop or a hexaethylene glycol linker were synthesized and studied for their binding affinity and specificity to their single-stranded DNA/RNA targets. Thermal denaturation studies revealed an increased affinity of FTFOs, due to addition of a Hoogsteen hydrogen bonding domain at the binding site, as the Watson-Crick domain forms a double helix with the target. DNase I hydrolysis and electrophoretic mobility shift analysis confirmed the formation of foldback triplexes relative to conventional double- and triple-stranded structures. The FTFOs showed increased sequence specificity mainly arising from their ability to recognize the target sequence twice. An FTFO with DNA components in both duplex- and triplex-forming domains showed preference for a DNA homopurine target strand.

Rumney and Kool studied the effects of oligoethylene glycol linkers in triplex DNAs by incorporating three to eight ethylene glycol units into linked all-pyrimidine structures targeting all-purine sequences (*12*). They have found that the optimum linker for blunt-ended triplexes is derived from octaethylene glycol giving complexes which are in some cases more stable than structures with natural pentanucleotide loops. As in the case of oligonucleotide duplexes, the optimum linker is ~12 Å longer than simple straight-line distances would indicate. For triplexes where the target is embedded in a longer strand, the optimum linker length is again octaethylene glycol, but the stability greatly depends on 5' or 3' loop orientation, with the latter being much more stable; both cases are less stable than the case with the natural pentanucleotide loop. The authors reported that most of this destabilization can be bypassed by simply extending one end of one of the binding domains near the loop in the 5'-direction (*12*).

The same group described a similar strategy targeting single-stranded pyrimidine sequences by triple helix formation using a double-length purine-rich oligonucleotide (*47*). To optimize binding properties, sequence and structural variations were made and the data were compared with the use of two separate strands. Variation of the lengths and sequences of loops bridging the binding domains demonstrates that dinucleotide loops composed of pyrimidines give the highest stability. Oligoethylene glycol-derived loop replacements were at least as stabilizing as the best dinucleotide linker, with hepta- and hexaethylene glycol-linked compounds having the highest T_m values (47).

Two Separate Binding Sites Linked by PEG. Kessler et al. described oligonucleotide hybrids which consisted of two triple helix forming oligonucleotides that had been connected by a flexible polymeric linker chain (48). Up to six triethylene glycol units were incorporated consecutively. Binding of this class of oligonucleotides to duplex DNA has been studied using a DNA segment, which possessed a pair of 12 base pair target sites for stable triple helix formation, separated by a duplex spacer region which was one helical turn long. Band shift and footprinting analysis showed that such hybrids can bind to both target sites simultaneously, if flexible linkers are included which are longer than 20-25 rotatable bonds. Molecular modeling confirmed that a flexible polymeric linker as short as 22 rotatable bonds is enough to link the two distant segments of triple helix, providing that the linker element travels a path which is external to the helix grooves and parallel to the long helix axis.

Schepartz and coworkers reported studies on a family of oligonucleotide-based molecules designed to recognize RNA in a way that acknowledges the complexity of RNA structure (23,49). Tethered oligonucleotide probes (TOPs) recognize RNA on the basis of both sequence and structure. Each TOP consists of two oligodeoxynucleotides joined by a flexible synthetic tether whose length and structure may be controlled by chemical synthesis. The two oligonucleotides in each TOP hybridize to two noncontiguous, single-stranded regions of a target RNA, and the tether traverses the distance between them. Cload et al. reported that TOPs containing neutral polyethylene glycol tethers are as effective as their counterparts containing polyanionic phosphodiester oligomers (49). TOPs were designed to recognize the spliced leader RNA (SL RNA) from *Leptomonas collosoma*. They contained DNA sequences that complement to noncontiguous regions of the SL RNA secondary structure. Although the TOPs were partially complementary to several other regions of the SL RNA, enzymatic mapping experiments indicated that the TOPs bound only to their targeted sites to form monomeric complexes when the SL RNA is in excess and that the two oligonucleotides within each TOP bind simultaneously. The kinetics and thermodynamics of TOP binding to the SL RNA were investigated. Chemical probing experiments revealed that TOPs can shift RNAs from one preferred secondary structure to another by causing "induced fit" in the target molecule.

In an extension of this strategy, Cload and Schepartz devised a single-step selection method to identify short sequences within the folded Rev response element (RRE) of the human immunodeficieny virus (HIV) RNA genome that are proximal and able to bind short oligonucleotides (50). The method employed a library of partially randomized TOPs complementary to all regions and an RNase H cleavage assay which identified those RRE regions preferred by the TOPs. Using this strategy, potent inhibitors of RRE function were identified. The random TOP selection procedure allows rational design of oligonucleotide-based ligands for RNA that exploit (rather than avoid) the complex structure of the RNA target.

Circular Oligonucleotides. Rumney and Kool reported the use of penta- and hexa-ethylene glycol chains in the linking of triplex-forming circular DNA oligonucleotides (*13*). They found that the longer linker of the two was the most stabilizing in triple-helical complexes, although the stability with a nucleotide loop was somewhat higher. These conjugates bound to complementary DNA and RNA sequences with high affinity, and they were highly resistant against exonuclease digestion.

Gao *et al.* bridged the ends of oligonucleotide duplexes with triethylene glycol loops (*51*). These bridged duplexes have high thermal dissociation temperatures, and the T_m for a triethylene glycol bridged 20 base pair duplex was higher than that for the corresponding pentathymidylate-bridged duplex. EcoR I endonuclease cleaved a ligated, bridged duplex at a slower rate than the corresponding unmodified duplex, whereas the unligated, bridged duplex was cleaved more rapidly. NMR spectroscopy indicated that the ligated duplex had a B-form conformation.

Nilsson *et al.* used so-called "padlock probes" for localized DNA detection by circularizing oligonucleotides (*52*). Oligonucleotide probes, consisting of two target-complementary segments, connected by a linker sequence, were designed. Upon recognition of the specific nucleic acid molecule the ends of the probes were joined through the action of a ligase, creating circular DNA molecules catenated to the target sequence. Hexaethylene glycol units were incorporated in the linker segments solely to reduce the number of synthetic steps required to span the ends of the two target-complementary segments. These probes provided highly specific detection with minimal background.

Loop Substitutions in Ribozymes. Three publications described the use of synthetic linkers to replace loops in hammerhead ribozymes (*20,21,53*). The hammerhead ribozyme is a short peace of RNA which can catalyze the intra- or intermolecular cleavage of RNA. Thomson *et al.* synthesized hammerhead ribozymes in which the tetranucleotide loop II was replaced by non-nucleotidic linkers of 7, 13, 17 and 19 atoms length, respectively, which were derived from oligoethylene glycols (*20*). Ribozymes with 17 and 19 atom linkers, in combination with a 4 base pair stem, had catalytic efficiencies which were 2 fold increased to that of the parent ribozyme with a tetranucleotide loop. Ribozymes with these linkers, but in combination with a shortened 2 base pair stem, showed a 2 fold decrease in catalytic efficiency when compared to the parent ribozyme. The other two groups came to similar conclusions, namely that loop II and parts of stem II can be replaced by synthetic linkers (*21,53*).

RNA-Protein Interactions. Ma *et al.* have studied the interactions between the TAR stem-loop structure of HIV-1 RNA and the Tat protein using nucleic acid duplexes of small size (miniduplexes) generated by automated synthesis (*17*). Four synthetic linkers of different length and hydrophobicity, including triethylene glycol and hexa-ethylene glycol, were incorporated into these model RNAs. These linker-derivatized RNA molecules were then assessed for their ability to bind to either a full-length protein or a short peptide (Tat-derived peptide) through RNA mobility shift assays. The results indicated that such modified miniduplex structures retain full binding activity, while T_m values were increased by 24-31 °C compared to an open duplex of the same length. In the second part, the authors extended their studies to the design of covalently closed cyclic RNA miniduplexes generated by T4 RNA ligase-catalyzed cyclization of linear precursors (*18*). When both ends of a shortened, wild-type TAR RNA stem (9 base pairs) were covalently linked through either nucleotidic loops (4-6 nucleotides) or synthetic linkers (derived from hexaethylene glycol), the resulting cyclic TAR RNA analogs were good substrates for binding by both Tat-derived peptide or full-length Tat protein. Interestingly, the cyclic TAR analogs failed to show any binding if the synthetic linker was reduced in length (e.g. derived from triethylene glycol), although such linkers were acceptable in the hairpin-shaped miniduplexes series.

Williams and Hall investigated, how the N-terminal RNA binding domain (RBD1) of the human U1A protein binds to two structurally very different RNA stem-loop structures (*54*). The experiments used RNA hairpins, in which the loop size was altered by deletion, insertion or substitution with oligoethylene glycol spacers, to determine what features of this RNA structure were critical for interaction with the RBD1. Substitution of the three nucleotides on the 3' side of the RNA hairpin loop by 6 to 18 ethylene glycol units did not significantly perturb the affinity, energetics or electrostatics of this RNA-protein association (*54*).

Effect of PEG Conjugation at the 2'-O-Position of Oligonucleotides

The 2'-O-methoxyethyl, the 2'-O-methyldi(oxyethyl) and the 2'-O-methyltri(oxyethyl) groups were recently introduced at several positions of antisense oligonucleotides (*29,30*). These 2'-modifications had a beneficial effect on the stability of duplexes with RNA target sequences; an increase in melting temperature T_m by about 1.1°C per modification was reported. With target DNA strands, however, duplex stability decreased. Additionally, specificity increased with RNA targets, measured as the T_m difference between the perfect duplexes and those containing a single mismatch. Data on nuclease stability were rather impressive giving increases by the factor 24 (for 2'-O-methoxyethyl) and >48 (for 2'-O-methyltri(oxyethyl)), compared to 2'-deoxyribonucleotides. After 96 hours, 60 and 90 %, respectively, undegraded oligonucleotide were measured in fetal calf serum (*29*). Chimeric oligonucleotides with 2'-O-methoxyethyl in the wings (terminal domains) and a central DNA phosphorothioate window were shown to efficiently downregulate C-'raf' kinase and PKC-α-messenger-RNA in tumor cell lines resulting in a profound inhibition of cell proliferation. The same compounds were able to effectively reduce the growth of tumors in animal models at low concentrations indicating the potential utility of these second generation antisense oligonucleotides for therapeutic applications (*30*).

Immobilized Conjugates

Oligoethylene glycols have been used as spacers for the immobilization of oligo-nucleotides on solid supports. Maskos and Southern described the synthesis and use of oligonucleotides immobilized on glass beads and microscope slides (*27*). These solid phases were utilized for the construction of oligonucleotide arrays to probe biological interactions. A device carrying all octapurine sequences was used to explore factors affecting molecular hybridization of the tethered oligonucleotides, to develop computer-aided methods for analyzing the data, and to test the feasibility of using the method for sequence analysis (*55*).

The specific detection of polymerase chain reaction (PCR) amplified sequences by hybridization of the PCR products to oligonucleotide probes immobilized via a poly-ethylene glycol spacer on carboxylated nylon membranes has been described by Zhang *et al.* (*19*). Using biotinylated PCR products in hybridization reactions and a chemi-luminescent detection system, high efficiency hybridization was obtained as well as a very good signal to noise ratio. The method has been applied successfully to the detection of RAS point mutations, cystic fibrosis deletion and point mutations and others.

PEG as a Matrix in Oligonucleotide Synthesis

Polyethylene glycol has found applications in two specific fields of oligonucleotide chemistry. The first one is PEG-supported liquid phase synthesis of oligonucleotides (*56-59*). Bonora *et al.* intended to use this methodology for large-scale synthesis of oligonucleotides. Monomethoxy-PEGs were used as soluble polymeric support where the starting nucleotide was coupled to. Coupling was carried out in homogeneous solution, and separation of the growing oligomer from excess reactants was achieved by precipitation. While the first reports used phosphotriester chemistry (*57*), the synthesis of 20mers by phosphoramidite chemistry (*58*), the application of H-phospho-nate chemistry (*60*) and the synthesis of phosphorothioate oligonucleotides (*59*) were described later. PEG-coupling of the starting nucleotide follows standard procedures of nucleotide chemistry, and oligonucleotide assembly is performed using commercially available reagents. With phosphoramidite chemistry, coupling yields of 99% were reported with 2.5 equivalents of phosphoramidite per support-bound oligonucleotide. After completion of the synthesis, the covalent bonds connecting the 3'-hydroxyl group with the PEG were hydrolyzed during standard deprotection, releasing a completely deprotected oligonucleotide.

Le Bec and Wickstrom applied this solution-phase strategy to the stereospecific synthesis of R_p and S_p methylphosphonate diastereomers by Grignard activated coup-ling, allowing the production of isomerically pure dinucleotides in contrast to common solid-phase protocols yielding racemic mixtures (*61*).

Polystyrene-PEG tentacle copolymers consisting of polyethylene glycol chains grafted onto insoluble polystyrene matrices have also been used in the automated synthesis of oligonucleotides and conjugates (*26,62*). Due to the excellent solvation properties and high mobility of the PEG, this system has been considered quasi-homogeneous. Reaction conditions had to be modified with respect to the standard protocols, and coupling yields of 98-99% were obtained. The synthesis of up to 200 μmole of RNA oligonucleotides was described (*63*). Gao *et al.* described the use of these supports in H-phosphonate synthesis (*64*), and Conte *et al.* reported their application in the automated synthesis of circular oligonucleotides by a combination of phosphoramidite and phosphotriester chemistry (*65*).

Affinity Partitioning of Nucleic Acids

The aqueous two-phase system PEG-dextran has been used quite often for the purification of proteins by liquid-liquid partitioning. The covalent attachment of an affinity ligand to one of the phase-forming polymers can strongly influence the partitioning behavior. Several enzymes were purified using PEG-coupled cofactors, substrate analogs or affinity dyes. For the fractionation of nucleic acids, base pair-specific ligands were coupled to the terminal positions of PEG molecules. Upon binding to DNA, these macromolecular ligands increase the affinity of the DNA for the PEG-rich phase (66). Jäschke et al. have developed a method for the sequence-specific isolation of single-stranded nucleic acids (67). Oligonucleotides-PEG conjugates were used to specifically increase the affinity of a target sequence for the PEG-rich phase. The affinity increases proportionally to the size of the coupled PEG. Model separations were described both in the batch mode and in a chromatographic system containing about 3000 theoretical plates. The statistical analysis of the data revealed, that for practically relevant separations in the batch mode PEGs with molecular weights above 30.000 would be required. For chromatographic applications, however, the described system using automatically synthesized oligonucleotide-PEG 4000 conjugates provides a suitable starting point.

Conclusions and Outlook

Polyethylene glycol and related polymers have found numerous applications in nucleic acid research and technology. The increasing use is due to a unique combination of physical, chemical and biological properties that already facilitated its use in the conjugation of polypeptides, lipids and low molecular weight drugs (5,6,68). In addition, the possibility to incorporate PEG during automated oligonucleotide synthesis allowed the easy preparation and investigation of various conjugates.

Based on the experience with protein-PEG conjugates, a number of future developments is conceivable:

Polyethylene glycols are currently being studied in several pharmaceutical companies to improve the efficacy and bioavailability of nucleic acid-based drugs. PEG is known to enhance the solubility and stability of proteins and to cause altered pharmakokinetics and systemic exposure. Altered immunogenicity and allergenicity have been reported. It is expected that oligonucleotides will benefit from the protection from enzymatic degradation, extended plamsa lifetime, reduced clearance and from reduction of other unwanted manifestations of of biological recognition as described for proteins (5). The work of Altmann et al. on the 2'-modified antisense oligo-nucleotides indicate the suitability of this approach for nucleic acids (30), and work in the ribozyme and aptamer field is underway.

Branched PEGs functionalized with different oligonucleotides could be a useful starting point for the development of novel biomaterials and for the systematic construction of supramolecular structures. The noncovalent linkage of complementary oligonucleotides by Watson-Crick base pairing would confer temperature sensitivity and specific enzymatic cleavability. Ternary conjugates, linking oligonucleotides via

PEG to signal peptides, oligosaccharides or lipids would combine different functions with only little interference and might be a feasable solution for improving cellular uptake or targeting specific cell types or organs with oligonucleotide drugs. The further development of chemical and enzymatic derivatization strategies and their systematic comparison is needed to explore these fields.

Acknowledgments

Drs. Martin Egli (Northwestern University Medical School, Chicago, IL) and Hugh Mackie (Glen Research Corp., Sterling, VA) are acknowledged for helpful suggestions, and Drs. W.A. Piecken and S.C. Gill (NeXstar Pharmaceuticals, Boulder, CO) for communicating unpublished data.

Literature Cited

1. De Mesmaeker, A.; Häner, R.; Martin, P.; Moser, H. E. *Acc. Chem. Res.* **1995,** *28,* 366-74.
2. Hélène, C. *Anticancer Drug Des.* **1991,** *6,* 569-84.
3. Marschall, P.; Thomson, J. B.; Eckstein, F. *Cell. Mol. Neurobiol.* **1994,** *14,* 523-38.
4. Gold, L.; Polisky, B.; Uhlenbeck, O.; Yarus, M. *Annu. Rev. Biochem.* **1995,** *64,* 763-97.
5. Zalipsky, S. *Adv. Drug Delivery Reviews* **1995,** *16,* 157-82.
6. *Poly(ethylene glycol) chemistry: biotechnical and biomedical applications;* Harris, J. M. (Ed.); Plenum: New York, 1992.
7. Goodchild, J. *Bioconjugate Chem.* **1990,** *1,* 165-87.
8. *Protocols for oligonucleotide conjugates;* Agrawal, S. (Ed.); Methods in Molecular Biology Vol. 26; Humana Press: Totowa, NJ, 1994.
9. Beaucage, S. L.; Iyer, R. P. *Tetrahedron* **1992,** *48,* 2223-2311.
10. Beaucage, S. L.; Iyer, R. P. *Tetrahedron* **1993,** *49,* 6123-94.
11. Durand, M.; Chevrie, K.; Chassignol, M.; Thuong, N. T.; Maurizot, J. C. *Nucleic Acids Res.* **1990,** *18,* 6353-9.
12. Rumney, S.; Kool, E. T. *J. Am. Chem. Soc.* **1995,** *117,* 5635-46.
13. Rumney, S.; Kool, E. T. *Angew. Chem. Int. Ed. Engl.* **1992,** *31,* 1617-9.
14. Jäschke, A.; Fürste, J. P.; Cech, D.; Erdmann, V. A. *Tetrahedron Lett.* **1993,** *34,* 301-4.
15. Jäschke, A.; Fürste, J. P.; Nordhoff, E.; Hillenkamp, F.; Cech, D.; Erdmann, V. A. *Nucleic Acids Res.* **1994,** *22,* 4810-7.
16. Jäschke, A.; Bald, R.; Nordhoff, E.; Hillenkamp, F.; Cech, D.; Erdmann, V. A.; Fürste, J. P. *Nucleosides Nucleotides* **1996,** *15,* 1519-29.
17. Ma, M. Y.-X.; Reid, L. S.; Climie, S. C.; Lin, W. C.; Kuperman, R.; Sumner-Smith, M.; Barnett, R. W. *Biochemistry* **1993,** *32,* 1751-8.
18. Ma, M. Y.-X.; McCallum, K.; Climie, S. C.; Kuperman, R.; Lin, W. C.; Sumner-Smith, M.; Barnett, R. W. *Nucleic Acids Res.* **1993,** *21,* 2585-9.

19.   Zhang, Y.; Coyne, M. Y.; Will, S. G.; Levenson, C. H.; Kawasaki, E. S. *Nucleic Acids Res.* **1991**, *19*, 3929-33.

20.   Thomson, J. B.; Tuschl, T.; Eckstein, F. *Nucleic Acids Res.* **1993**, *21*, 5600-3.

21.   Benseler, F.; Fu, D.-J.; Ludwig, J.; McLaughlin, L. W. *J. Am. Chem. Soc.* **1993**, *115*, 8483-4.

22.   Altmann, S.; Labhardt, A. M.; Bur, D.; Lehmann, C.; Bannwarth, W.; Billeter, M.; Wüthrich, K.; Leupin, W. *Nucleic Acids Res.* **1995**, *23*, 4827-35.

23.   Cload, S. T.; Schepartz, A. *J. Am. Chem. Soc.* **1991**, *113*, 6324-6.

24.   Koroleva, O. N.; Volkov, E. N.; Drutsa, V. L. *Bioorg. Khim.* **1994**, *20*, 420-32.

25.   Efimov, V. A.; Pashkova, I. N.; Kalinkina, A. L.; Chakhmakhcheva, O. G. *Bioorg. Khim.* **1993**, *19*, 800-4.

26.   Bayer, E.; Maier, M.; Bleicher, K.; Gaus, H. J. *Z. Naturforsch.* **1995**, *50 b*, 671-76.

27.   Maskos, U.; Southern, E. M. *Nucleic Acids Res.* **1992**, *20*, 1679-84.

28.   Efimov, V. A.; Kalinkina, A. L.; Chakhmakhcheva, O. G. *Nucleic Acids Res.* **1993**, *21*, 5337-44.

29.   Martin, P. *Helv. Chim. Acta* **1995**, *78*, 486-504.

30.   Altmann, K. H.; Dean, N. M.; Fabbro, D.; Freier, S. M.; Geiger, T.; Häner, R.; Hüsken, D.; Martin, P.; Monia, B. P.; Müller, M.; Natt, F.; Nicklin, P.; Phillips, J.; Pieles, U.; Sasmor, H.; Moser, H. E. *Chimia* **1996**, *50*, 168-76.

31.   Livak, K. J.; Hobbs, F. W.; Zagursky, R. J. *Nucleic Acids Res.* **1992**, *20*, 4831-7.

32.   Pitulle, C.; Kleineidam, R. G.; Sproat, B.; Krupp, G. *Gene* **1992**, *112*, 101-5.

33.   Tong, G.; Lawlor, J. M.; Tregear, G. W.; Haralambidis, J. *J. Org. Chem.* **1993**, *58*, 2223-31.

34.   Igloi, G. *Anal. Biochem.* **1996**, *233*, 124-9.

35.   Moore, M. J.; Sharp, P. A. *Science* **1992**, *256*, 992-7.

36.   Oh, B.-K.; Pace, N. R. *Nucleic Acids Res.* **1994**, *22*, 4087-94.

37.   Chu, B. C. F.; Wahl, G. M.; Orgel, L. E. *Nucleic Acids Res.* **1983**, *11*, 6513-29.

38.   Kawaguchi, T.; Asakawa, H.; Tashiro, Y.; Juni, K.; Sueishi, T. *Biol. Pharm. Bull.* **1995**, *18*, 474-6.

39.   Manoharan, M.; Tivel, K. L.; Andrade, L. K.; Mohan, V.; Condon, T. P.; Bennett, C. F.; Cook, P. D. *Nucleosides Nucleotides* **1995**, *14*, 969-73.

40.   Jones, D. S.; Hachmann, J. P.; Osgood, S. A.; Hayag, M. S.; Barstad, P. A.; Iverson, G. M.; Coutts, S. M. *Bioconjugate Chem.* **1994**, *5*, 390-9.

41.   Grossman, P. D.; Bloch, W.; Brinson, E.; Chang, C. C.; Eggerding, F. A.; Fung, S.; Iovannisci, D. A.; Woo, S.; Winn-Deen, E. S. *Nucleic Acids Res.* **1994**, *22*, 4527-34.

42.   Baron, H.; Fung, S.; Aydin, A.; Bähring, S.; Luft, F. C.; Schuster, H. *Nature Biotechnology* **1996**, *14*, 1279-82.

43.   Durand, M.; Peloille, S.; Thuong, N. T.; Maurizot, J. C. *Biochemistry* **1992**, *31*, 9197-9204.

44.   Dagneaux, C.; Liquier, J.; Taillandier, E. *Biochemistry* **1995**, *34*, 14815-8.

45.   Giovannangeli, C.; Thuong, N. T.; Hélène, C. *Proc. Natl. Acad. Sci. USA* **1993**, *90*, 10013-7.

46.   Kandimalla, E. R.; Agrawal, S. *Gene* **1994**, *149*, 115-21.
47.   Vo, T.; Wang, S.; Kool, E. T. *Nucleic Acids Res.* **1995**, *23*, 2937-44.
48.   Kessler, D. J.; Pettitt, B. M.; Cheng, Y.-K.; Smith, S. R.; Jayaraman, K.; Vu, H. M.; Hogan, M. E. *Nucleic Acids Res.* **1993**, *21*, 4810-5.
49.   Cload, S. T.; Richardson, P. L.; Huang, Y.-H.; Schepartz, A. *J. Am. Chem. Soc.* **1993**, *115*, 5005-14.
50.   Cload, S. T.; Schepartz, A. *J. Am. Chem. Soc.* **1994**, *116*, 437-42.
51.   Gao, H.; Chidambaram, N.; Chen, B. C.; Pelham, D. E.; Patel, R.; Yang, M.; Zhou, L.; Cook, A.; Cohen, J. S. *Bioconjugate Chem.* **1994**, *5*, 445-53.
52.   Nilsson, M.; Malmgren, H.; Samiotaki, M.; Kwiatkowski, M.; Chowdhary, B. P.; Landegren, U. *Science* **1994**, *265*, 2085-8.
53.   Hendry, P.; Moghaddam, M. J.; McCall, M. J.; Jennings, P. A.; Ebel, S.; Brown, T. *Biochim. Biophys. Acta* **1994**, *1219*, 405-12.
54.   Williams, D. J.; Hall, K. B. *J. Mol. Biol.* **1996**, *257*, 265-75.
55.   Southern, E. M.; Maskos, U.; Elder, J. K. *Genomics* **1992**, *13*, 1008-17.
56.   Brandstetter, F.; Schott, H.; Bayer, E. *Tetrahedron Lett.* **1973**, *32*, 2997-3000.
57.   Bonora, G. M.; Scremin, C. L.; Colonna, F. P.; Garbesi, A. *Nucleic Acids Res.* **1990**, *18*, 3155-9.
58.   Bonora, G. M.; Biancotto, G.; Maffini, M.; Scremin, C. L. *Nucleic Acids Res.* **1993**, *21*, 1213-7.
59.   Scremin, C. L.; Bonora, G. M. *Tetrahedron Lett.* **1993**, *34*, 4663-6.
60.   Zaramella, S.; Bonora, G. M. *Nucleosides Nucleotides* **1995**, *14*, 809-12.
61.   Le Bec, C.; Wickstrom, E. *Tetrahedron Lett.* **1994**, *35*, 9525-8.
62.   Bayer, E.; Bleicher, K.; Maier, M. *Z. Naturforsch.* **1995**, *50 b*, 1096-1100.
63.   Tsou, D.; Hampel, A.; Andrus, A.; Vinayak, R. *Nucleosides Nucleotides* **1995**, *14*, 1481-92.
64.   Gao, H.; Gaffney, B. L.; Jones, R. A. *Tetrahedron Lett.* **1991**, *32*, 5477-80.
65.   Conte, M. R.; Mayol, L.; Montesarchio, D.; Piccialli, G.; Santacroce, C. *Nucleosides Nucleotides* **1993**, *12*, 351-8.
66.   Müller, W. In *Partitioning in aqueous two-phase systems;* Walter, H.; Brooks, D. E.; Fisher, D. (Eds.) Academic Press: New York, 1985, , pp. 227-66.
67.   Jäschke, A.; Fürste, J. P.; Erdmann, V. A.; Cech, D. *Nucleic Acids Res.* **1994**, *22*, 1880-4.
68.   Katre, N. V. *Adv. Drug Delivery Rev.* **1993**, *10*, 91-114.

Chapter 19

Design of Antitumor Agent-Terminated Poly(ethylene glycol) Conjugate as Macromolecular Prodrug

Tatsuro Ouchi, Hidetoshi Kuroda, and Yuichi Ohya

Department of Applied Chemistry, Faculty of Engineering, Kansai University, Suita, Osaka 564, Japan

Since poly(ethylene glycol) (PEG) is a water-soluble and biocompatible polymer, PEG is a very interesting material as a carrier of macromolecular prodrug. Although doxorubicin (DXR) is one of the most prominent antitumor agents in cancer chemotherapy, its very strong side-effects, its poor water-solubility, and its instability have been cited as unsolved problems. In order to solve the defects of DXR, the design of a DXR-terminated PEG conjugate as a macromolecular prodrug of DXR was investigated. The MeO-PEG/Schiff's base/DXR conjugate showed a lysosomotropic release behavior of free DXR, very good stability in PBS for a long period, and exhibited a significant cytotoxic activity against p388D$_1$ *lymphocytic leukemia* cells *in vitro*. Moreover, this conjugate exhibited stronger cytotoxic activity than free DXR against DXR-resistant p388D$_1$ *leukemia* cells. Furthermore, the Lactose/PEG/amide/DXR conjugate showed stronger cell-specific cytotoxic activity against HLE *human hepatoma* cells than MeO-PEG/amide/DXR conjugate. The cytotoxic activity of Lactose/PEG/amide/DXR conjugate against HLE cells was inhibited by addition of galactose.

In cancer chemotherapy, the side-effects of antitumor agent are serious problems. In comparison with the low molecular prodrug, the polymer/drug conjugate can generally be expected to overcome the problem of side-effects by improving the body distribution of drug and to have a prolonged duration of activity. Since poly(ethylene glycol) (PEG) is a water-soluble and biocompatible polymer with modifiable hydroxyl group at the end group, it has been utilized as a modifier of drugs or proteins. In a previous paper (*1*), we reported that the conjugate of monomethoxy-poly(ethylene glycol) (MeO-PEG) bound to 5-fluorouracil (5FU) *via* urethane or urea bond produced the remarkable survival effect in mice bearing p388D$_1$ *lymphocytic leukemia* cells. These results showed the potential of PEG as a drug carrier. Doxorubicin (DXR) is one of the best clinical antitumor agents. However, its very strong side-effects, its poor water-solubility, and its instability have also been cited. In order to solve the defects of DXR, some kinds of PEG/DXR conjugate have been synthesized as macromolecular prodrugs (*2, 3*). The antitumor activity of a polymer/drug conjugate is influenced by the hydrolysis rate of the bond between drug and carrier polymer (*4-7*). So, we employed ester, amide and Schiff's base bonds as the modes of bond between carrier and drug to achieve the intracellular release of DXR from conjugates under lysosomal acidic condition after their uptake into tumor cells *via* endocytocis (*8*). The formation of polymer micelle is of interest in the drug delivery system and the field of biomedical polymer (*9, 10*). Since the PEG/DXR conjugates designed in this study consist of the hydrophobic part of DXR and the hydrophilic part of PEG, they have the ability to form the intermolecular aggregates in aqueous solution. The formation of aggregates of the conjugates might influence the release behavior of DXR from themselves, and their stability and cytotoxic activity. Therefore, we investigated the state of the conjugates in aqueous solution. Moreover, to achieve receptor-mediated drug delivery to specific cells, we employed the lactose residue as a targeting moiety to *hepatoma* cells. The present paper concerns the release behavior of DXR from PEG/DXR conjugates, their stability in aqueous solution at 37°C *in vitro*, the effect of conjugates on cytotoxic activity against p388D$_1$ *lymphocytic leukemia* cells and HLE *human hepatoma* cells *in vitro* (*11*), and the possibility of overcome of resistance of DXR by PEG/DXR conjugation technique.

Experimental

Materials. Monomethoxy-poly(ethylene glycol) (MeO-PEG; n=20) was provided by Toho Chemical Industry Co. Ltd. Monomethoxy-poly(ethylene glycol) acid (MeO-PEG acid; n=20) and poly(ethylene glycol) diacid (PEG diacid; n=20) were provided by Kawaken Fine Chemical Co. Ltd. Doxorubicin hydrochloride (DXR•HCl) and 14-bromodaunorubicin (14Br-DXR) were obtained from Meiji Seika Co. Ltd.

Synthesis of MeO-PEG/ester/DXR conjugate 1. The fixation of DXR to MeO-PEG *via* ester bond was carried out according to the coupling reaction shown in Scheme I (*11*). The MeO-PEG/ester/DXR conjugate **1** was isolated by gel filtration chromatography (Sephadex LH-20 column, 1.0×100 cm; eluent: DMF). Yield 75%. The characterization of the conjugate obtained was carried out by IR, UV spectra and thin layer chromatography (TLC) (plate: Merck Kieselgel 60 F$_{254}$, eluent: chloroform/methanol, v/v 5:1). IR(KBr): 2887(CH$_2$), 1734(COO), 1113 cm^{-1} (OCH$_2$CH$_2$). The degree of end-capping of DXR per PEG molecule (DDXR) was estimated from the UV absorbance at 495 nm in DMF by using the ε_{495} value of 11700 mol^{-1} dm^3 cm^{-1} of DXR•HCl as a standard.

Synthesis of MeO-PEG/amide/DXR conjugate 2. The fixation of DXR to MeO-PEG *via* amide bond was performed by water-soluble carbodiimide (WSC) method according to the reaction shown in Scheme II (*11*). The MeO-PEG/amide/DXR conjugate **2** was isolated by gel filtration chromatography (Sephadex LH-20 column, 1.0×100 cm; eluent: DMF). Yield 70%. The characterization of the conjugate obtained was carried out by measurement of IR, UV spectra and TLC analysis. IR(KBr): 2897(CH$_2$), 1638(CONH), 1113 cm^{-1}(OCH$_2$CH$_2$).

Synthesis of MeO-PEG/Schiff's base/DXR conjugate 3. The fixation of DXR to MeO-PEG *via* Schiff's base bond was carried out according to Scheme III (*11*). The MeO-PEG aldehyde was prepared according to a modification of the method reported by Harris *et al.* (*12*). The MeO-PEG aldehyde was reacted with DXR to give MeO-PEG/Schiff's base/DXR **3** conjugate at 0°C for 12h. The conjugate **3** was isolated by

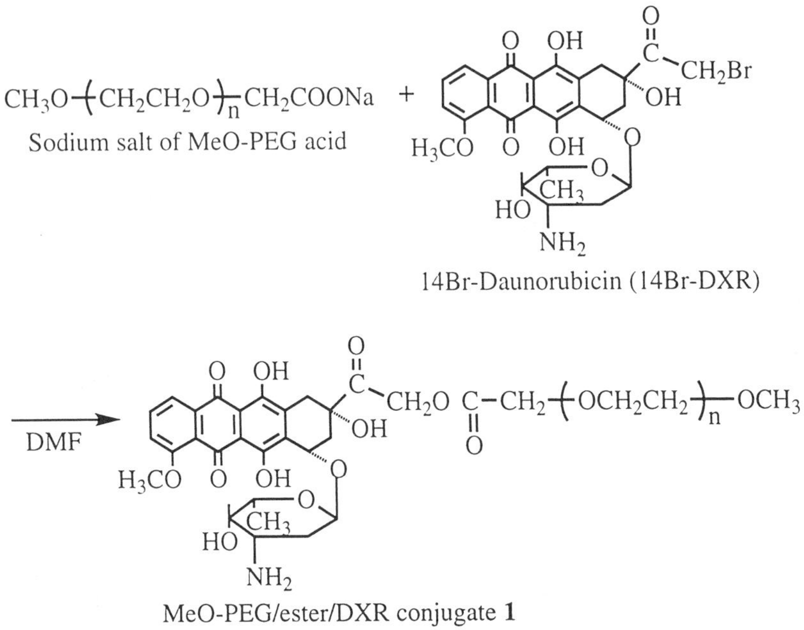

CH₃O—(CH₂CH₂O)ₙ—CH₂COONa   +

Sodium salt of MeO-PEG acid

14Br-Daunorubicin (14Br-DXR)

MeO-PEG/ester/DXR conjugate **1**

Scheme I. Synthetic route of MeO-PEG/ester/DXR conjugate **1**.

CH₃O—(CH₂CH₂O)ₙ—CH₂COOH   +

MeO-PEG acid

Doxorubicin (DXR)

MeO-PEG/amide/DXR conjugate **2**

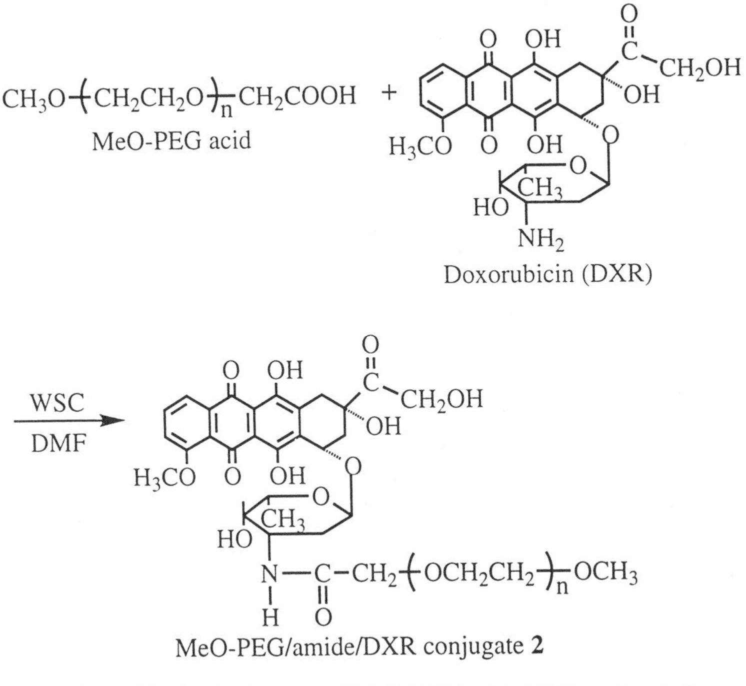

Scheme II. Synthetic route of MeO-PEG/amide/DXR conjugate **2**.

gel filtration chromatography (Sephadex LH-20 column, 1.0×100 cm; eluent: DMF). Yield 60%. The characterization of the conjugate obtained was carried out by measurement of IR, UV spectra and TLC analysis. IR(KBr): 2889(CH_2), 1654(CH=N), 1113cm^{-1}(OCH_2CH_2).

Synthesis of Lactose/PEG/amide/DXR conjugate 4. The synthesis of Lactose/PEG/amide/DXR conjugate 4 was performed according to the reaction steps shown in Scheme IV. The PEG acid mononitrobenzyl ester was reacted with *N*-lactonyl ethylenediamine by N,N'-dicyclohexylcarbodiimide (DCC)/N-hydroxysuccinimide (HOSu) method to afford the PEG nitrobenzyl ester end-capped with lactose. The deprotected lactose end-capped PEG acid was reacted with DXR by WSC/1-hydroxybenzotriazole (HOBt) method to give the Lactose/PEG/amide/DXR conjugate 4. The conjugate 4 was isolated by gel filtration chromatography (Sephadex LH-20 column, 1.0×100 cm; eluent: DMF). Yield 55%. The characterization of conjugate 4 was carried out by measurement of IR, UV spectra and TLC analysis. IR(NaCl): 2815(CH_2), 1635(CONH), 1113 cm^{-1} (OCH_2CH_2).

Determination of release rate of DXR from conjugate. The release behavior of DXR from the conjugates *in vitro* was investigated in 1/15M KH_2PO_4-Na_2HPO_4 buffer solution (pH=7.4) and 1/5M Na_2HPO_4-1/10M citric acid buffer solution (pH=4.0) at 37°C. The amount of DXR released was estimated by HPLC (column: TSK gel ODS-120; eluent: acetonitrile/1/100M $NH_4H_2PO_4$ aqueous solution, v/v 65:35; fluorescence detector with excitation at 480nm and emission at 590nm) (*13-15*).

Measurement of stability of conjugate. Since a free DXR is very unstable, it is converted gradually to a dark red precipitate in aqueous solution. So, the stability of conjugates was estimated by measurement of decrease of absorbance at 495 nm in aqueous solution with the passage of time. The conjugates and free DXR were dissolved in 1/15 KH_2PO_4-Na_2HPO_4 buffer solution (pH=7.4) at 37°C and the absorbance at 495 nm of samples was followed, respectively.

Observation of aggregation of conjugates in aqueous solution. The conjugates obtained consist of the hydrophilic part (PEG) and the hydrophobic part (DXR). Therefore, the conjugates have the possibility to form the aggregates in aqueous solution. The formation ability of the intermolecular aggregation of the conjugates in aqueous solution was investigated by the measurement of fluorescence. The conjugates (1 wt%) were dissolved in PBS (pH=7.4) as well as 1 wt% sodium dodecyl sulfate (SDS) solution at 37°C and diluted gradually two-fold with each solution. The fluorescence intensity of the solution at each concentration was measured by the fluorescence detector with excitation at 480 nm and emission at 590 nm.

Assay of cytotoxic activity of conjugate. The cytotoxic activity of the conjugates obtained was measured against p388D$_1$ *lymphocytic leukemia* cells or DXR-resistant p388D$_1$ *lymphocytic leukemia* cells *in vitro* (*4*). The tumor cell suspension(100 μl) containing 1×10^4 p388D$_1$ *leukemia* cells in culture medium containing 10% FCS was distributed in a 96-wells multi-plate (Corning 25860MP), and then incubated with conjugates or free DXR in a humid atmosphere containing 5 % CO_2 at 37°C for 48 h. The number of viable cells was determined by means of the MTT (3-(4,5-dimethylthazol-2-yl)-2,5-dephenyl tetrazolium bromide) assay using a microplate reader (MTP-120, Corona Electric Co.). The cytotoxic activity was estimated by the

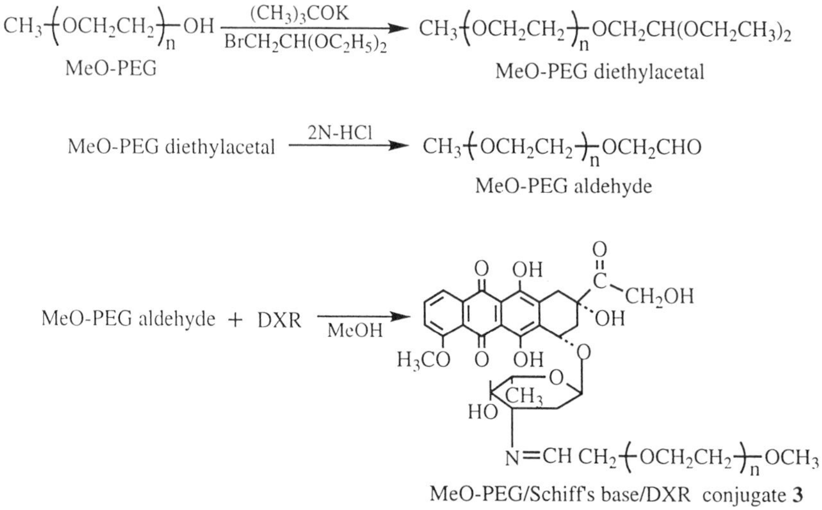

Scheme III. Synthetic route of MeO-PEG/Schiff's base/DXR conjugate **3**.

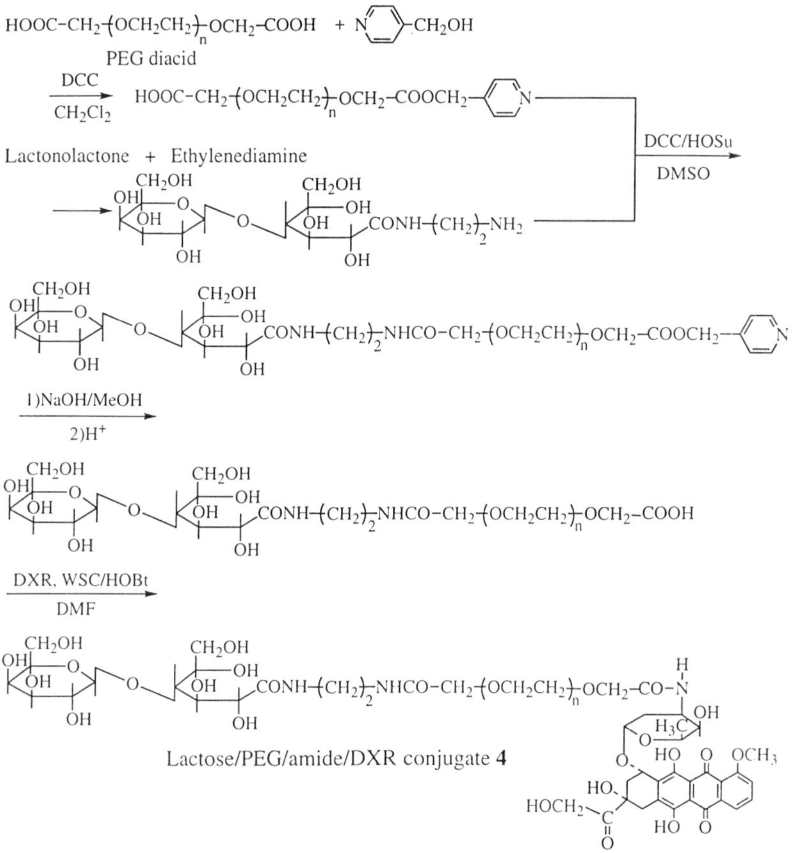

Scheme IV. Synthetic route of Lactose-PEG/amide/DXR conjugate **4**.

following equation:

$$\text{Cytotoxic activity } (\%) = (C - T) / C \times 100$$

C: number of viable cells after 48 h incubation without drug
T: number of viable cells after 48 h incubation with drug

The galactose receptor-mediated cytotoxic activity against HLE *human hepatoma* cells *in vitro* was evaluated by the following two method. One is the measuremant of cytotoxic activity of the Lactose/PEG/amide/DXR conjugate 4 binding on HLE cellular surface (Method 1). Aliquots of tumor cell suspension ($100\mu l$) containg 1×10^4 cells in culture medium were distributed in a 96-wells multi-plate and incubated in a humid atmosphere containing 5% CO_2 at 37°C for 48 h. The cells washed with culture medium were distributed in $100\mu l$ of fresh culture medium containing $20\mu l$ of PBS solution of the conjugate 2, 4 or free DXR (7.0×10^{-4} DXR mol/l) and preincubated at 4°C for 1 or 2.5 h. In order to remove the conjugates not bound to the cells and the free DXR not incorporated into the cells, the cells were washed several times with culture medium. The washed cells were added to $100\mu l$ of fresh culture medium, and then cultured in a humid atmosphere containing 5% CO_2 at 37°C for 48 h. The number of viable cells was determined by means of MTT assay. The other is the test of blocking effect of the addition of galactose on the uptake of conjugate 4 into HLE cells *in vitro* (Method 2). Aliquots of tumor cell suspension ($100\mu l$) containing 1×10^4 cells in culture medium were distributed in a 96-wells multi-plate and incubated in a humid atmosphere containing 5% CO_2 at 37°C for 48 h. The cells were distributed in $100\mu l$ of culture medium containing PBS solution of the conjugate and galactose (7.0×10^{-4} DXR mol/l and 1.0×10^{-3} mol/l, respectively) and preincubated at 4°C for 2.5 h. The cells washed several times with culture medium were added to $100\mu l$ of fresh culture medium and then cultured in a humid atmosphere containing 5% CO_2 at 37°C for 48 h. The cytotoxic activity was calculated by the above equation.

Results and Discussion

Synthesis of PEG/DXR conjugates. The conjugates of MeO-PEG end-capped with DXR *via* ester, amide or Schiff's base bonds and the conjugate of lactose-PEG end-capped with DXR *via* amide bond were synthesized through the reaction steps shown in Schemes I~IV. All conjugates obtained were water-soluble. It was confirmed by TLC that free DXR and adriamycinon as a degradation product were not contained in the conjugates obtained. DXR•HCl and 14Br-DXR showed the red spots at R_f of 0-0.05, whereas the conjugates showed broad red spots at R_f of 0.21-0.62. These conjugates could be easily separated from non-immobilized free 14Br-DXR, DXR and adriamycinon by gel filtration chromatography. Although the further purification was tried by other column chromatography techniques, we could not succeed to remove perfectly the non-reacted MeO-PEG, MeO-PEG derivatives or Lactose end capped PEG acid. The results of degree of end-capping of DXR per PEG molecule (DDXR) for the conjugates are summarized in Table I. The maximum value of DDXR obtained was 86 mol%.

Release behavior of DXR from conjugate. The release behavior of DXR from the conjugates was investigated in 1/15M KH_2PO_4-Na_2HPO_4 buffer solution of pH 7.4 and 1/5M Na_2HPO_4-1/10M citric acid buffer solution of pH 4.0 at 37°C *in vitro*.

The results of release rate of DXR from the conjugates are shown in Figure 1 (*11*). The release rate of DXR from MeO-PEG/ester/DXR conjugate **1** was very fast in both pH aqueous solutions. The release rates of the DXR from MeO-PEG/amide/DXR conjugate **2** and MeO-PEG/Schiff's base/DXR conjugate **3** were very slow in the pH7.4 buffer solution; these results show that amide and Schiff's base bonds between DXR and MeO-PEG are hardly hydrolyzed to release free DXR in the medium outside the cells. On the contrary, in the pH 4.0 buffer solution which approximated the lysosomal acidic condition in the cells, the release rate of DXR from MeO-PEG/Schiff's base/DXR conjugate **3** was faster than that of MeO-PEG/amide/DXR conjugate **2**. These pH-sensitive release behavior of DXR from PEG/Schiff's base/DXR conjugate **3** suggested that the conjugate **3** released hardly free DXR outside cells and then released easily free DXR in the lysosomes after its uptake into cells *via* endocytocis. On the other hand, Prof. Kopecek and Prof. Duncan *et al.* succeeded in the lysosomal digestion of linkage between polymer and drug by using the Gly-Phe-Leu-Gly tetrapeptide spacer group (*7*). The Schiff's base linkage had the advantage of the tetrapeptide spacer group in the ease of its synthesis.

Stability of DXR residue in conjugate in aqueous solution. A free DXR is found to be easily destroyed in water by the attack of hydroxyl ion on the anthracycline ring. Since its absorbance at 495 nm is found to decrease with the degradation of DXR, the stability of DXR residue in conjugates in aqueous solution could be evaluated by the measurement of the absorbance at 495 nm. The results of change of absorbance at 495 nm of the conjugates and free DXR in PBS (pH=7.4) with the passage of time are shown in Figure 2 (*11*). While the absorbance of free DXR retained only about 55 % of initial absorbance at 495 nm after 4 days, the absorbance of MeO-PEG/amide/DXR conjugate **2** and MeO-PEG/Schiff's base/DXR conjugate **3** retained over 90 % of initial absorbance at 495 nm after 4 days. On the contrary, the stability of MeO-PEG/ester/DXR conjugate **1** was no good similarly to that of free DXR. The very fast release rate of DXR from the MeO-PEG/ester/DXR conjugate **1** was suggested to reflect in no good stability of DXR residue in conjugate **1.** As the DXR residue in conjugate **1** was bound to PEG at its 14-position *via* ester bond, the DXR residue in conjugate **1** had a free amino group in its sugar moiety. It was observed by Masuike *et al.* that the free DXR was easily destroyed under the basic condition (*16*). Therefore, it was presumed that the coexistence of a free amino group would influence the stability of the DXR residue in conjugate **1** in aqueous solution. The good stabilization of MeO-PEG/amide/DXR conjugate **2** and MeO-PEG/Schiff's base/DXR conjugate **3** in aqueous solution can be explained owing to the protection of free amino groups of their DXR residues by conjugation technique and the formation of aggregates in both MeO-PEG/DXR conjugates in aqueous solution as described below. The DXR residues in conjugates **2** and **3** were considered to be able to escape from the attack by hydroxyl ion because the structural barriers against hydroxyl ion were formed by the attachments of DXR to PEG *via* covalent bonds.

Formation of aggregates of conjugate in aqueous solution. The state of aggregation of conjugates was estimated by measurement of fluorescence spectrum. The results for MeO-PEG/amide/DXR conjugate **2** as one example are shown in Figure 3 (*11*). It was observed in Figure 3 that the values of fluorescence intensity of all samples used were quenched at the same concentration. In comparison with the fluorescence intensity of free DXR or the mixture of free DXR and MeO-PEG, that of MeO-PEG/amide/DXR conjugate **2** was very low. These results suggested that the MeO-PEG/DXR conjugate formed a certain state of aggregation in aqueous solution. Since the fluorescence of MeO-PEG/DXR conjugate was not completely

Table I. Values of R_f and degree of end-capped DXR
for PEG/DXR conjugates

Conjugate	R_f^a	DDXR^b (mol%)
MeO-PEG/ester/DXR	0.60	75
MeO-PEG/amide/DXR	0.61	85
MeO-PEG/Schiff's base/DXR	0.62	86
Lactose/PEG/amide/DXR	0.21	83

^{a} Eluent of TLC: chloroform/methanol, v/v 5:1.
^{b} Degree of end-capping of DXR per PEG molecule.

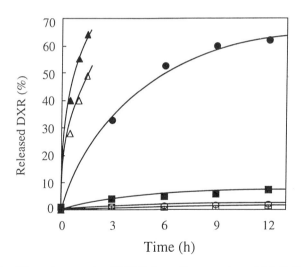

Time (h)

Figure 1. The release rate of DXR from three modes of MeO-PEG/DXR conjugate in 1/15M KH$_2$PO$_4$-Na$_2$HPO$_4$ buffer solution (pH=7.4) or 1/5M Na$_2$HPO$_4$-1/10M citric acid buffer solution (pH=4.0) at 37°C. (▲) conjugate **1**, pH=4.0; (△) conjugate **1**, pH=7.4; (■) conjugate **2**, pH=4.0; (□) conjugate **2**, pH=7.4; (●) conjugate **3**, pH=4.0; (○) conjugate **3**, pH=7.4. (Reproduced with permission from ref. 11. Copyright 1995 Technomic Publishing Co., Inc.)

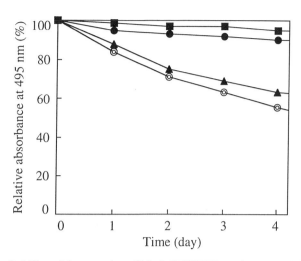

Figure 2. Stability of three modes of MeO-PEG/DXR conjugate and free DXR in PBS (pH=7.4) at 37°C. (▲) conjugate **1**; (■) conjugate **2**; (●) conjugate **3**; (◎) free DXR. (Reproduced with permission from ref. 11. Copyright 1995 Technomic Publishing Co., Inc.)

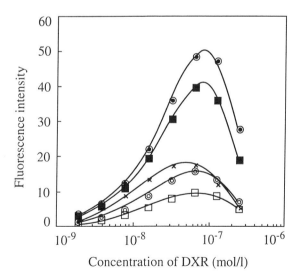

Figure 3. Fluorescence intensity of free DXR and MeO-PEG/amide/DXR conjugate **2** in PBS (pH=7.4) at 37°C. (□) conjugate **2** and (■) conjugate **2** plus 1 wt% SDS; (◎) free DXR; (●) free DXR plus 1 wt% SDS; (×) free DXR plus MeO-PEG. (Reproduced with permission from ref. 11. Copyright 1995 Technomic Publishing Co., Inc.)

quenched, it was suggested that the aggregation was not in the complete micellar state. Therefore, it was considered that the stability of the conjugate in aqueous solution might depend on the structural barrier by PEG based on the formation of aggregates of conjugate in aqueous solution.

Cytotoxic activity of conjugate against tumor cells. The effects of dose of MeO-PEG/ester/DXR conjugate **1**, MeO-PEG/amide/DXR conjugate **2** and MeO-PEG/Schiff's base/DXR conjugate **3** on the cytotoxic activity against p388D$_1$ *lymphocytic leukemia* cells *in vitro* are shown in Figure 4 (*11*). Although the MeO-PEG/Schiff's base/DXR conjugate **3** did not show the stronger cytotoxic activity than the free DXR, the MeO-PEG/Schiff's base/DXR conjugate **3** exhibited the stronger cytotoxic activity than the PEG/amide/DXR conjugate **2**. Since the MeO-PEG/Schiff's base/DXR conjugate **3** could be presumed to be uptaken into lysosome in tumor cells *via* endocytosis (*17*) and to be hydrolyzed easily in the acid environment of lysosome to release free DXR (*8*), the MeO-PEG/Schiff's base/DXR conjugate **3** might show the relatively strong cytotoxic activity. Although the MeO-PEG/amide/DXR conjugate **2** was also presumed to be uptaken into lysosome through the same route, the cytotoxic activity of MeO-PEG/amide/DXR conjugate **2** might be weaker than MeO-PEG/Schiff's base/DXR conjugate **3** because of the too slow release rate of DXR from the conjugate **2** in lysosome. Namely, it was suggested the release rate of DXR from the conjugate after uptake into tumor cells influenced directly the cytotoxic activity. Thus, the Schiff's base was confirmed to be available for a lysosomally digestible linkage between polymer and drug in the macromolecular prodrug. Accordingly, MeO-PEG/Schiff's base/DXR conjugate **3** can be expected to show the good therapeutical results against animals bearing tumor cells.

On the other hand, the evaluation results of galactose receptor-mediated cytotoxic activity by the Lactose/PEG/amide/DXR conjugate **4** tested by Method 1 and Method 2 are shown in Figures 5 and 6. respectively. The Lactose/PEG/amide/DXR conjugate **4** was found to show stronger cell-specific cytotoxic activity against HLE *human hepatoma* cells, compared with the MeO-PEG/amide/DXR conjugate **2** (Figure 5). Moreover, the cytotoxic activity of conjugate **4** against HLE *human hepatoma* cells was confirmed to be inhibited by the addition of galactose (Figure 6). These results suggested that the introduction of lactose residue into the conjugate could accelerate the galactose receptor mediated uptake of the conjugate *via* endocytosis into HLE cells and derive the specific cytotoxic activity against *hepatoma* cells.

Overcome of resistance by PEG/DXR conjugation technique. The results of cytotoxic activity against DXR-resistant p388D$_1$ lymphocytic *leukemia* cells are shown in Table II. The MeO-PEG/Schiff's base/DXR conjugate **3** was found to exhibit 22 times as strong cytotoxicity as free DXR against DXR-resistant p388 *leukemia* cells. Many P-glycoproteins are known to appear in the surface of cells which gain the resistance to some drug. It has been reported to be the main cause of gain of drug resistance that these P-glycoproteins recognize and expel the drug. The parent drug released lysosomally from a macromolecular prodrug uptaken into the cell *via* endocytosis is liberated in a central region of the cell, while the low molecular weight free drug uptaken into the cell *via* diffusion is located near the inside of cell membrane. So, the probability of expulsion of drug from the cell for the macromolecular prodrug should be smaller than that for the free drug. The appearance of significant strong cytotoxic activity of the MeO-PEG/Schiff's base/DXR conjugate **3** against DXR-resistant p388D$_1$ *leukemia* cells can be explained by such a reason.

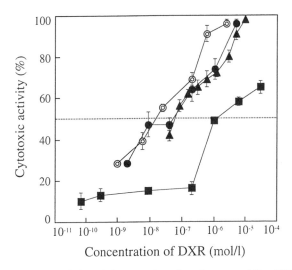

Figure 4. Cytotoxic activity of three modes of conjugate and free DXR against p388D₁ *lymphocytic leukemia* cells *in vitro*. (▲) conjugate **1**; (■) conjugate **2**; (●) conjugate **3**; (◎) free DXR. (Reproduced with permission from ref. 11. Copyright 1995 Technomic Publishing Co., Inc.)

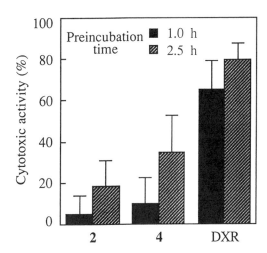

Figure 5. Cytotoxic activity of MeO-PEG/amide/DXR conjugate **2**, Lactose/PEG/amide/DXR conjugate **4** and free DXR against HLE *human hepatoma* cells *in vitro*.

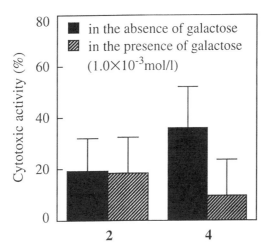

Figure 6. Effect of the addition of galactose on cytotoxic activity of MeO-PEG/amide/DXR conjugate **2** and Lactose/PEG/amide/DXR conjugate **4** against HLE *human hepatoma* cells *in vitro*.

Table II. Cytotoxic activity of MeO-PEG/Schiff's base/DXR conjugate **3** and free DXR against common p388D$_1$ *leukemia* and DXR-resistant p388 D$_1$ *leukemia* cells

Sample	IC$_{50}$ against common p388D$_1$		IC$_{50}$ against DXR-resistant p388D$_1$	
	IC$_{50}$ (μg/ml)	IC$_{50}$/DXR molecule (mol/l)	IC$_{50}$ (μg/ml)	IC$_{50}$/DXR molecule (mol/l)
Conjugate **3**	3.8×10^{-2}	2.4×10^{-8}	1.2×10^{-1}	7.6×10^{-8}
Free DXR	1.4×10^{-2}	2.4×10^{-8}	1.0	1.7×10^{-6}

Acknowledgements

The authors wish to express their sincere appreciations to Meiji Seika Co. Ltd. for providing DXR•HCl and 14Br-DXR. They express thanks to Toho Chemical Industry Co. Ltd. and Kawaken Fine Chemical Co. Ltd. for providing MeO-PEG and MeO-PEG acid, PEG diacid, respectively.

This work was supported by a Grant-in-Aid for Scientific Research from the Ministry of Education, Science and Culture, Japan.

References

1. Ouchi, T.; Hagihara, Y.; Takahashi, K.; Takano, Y.; Igarashi, I. *Drug Design and Discovery* **1992**, 9, 93.
2. Caliceti, P.; Monfardini, C.; Sartore, L.; Schiavon, O.; Baccichetti, F.; Carlassare, F.; Veronese, F. M. *Farmaco.* **1993**, 48, 919.
3. Nathhan, A.; Zalipsky, S.; Kohn, J. *J. Bioact. Compat. Polym.* **1994**, 9, 239.
4. Ohya, Y.; Hirai, K.; Ouchi, T. *Makromol. Chem.* **1992**, 193, 1881.
5. Ouchi, T.; Fujino, A.; Tanaka, K.; Banba, T. *J. Control. Rel.* **1990**, 12, 143.
6. Ouchi, T.; Yuyama, H.; Vogl, O. *J. Macromol. Sci.-Chem.* **1987**, A24, 1101.
7. Duncan, R.; Cable, H. C.; Lloyd, J. B.; Rejmanova, P.; Kopecek, J. *Makromol. Chem.* **1983**, 184, 1997.
8. Galloway, C. J.; Dean, G. E.; Marsh, M.; Rundnick, G.; Mellman, I. *Proc. Acad. Sci. USA.* **1983**, 80, 3334.
9. Yokoyama, M.; Miyauchi, M.; Yamada, N.; Okano, T.; Sakurai, Y.; Kataoka, K.; Inoue, S. *Can. Res.* **1990**, 50, 1693.
10. Nukui, M.; Hoes, K.; Berg, H.; Feijen, J. *Makromol. Chem.* **1991**, 192, 2925.
11. Ohya, Y.; Kuroda, H.; Hirai, K.; Ouchi, T. *J. Bioact. Compat. Polym.* **1995**, 10, 51.
12. Harris, J. M.; Struck, E. C.; Case, M.G.; Paley, M. S. *J. Polym. Sci.: Polym. Chem. Ed.* **1984**, 22, 341.
13. Hirano, T.; Ohashi, S.; Morimoto, S.; Tsuda, K.; Kobayashi, T.; Tsukagoshi, S. *Makromol. Chem.* **1986**, 187, 2815.
14. Duncan, R.; Seymour, L. W.; O'Hare, K. B.; Flanagan, P. A.; Wedge, S. J. *Control. Rel.* **1992**, 19, 331.
15. Haneke, A. C.; Crawford, J.; Aszalos, A. *J. Pharm. Sci.* **1981**, 70, 1112.
16. Masuike, T.; Odake, J.; Takemoto, Y. *Yakugaku Zasshi* **1984**, 104, 614.
17. Silverstein, S. C.; Steinman, R. M.; Cohn, Z. A. *Ann. Rev. Biochem.* **1977**, 46, 667.

Chapter 20

Poly(ethylene glycol)-Grafted Polymers as Drug Carriers

E. H. Schacht and K. Hoste

Polymers Materials Research Group, Institute for Biomedical Technologies (IBITECH), University of Ghent, Krijgslaan 281, S4 bis, Ghent, Belgium

This paper describes the synthesis and evaluation of polyethylene glycol modified dextran and poly[N-(2-hydroxyethyl)-L-glutamine] (PHEG). The graft copolymers show aggregate formation in the liquid and solid state. In vitro and in vivo biological evaluation revealed that the PEG-modified polymers are interesting as potential drug carriers.

Chemotherapy often has limited success because of a lack in cell selectivity for the conventional dosage forms. Undesirable interaction with non-target cells results in unwanted side effects. This phenomenon is severely hampering cancer chemotherapy.

Over the past two decades extensive research has been devoted to the design of advanced drug delivery systems that can deliver the active agent to the preferred site of action. Among the various concepts that have been proposed to achieve a more efficient drug delivery is the macromolecular prodrug approach. In this concept, a drug is linked onto a polymeric carrier. This carrier can be inert or biodegradable. The polymer can be designed to contain structural elements that can modify the water/lipid solubility. In addition, interaction with target cells can be promoted by introducing so called targeting groups onto the polymer backbone*(1-2)*.

An additional variable is the linkage inbetween the carrier and the drug moiety. Spacer groups can be introduced aiming to provide site selective drug release. Oligopeptides which are a good substrate for target associated enzymes are attractive spacer candidates*(3)*. This polymeric prodrug concept was first presented in a comprehensive model by Ringsdorf in 1975*(4)*.

In designing proper macromolecular prodrugs the solubilizing component is an important contributor. Since a lot of drugs and peptidic spacer groups are hydrophobic, attachment of such groups onto hydrophilic polymers seriously reduces the water solubility and limits their parental applicability.

Hydrophobic side groups tend to aggregate and eventually cause precipitation. Water solubility of the conjugates can be enhanced by introducing hydrophilic solubilizers which can prevent aggregation or prevent aggregates from precipitation. A suitable candidate for promoting polymer-drug conjugate solubility is polyethylene glycol (PEG). PEG is known to be a non-toxic and non-immunogenic water-soluble polymer.

Abuchowski and co-workers demonstrated that substitution of proteins with PEG makes the proteins less immunogenic and more stable*(5-7)*. At present, PEG-enzyme conjugates are accepted for clinical application*(8)*. Considering the substantial amount of biological data, PEG is an attractive polymer for modification of biologically active polymers. In the macromolecular prodrug approach PEG-ylated polymers, block copolymers or graft copolymers, are attractive carriers.

PEG-containing Block Copolymers

A first example of a PEG-containing block copolymer used as drug carrier was reported by Ringsdorf et al *(9)*. Amino terminated PEG was used as initiator for the ring opening polymerization of the N-carboxyanhydride (NCA, Leuch's anhydride) of L-lysine. In the block copolymer the ε-amino groups were partially acylated with palmitoyl groups in order to promote aggregation. Sudan Red 7B solubilisation studies revealed that the block copolymers formed micelles with PEG as a hydrophilic shell and palmitoyl substituted poly(L-lysine) (PLL) as the inner core. Differential Scanning Calorimetry (DSC) with liposomes indicated that the block copolymer interacts with and probably penetrates lipid membranes.

Furthermore, the PEG-PLL block copolymers were substituted with sulfidoderivatives of cyclophosphamide (CP), an alkylating antitumor agent*(10-11)*. Normally, such cyclophosphamide derivatives are hydrolyzed rapidly to give the active metabolite 4-hydroxycyclophosphamide. With the PEG-PLL derivatives however, in vitro studies indicated that the cyclophosphamide-block copolymer acts as an intracellular depot for the active metabolite of cyclophosphamide. Cellular uptake of the block copolymer occurs prior to sustained release of the active drug*(12)*.

Another block copolymer, containing PEG, which has been used as a drug carrier is described elsewhere in this book by Kataoka. Kataoka et al. synthesized block copolymers of PEG and poly(aspartic acid) and used them as carriers for the hydrophobic anti-cancer drug adriamycine*(13)*. In vitro and in vivo micelle formation of PEG-P(Asp(ADR)) was confirmed by different techniques*(14)*. Again, PEG formed the hydrophilic shell of the micellar structure. Furthermore, it was shown that free adriamycine could be physically trapped into the PEG-P(Asp(ADR)) micelle*(15)*. It was demonstrated that micelles formed by block copolymers can act as an efficient reservoir for free drugs.

The above described examples illustrate very well that PEG-containing copolymers can form micellar structures which are capable of carrying hydrophobic drugs. The drug might be covalently linked to the carrier or can be entrapped physically into the micelles. Both cases show the same benefit : introducing PEG results in improved solubilisation of hydrophobic drugs.

In addition to PEG containing block copolymers, PEG-grafted copolymers are an interesting alternative as drug carrier. Examples are described in the next section.

PEG-containing Graft Copolymers

A large number of synthetic polymers have been proposed as carriers for preparing macromolecular prodrugs*(16-23)*. Among them are the polysaccharide dextran (Figure 1) and the poly(α-amino acid)derivative, poly-[N-(2-hydroxyethyl)-L-glutamine] (PHEG) (Figure 2).

We have selected both polymers for the preparation of macromolecular derivatives of cytotoxic and anti-bacterial agents*(24-26)*. It was observed that the maximal acceptable drug content in the conjugate was limited by a lack of water solubility. In order to improve the solubilizing ability of the carriers, polyethylene glycol was grafted onto the polymer backbone.

Synthesis. Dellacherie reported before the preparation of PEG-grafted dextran by reaction of α-methyl-ω-amino-polyethylene glycol with epichlorohydrin activated dextran (Figure 3)*(27)*.

Gel permeation chromatography (GPC) indicated the presence of a high molecular weight polydisperse polymer. A drawback of this preparation method is the formation of less defined intermediates during epichlorohydrin activation . Moreover, the alkaline conditions (pH 10-11) required for coupling can induce polysaccharide depolymerization.

A more elegant method for synthesizing PEG-substituted dextran was developed in our research group*(28)*. Dextran-PEG was easily obtained by reaction of 4-nitrophenyl chloroformate activated dextran with an equivalent amount of α-methyl-ω-amino-polyethylene glycol (PEG-NH$_2$) (Figure 4).

PEG-amine was prepared by quantitative conversion of the PEG-hydroxyl group into the tosylate and subsequent amination with aqueous ammonia (Figure 5)*(29)*.

In order to remove non-reacted PEG-NH$_2$, the reaction mixture was passed over a strong acid ion exchange resin. The reaction product was finally isolated by preparative (GPC). Direct isolation via preparative GPC was not feasible since the PEG-modified polymers tend to solvate free PEG-NH$_2$. This was clearly demonstrated by analytical GPC analysis of a mixture PEG-dextran and FITC-labeled PEG(Figure 6). In analytical GPC, PEG-FITC appears as a single peak that can be detected by UV-detection at 492 nm. At this detection wavelength no signal is observed for PEG-dextran. However, analysis of a mixture of dextran-PEGand PEG-FITC gives one broad UV-sensitive signal corresponding with the elution peak observed by refractive index detection of dextran-PEG alone.

Phase separation of dextran and PEG on a molecular level in a dextran-PEG conjugate can result in the formation of a PEG-core in which free PEG-NH$_2$ can be trapped. This implies that conjugates isolated by precipitation or by preparative GPC may be contaminated. In such case NMR analysis can give erratic data concerning conjugate composition.

The method described above allows the preparation of PEG-grafted dextrans with well controlled degree of substitution and different PEG lengths.

In a similar way, a series of PEG-grafted PHEG-derivatives were prepared*(30)*.

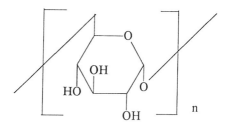

Figure 1. Dextran

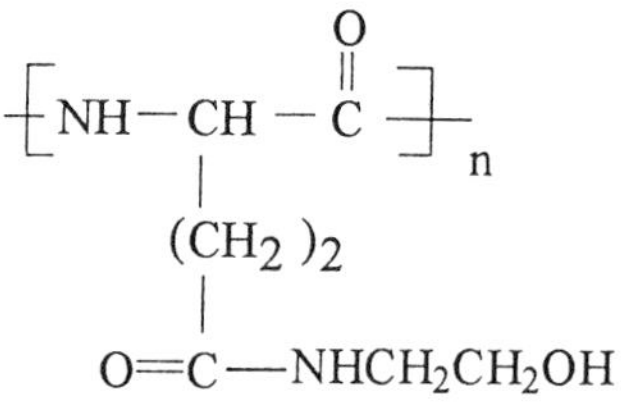

$$\left[\!\!\left[NH\!-\!CH\!-\!\overset{\displaystyle \overset{O}{\|}}{C} \right]\!\!\right]_n$$
$$\underset{\displaystyle \underset{\displaystyle O=C-NHCH_2CH_2OH}{|}}{(CH_2)_2}$$

Figure 2. Poly-[N-(2-hydroxyethyl)-L-glutamine]

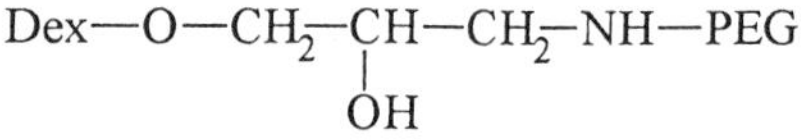

$$Dex\!-\!O\!-\!CH_2\!-\!\underset{\displaystyle \underset{OH}{|}}{CH}\!-\!CH_2\!-\!NH\!-\!PEG$$

Figure 3. Dextran-PEG prepared by epichlorohydrin activated dextran

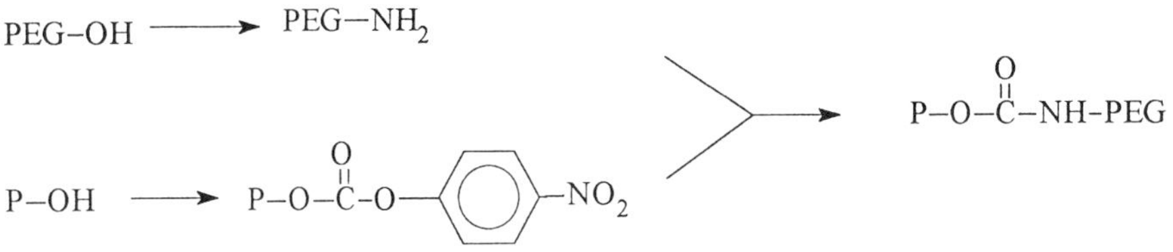

Figure 4. Synthesis of PEG-grafted dextran by 4-nitrophenyl chloroform activated dextran

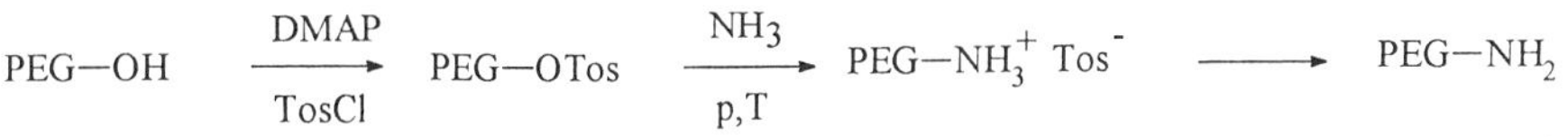

Figure 5. Synthesis of α-methyl-ω-amino-polyethylene glycol

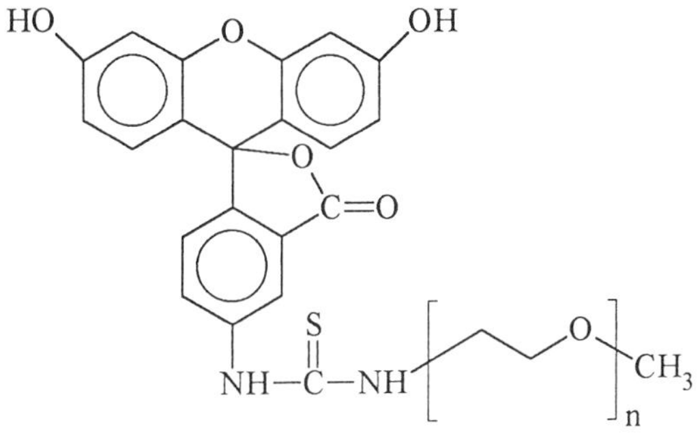

Figure 6. FITC-labeled PEG

Physico-chemical Properties of dextran-PEG and PHEG-PEG. Properties of the graft copolymers were investigated by demixing experiments, DSC analysis, analytical GPC and determination of the 'Critical Aggregation Concentration'.

Emulsifying Properties. During isolation of the dextran-PEGs by precipitation it was observed that the PEG-grafts significantly change the solubility properties of the polymer.

PEG, although water soluble is also soluble in many organic solvents. Dextran and PHEG however display good water solubility and dissolve in some aprotic polar solvents like dimethyl sulfoxide and N,N'-dimethyl formamide. Therefore, it is anticipated that the conjugates may act as tensioactives. In a water/organic two phase system they will be positioned at the interface and be able to act as an emulsifier. The emulsifying ability of dextran-PEG and PHEG-PEG conjugates was evaluated for a water/octanol biphasic system. Demixing times are given in Table I.

Table I : Demixing times of the PEG-grafted polymers in water/n-octanol.

Product	Mw carrier	DS (%)	Concentration water phase (mg/ml)	Demixing time (min)
Dex-PEG(750)	31210	5	1.05	1
Dex-PEG(750)	31210	28	1.10	5
Dex-PEG(2000)	31210	4	1.10	8
Dex-PEG(2000)	31210	9.5	1.20	30
Dex-PEG(5000)	31210	3	1.10	>720
PHEG-PEG(750)	27100	11	1.20	20
PHEG-PEG(2000)	27100	6	1.00	25
PHEG-PEG(5000)	27100	1.6	1.00	36 uur
PHEG-PEG(5000)	27100	6	1.00	>48 uur
Dextraan	31210		1.15	0.75
PHEG	27100		1.00	2
Dex/PEG(2000)			0.43/0.60	3
Dex/PEG(5000)			0.35/0.75	6

SOURCE : adapted from ref. 29 and 31.

As indicated in table I a small amount of conjugate allows to obtain good emulsions. The emulsifying capacity obviously depends on the length of the PEG-substituents and the degree of substitution. This data further support the above suggested hypothesis that the conjugates may form colloidal solutions.

DSC Analysis of the Conjugates. PEG of molecular weight below 2000 is a viscous liquid at room temperature whereas polymers of molecular weight 2000 and higher are semicrystalline materials with a melting point in the range 30°C to 60°C depending on the molecular weight. If phase separation occurs in the solid state, the

This was confirmed by DSC analysis of solid samples of the conjugates. The area under the melt endotherm per unit sample weight is a measure for the degree of crystallinity. Results are given in Table II.

Table II : DSC results of PEG-grafted polymers

Product	Mw carrier	DS (%)	Heat of melting of the conjugate (J/g)	Heat of melting of PEG (J/g)
Dex-PEG(5000)	31210	3.5	44.87	89.13
Dex-PEG(5000)	31210	8	80.99	114.07
Dex-PEG(5000)	31210	12.5	100.68	128.75
PHEG-PEG(2000)	27100	6	28.14	68.80
PHEG-PEG(2000)	27100	11	44.85	82.29
PHEG-PEG(5000)	27100	1.6	18.74	59.12
PHEG-PEG(5000)	27100	6	80.23	126.75
Dextran	31210		/	/
PHEG	27100		/	/
PEG(2000)				187.83
PEG(5000)				189.5

SOURCE : Adapted from ref. 29 and 30.

As expected, conjugates with short PEG side groups (MW = 750) do not crystallize. Polymers with PEG length 2000 or higher give a melt endotherm. The heat of melting increases with increasing length of the PEG side group and with increasing degree of substitution. The data obtained for the PHEG-PEG$_{2000}$ conjugates are in contrast with the results obtained for the dextran-PEG$_{2000}$ conjugates where no melting endotherm was observed. It is known that dextran is a rigid polymer(31). The above mentioned results suggest that PHEG has a more flexible chain, so that phase separation may occur, resulting in the formation of crystalline PEG domains.

Analytical GPC Analysis. Further evidence for the occurence of association phenomena in PEG-ylated polymers was obtained by aqueous gel permeation chroma-tography. The molecular weight was determined using dextran standards for calibration. Weight averaged and number averaged molecular weight data for a series of PEG grafted polymers is given in Table III. The same table also contains the expected molecular weight calculated on the basis of the molecular weights of the backbone polymer and the PEG grafts and the degree of substitution.

From this table it is clear that the molecular weight of the conjugates with PEG$_{750}$-side groups agrees well with the calculated value except for high degrees of substitution where lower values are noticed. For the conjugates with a high degree of substitution of PEG$_{2000}$ the apparent molecular weight is higher than calculated. The conjugates even eluted in the void volume of the column.

Conjugates with PEG_{5000}-side groups show very low apparent polecular weights for a low degree of substitution. At high degree of substitution a higher molecular weight is observed. Again the conjugates eluted in the void volume of the column.

This indicates that for a low degree of substitution, conjugates with PEG_{5000} form intramolecular aggregates which have a small hydrodynamic volume, whereas a high degree of substitution gives rise to intermolecular aggregates with a large hydrodynamic volume. This causes the conjugates to elute in the void volume of the column. Similar results were obtained for the PHEG-PEG-conjugates.

Table III : Molecular weights determined by analytical GPC.

Product	Degree of substitution (%)	M_w	M_n	Calculated molecular weight
Dex-PEG(750)	5	39 550	22 620	38 760
Dex-PEG(750)	10	46 470	17 590	45 510
Dex-PEG(750)	20	43 140	14 200	59 010
Dex-PEG(750)	28	38 380	14 020	71 760
Dex-PEG(2000)	3	51 330	28 830	43 300
Dex-PEG(2000)	9	*	*	65 260
Dex-PEG(2000)	20	*	*	109 260
Dex-PEG(5000)	3	12 650	3 930	56 260
Dex-PEG(5000)	9	*	*	184 320
Dex-PEG(5000)	20	*	*	209 320

* = elutes in the void volume of the column (the exclusion limit is 200 000)

Determinaton of the "Critical Aggregation Concentration" (CAC).   As discussed before GPC analysis revealed that free $PEG-NH_2$ can be trapped in PEG-dextran aggregates.Since dextran and PHEG are water soluble, but PEG is soluble in water and organic solvent, one can anticipate that, in aqueous solution, dextran-PEG and PEG-PHEG act as tensioactive and form aggregates with the "more" hydrophilic polymeric carrier as the outer shell and the "less" hydrophilic PEG as the core.

Tensioactive molecules tend to organize at the water-air interface and decrease the surface tension of the water. At a certain concentration, when the surface is saturated, they form aggregates (micelles). From that concentration on the surface tension remains constant.

For the PEG-substituted polymers, aggregation is likely to occur as well. The "Critical Aggregation Concentration" (CAC) was determined by measuring the surface tension of the solution as a function of the PEG-polymer concentration using a Cahn Dynamic Contact Angle Analyzer. An example is given in Figure 7.

For all polymers included in this study the CAC's ranged between 1 and 3.5 mg/l. These concentrations are very low but indicate that even at very low concentrations stable aggregates can be formed.

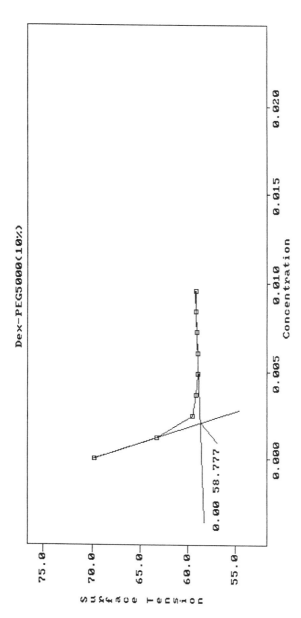

Figure 7. Surface tension of a solution of Dex-PEG$_{5000}$(10%) as a function of the concentration

Biological Evaluation of the PEG-grafted Polymers. In vitro and in vivo biological experiments were performed. The influence of PEG-substitution on the biodegra-dability of PHEG was investigated in vitro. In vivo experiments consisted of blood clearance and body distribution studies.

Biodegradation Studies for PEG-PHEG Conjugates in Presence of lysosomal Enzymes. The degradation of PHEG in presence of enzymes and enzyme mixtures has been reported by several authors*(32-37)*. Pytela and coworkers have evidenced the in vitro degradation of PHEG derivatives by Cathepsin B*(35)*. Moreover, Vercauteren described the degradation of PHEG in presence of isolated rat lysosomal enzymes *(38)*. Both authors observed random degradation of the poly-aminoacid using GPC analysis. This method is particularly advantageous for the evaluation of the endopeptidase activity, since cleavage in one macromolecule reduces the average M_n significantly.

It can be anticipated that chemical modification of PHEG may alter the susceptibility towards degrading enzymes. Therefore, the effect of PEG grafting on the enzymatic degradabilityof PHEG, was investigated.

In this study, the homopolymer and the graft copolymer were incubated in presence of rat liver lysosomal enzymes (Tritosomes) at 37°C. At regular time intervals samples of the incubation medium were analyzed by analytical GPC using a Viscometer-Refractometer detection system (type Viscotek 250 Dual Detector). An overlay of PHEG chromatograms at different incubation times (Figure 8) (Reported with permission from ref. 31 Copyright 1996) demonstrates a shift of the peaks towards the low molecular weight range with increasing degradation time.The observed M_n values at different incubation times are summarized in Table IV.

Table IV : Change in M_n values of PHEG and PHEG-PEG conjugates during incubation in presence of rat lysosomal enzymes.

Product	Degree of Substitution	t = 0 hours	t = 3 hours	t = 5 hours	t = 8 hours
PHEG		27100	7860	2860	2700
PHEG-PEG(750)	6 %	26850	8020	5950	4050
PHEG-PEG(750)	11 %	42550	18400	10950	6650
PHEG-PEG(2000)	6 %	41640	18280	11080	6020
PHEG-PEG(2000)	11 %	61400	22500	10280	6500
PHEG-PEG(5000)	1.6 %	46000	12300	8400	5930
PHEG-PEG(5000)	6 %	79000	30390	15900	7400

SOURCE : Reprinted with permission from ref. 31 Copyright 1996

These data suggest that substitution of PHEG with PEG chains only slightly decreases the enzymatic degradability of the backbone. Increasing the molecular weight or the degree of substitution decreases the initial rate of degradation.

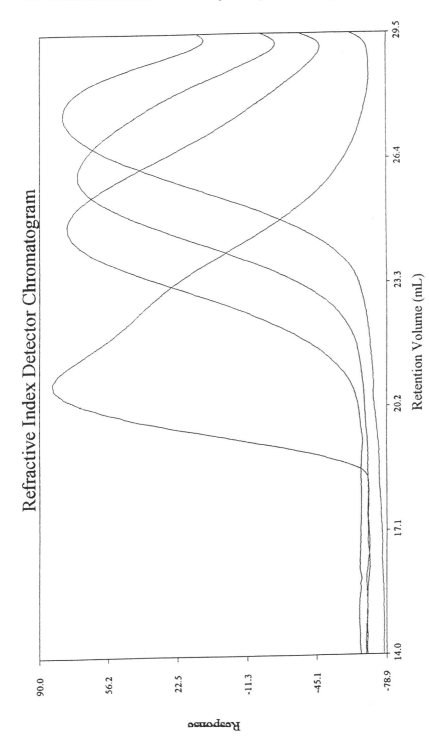

Figure 8. Overlay of PHEG chromatograms at different incubation times

However, 8 hours after incubation only a minimal difference in M_n-value could be observed for the different copolymers. It is clear that all PHEG-PEG conjugates included in this study are subject to lysosomal degradation. In absence of lysosomal enzymes no degradation was observed (GPC data not shown).

In Vivo Evaluation of the Graft Copolymers. Since the PEG-substituted polymers are intended to be used as drug carriers, it is important to study their body distribution and blood clearance. For in vivo detection, polymer derivatives were radiolabeled by substituting the polymer with tyrosinamide (Figure 9).

Blood clearance. A solution of the polymer was injected intravenously in female BALB/c mice. At regular time intervals blood samples were taken and assayed for radioactivity. The amount of radioactivity in the blood after two minutes was taken as 100%. Results for the PHEG conjugates shown in Figure 10, indicate that the PHEG-PEG conjugates are cleared from the blood more slowly than unmodified PHEG. After 3 hours, there is still 50 % in the blood, whereas for unmodified PHEG the concentration is already reduced to 25 % of the initial value.

The results for PEG-grafted dextrans are shown in Figure 11. Dex-PEG$_{750}$ (19 %) shows the same behaviour as the PHEG-PEG conjugates. After 3 hours there is still 34 % left in the blood. However, Dex-PEG$_{5000}$ (DS = 3 %) behaves quite differently. It is cleared much faster than dextran. As will be discussed later this conjugate is also rapidly excreted from the body. These results can be explained by the small hydrodynamic volume observed during GPC analysis of Dex-PEG$_{5000}$ (M_W = 20730, M_n = 18570). Dependent on the kind polymer, the renal excretion limit varies between 40000 for dextran and 60000 for PHEG.

The molecular size of Dex-PEG$_{5000}$ (3 %) is apparently below the renal barrier. This explains its rapid blood clearance and excretion. Blood clearance studies show that PEG-substituted polymers with a hydrodynamic volume above the renal excretion limit, circulate longer in the blood than the parent carriers. This feature makes them promising as new drug carriers.

Body distribution. A solution of the PEG-ylated polymer was injected intravenously. At regular times the animals were killed and dissected. Liver, kidneys and spleen were removed and urine was collected. The organs were homogenized and the radioactivity was measured.

Results obtained for the PHEG-conjugates are shown in Figure 12 (a,b,c). It is clear that for all PHEG conjugates spleen uptake is negligable. The amounts found in liver and kidneys are also very low. It should be noted that liver was not perfused. Hence, the radioactivity found in liver is mainly due to the blood retained within the organ. It is clear that PHEG is excreted much faster than the PHEG-PEG conjugates. After 24 hours, 85 % of PHEG is excreted, whereas for PHEG-PEG$_{2000}$ and PHEG-PEG$_{5000}$ this is 50 %, respectively 5 %. What is not excreted remains mainly in the blood.

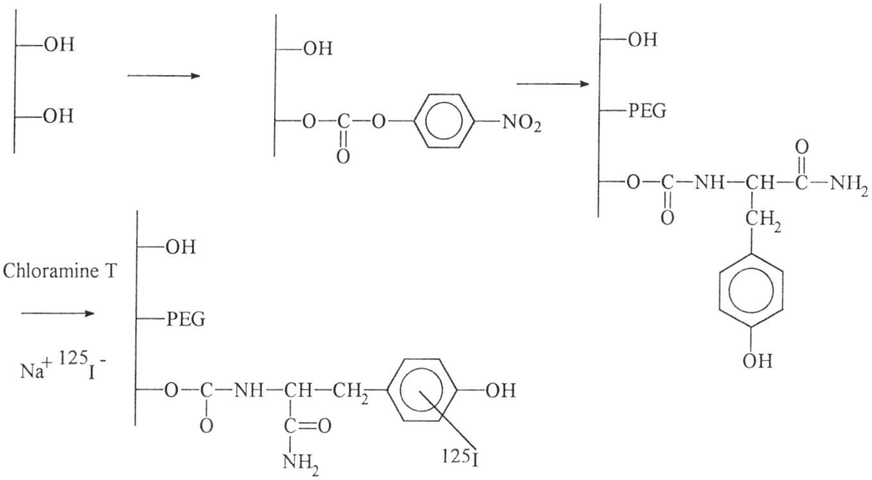

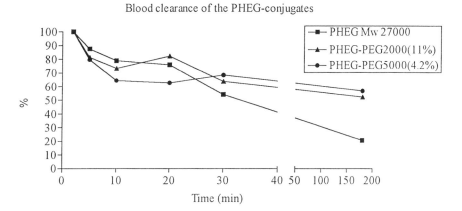

Figure 9. Substitution of the polymer with tyrosinamide

Blood clearance of the PHEG-conjugates

Figure 10. Blood clearance of the PHEG-PEG-conjugates

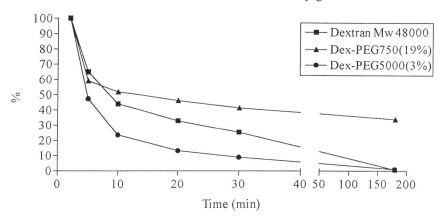

Figure 11. Blood clearance of the dex-PEG-conjugates

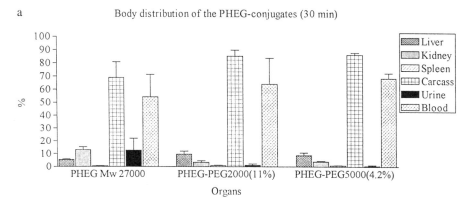

Figure 12 (a,b,c). Body distribution of the PHEG-PEG-conjugates

b

Body distribution of the PHEG-conjugates (3 h)

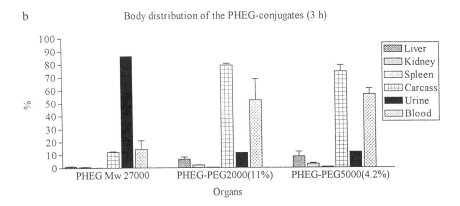

c

Body distribution of the PHEG-conjugates (24 h)

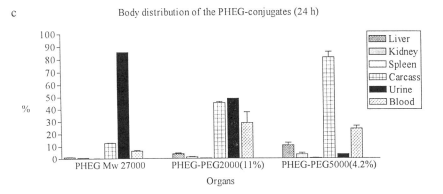

Figure 12. *Continued.*

A body distribution study for dextran and PEG-ylated dextrans (Figure13 a,b,c) reveals that dextran is captured in large amounts by the liver. Earlier studies have shown that dextran is recognized by liver recceptors *(39)*. In contrast, liver uptake of PEG-grafted dextrans is significantly lower. The liver uptake after 30 minutes is 10 % or 15 % for dextran-PEG$_{750}$, respectively dextran-PEG$_{5000}$. After 24 hrs, liver content was very low. This indicates that the PEG-ylated dextrans are not internalized by liver cells. The measured organ content is due to blood contained within the isolated organ.

Unmodified dextran (MW 48,000) is not rapidly excreted from the body, since a large fraction can be found in liver even after 24 hrs. dextran-PEG$_{750}$ is excreted faster than dextran. On the other hand dextran-PEG$_{5000}$ is rapidly blood cleared and excreted. As discussed before, the latter is due to intramolecular association.

a

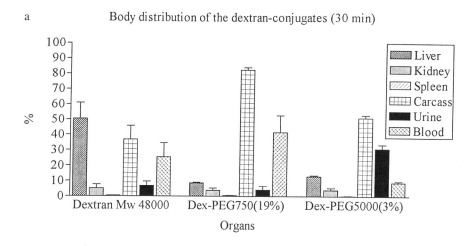

b

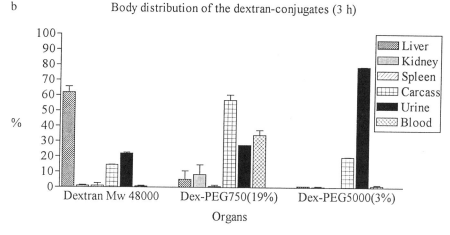

Figure 13 (a,b,c). Body distribution of the dex-PEG-conjugates

c   Body distribution of the dextran-conjugates (24 h)

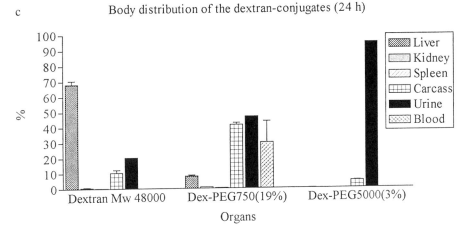

Figure 13. *Continued.*

Conclusion.

In vitro and in vivo biological studies have demonstrated the potential of PEG-grafted dextran and PHEG as drug carriers.

Introducing PEG side groups results in a reduced liver uptake. Blood clearance is remarkably slower except for conjugates which, due to intra molecular association, have size below renal threshold value.

For the range of substitution and PEG length used in this study, PEG-grafted PHEG is still susceptible to lysosomal degradation. Hence, if during extended blood circulation polymer may be captured by cells, it could still be processed in the lysosomal compartment.

Acknowledgements.

This work was financially supported by the IWT (Vlaams Instituut voor de Bevordering van het Wetenschappelijk-Technologisch onderzoek in de Industrie) and the Belgian 'Dienst voor Programmatie van het Wetenschapsbeleid', research grant PAI III, No. 40.

Literature Cited.

(1) Duncan, R.; Seymour, L. W.; Scarlett, L.; Lloyd, J. B.; Rejmanova, P.; Kopecek, J. *Biochem. Biophys. Acta* **1986**, *886*,62.
(2) Tsukada, Y.; Okkawa, K.; Hibi, N. *Cancer Res.* **1987**, *47*, 4293.
(3) Subr, V.; Kopecek, J.; Pohl, J.; Bandys, M.; Kostka, V. *J. Contr. Rel.* **1988**, *8*, 133.
(4) Ringsdorf, H. *J. Polymer Sci. : Symposium* **1975**, *51*, 135.

(5) Abuchowski, A.; van Es, T.; Palczuk N. C.; Davis, E. F. *J. Biol. Chem.* **1977**, *252*, 3582.

(6) Abuchowski, A.; Kazo, G. M.; Verhoest, C.R.Jr.; van Es, T.; Kafkewitz, D.; Vian, A. T.; Davis, F. F. *Cancer Bio. Chem. Biophys.* **1984**, *7*,175.

(7) Abuchowski, A.; Mc Coy, J. R.; Palczuk, N. C.; van Es, T.; Davis, F. F. *J. Biol. Chem.* **1977**, *252*, 3582.

(8) Fuertges, F.; Abuchowski, A. *J. Contr. Rel.* **1990**, *11*, 1990.

(9) Pratten, M. K.; Lloyd, J. B.; Hörpel, G.; Ringsdorf, H. *Makromol. Chem.* **1985**, *186*, 725.

(10) Peter, G.; Wagner, T.; Hohorst, H. J. *Cancer Treat. Rep.* **1976**, *60*, 429.

(11) Hirano, T.; Klesse, W.; Ringsdorf, H. *Makromol. Chem.* **1979**, *180*, 1125.

(12) Bader, H.; Ringsdorf, H.; Schmidt, B. *Angew. Makromol. Chem.* **1984**, *123/124*, 457.

(13) Yokoyama, M.; Miyouch, M.; Yamada, N.; Okano, T.; Sakurai, Y.; Kataoka, K.; Inoue, S. *J. Contr. Rel.* **1990**, *11*, 269.

(14) Yokayama, M. *Crit. Rev. Ther. Drug* **1994**, *9* , 213.

(15) Yokayama, M.; Okano, T.; Sakurai, Y.; Kataoka, K. *J. Contr. Rel.* **1994**, *32*, 269.

(16) Schacht, E. *Ann. Natl. Acad. Sci.* **1985**, *446*, 199.

(17) Kwon, G.; Kataoka, K. *Adv. Drug Deliv. Rev.* **1995**, *16*, 295.

(18) Matsumoto, S.; Yamamoto, A.; Takakura, Y.; Hashida, M.; Sezaki, H. *Cancer Res.* **1986**, *46*, 4463.

(19) Duncan, R.; Kopecek, J.; Lloyd, J. B. In *Polymers in medicine : Biomedical and pharmalogical applications*; Chiellini, E.; Guisti, P., Eds.; Plenum Press : New York, 1983, Vol. 23, pp 97-114.

(20) Maeda, H. *Adv. Drug Deliv. Rev.* **1991**, *6*, 181.

(21) Sela, M.; Katachalski, E. *Adv. Protein Chem.* **1987**, *14*, 391.

(22) Mosigny, M.; Roche, A. C.; Midoux, P.; Mayer, R. *Adv. Drug Deliv. Rev.* **1994**, *14*, 1.

(23) Yokoyama, M.; Okano, T.; Sakurai, Y.; Ekimoto, H.; Shibazaki, C.; Kataoka, K. *Cancer Res.* **1991**, *51*, 3229.

(24) De Marre, A.; Soyez, H.; Schacht, E. *J. Contr. Rel.* **1994**, *32*, 129.

(25) Soyez, H.; Schacht, E.; De Marre, A.; Seymour, L. *Macromol. Symp.* **1996**, *103*, 163.

(26) De Marre, A.; Seymour, L.; Schacht, E.; *J. Contr. Rel.* **1994**, *31*, 89.

(27) Coessens, V.; Schacht, E.; Domurado, D. *J. Contr. Rel.* **1996**, *38*, 141.

(28) Duval, J. M.; Delestre, C.; Carré, M. C.; Hubert, P.; Dellacherie, E. *Carbohyd. Polym.* **1991**, *15* ,233.

(29) Hoste, K.; Bruneel, D.; De Marre, A.; De schrijver, F.; Schacht, E. *Macromol. Chem. Rapid Commun.* **1994**, *15* , 697.

(30) Loccufier, J.; Crommen, J.; Vandorpe, J.; Schacht, E. *Makromol. Chem. Rapid Commun.* **1991**, *12*, 159.

(31) De Marre, A.; Hoste, K.; Bruneel, D.; Schacht, E.; De Schrijver, F. *J. Bioact. Compat. Pol.* **1996**, *11*, 85.

(32) Basedow, A.; Ebert, K. *J. Polym. Sci. Polym. Symp.* **1979**, *66*, 101.

(33) Dickinson, H.; Hiltner, A. *J. Biomat. Res.* **1981**, *15*, 591.

(34) Hayashi, T.; Tabata, Y.; Nakajima, A. *Rep. Prog. Phys. Jap.* **1993**, *26*, 591.

(35) Rypacek, V.; Saudek, V.; Pytela, J.; Skarda, V. *Makromol. Chem. Suppl.* **1985,** *9*, 129.

(36) Pytela, J.; Skarda, V.; Drobnik, J.; Rypacek, F. *J. Contr. Rel.* **1989,** *10*, 17.

(37) Pytela, J.; Kotva, R.; Metalova, M.; Rypacek, F; *Int. J. Biol. Macromol.* **1990,** *12*, 241.

(38) Vercauteren, R., University of Gent, *PhD Thesis*, **1992.**

(39) Vansteenkiste, S., University of Gent, *PhD Thesis*, **1992.**

New Functional Polymers, Surfaces, and Hydrogels

Chapter 21

Hydrazide Derivatives of Poly(ethylene Glycol) and Their Bioconjugates

Samuel Zalipsky[1] and Sunitha Menon-Rudolph[2]

[1]SEQUUS Pharmaceuticals, Inc. 960 Hamilton Court, Menlo Park, CA 94025
[2]The R. W. Johnson Pharmaceutical Research Institute, 1000 Route 202, Raritan, NJ 08869

Hydrazide derivatives of poly(ethylene glycol) (PEG-Hz) have a number of attributes making them useful for preparation of conjugates, particularly of polypeptides and glycoproteins. They form conjugates in mildly acidic aqueous solutions via two modes of reactivity. The first one involves hydrazone formation with reactive carbonyls generated on the substrate molecule by several different methods. These include oxidation of oligosaccharide residues of glycoproteins, glyoxylate / Cu^{2+} -mediated transamination of the N-terminal residue of polypeptides, periodate oxidation of N-terminal Ser or Thr residues. The second mode involves coupling with carbodiimide-activated carboxyl groups forming diacylhydrazide linkages with PEG. Synthesis of PEG-Hz is straightforward by hydrazinolysis of esters of either carboxymethylated PEG or urethane-linked amino acid. Having an unusual amino acid, e.g. β-Ala, as part of the linker offers a convenient way for composition determination of protein conjugates, particularly those containing multiple chains of mPEG-O(C=O)-β-Ala-Hz, by amino acid analysis. Our work involving PEG-Hz conjugation, including examples of preparation of N-terminally modified polypeptides, oligosaccharide-linked glycoproteins, polypeptides modified on their carboxyl groups, and immunoconjugates of enzymes and liposomes is discussed in this review.

There are a number of reasons why polymers are conjugated to biologically-relevant molecules. In the most general sense, polymers are very effective modifiers of physicochemical properties such as molecular size, solubility, permeability, diffusion rate. To achieve significant alteration of properties, macromolecular substrates , e.g. proteins, are often modified by multiple attachments of polymer chains, while in the case of low molecular weight molecules, one polymer chain of an appropriate length can often suffice. Among the polymers used for covalent modification of biologicals, PEG* is one of the most widely used. This is mainly due to the array of useful properties that PEG possesses. These include a wide range of solubilities in both aqueous and organic solutions, lack of toxicity and immunogenicity, non-biodegradability yet ease of excretion from living organisms. Conveyance of some of these properties onto PEG conjugates is often the principal motivation for their development (*1*). Another reason for extensive use of PEG lies in its availability in the form of well defined molecular weight materials of low dispersity. PEGs are usually available as bis-hydroxyls (HO-PEG-OH) or as monohydroxyls (CH$_3$O-PEG-OH, mPEG-OH). The art of transformation of these functionalities into a wide array of useful reactive groups suitable for bioconjugation reactions was perfected over the last two decades (*2*).

Most of the known methods for preparation of PEG conjugates rely on modification of amino groups of biologically active molecules with PEG-based electrophiles, usually an active ester or a reactive alkylating group (*1,3*). In the case of multifunctional biological molecules, e.g. polypeptides and proteins, other functional residues sometimes are desirable targets for modification with PEG. There is a great deal of interest in the development of site-specific reagents for protein conjugation. Hydrazides are attractive derivatives in relation to the latter two aspects of bioconjugation. The versatility of hydrazide functionality has been long known in the field of bioconjugates [see reviews (*4-6*)]. Hydrazides in general are good nucleophiles, yet relatively weak bases (pK$_a$ ≈ 3). Hence they are prone to many similar reactions that amino groups are known to undergo. Yet under mildly acidic conditions maintained in aqueous buffers, hydrazides are often more reactive than amines. There are two principal modes of reactivity of hydrazides that we were interested in, both of which are highly selective and useful for conjugation with peptides, proteins or glycoproteins (Scheme I). (A) Hydrazides react with carbonyls, which as will be discussed below, can be introduced onto various peptide ligands often in a site-specific manner. This reaction results in relatively stable hydrazone linkages that usually, in contrast to Schiff base analogs, do not require further stabilization by reduction. (B) Carboxyl groups of proteins, which are readily activated with water-soluble carbodiimide (e.g. EDC) at mildly acidic pH, react readily with hydrazides, while the amino groups on the same proteins remain deactivated due to protonation.

During the last decade we have used a number of hydrazide derivatives of PEG that were utilized in various conjugation protocols including protein modifications, preparation of immunoconjugates, and attachment of ligands to surfaces of PEG-grafted liposomes. While some of the methods developed were presented at various meetings

---

* See abbreviations list at the end of the chapter.

and were part of a few patent applications (7-9), only a few complete reports on this subject have been published (10-14). It is the purpose of this paper to summarize and to review our results with hydrazide derivatives of PEG and their bioconjugates with some emphasis on the previously unpublished aspects of this effort.

Results and Discussion

Background. The earlier literature made mention of hydrazide derivatives of PEG (PEG-Hz) only in a few places (15,16). Andresz et al. used a hydrazide derivative of PEG for preparation of mono- and oligo-saccharide conjugates interlinked through hydrazone linkages (15). The synthesis of PEG-Hz was accomplished in high yield by hydrazinolysis of esters of carboxymethylated-PEG. The same process was recently suggested as a convenient approach to PEG-hydrazides (17). The carboxymethyl hydrazides of PEG were mentioned in a few patents (18-20). In these cases, by a reaction with nitrous acid, the hydrazides were converted *in situ* into acyl azides and then used to acylate amino-containing residues of proteins. Reaction of mPEG-tresylate with an excess of adipic dihydrazide (ADH) was also utilized for introduction of hydrazido groups. The resulting product was used for conjugation with periodate oxidized ATP (16). The very low content of ATP in the final conjugate (0.06 mole ATP / mole PEG) indicates that this functionalization and conjugation approach was not efficient. Note that the tresylate chemistry and its use in bioconjugation protocols has been subjected to a serious revision (21,22), which makes the latter approach to PEG-hydrazide even more dubious.

Synthesis of PEG-Hz derivatives. Our approach to hydrazide derivatives utilized PEG-linked amino acid esters as starting materials. These materials are readily attainable by a few different pathways shown in Scheme II. They undergo hydrazinolysis without any significant cleavage of the urethane linkage to yield PEG-amino acid-hydrazides. Direct endcapping of the terminal hydroxyl groups of PEG with commercially available ethyl isocyanato-acetate or -propionate derivatives provide for a convenient approach to PEG-linked amino acid esters. (10,23,24). Alternatively PEG-chloroformate, easily generated by treatment of the polymer with phosgene, reacts with amino acid esters in high yield (7). Succinimidyl carbonate (25) or other active carbonates of PEG are also suitable for urethane linking amino acid residues to the polymer under mild conditions (26,27). The carboxyl group can be then readily converted into the hydrazide, for example by coupling *tert*-butyl carbazate followed by acidolytic removal of the Boc protecting group (10,28).

Most of the transformations shown in Scheme II are clean and high yielding reactions, easily amenable to scale up. Although all the reactions shown are for the β-

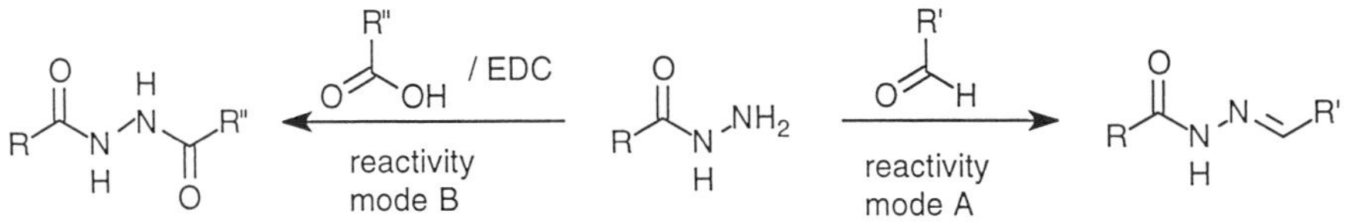

Scheme I.   Two reactivity modes of hydrazides discussed in this review

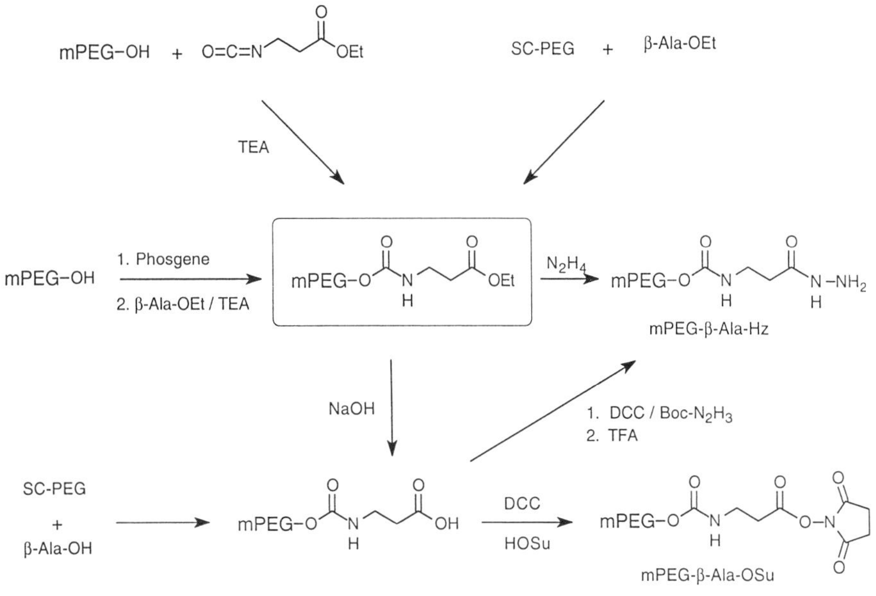

Scheme II.   Synthesis of mPEG-β-Ala hydrazide and succinimidyl ester derivatives.

alanine derivatives, we have had equal success with glycine and some limited yet positive experience with other amino acid derivatives. Simpler Hz derivatives, PEG-carboxymethyl-Hz and PEG-carbazate, were synthesized by hydrazinolysis of the appropriate ethyl esters (29) or active carbonates (25) respectively.

Using α-hydroxy-ω-carboxy-PEG as a starting material, we prepared a new derivative containing the Boc-protected Hz-group and the amino reactive succinimidyl carbonate at the opposite terminals of the PEG chains (Scheme III). This heterobifunctional polymer was urethane-linked to DSPE, followed by acidolytic exposure of the Hz group (10). The resulting Hz-PEG-DSPE was used in preparation of PEG-grafted liposomes containing bioconjugation-prone Hz groups at the extremities of the polymer chains (11-14,30).

Amino acid as part of the linker: analytical implications. There are two principal reasons for choosing urethane-linked amino acid hydrazides. The ease of preparation is one of them. The second reason relates to the use of the amino acid linker as a marker for characterization of conjugates. Although the urethane linkage between PEG and an amino acid residue is stable to all the conditions to which the biological conjugate may be exposed (31), which includes in vivo or bioreactor applications, it can be quantitatively cleaved under conditions used for acid hydrolysis of proteins as performed for amino acid analysis (usually 6 N HCl, 110 °C, 24 h). Amino acid analysis (AAA), in general, can be used with high precision for protein quantitation. The positioning of an unusual amino acid that does not appear in natural polypeptide sequences, like β-Ala or Nle (26), as part of the linker between PEG and protein allows for convenient determination of the composition of such conjugates by AAA. In fact, knowing the amino acid composition of a protein, it is possible to determine both concentration and composition of its PEG conjugate in just one AAA run consuming only a small aliquot of the conjugate preparation, < 50 μg. This is particularly attractive for exploratory work when limited quantities of the protein are available for modification.

The traditional method for characterization of amino group-linked PEG-proteins (32) involved determination of the percentage of unreacted amino groups on the modified protein with 2,4,6-trinitrobenzenesulfonate (TNBS) (33). With all its shortcomings this method was widely used, and is still in use, in conjunction with a variety of PEG-based amino-reactive reagents (3). Note that for both modes of protein or glycoprotein modifications with hydrazide reagents, that we were interested in, there is no comparable analytical methodology for determination of reacted sites. As an initial validation exercise we decided to compare the results of composition analysis performed by both amino group determination and AAA on a series of BSA conjugates obtained using the succinimidyl ester of urethane-linked mPEG-β-Ala (see Scheme II). The results of these analyses, calculated in terms of the number of mPEG residues per BSA molecule, are summarized in Table I. We compared both fluorescamine (34) and

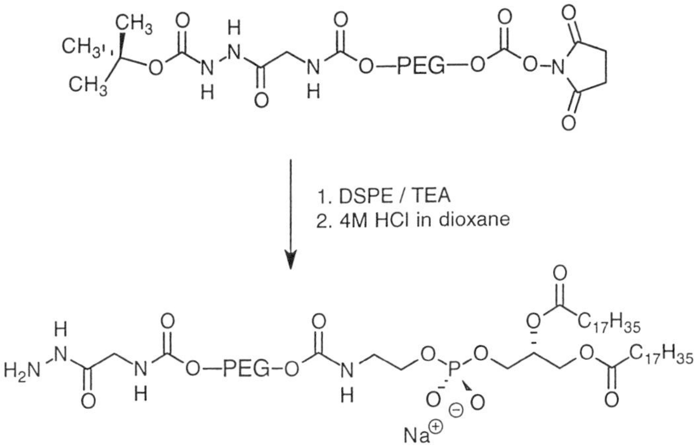

Scheme III.   Preparation of Hz-PEG-DSPE (*10*).

Table I. Determination of compositions (k values) of [mPEG]$_k$-BSA conjugates, prepared as shown below, by amino acid analysis and by amino group quantitation.

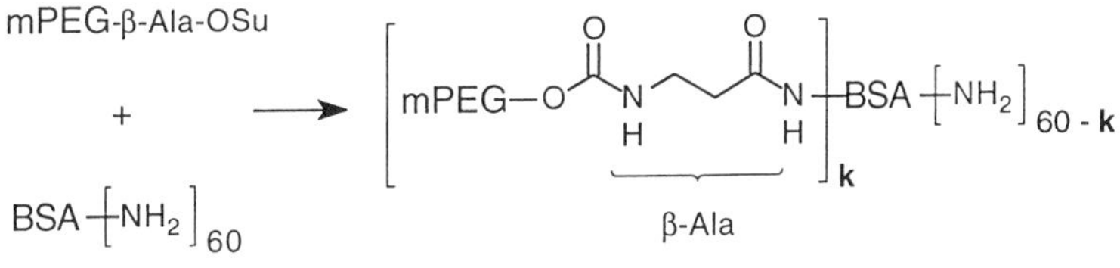

Sample	β-Ala[a]	TNBS[b]	Fluorescamine[c]
1	44.4	42.6	37.6
2	28.3	29.8	26.6
3	18.0	19.7	13.3
4	12.0	10.4	9.5
5	6.2	5.5	3.4

[a] Determined by AAA of the hydrolysates.

[b] Determined by TNBS assay according to Habeeb (*33*).

[c] Determined by Fluorescamine assay according to Stocks *et al.* (*34*).

TNBS methods (33) which determine the fraction of unreacted amino groups on the modified protein. The results of the latter method were consistently in closer agreement with AAA through the entire range of the extent of BSA modification. These data were consistent with our previous observations that succinimidyl esters of PEG are selective in their reactivity toward amines (25). Further the TNBS procedure of Habeeb (33) results in a more accurate determination of amino groups on proteins. Note that the fluorescence method consistently underestimated the extent of modification, probably due to the incomplete reaction of fluorescamine with the buried amino groups of BSA, and also possibly because of quenching effects. Sartore *et al.* published results of similar measurements performed on a few mPEG-proteins modified to various degrees with the succinimidyl ester of urethane-linked mPEG-Nle (26). The AAA and TNBS results were often markedly different, with the latter method usually overestimating the extent of modification. The results summarized in Table I suggest two conclusions. (1) The composition of mPEG-β-Ala-protein can be accurately determined by quantitation of β-Ala by AAA. (2) Fluorescamine is not a very accurate method for determination of the extent of modification of proteins.

Based on the first conclusion, quantitation of β-Ala by AAA has proven to be quite useful in characterization of conjugates derived from mPEG-β-Ala-Hz. One of the most useful conjugation reactions of this reagent involves modification of oxidized glycoproteins. Oxidation of vicinal diols of the oligosaccharide moiety with periodate was utilized for generation of reactive aldehyde groups, which were then converted into hydrazone attachments with mPEG-β-Ala-Hz (Scheme I, mode A). For determination of the optimal conditions for this conjugation reaction we used glucose oxidase (GO) as a model substrate. This enzyme is readily available and inexpensive. It is a glycoprotein containing 18% carbohydrate, which makes it possible to achieve a broad range of modifications. By performing conjugation reactions at various pH values, other than the standard conditions, a series of PEG-GO hydrazones was generated. The purified conjugates were analyzed by SEC-HPLC and also subjected to AAA for quantitation of the incorporated β-Ala residues. As illustrated in Figure 1 the conjugate obtained at pH 5.5 eluted with the shortest retention time on SEC-HPLC, indicating that this was the largest molecular size with the highest extent of modification. Accordingly, the largest number of mPEG-β-Ala residues was detected in this conjugate (Figure 1). Although this conjugation reaction generally can be performed in a relatively broad pH range, the best results are obtained at pH 5 - 6.5, provided the protein is not inactivated.

The two modes of hydrazide reactivity. Utilizing these conjugation conditions we prepared hydrazone conjugates of PEG-hydrazides and various glycoproteins (7) (see Table II). Immunoglobulins are in particular attractive as targets of this approach. These glycoproteins contain ≈ 4% carbohydrate located on the Fc region of the molecule far from the antigen binding sites (6). Therefore the site specificity is the main

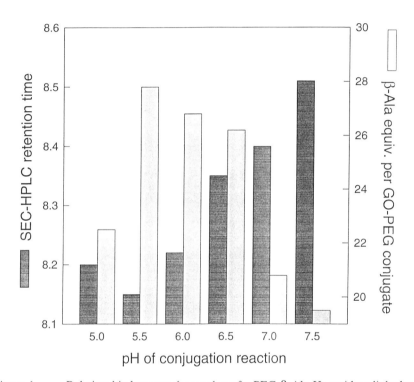

Figure 1.   Relationship between the number of mPEG-β-Ala-Hz residues linked per glycoprotein molecule of glucose oxidase and pH of the conjugation reaction.
GO (3 mg/ml) was oxidized with 20 mM $NaIO_4$ for 1 h and then treated with ≈ 2000-fold molar excess of mPEG-5k-β-Ala-Hz. The purified conjugates obtained at various pH values, under otherwise identical conditions, were analyzed by SEC-HPLC and β-Ala content determined by AAA.

Table II. Summary of PEG-hydrazones discussed in this manuscript of general formula[a]

R	X	Ligand	Comments
CH$_3$O-	β-Ala deriv.	IgG	Periodate-oxidized IgG reacted with an excess mPEG-Hz, 2-9 mPEG chains / IgG.
CH$_3$O-	β-Ala deriv.	Glucose Oxidase	Used as a model glycoprotein to determine optimal conditions of conjugation (Fig. 1).
CH$_3$O-	β-Ala deriv.	Ovalbumin	Periodate-oxidized glycoprotein was modified as a model, 2-5 mPEG residues / ovalbumin.
CH$_3$O-	OCH$_2$ O NH	Hormone glycoprotein of 29 kDa	Site-specific PEGylation on N-terminal. Protein transaminated by glyoxalate / Cu^{2+}, and site-specifically coupled with Hz derivatives of mPEG-5k. Best yield was obtained with the carbazate derivative (X = O).
HRP-	Gly deriv.	IgG	HRP modified on its carboxyl groups with Hz-PEG-Hz, 3-4 residues / HRP. Periodate-treated IgG was labeled with HRP-PEG-Hz.
DSPE anchored in liposome bilayer	Gly or β-Ala derivs	IgG	Immunoliposomes were prepared by coupling various periodate-treated IgGs with Hz-PEG-liposomes. The number of IgG per vesicle can be controlled by changing conditions, e.g. ratio of the reactants.
DSPE anchored in liposome bilayer	Gly or β-Ala derivs	YIGSR-NH$_2$	Site-specific conjugation to N-terminal. TYIGSR-NH$_2$ was converted into N-α-glyoxylyl-YIGSR-NH$_2$ with periodate and then coupled with Hz-PEG-liposomes. Ligand densities of hundreds per vesicle.

[a] Z is usually H or an alkyl residue of a reactive ketone.

advantage of this approach to modification of IgGs and other glycoproteins (*5*). Due to the usually low carbohydrate content, however, we found it difficult to attach multiple chains of PEG per IgG. Note that the generation of aldehyde groups on many glycoproteins can be achieved enzymatically by utilizing neuraminidase and galactose oxidase (*35*). These conditions are considerably milder than periodate treatment, which is known to affect several oxidation-prone amino acid residues (Trp, Tyr, sulfur-containing amino acids, N-terminal Ser and Thr). Occasionally we have encountered IgG that was inactivated by a periodate treatment, presumably by oxidation of one or more of these sensitive residues (*36*).

Oxidation of the oligosaccharide moiety of glycoproteins is only one of several useful methods to generate reactive carbonyls on polypeptides. In a more general sense polypeptides bearing reactive carbonyl(s) are suitable for conjugation with PEG-Hz derivatives via pathway A of Scheme I. For example, King *et al.* (*37*) described utilization of the succinimidyl ester of formyl benzoic acid as a reagent for introduction of aromatic aldehydes onto proteins by substituting some of the amino groups. Although this protein modification is random, the aryl aldehydes were shown to possess good reactivity toward hydrazides and excellent selectivity over amines.

A few very useful methods for the introduction of reactive carbonyls onto peptides, selectively on their N-terminal residues, were described by Dixon and Fields (*38,39*). The most general of these reactions, transamination, is glyoxylate mediated and catalyzed by bivalent metal ions, usually Cu^{2+}. The transamination reaction results in conversion of the N-terminal residue of the peptide into a α-keto acid residue. As will be illustrated by a specific example below, the transaminated polypeptide can react with hydrazide derivatives of mPEG producing well defined conjugates containing only one polymer chain linked to the N-terminus. Conjugation of transaminated interleukin-1 receptor antagonist to aminooxy-PEG derivatives via an oxime linkage was recently reported (*40*). The peptides containing N-terminal Ser or Thr residues undergo very rapid periodate oxidation ($\approx$ 1000 fold faster than 1,2 diol) resulting in destruction of these amino acid residues and formation of N-α-glyoxylyl-peptides (*38,39*). This very mild approach to aldehyde-bearing peptides can be easily followed by hydrazone formation with PEG-hydrazide (*14*).

Covalent attachment of PEG-Hz to carbodiimide-activated carboxyl groups of proteins (Scheme I, mode B) proved useful in a number of cases (*7*). This approach is particularly attractive as an alternative to a more common amino group modification when the protein has very few amines or it is inactivated by their modification. This mode of reactivity of PEG-Hz results in a random modification of available Asp and Glu residues via diacyl hydrazide attachments of the polymer chains. Under conditions optimal for water-soluble carbodiimide reactivity, around pH 5 (*41*), the conjugation of proteins with mPEG-β-Ala-Hz proceeds cleanly without any detectable modification of the amino groups (see the experimental procedure for mPEG-GCSF preparation). As would be expected, and judging from the isoelectric focusing analysis, the product of such a protein modification consists of multiple components with higher pI values than the parent protein. While using this mode of hydrazide reactivity, one has to be aware of carbodiimide-associated side reactions (42), and try to suppress or minimize them depending on their adverse effect on the final conjugate. Among the known protein modification reactions of carbodiimides are formation of O-isourea and S-isourea with Tyr and Cys residues respectively, and particularly troublesome side reaction of N-acylurea formation with Asp and Glu derived carboxyls (42). Consistent with our own observations, suppression of N-acylurea formation by carbodiimide / hydrazide combination was reported under conditions that predominantly yield this side-product with carbodiimide / amine (43).

Specific examples of PEG-Hz conjugates. Wilchek and coworkers described preparation of the Hz-containing enzymes (horseradish peroxidase and alkaline phosphatase) which were used for labeling glycoproteins (44). In analogy with this strategy, we reacted a large excess of dihydrazide of PEG-4k with HRP, utilizing the carboxyl groups of the enzyme for the attachment of the homobifunctional PEG (Scheme I, B). The HRP-PEG-Hz, produced in this reaction contained 3-4 PEG-Hz chains per molecule and retained over 90% of the specific activity of the parent enzyme, as was determined using o-phenylenediamine as the substrate. The Hz-groups on the far ends of the HRP adduct were utilized for conjugation with periodate-oxidized IgG (α-HCG) utilizing the specific reactivity of its oligosaccharide moiety (Table II). Despite the presence of several PEG chains on the surface of the enzyme, and in contrast to some previously reported observations (45), the HRP-PEG-Hz conjugate bound readily, just like the parent HRP, to an affinity column of Concanavalin A. We took advantage of this binding for purification of the HRP-PEG-Hz conjugate from the excess of the PEG-Hz reagent and also for purification of the final HRP-PEG-IgG product.

N-terminal protein PEGylation. A glycoprotein hormone of 29 kDa (apparent molecular weight 35 kDa by SDS-PAGE) was site-specifically PEGylated on its N-terminal in two steps (see Table II). The protein was first transaminated according to Dixon and Fields (38) and then covalently modified on the freshly exposed α-keto acyl residue with either mPEG-hydrazide or mPEG-carbazate derivatives of 5 kDa. A single conjugate of 34 kDa (46 kDa by SDS-PAGE) was formed in both cases in 25 and 95 % yield, respectively. The conjugation reaction was enhanced by the presence of SDS in the reaction mixture, probably by partial unfolding of the protein and enhanced exposure of the N-terminal residue. The mPEG-glycoprotein was characterized by electrophoresis methods on 4-15% SDS-PAGE gradient gels. Figure 2 shows a Coomassie stained gel of the native, transaminated, and mPEG-5k-conjugated derivatives of the glycoprotein. Figure 3 shows the Western blot experiment with the

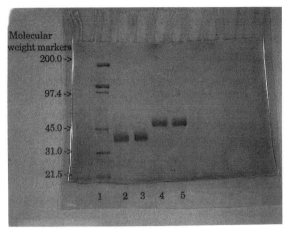

Figure 2.   SDS-PAGE (gradient 4-15%, Coomassie blue staining) of a glycoprotein hormone and its N-terminal conjugate with mPEG-5k. Lane 1: molecular weight markers from top to bottom: myosin (200 kDa), phosphorylase B (97.4 kDa), ovalbumin (45 kDa), carbonic anhydrase (31 kDa), trypsin inhibitor (21.5 kDa). Lane 2: native glycoprotein. Lane 3: transaminated glycoprotein. Lane 4: glycoprotein modified with mPEG-carbazate; lane 5, glycoprotein modified with mPEG-OCH₂-Hz.

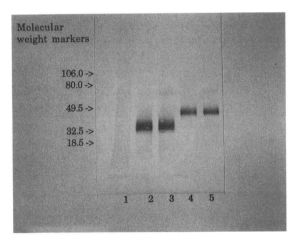

Figure 3.   Western blot analysis of glycoprotein hormone and its N-terminal conjugate with mPEG-5k. The same lane assignments as in Fig. 2. The following prestained molecular weight markers (and their apparent molecular weights) were used (lane 1): Phosphorylase B (106 kDa), bovine serum albumin (80 kDa), ovalbumin (49.5 kDa), trypsin inhibitor (32.5 kDa), lysozyme (18.5 kDa). Electrophoresis was performed on 4-15% gradient gels. Proteins were electrophoretically transferred to nitrocellulose membrane. The membrane was first incubated with the primary antibody recognizing the glycoprotein, and then with a secondary antibody-AP conjugate recognizing the Fc portion of the primary antibody. AP color developing substrate was used for visualization.

same lane assignments for the samples as in Figure 2. Both figures show that the molecular weight of the transaminated protein in lane 3 is essentially the same as that of the native protein. Disapearance of one Ala residue on the transaminated derivative was independently confirmed. Upon PEG conjugation with the mPEG-Hz linkers (lanes 4 and 5) the molecular weight of the PEG-conjugate is shifted to a higher molecular weight. The results obtained from the gel electrophoresis and Western blot experiments demonstrated that PEG-conjugation was complete after transamination, giving rise to a unimolecular mPEG-hydrazone conjugate. The detection of only one molecular-weight species of the conjugate demonstrates the specificity of the reaction. These observations were further corroborated by matrix-assisted laser desorption ionization mass spectrometry (MALDI-MS). The mPEG-5k-carbazate yielded an experimental molecular mass of 5157 as shown in Figure 4A. For the native glycoprotein in Figure 4B the peak labelled at m/z 28570 represents the singly charged monomer. For the mPEG-5k-glycoprotein shown in Figure 4C the peak for the singly charged monomer is seen at m/z 34280, confirming the addition of one mPEG chain of 5 kDa onto the glycoprotein. The stability of the mPEG-carbazate-glycoprotein was probed by incubating the conjugate in acetate buffer (0.1 M, pH 4.5) over a period of 4 days. No breakdown or degradation was observed with SDS-PAGE despite the acidic conditions. Although the *in vitro* activity of this particular conjugate was decreased, the *in vivo* circulating time and hence the bioavailability was enhanced.

Note that the issue of composition determination by AAA is more relevant to conjugates composed of multiple mPEG-chains. For the conjugates containing only one polymer chain per protein, as discussed above for the N-terminal modification, it is sufficient to demonstrate completion of the reaction and consumption of the starting materials. In these cases there is little advantage to having a reference amino acid as part of the linker, and simpler derivatives like the hydrazide of carboxymethylated PEG (*15,17*) or PEG-carbazate are sufficient. It is intriguing, that the latter derivative was the most effective PEGylating reagent in the last example.

Ligand-bearing PEG-grafted liposomes.   Long-circulating, RES-evading liposomes possess these valuable properties due to external surface-grafted PEG chains, usually of 1-5 kDa (*46*). We found that recently-introduced Hz-PEG-DSPE (Scheme III) can completely replace the mPEG-DSPE conjugate, originally used for preparation of such liposomes, without any measurable effect on the pharmacokinetic behavior (Figure 5) or tissue distribution. However, in contrast to the inert methoxy end groups of mPEG, the Hz-terminated polymer chains find use for attachment of various biologically relevant ligands. The most obvious application of such systems is in targeted delivery of liposomal drugs, and IgG is most commonly used as the targeting moiety. A variety of immunoglobulins were conjugated to Hz-PEG-2k-liposomes through their carbohydrate residues, taking advantage of the previously developed methods for coupling mPEG-Hz (Table II). The antigen binding activity of the immunoliposomes prepared by this methodology was usually well preserved, although the pharmacokinetic behaviour typically required some fine-tuning to balance the beneficial properties of the liposomes and to optimize the density of the attached IgG residues (*11,30,36,47*). As illustrated in Figure 5 immunoliposomes containing on average 10 IgG residues per lipid vesicle exhibited prolonged systemic circulation although their clearance rate was noticeably faster than the parent liposomes and the free IgG (*11*). Note that both ^{125}I and ^{67}Ga labels placed on the protein and liposome, respectively, exhibited the same pharmacokinetic curves, indicating that the components of the immunoliposome do not dissociate while in blood circulation.

Another important application of ligand-bearing, long-circulating liposomes pertains to their use as platforms for presentation of various small ligands. Such low molecular weight molecules (peptides, oligosaccharides, etc.) in their free form are promptly cleared from the systemic circulation. We devised a convenient method for attaching peptides to the Hz-extremities of the liposome-grafted PEG chains. Peptides containing Ser or Thr residues at their N-termini can be readily obtained by modern peptide synthesis methods. Brief treatment of such peptides with periodate cleaves the 1-amino-2-hydroxyl residue resulting in N-α-glyoxylyl peptide, which reacts with Hz-PEG-liposomes producing hydrazone attachments. The conjugation is site-specific to the N-terminal residue of the peptide. It is performed in a one-pot procedure, and requires only one chromatography or dialysis step for the final product purification (*14*). We demonstrated the feasibility of this approach to peptidoliposomes with the cell adhesive peptide sequence, TYIGSR-NH$_2$. It was possible to attach several hundreds of the peptide residues per liposome (100 nm) without any difficulties. Double-label pharmacokinetic experiments performed in rats with peptidoliposomes (200 peptide residues per lipid vesicle), containing ^{125}I-YIGSR-NH$_2$ and ^{67}Ga-desferal loaded in the internal aqueous compartment, showed that both labels exhibited identical clearance characteristics. Thus the peptide and the liposome remained conjugated in the blood plasma and remained there for only slightly shorter periods of time than the parent Hz-PEG-liposomes. This indicates that the hydrazone attachments formed by liposome-grafted Hz-PEG and N-α-glyoxylyl-YIGSR-NH$_2$ were stable under the conditions existing *in vivo* (*14*). These experiments suggested that this general approach might be useful for increasing the systemic exposure of biologically active short peptide sequences.

Concluding Remarks

As illustrated by the examples above, hydrazide derivatives of PEG are useful for preparation of peptide and protein adducts via a number of conjugation methods. Mild conjugation conditions, high selectivity, and site-specificity of the coupling reactions are among the beneficial characteristics of these methods. The latter feature is of particularly

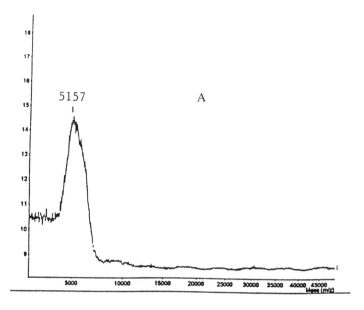

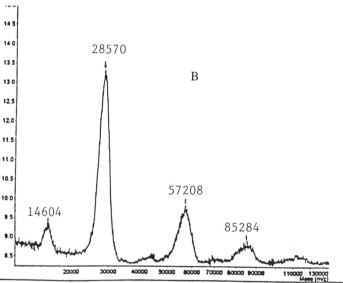

Figure 4.   MALDI-MS of an N-terminally PEGylated glycoprotein hormone and the
individual components of the conjugate. (A) mPEG-carbazate, m/z 5157; (B) Singly
charged monomer of the native glycoprotein, m/z 28570. The signals at m/z 57208 and
85284 represent the dimer and trimer. Doubly charged ion, m/z 14604. (C) Singly
charged monomer of mPEG-glycoprotein hydrazone, m/z 34280. The signals at m/z
69071 and 102955 represent the dimer and trimer. Doubly charged ion, m/z 17092.
The spectra were acquired on a Finnigan-MAT time of flight LaserMAT 2000 mass
spectrometer.

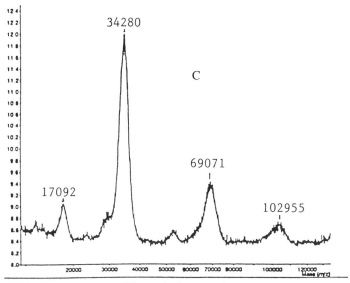

Figure 4. *Continued.*

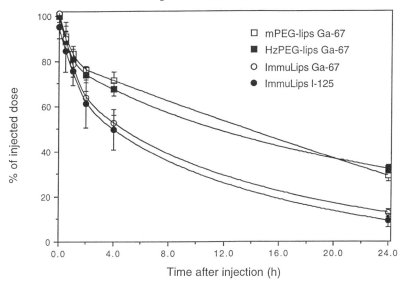

Time after injection (h)

Figure 5. Blood lifetimes of liposomes prepared from hydrogenated soy phosphatidyl choline, cholesterol, mPEG-DSPE (mole ratio 1.85:1:0.15), same formulation containing Hz-PEG-DSPE (5 mole %) instead of the mPEG analog, and immunoliposomes prepared by coupling oxidized IgG to the Hz-PEG-liposomes. The immunoliposomes were labeled with [125]I (protein label) and ^{67}Ga-desferal in the internal aqueous compartment. Four Sprague-Dawrey rats were injected with each liposomal formulation via a tail vein. Samples collected at various times by retro-orbital bleeding were used to determine radioactivity in blood.

great utility for preparation of N-terminally modified polypeptides or glycoproteins carrying PEG chains only on their oligosaccharide residues. For both modes of reactivity discussed here, hydrazone formation with reactive carbonyls or modification of carboxyl groups forming diacylated hydrazine attachments, excellent selectivity against the amino groups was observed. Considering that most of the commonly used PEG-based reagents are designed for modification of amino groups in a rather random fashion, addition of the PEG-hydrazides to the arsenal of polypeptide reacting agents markedly expands the bioconjugate design options. Since in many instances optimal preservation of biological activity of the PEGylated polypeptides is dependent on the total number and the positioning of the polymer chain attachment sites (*1,3*), use of site-specific methods for conjugate preparation is likely to yield well defined adducts with better preservation of biological activity. It can be concluded with confidence that many more useful bioconjugates will be constructed utilizing the chemical versatility of PEG-hydrazides.

Experimantal Procedures

TNBS determination of Hz groups. TNBS assay as described for determination of amino groups (*48*) was used with the following modifications. ADH purified by recrystalization was used to prepare standard solutions with concentrations in the range 0 - 0.1 mequiv. Hz/l. Potassium borate (0.2 M, pH 9.3) was used for preparation of the standard and the sample solutions. A sample of Hz-containing material (1 ml solution, < 0.1 mequiv. Hz/l) in the borate buffer was treated with TNBS (25 μl, 30 mM). After $\approx$ 45 min incubation at 25 °C the maroon-red color was measured at 500 nm. A calibration curve was obtained with the standard ADH solutions. Typically ε_{500} = 16,500 - 17,000 $M^{-1} \cdot cm^{-1}$ was obtained by this procedure.

Test for presence of PEG in aqueous solutions (PMA test). The presence of PEG was detected by mixing a drop of a sample solution with a drop of PMA solution (1% in 1 N HCl). Immediate appearance of white precipitate of the interpolymer PEG-PMA complex indicated the presence of PEG. This test proved to be very specific to the polyether backbone of PEG, and was not influenced by presence of salts in the solution or derivatization / conjugation of PEG, as long as pH < 4 was maintained (*24,29,49*).

Preparation of mPEG-O(C=O)-β-Ala-OEt (7). mPEG (100 g, 20 mmol) was dissolved in toluene (250 ml) and azeotropically dried. The solution was brought to 25 °C, diluted with methylene chloride (50 ml) and then treated with phosgene (30 ml of 20% toluene solution, 56 mmol) overnight. The solvents and the excess of phosgene were removed by rotary evaporation under vacuum. The solid residue of polymeric chloroformate was dissolved in methylene chloride (90 ml) and treated with β-Ala-OEt hydrochloride (6.1 g, 40 mmol) predissolved in methylene chloride (total vol. 30 ml) followed by TEA (8.4 ml, 60 mmol). Approximately 30 min later the solution was diluted with toluene (50 ml), filtered and evaporated to dryness. The crude product was dissolved in warm (50 °C) ethyl acetate (500 ml) and filtered through celite. The filtrate was diluted with *iso*-propanol to a total volume of 1000 ml and left overnight at 25 °C to facilitate precipitation of the product. Another recrystallization of the product from *iso*-propanol was performed. The yield of the dried mPEG-β-Ala-OEt was 98 g (95 %). I.R. (neat): 3341 (N-H), 1723 (C=O,urethane) cm^{-1}. ^{1}H-NMR (CDCl$_3$): δ 1.27 (t, J = 7 Hz, C$\underline{H}_3$CH$_2$, 3H), 2.53 (t, J = 6 Hz, CH$_2$ of β-Ala, 2H), 3.38 (s, CH$_3$O, 3H), 3.65 (s, PEG), 4.15 (q, J = 7 Hz, CH$_3$C$\underline{H}_2$O, 2H), 4.21 (t, J = 5 Hz, CH$_2$C$\underline{H}_2$OC=ON, 2H), 5.3 (broad, NH, 1H). Alternatively ethyl esters of urethane-linked Gly or β-Ala were prepared by reacting azeotropically dried dihydroxy-PEG or mPEG-OH with ethyl isocyanatoacetate or ethyl isocyanatopropionate (1.5 equiv. / OH) in presence of TEA (1 equiv.). Overnight reactions at 25 °C routinely resulted in quantitative conversion of the terminal hydroxyl groups (*23*).

Preparation of mPEG-O(C=O)-β-Ala-OH and its succinimidyl ester.
Hydrolysis of mPEG-O(C=O)-β-Ala-OEt was performed in 2N NaOH at 25 °C for 2 h (*23,24*), or alternatively by adjusting the pH to 11 - 12 and maintaining it with dropwise addition of aqueous NaOH until the pH stabilized (*10*). The solution was acidified and the mPEG-O(C=O)-β-Ala-OH recovered by standard procedures. The succinimidyl ester was prepared using several known procedures (*25,28,50*).

Preparation of mPEG-O(C=O)-β-Ala-Hz (*7*). mPEG-β-Ala-OEt (62 g, 12 mmol) was dissolved in pyridine (120 ml) and treated with hydrazine (12 ml, 0.375 mole) under reflux for 6 h. The solution was rotary evaporated to dryness and the residue crystallized twice from *iso*-propanol and dried *in vacuo* over P$_2$O$_5$. The yield was 60 g (97 %). Colorimetric assay of hydrazide groups using TNBS gave 0.2 mmol/g (103 % of theoretical). The β-Ala content of the polymer was 0.205 mmol/g (105 % of theoretical) as determined by AAA of a completely hydrolysed (6N HCl, 110 °C, 24 h) aliquot of the product. ^{1}H-NMR (CDCl$_3$): δ 2.43 (br t, CH$_2$ of β-Ala, 2H), 3.38 (s, CH$_3$O, 3H), 3.64 (s, PEG), 4.21 (br t, CH$_2$C$\underline{H}_2$OC=ON, 2H), 5.3 (broad, NH, 1H). ^{13}C-NMR (CDCl$_3$): δ 171.2 (C=O, of Hz); 156.4 (C=O, urethane); 71.8 (CH$_3$O$\underline{C}$H$_2$); 70.0 (PEG); 68.5 ($\underline{C}$H$_2$CH$_2$OC=O); 63.7 (CH$_2$$\underline{C}H_2$OC=O); 58.9 (CH$_3$O); 37.1 (NHCH$_2$$\underline{C}H_2$); 33.9 (NH$\underline{C}H_2CH_2$) ppm. IR (neat): 3328 (NH); 1719 (C=O, urethane); 1671 (C=O, Hz) cm^{-1}.

Preparation of mPEG-carbazate. Active carbonates (nitrophenyl carbonate or succinimidyl carbonate) or mPEG-chloroformate reacted cleanly with hydrazine to yield mPEG-O(C=O)N_2H_3, which was then purified from excess low molecular weight reagents by repeated recrystallization from ethanol.

Attachment mPEG-β-Ala-Hz to carbodiimide-activated carboxyl groups of GCSF (7). The mPEG-β-Ala-Hz (15.0 g, 2.9 mmol) was added to a solution of GCSF (86 mg, 4.78 mmole) in 1 mM HCl (86 ml) , followed by EDC (128 mg, 0. 667 mmol). The reaction mixture was stirred at 25 °C for 90 minutes while maintaining the pH at 4.7 - 5.0 range. Excess reagents were removed by extensive diafiltration of the reaction solution at 4 °C against 1 mM HCl. Comparison of the PEG-conjugate and native GCSF using SEC-HPLC (Zorbax GF-450 column, mobile phase was 0.2 M phosphate buffer pH 7.5) showed a substantially increased molecular weight of the conjugate. The average number of mPEG residues in the PEG-GCSF was 5.8, as determined by β-Ala quantitation by AAA. TNBS assay (*33*) confirmed that both native and PEG-modified GCSF had the same number of amino groups, indicating that the EDC activated carboxylic acid groups of the protein did not react with amino groups of the protein. The conjugate of mPEG-GCSF gave four separate bands on SDS-PAGE (PhastGel, Homogenous 12.5, Pharmacia) in the range from 29 to 67 kDa. Isoelectric Focusing (PhastGel-IEF 3-9, Pharmacia) of the conjugate resulted in the separation of six bands with pI values arranging between 6.8 and 9.0, noticeably higher than the native protein (pI 5.2; 5.9).

Preparation of HRP-PEG-Hz. HRP (40 mg, 1 μmol) and PEG-dihydrazide (320 mg, 160 μmol Hz) were dissolved in water (1 ml). EDC (50 mg) was added. Aqueous HCl (1 N) was used to maintain pH ≈ 5 for 60 min. The solution was adjusted to pH 7.5 and incubated overnight at 25 °C. The reaction solution was concentrated on Centricon-10 (Amicon), diluted with water and concentrated again. The dilution / concentration cycle was repeated until the filtrates became negative to PMA test. This indicated that all the unreacted PEG-dihydrazide was removed. When an aliquot of the purified conjugate solution was treated with PMA solution at pH 3, all the brown color of the enzyme was completely precipitated with the PEG-PMA interpolymer complex forming a clear colorless solution, suggesting that all the HRP was bound to PEG. Analysis of the HRP-PEG-Hz on SEC-HPLC (Zorbax GF-250) indicated one peak of product comparable in its elution time to a dimer of HRP (9.2 min). For comparison (HRP, MW 40kDa -10.3 min, BSA, 68kDa - 9.6 min, PEG-4kDa - 10 min). The conjugate contained on average of 3.5 Hz groups per HRP molecule as determined by TNBS assay. The conjugate exhibited binding to a Con A-Sepharose column and eluted under the same conditions as the starting HRP (15 - 20 mM methyl-α-D-mannoside). This finding was utilized in the later preparations for purification of HRP-PEG-Hz.

Conjugation of HRP-PEG-Hz to IgG. IgG (α-HCG, 3 mg) in phosphate saline buffer (0.5 ml, 50 mM Na_2HPO_4, 0.15 M NaCl, pH 5.5 adjusted with acetic acid) was treated with $NaIO_4$ (1.5 - 1.7 mg) predissolved in the same buffer (50 µl) in the dark at 4 °C for 2.5 h. The concentration of periodate in the reaction solution was 13 -15 mM. The oxidized IgG was purified by passing the reaction solution through a PD-10 column preequilibrated with the same buffer, and quickly mixed with HRP-PEG-Hz (3 mg). The conjugation solution was concentrated to $\approx$ 1 ml and incubated overnight at 4 °C. The reaction solution was injected into preparative SEC-HPLC (Zorbax GF-250 XL) and fractions containing the conjugate collected. At this stage the unreacted HRP-PEG-Hz was completely separated. The collected fractions were pooled and applied onto a Con A-Sepharose column equilibrated with the loading buffer (20 mM Tris · HCl, 0.3 M NaCl, 1 mM $CaCl_2$, 1 mM $MnCl_2$, pH 7.1). if any unconjugated IgG was present, it did not bind to the affinity column. The pure IgG-PEG-HRP was eluted from the Con A column with a gradient of methyl-α-D-mannoside (0 to 25 mM in 50 min at 1 ml/min). The ratio of HRP / IgG in the product was determined by UV / Vis calculating A_{403} / A_{280}. Typically 0.3 - 0.9 A_{403} / A_{280} values were observed indicating 1-3 HRP residues per IgG.

mPEG-hydrazone of a transaminated glycoprotein hormone. General methods for transamination of polypeptides have been described in detail elsewhere (*38,39*). For transamination of the glycoprotein hormone, the concentration of the protein was first adjusted to 1 mg/ml with 2 M sodium acetate in 0.4 M acetic acid, 0.1 M glyoxalic acid and 10 mM cupric sulfate (pH 5.5). Alternatively, the protein can be dialyzed or diafiltred into the above buffer. The reaction was allowed to proceed for two hours at room temperature. The extent of transamination was monitored by the method described by Fields and Dixon (*51*). The transaminated protein sample solution in 1 M HCl was placed in a cuvette and reacted with 2,4 - dinitrophenylhydrazine in 1 M HCl. The reaction was continuously followed by measuring the absorbtion at 370 nm until no further change was observed. The initial reading was subtracted from the final reading. The difference in absorbance was proportional to the amount of carbonyl groups on the protein molecule. The transaminated glycoprotein was subjected to AAA on an ABI 420H system (Applied Biosystems, Foster City, CA) using pre-column phenyl isothiocyanate derivatization. The transaminated protein was separated from the low molecular weight reagents by gel filtration chromatography using 100 mM sodium acetate at pH 4.5. In the second step of the reaction the transaminated protein in the elution buffer was reacted with mPEG-5k-Hz (10 to 20 mg). The PEG coupling reaction was carried out for 40 hours at room temperature. The conjugated product was purified by Sephacryl S-200 gel filtration. Three types of reactive mPEG-5k-Hz derivatives were evaluated (Table II). Of the three, the reaction between the transaminated glycoprotein and mPEG-5k-carbazate produced the best results. It was more than 95% complete. In the case of the transaminated protein coupling reaction with mPEG-5k-hydrazide or mPEG-5k-semicarbazide the coupling reactions were only about 25% complete. The yields were improved when 0.1 % SDS was used in the reaction mixture. Gel filtration chromatography (TSK G3000SWXL column) of mPEG-carbazate-glycoprotein showed clear increase in molecular size (17.2 min) over the parent glycoprotein (18.3 min).

Preparation of immunoliposomes and peptidoliposomes. Liposomes were prepared and radiolabeled as described elsewhere (11-13,36). Briefly, lipid films were formed by evaporation of lipid mixtures from chloroform and hydrated with the desired aqueous phase, followed by vigorous shaking. For the ^{67}Ga-labeling the hydrating buffer contained 10 mM desferoxamine mesylate in 0.9% saline (isotonic). The resulting liposomes were extruded through polycarbonate defined pore filters to give a mean particle size of 95±15 nm, as determined by dynamic light scattering (Coulter N4SD). The typical lipid formulation contained: Hz-PEG-2k-DSPE (1.6 -5 mole %), mPEG-2k-DSPE (0-5 mole %), phosphatidyl choline (50 -60 mole %), and cholesterol (30-40 mole%). Antibodies (1-5 mg/ml), if necessary spiked with the appropriate amount of ^{125}I-IgG, were first incubated in acetate buffer (0.1M, pH 5.6) with periodate (2-20 mM) for 20-120 min. The excess of periodate was quenched by addition of NAM solution (0.1 ml at 0.5 M). Quick consumption of NAM commensurate with the amount of periodate present, presumably by conversion into Met-sulfoxide and -sulfone derivatives, was confirmed by HPLC. Finally, Hz-PEG-liposomes (1.2 ml, 20-50 μmole phospholipid / ml) were added. The mixture was left at room temperature for 1 h and then refrigerated (6 °C) until the following day. Samples were assayed for conjugation efficiency by chromatography on Sepharose 4B to separate liposome bound antibody from free IgG in 25 mM HEPES containing 15 mM NaCl, pH 7.2. The observed increase in the particle size of the immunoliposomes (diameter ≈ 120 nm), relative to the parent liposomes (≈ 100 nm) was consistent with the approximate size of IgG molecules (100Å) (11). Peptides containing N-terminal Ser or Thr residues were oxidized with periodate at pH 7.2 for 5 min producing N-α-glyoxylyl peptides, which were then coupled to Hz-PEG-liposomes in a similar manner (14). The composition of immuno- and peptido-liposomes was determined by AAA and phosphate analysis.

Abbreviations
ADH, adipic acid dihydrazide; AAA, amino acid analysis; ATP, adenosine triphosphate; AP, alkaline phosphatase; Boc, *tert*-butyloxycarbonyl; BSA, bovine serum albumine; DCC, N,N'-dicyclohexylcarbodiimide; DSPE, distearoylphosphatidylethanolamine; EDC, N-ethyl-N'-(3-dimethylamino-propyl)carbodiimide; GO, glucose oxidase; HCG, human chorionic gonadotropin; HRP, horseradish peroxidase; Hz, hydrazide; MALDI-

MS, matrix-assisted laser desorption ionization mass spectrometry ; NAM, N-acetyl methionine; PAGE, poly(acrylamide) gel electrophoresis; PEG-Nk, poly(ethylene glycol)-molecular weight in kiloDaltons; mPEG, monomethoxy-PEG; PMA, poly(methacrylic acid); RES, reticuloendothelial system; SC, succinimidyl carbonate; SDS, sodium n-dodecyl sulfate; SEC, size exclusion chromatography; Su, succinimide; TEA, triethylamine; TFA, trifluoroacetate; TNBS, 2,4,6-trinitrobenzene sulfonate.

Acknowledgements
The authors would like to thank the following individuals who contributed to the various projects mentioned in this review: C. Asmus, C. Engbers, J. Harding, C. Lee, B. Puntambekar, Z. Wei, C. Wogheren, M. Woodle.

Literature Cited

1. Zalipsky, S. *Adv. Drug Delivery Rev.* **1995**, *16*, 157-182.
2. Zalipsky, S. *Bioconjugate Chem.* **1995**, *6*, 150-165.
3. Zalipsky, S.; Lee, C. In *Poly(ethylene glycol) Chemistry: Biotechnical and Biomedical Applications*; J. M. Harris, Ed.; Plenum Press: New York, 1992; pp 347-370.
4. Inman, J.K. *Methods Enzymol.* **1974**, *34*, 30-58.
5. Wilchek, M.; Bayer, E.A. *Methods Enzymol.* **1987**, *138*, 429-442.
6. O'Shannessay, D.J.; Quarles, R.H. *J. Immunol. Meth.* **1987**, *99*, 153-161.
7. Zalipsky, S.; Lee, C.; Menon-Rudolph, S. *Int. Patent Appl.* **1992**, WO 92/16555.
8. Zalipsky, S.; Woodle, M.C.; Martin, F.J.; Barenholz, Y. *Int. Patent Appl.* **1994**, WO 94/21235.
9. Zalipsky, S.; Martin, F. *Int. Patent Appl.* **1994**, WO 94/21281.
10. Zalipsky, S. *Bioconjugate Chem.* **1993**, *4*, 296-299.
11. Harding, J.A.; Engbers, C.M.; Newman, M.S.; Goldstein, N.I.; Zalipsky, S. *Biochim. Biophys. Acta* **1996**, in press.
12. Goren, D.; Horowitz, A.T.; Zalipsky, S.; Woodle, M.C.; Yarden, Y.; Gabizon, A. *Br. J. Cancer* **1996**, *74*, 1749-1756.
13. Zalipsky, S.; Newman, M.; Punatambekar, B.; Woodle, M.C. *Polym. Materials: Sci. & Eng.* **1993**, *67*, 519-520.
14. Zalipsky, S.; Puntambekar, B.; Bolikas, P.; Engbers, C.M.; Woodle, M.C. *Bioconjugate Chem.* **1995**, *6*, 705-708.
15. Andresz, H.; Richter, G.C.; Pfannemuller, B. *Makromol. Chem.* **1978**, *179*, 301-312.
16. Persson, L.-O.; Olde, B. *J. Chromatogr.* **1988**, *457*, 183-193.
17. Kogan, T.P. *Synth. Commun.* **1992**, *22*, 2417-2424.

18. Davis, F.F.; Van Es, T.; Palczuk, N.C. *U.S. Patent 4,179,337* **1979**,

19. Rubinstein, M.; Simon, S.; Bloch, R. *U.S. Patent 4,101,380* **1978**,

20. Shimizu, K.; Nakahara, T.; Kinoshita, T. *U.S. Patent 4,495,285* **1985**,

21. King, J.F.; Gill, M.S. *J. Org. Chem.* **1996**, *61*, 7250-7255.

22. Gais, H.-J.; Ruppert, S. *Tetrahedron Lett.* **1995**, *36*, 3837-3838.

23. Zalipsky, S.; Albericio, F.; Slomczynska, U.; Barany, G. *Int. J. Pept. Prot. Res.* **1987**, *30*, 740-783.

24. Zalipsky, S.; Chang, J.L.; Albericio, F.; Barany, G. *Reactive Polym.* **1994**, *22*, 243-258.

25. Zalipsky, S.; Seltzer, R.; Menon-Rudolph, S. *Biotechnol. Appl. Biochem.* **1992**, *15*, 100-114.

26. Sartore, L.; Caliceti, P.; Schiavon, O.; Monfardini, C.; Veronese, F.M. *Appl. Biochem. Biotechnol.* **1991**, *31*, 213-222.

27. Chiu, H.-C.; Zalipsky, S.; Kopeckova, P.; Kopecek, J. *Bioconjugate Chem.* **1993**, *4*, 290-295.

28. Nathan, A.; Bolikal, D.; Vyavahare, N.; Zalipsky, S.; Kohn, J. *Macromolecules* **1992**, *25*, 4476-4484.

29. Zalipsky, S.; Barany, G. *J. Bioact. Compatible Polym.* **1990**, *5*, 227-231.

30. Zalipsky, S.; Hansen, C.B.; Lopes de Menezes, D.E.; Allen, T.M. *J. Controlled Release* **1996**, 153-161.

31. Larwood, D.J.; Szoka, F.C. *J. Labelled Compd. Radiophrm.* **1984**, *21*, 603-614.

32. Abuchowski, A.; Van Es, T.; Palczuk, N.C.; Davis, F.F. *J. Biol. Chem.* **1977**, *252*, 3578-3581.

33. Habeeb, A.F.S.A. *Anal. Biochem.* **1966**, *14*, 328-336.

34. Stocks, S.J.; Jones, A.J.M.; Ramey, C.W.; Brooks, D.E. *Anal. Biochem.* **1986**, *154*, 232-234.

35. Solomon, B.; Koppel, R.; Schwartz, F.; Fleminger, G. *J. Chromatogr.* **1990**, *510*, 321-329.

36. Hansen, C.B.; Kao, G.Y.; Moase, E.H.; Zalipsky, S.; Allen, T.M. *Biochim. Biophys. Acta* **1995**, *1239*, 133-144.

37. King, T.P.; Zhao, S.W.; Lam, T. *Biochemistry* **1986**, *25*, 2774-5779.

38. Dixon, H.B.F.; Fields, R. *Methods Enzymol.* **1972**, *25*, 409-419.

39. Dixon, H.B.F. *J. Protein Chem.* **1984**, *3*, 99-108.

40. Gaertner, H.F.; Offord, R.E. *Bioconjugate Chem.* **1996**, *7*, 38-44.

41. Papisov, M.I.; Maksimenko, A.V.; Torchilin, V.P. *Enzyme Microb. Technol.* **1985**, *7*, 11-16.

42. Carraway, K.L.; Koshland, D.E. *Methods Enzymol.* **1972**, *25B*, 616-623.

43. Pouyani, t.; Prestwich, G.D. *Bioconjugate Chem.* **1994**, *5*, 339-347.

44. Gershoni, J.M.; Bayer, E.A.; Wilchek, M. *Anal. Biochem.* **1985**, *146*, 59-63.

45. Wieder, K.J.; Davis, F.F. *J. Appl. Biochem.* **1983**, *5*, 337-347.

46. *Stealth® Liposomes* ; Lasic, D.; Martin, F., Eds.; CRC Press: Boca Raton, FL, 1995.
47. Allen, T.M.; Agrawal, A.K.; Ahmad, I.; Hansen, C.B.; Zalipsky, S. *J. Liposome Res.* **1994**, *4*, 1-25.
48. Snyder, S.L.; Sobocinski, P.Z. *Anal. Biochem.* **1975**, *64*, 284-288.
49. Oto, E.K.; Zalipsky, S.; Quinn, Y.P.; Zhu, G.Z.; Uster, P.S. *Anal. Biochem.* **1995**, *229*, 106-111.
50. Buckman, A.; Morr, M.; Johansson, G. *Makromol. Chem.* **1981**, *182*, 1379-1384.
51. Fields, R.; Dixon, H.B.F. *Biochem. J.* **1971**, *121*, 587-589.

Chapter 22

Protein Adsorption on Poly(ethylene oxide)-Grafted Silicon Surfaces

Susan J. Sofia and Edward W. Merrill

Department of Chemical Engineering, Massachusetts Institute of Technology, 77 Massachusetts Avenue, Room 56–291, Cambridge, MA 02139

Poly(ethylene oxide), in linear and star form, was covalently grafted to silicon surfaces in a range of grafting densities and the surfaces tested for their ability to adsorb proteins (cytochrome-c, albumin, fibronectin). The surfaces were analyzed by X-ray photoelectron spectroscopy (XPS) and ellipsometry, where both PEO content and amount of adsorbed protein were determined. All three proteins were found to reach zero adsorption at the highest grafting densities on all three linear PEG surfaces (PEG MW 3400, 10,000, and 20,000 g/mol). On both star PEO surfaces, albumin and fibronectin decreased to zero adsorption at intermediate grafting densities, whereas cytochrome-c continued to adsorb at all grafting densities. It was found that grafting densities of linear PEG in a brush regime were necessary to prevent protein adsorption. On PEO star surfaces, however, it was the decrease in space between grafted molecules to less than the size of the protein that was necessary in preventing the protein from adsorbing to the surface.

Increasing the biocompatibility of surfaces through the incorporation of PEO has been extensively investigated over the years. Two widespread ways of achieving this are: 1) by covalently binding of one or both ends of the PEO chain to an insoluble substrate surface, or 2) by incorporating PEO into a network with another polymer, either as an interpenetrating network (IPN) or covalently through a block copolymer network. Nagoaka and coworkers (1,2) end-linked one end of PEO via a methacrylate to various insoluble polymers. PEO end-linked at one or both ends, either to a surface or in a network, has been achieved with a wide variety of materials: segmented polyurethanes (where the PEO constitutes the "soft segment" continuous phase) (3-8), polysiloxanes (9-11) polystyrenes (12-15), polyfunctional isocyanates (16), poly(ethylene terephthalates) (6,17), poly(methyl methacrylates) (6,18,19), poly(lactic-glycolic acids) (20), poly(tetramethylene oxide) (21), and poly(vinyl chloride) (22). An important outcome of these studies is the common trend of decreasing protein and platelet adsorption and biological activity with increasing PEO content, leading to the conclusion that the highest level of biocompatibility would be achieved when the surface consists solely of amorphous PEO polymer (*4,23*).

The accurate experimental characterization of PEO chains grafted to a polymer surface has traditionally been difficult to accomplish. This is mainly due to the fact that polymers can have very rough, non-uniform, and dynamic surfaces that make their accurate characterization quite a challenge. The use of X-ray photoelectron spectroscopy (XPS) or secondary ion mass spectrometry (SIMS) have traditionally been used, but these techniques primarily determine qualitatively the PEO content on the surface. The direct correlation of biological interactions with PEO surface content had not been achieved. At best, the efficacy of these surfaces was usually determined by observing the extent of biological interaction with them, and then correlating these interactions with the polymer characteristics, such as PEO molecular weight and PEO content in the material, but not specifically with the amount of PEO present on the surface.

There are several studies that have lent some insight into the PEO surface requirements needed for the prevention of protein adsorption. Prime et al. (*24*) studied protein adsorption onto surfaces containing PEG oligomers of varying content and molecular weight, which was achieved with self-assembled-monolayers (SAMs) on gold substrates (*24,25*). (We use PEG in place of PEO to refer to molecular weights of less than 25,000 g/mol.) The study showed convincingly that only very short lengths of PEO (PEGs of only 2 to 6 monomer units) can succeed, at sufficient densities, in preventing protein adsorption. This dispels the notion that very small molecular weight PEG loses its "ability" to repel proteins (*26*). Jeon et al. (*27,28*), in a theoretical study, modeled the protein resistance of a grafted PEO surface through a combination of hydrophobic interactions, Van der Waals attraction between a protein and the PEO/substrate surface, and steric repulsion of the grafted PEO chains as they are compressed by an approaching protein. The authors concluded that high surface densities and long chain lengths of PEO are desirable for resistance to protein adsorption, with surface density being the more important of the two parameters. However, the surface conditions considered by Jeon et al. were a dense, stretched-chain brush regime, which would be extremely difficult to achieve experimentally. It is generally found that the maximum grafting densities achieved from grafting solvated chains usually correlate to chain spacings on the order of the radius of gyration of the molecule (*29-32*). In addition, the authors state that certain assumptions and approximations that were made in their model lend the qualitative trends seen in their results to be valid, but the exact calculated values may not be entirely accurate. In the following experimental study, we hoped to achieve two things: first, to give a more realistic view of the dependence of protein adsorption on the surface content of grafted, randomly coiling PEO molecules, and second, to study protein adsorption on surfaces grafted with PEO star molecules, and compare the adsorption behavior with that found on linear PEO grafted surfaces.

PEO Grafted to Silicon Surfaces

Linear and star-shaped PEO molecules were grafted to silicon surfaces at varying grafting desities. In the case of the linear PEG surfaces, the range of grafting desities spanned from a mushroom (individual chain) regime to the onset of a brush (overlapping, stretched chain) regime. Star surfaces consisted of individual star molecules grafted in increasing density up to the point of near chain overlap. Silicon wafers were used as a model substrate surface, where the surface is stable and extremely flat. A triaminosilane was covalently coupled to the surface, which provided a high concentration of grafting sites (both primary and secondary amines) for the attachment of the PEO (33). Tresyl chloride was used to couple the PEO hydroxyl chain ends to the amines on the surface (34,35), with a range of grafting densities being achieved by varying the concentration of tresylated PEO in the coupling solution (0.01 to 15 wt. %) (36).

Both linear and star-shaped PEO molecules were grafted to the silicon surfaces in order to study the differences in their grafting densities and effectiveness at preventing protein adsorption. Linear PEO random coils have a very loose, non-dense structure with only two chain ends per molecule. These chain ends could be located anywhere within the volume of the molecule, thus causing its grafting probability to decrease with increasing molecular weight. Their non-dense structure leads the chains to easily overlap with one another, even at low solution concentrations. PEG molecules of three molecular weights were grafted: 3400, 10,000, and 20,000 g/mol. Star molecules, on the other hand, having a central core region of poly(divinyl benzene) with PEO arms extending radially from that core (37,38), have a very dense, hard-sphere character which renders these molecules difficult to overlap except at extremely high concentrations (39). They have a large number of chain ends per molecule that are located at the outer regions of the molecule due to the steric hindrance within the core region (40,41). Thus, their probability of grafting to a surface is extremely high and would only decrease with a return to more random-coil characteristics (i.e., with increasing arm molecular weight or decreasing functionality). Two types of star molecules, reference numbers 228 and 3510, were bound to the silicon surface and have the following properties: #228: $\overline{M}w = 200,000$ g/mol, $\overline{M}_{arm} = 10,000$ g/mol, $\bar{f}$ = 20; #3510,: $\overline{M}w = 350,000$ g/mol, $\overline{M}_{arm} = 5200$ g/mol, $\bar{f} = 70$ ($\bar{f}$ is the average number of arms per molecule, or its functionality).

The PEO surfaces were contacted with protein solutions of cytochrome-c, albumin, and fibronectin for 24 hours and the respective adsorptions measured. These three proteins were chosen because they span a wide range of sizes, from 12 kD for cytochrome-c (spherical in shape, diameter 34Å), 68 kD for albumin (spherical in shape, diameter 72Å), to 500 kD for fibronectin (rod-like shape, 600Å long and 25Å wide). This was so adsorption as a function of protein size could also be studied.

Analysis of the surfaces was performed using ellipsometry and X-ray photoelectron spectroscopy (XPS). This included a new method for measuring protein adsorption using XPS by measuring the attenuation of the underlying silicon signal and calculating the thickness of the adsorbed protein layer.

Analysis of PEO Surfaces

Thickness of Dried PEO Layer. XPS high resolution scans of the carbon 1s photoelectron were used to quantify the presence of the PEO on the surface using an aminosilane surface as a reference. Figure 1 shows an example of a high resolution

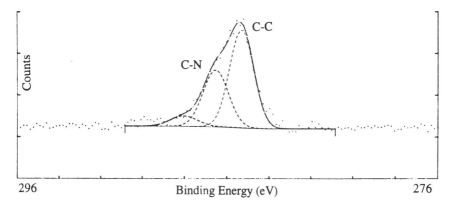

Figure 1. High resolution carbon scan of an aminosilane-coupled silicon surface.

carbon scan for an aminosilane control surface. The largest peak at lowest binding energy is indicative of alkane carbon (C-C), and the smaller, intermediate peak is the amine peak (C-N). The small, highest energy peak is probably due to the presence of carbon dioxide complexed onto the surface. Figures 2 and 3 show the scans for linear PEG 20k and star 3510 coupled surfaces at two different coupling concentrations of tresylated PEO. These scans clearly show the growing intensity of the ether peak of PEO (C-O, shifted 1.5 eV from the C-C peak) as the coupling concentration increases, indicating the increasing PEO content on the surface. It is important to remember that it is the <u>dried</u> (dehydrated) PEO layer that is being analyzed in XPS and ellipsometry, where any inhomogeneities on the surface are averaged out as if the surface had a uniform layer.

The intensity of the ether peak can be used to estimate the dried thickness of the PEO layer through the relation:

$$\frac{I}{I_0} = 1 - \exp\left(\frac{-d}{\lambda \sin(\theta)}\right) \tag{1}$$

where I is the intensity of the ether peak from a certain PEO layer thickness,

I_0 is the intensity from an "infinitely" thick PEO layer,

d is the thickness of the PEO layer (Å),

λ is the attenuation length of carbon 1s photoelectron through an organic layer (Å), and

θ is the take-off-angle used when taking the XPS measurements.

The attenuation length λ was found using the results of Laibinis et al. (42). Knowing θ and λ, and measuring I and I_0, values of the PEO thickness were calculated. These thicknesses were confirmed by ellipsometry, and PEO thickness as a function of tresylated PEO coupling concentration is shown in Figure 4.

Calculation of PEO Grafting Density. The values of the dry thickness of PEO were used to calculate the grafting density, σ , of the PEO molecules on the surface. The definition of grafting density for linear molecules is defined to be (43)

$$\sigma(\text{linear}) = \left(\frac{a}{L}\right)^2 \tag{2}$$

where a is the size of a monomer unit (a $\approx$ 3 Å) (24,27), and L is the average distance between grafted, hydrated chains on the surface. The parameter L can be estimated from knowing the molecular weight, M, and the dry, grafted thickness, h, of PEO on the surface

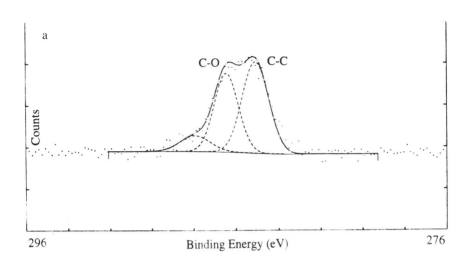

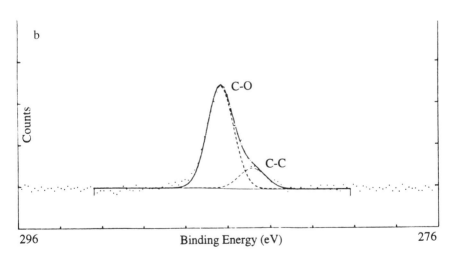

Figure 2. High resolution carbon scans of PEG 20k grafted surfaces from two different coupling concentrations (w/v): (a) 0.5%, (b) 15%.

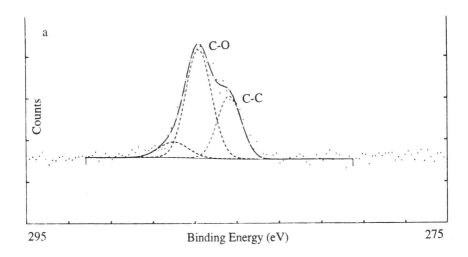

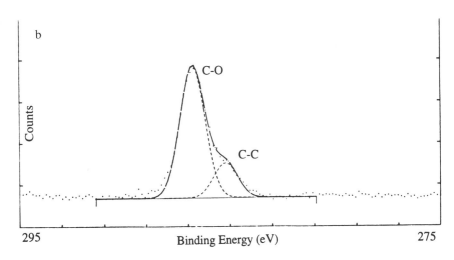

Figure 3. High resolution carbon scans of star 3510 grafted surfaces from two different coupling concentrations (w/v): (a) 0.05%, (b) 12%.

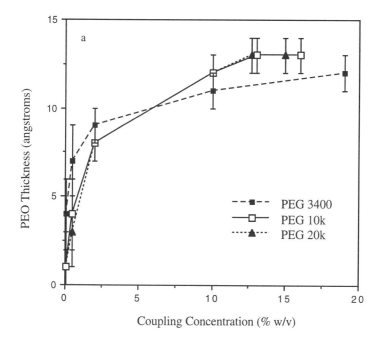

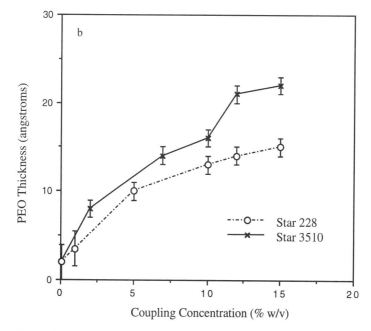

Figure 4. Dry PEO thickness as a function of tresylated PEO coupling concentration for: (a) linear PEG, and (b) star PEO.

$$L(\text{Å}) = \left(\frac{M}{\rho_{dry} \, h \, N_A} \right)^{1/2} \quad (3)$$

where ρ_{dry} is the density of the dry PEG layer and N_A is Avogadro's number.

Due to the large structural difference between linear and star PEO molecules, the size of a monomer unit is not appropriate to use in the calculation of grafting density for stars. Therefore, in place of a in equation 2, we use the radius of gyration, R_G^{star}, of the star molecules, since radius of gyration is a convenient parameter that can be measured or obtained (*39*). Grafting density for a star PEO surface is then

$$\sigma(\text{star}) = \left(\frac{R_G^{star}}{L} \right)^2 \quad (4)$$

PEO Chain Overlap. The relationship between attained grafting density and tresylated PEO coupling concentration for the three linear PEG molecules is shown in Figure 5. All three PEG molecules show the same general behavior of a rapid rise in grafting density at low coupling concentrations, with a leveling off such that a maximum grafting density is attained. This asymptotic behavior has been observed previously (*44,45*) and is mainly due to the interaction between grafted PEO chains on the surface. If we look at a calculation of the critical concentration of PEO, c_{crit}, marking the onset of chain overlap in solution, it can be estimated by

$$c_{crit} = \left(\frac{M}{N_A \, \frac{4}{3} \pi \, R_G^{\,3}} \right) \quad (5)$$

where R_G is the radius of gyration of the PEO molecule. For linear PEO molecules, this can be calculated from Flory (*46*); for stars it can be estimated from the results of Bauer et al. (*39*) or measured from light scattering experiments. Therefore, the critical concentrations for the three linear PEG solutions are $c_{crit}(3400) \approx 8\%$, $c_{crit}(10k) \approx 5\%$, and $c_{crit}(20k) \approx 3\%$ (w/v). The highest coupling concentrations used in the experiments were 2.5 to 5 times larger than these c_{crit} values, indicating there was significant chain overlap in solution. The concentration of PEO at the surface is not necessarily the same as that in the bulk solution, but it is likely that significant chain overlap on the surface was also achieved. It is evident, therefore, that the surface becomes saturated in PEO such that steric hindrance and excluded volume effects have a strong influence on the chains binding to the surface, as expected. There is an increased resistance for additional PEO chains to penetrate the PEO layer and bind to the surface (*47*).

Figure 6 shows the attained grafting density as a function of tresyl-star coupling concentration for both star 228 and star 3510. Unlike the behavior of linear PEO, there

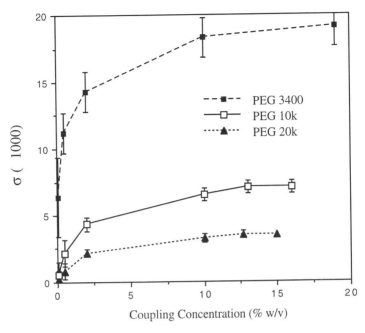

Figure 5. Grafting density as a function of tresylated PEO coupling concentration for linear PEG 3400, 10k, and 20k.

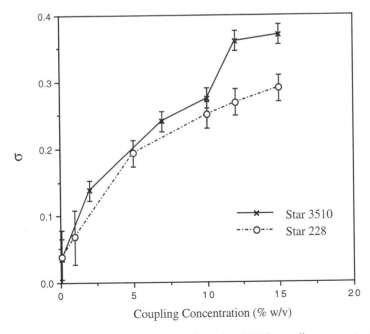

Figure 6. Grafting density as a function of tresylated PEO coupling concentration for stars 228 and 3510.

is no leveling off at a maximum grafting density at large concentrations. Instead, σ continues to increase even at the highest coupling concentrations. Again, we can explain these results by looking at the critical concentration of the molecules in solution: $c_{crit}(228) \approx 15\%$ and $c_{crit}(3510) \approx 13.5\%$. At the highest coupling concentration of 15%, both stars have just reached the point of overlap in solution. Therefore, from this point onward is where we would expect grafting density to reach a maximum due to steric repulsion and excluded volume effects on the surface. The extremely large values of c_{crit} for the stars are testimony to how dense the stars are, where such high concentrations are needed before the stars are forced to interpenetrate one another.

Protein Adsorption on PEO Surfaces

Measurement of Adsorbed Protein. The thickness of the layer of adsorbed protein can be determined by analyzing the attenuation of the silicon 2p photoelectron signal in an XPS scan, as referenced to the equivalent PEO surface not contacted with protein. The attenuation of a photoelectron signal due to an overlayer is described by the equation

$$\frac{I}{I_o} = \exp\left(\frac{-d}{\lambda \sin\theta}\right) \tag{6}$$

where I is the intensity of the Si 2p photoelectron with a protein overlayer,
   I_o is the intensity of the Si 2p with no protein overlayer,
   d is the thickness of the protein overlayer (Å),
   λ is the attenuation length of Si 2p through an organic overlayer (Å), and
   θ is the take-off-angle used in taking the measurements.
Therefore, the protein thickness, d, can be calculated by a measurement of I and I_0 in an XPS scan. This dry protein thickness can be converted to the common measure of adsorbed protein, ng/cm^2 or $\mu g/cm^2$, by estimating that the density of the adsorbed layer is ~1 g/cm^3. This is a safe assumption as, in general, most organic layers have a density close to 1 g/cm^3. In addition, proteins in aqueous solution have a fixed conformation such that they are not hydrated or swelled by water, and are therefore also thought to have a density close to 1 g/cm^3. Therefore, every Å thickness equals ca. 10 ng/cm^2 protein, with zero thickness being $\leq$ 1 ng/cm^2 of adsorbed protein.

Protein Adsorption on Linear PEG Surfaces. Figure 7 shows the adsorption as a function of PEO grafting density for cytochrome-c, albumin, and fibronectin on surfaces grafted with linear PEG 3400, 10k, and 20k, as measured by XPS. Again, these three proteins were chosen because they span a wide range of protein size, such that adsorption as a function of protein size could also be studied. At the lowest grafting densities where grafted PEO chains are sparse, there is maximum adsorption. As grafting density increases, the adsorption of all three proteins declines until it reaches zero adsorption at the highest grafting densities. This general behavior is the same on all three PEG surfaces. There are several important conclusions to draw from these results. First, there is no specific PEO molecular weight, nor universal range of PEO grafting densities, that are necessary for the prevention of protein adsorption, at least in the ranges of protein size and PEG molecular weight studied. Rather, there is a defined minimum in grafting density for a given PEG molecular weight above which significant prevention of protein adsorption is achieved. This minimum correlates exactly with the onset of maximum achievable PEO grafting density (the "knee" of the curves) as shown in Figure 5. If we look more closely at the range of PEG grafting densities, we see that they span a range of PEG chain spacings of $L > R_G^{linear}$ to $L \leq R_G^{linear}$, i.e., from a mushroom to brush regime (43). Of significance is that it is at the transition point of $L \approx R_G^{linear}$, the onset of the brush regime, where the minimum in protein adsorption is reached. The brush region in good solvent is described as being when grafted chains repulse one another due to excluded volume and thus stretch away from the surface and from each other. This results in a more favorable enthalpic interaction of the polymer with the solvent (since $\chi < 0.5$) while sacrificing a loss of entropy due to the chains being more stretched (30,48,49). When a protein interacts with a PEO surface, it is rejected due to a steric repulsion force that is generated from the loss of configurational entropy as the PEO chains are compressed by the approaching protein (27,31,50). This repulsion force arises from a disruption of the PEO chain conformation and its interaction with surrounding water molecules. At high PEO segment densities, as in the brush regime, this repulsion force is then that much greater (27), and therefore we find that the proteins are rejected from the surface. The adsorption behavior seen with our PEG surfaces are in accordance with the findings of Malmsten and van Alstine (45), where PEO amphiphiles were adsorbed in varying densities to methylated silica and phosphatidic acid surfaces and the adsorption of several serum proteins was studied.

    Another important finding from the above results is the fact that protein adsorption behavior on linear PEG surfaces seems to be independent of protein size, at least in the range of protein sizes studied, in that the adsorption of all three proteins decreases from maximum to minimum in the same range of grafting density for a given PEG surface. This can be explained through a comparison of the relative size of the proteins, d_p, compared to the size of the PEO chains, twice R_G. As found by Abbott et al. (51-54), the size of the protein relative to the PEO chain size has a significant effect on the partition coefficient of proteins in PEO/dextran two phase systems. If $d_p \ll 2R_G$, then the protein is small enough to enter the volume of the PEO chain with minimum disruption of the PEO chain, and the partition coefficient is increased. For $d_p \geq 2R_G$, the protein is excluded from the PEO chain volume and thus can only reside in the spacing between hydrated chains. When the PEO chains are overlapped in solution ($c > c_{crit}$), proteins become nearly completely excluded from the PEO rich phase, resulting in a significant decrease in the partition coefficient. The same phenomenon can be said to occur on PEO grafted surfaces, where we found that PEO overlap on the surface in a brush regime is very effective at preventing proteins ($d_p \geq 2R_G$) from reaching and adsorbing to the underlying surface.   In addition, one recent theoretical study has evidenced that the kinetics of protein adsorption can be strongly influenced by not only PEO grafting density, but also on the molecular weight of the grafted PEO chains, in that grafted layers of higher molecular weight PEO, giving thicker PEO layers, can retard the rate of protein adsorption (55).

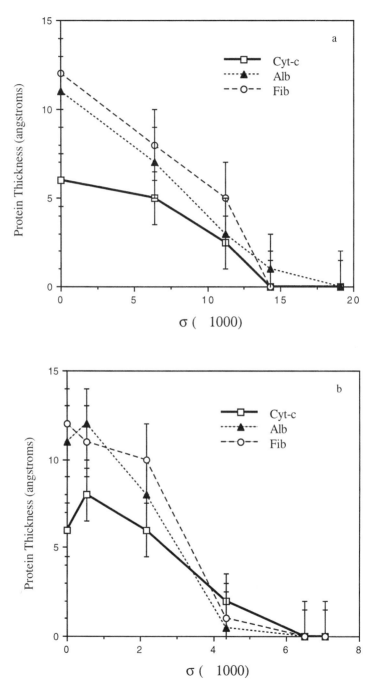

Figure 7. Protein adsorption as a function of linear PEG grafting density:
(a) 3400, (b) 10k, and (c) 20k.

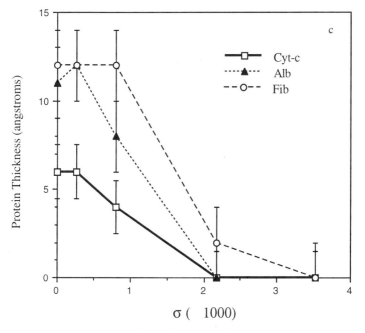

Figure 7. *Continued.*

The above findings are of particular significance when applied to the development of a polymer material having PEO grafted to or incorporated into its surface for the purpose of improving biocompatibility. In order to achieve sufficient PEO surface contents (i.e., a brush regime), the required grafting density increases as the molecular weight of the PEO decreases. In particular, for PEG molecular weights less than 1000-2000 g/mol, the required chain spacings would potentially need to be less than R_G to achieve a brush regime since they are no longer randomly coiling molecules, but rather have a more extended chain conformation when free in solution. This explains why in many of the studies cited in the beginning of this chapter, small PEO molecular weights and contents used in the polymer materials generally exhibited greater protein and platelet adsorption and thus poor enhancement in biocompatibility. Their surface content was too small (i.e., low grafting densities) to be very effective at preventing protein and platelet adsorption.

Protein Adsorption on Star PEO Surfaces. Figure 8 shows the protein adsorption of cytochrome-c, albumin, and fibronectin as a function of grafting density on each of the star PEO surfaces. Albumin and fibronectin show similar adsorption behavior, with a decline from maximum to minimum adsorption occurring in the range of low to intermediate star grafting densities. Cytochrome-c, in contrast, shows markedly different behavior. Rather than declining to zero adsorption, cytochrome-c continued to adsorb at all grafting densities, although with a decreasing trend. We can again look to the spacing of PEO chains to explain this behavior. The fact that the highest coupling concentrations used to bind the PEO stars to the surface was just at the point of chain overlap in solution leads us to conclude that there was also no overlap of star molecules on the surface. This then means that there are open spaces between grafted molecules on the surface. Star molecules are more dense in polymer segments than a linear molecule of equivalent hydrodynamic size, generally 2 to 3 times more dense or greater, as well as being almost melt-like in the core region of the molecule, leading stars to behave as hard spheres in solution. The steric repulsion force caused by an approaching protein would be tremendous, with little probability of a protein compressing or diffusing into a star molecule. Therefore, the only place for proteins to adsorb is in the spaces between star molecules. Protein size as it relates to the size of the open spaces on the surface thus plays a key role in adsorption. From the data shown in Figure 8, we can hypothesize that the grafting density of star molecules reached a point such that the open spaces became small enough to exclude albumin and fibronectin from the surface but not cytochrome-c. Taking into account the size of a hydrated star molecule, the average spacing between molecules on the surface, and the size of the protein, it was indeed found that albumin and fibronectin were excluded from the surface when the spaces between star molecules decreases to less than the size of the protein, but that cytocrhome-c remained smaller than these spaces even at the highest star grafting densities and thus continued to adsorb. Higher grafting densities resulting from higher coupling concentrations (forcing the star molecules into closer contact to overlap) would need to be achieved before the exclusion of small proteins like cytochrome-c could be realized. Overall, a surface fully covered with PEO star molecules has the potential to be much more effective than a linear PEO surface at

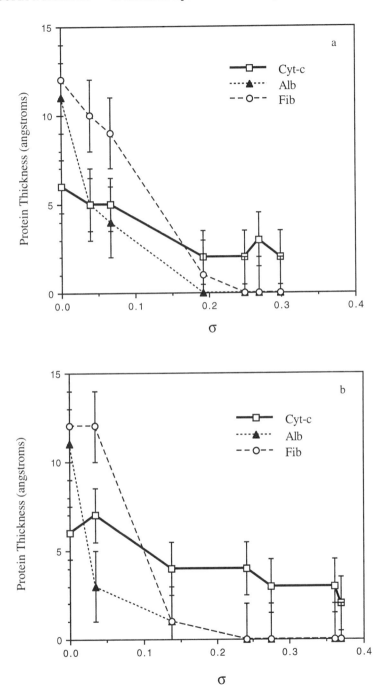

Figure 8. Protein adsorption as a function of PEO star grafting density:
(a) star 228, and (b) star 3510.

preventing protein and cell adsorption, as star molecules, due to their hard-sphere character, are evidenced here to be essentially impenetrable by proteins.

Ellipsometry Results. The protein adsorption of the three proteins on each of the three PEG surfaces and two star surfaces were also measured by ellipsometry (data not shown). Due to the fact that the index of refraction of a protein layer, n_p, can vary by as much as $\pm25\%$ depending on the specific protein that adsorbs, how much protein adsorbs, and possibly on the extent of denaturing, the results from ellipsometry can only be accepted for the general trend they reveal and not for the specific values of layer thickness measured (24). In this regard, the ellipsometry results correlated very well with the results obtained from XPS on all linear and star PEO surfaces.

Conclusions

In this chapter, we presented the key results from a study whose purpose was to investigate the dependence of protein adsorption on PEO surface content, molecular weight, PEO molecular type, and protein size. Silicon wafers provided a model surface on which accurate measurements with XPS and ellipsometry could be made.

The protein adsorption on the linear PEG surfaces was found to be independent of protein size, as $d_p \geq R_G^{linear}$. There was a decline from maximum to minimum adsorption of all three proteins in the same range of grafting density for a given PEG surface. What was found to be of main importance in the prevention of protein adsorption was the transition to a brush regime of the grafted PEO chains on the surface. Protein adsorption dropped to zero when chain spacings became smaller than R_G^{linear} (the chains being at least half-overlapped with each other). These findings are important in that they can be used as a guideline for the development of a material where increased biocompatibility through the grafting or incorporation of PEO into the surface is desired.

The protein adsorption behavior on star PEO surfaces was found to be markedly different than that on linear PEG surfaces. Adsorption on star surfaces is highly dependent on protein size. This is because star molecules are more dense in PEO than a linear molecule of equivalent hydrodynamic size such that they exhibit hard-sphere character in solution. Their dense structure results in the adsorption of proteins to occur only in the available spaces between grafted molecules, thus making the adsorption dependent on the relative sizes of the protein and the open spaces on the surface. Large coupling concentrations of PEO star molecules to greater than c_{crit} are needed to achieve overlap of PEO star molecules in solution as well as on the surface, so that a densely packed layer of star molecules on the surface can be realized.

Literature Cited

1. Mori, Y.; Nagaoka, S.; Takiuchi, H.; Kikuchi, T.; Noguchi, N.; Tanzawa, H.; Noishiki, Y. *Trans, Am. Soc. Artif. Internal Organs* **1982**, *28*, 459.
2. Nagaoka, S.; Mori, Y.; Takuchi, H.; Yokota, K.; Tanyawa, H.; Nishiumi, S. *Polymer Preprints* **1983**, *24*, 67.
3. Brash, J.L.; Uniyal, S. *J. Polym. Sci.* **1979**, *66*, 377.

4. Sa da Costa, V.; Brier-Russell, D.; Trudell, G.; Waugh, D.F.; Salzman, E.W.; Merrill, E.W. *J. Coll. Interface Sci.* **1980**, *76*, 594.
5. Mahmud, N.; Wan, S.; Sa da Costa, V.; Vitale, V.; Brier-Russell, D.; Kuchner, L.; Salzman, E.W.; Merrill, E.W. In *Physical Chemical Aspects of Polymer Surfaces,* Mittall, K.L., Ed., Plenum Press: New York, New York, 1983, Vol. 2.
6. Desai, N.P.; Hubbell, J.A. *Biomaterials* **1991**, *12*, 144.
7. Bots, J.G.F.; van der Does, L.; Bantjes, A. *Biomaterials* **1986**, *7*, 393.
8. Silver, J.H.; Myers, C.W.; Lim, F.; Cooper, S.L. *Biomaterials* **1994**, *15*, 695.
9. Verdon, S.L.; Chaikof, E.L.; Coleman, J.E.; Hayes, L.L.; Connolly, R.J.; Ramberg, K.; Merrill, E.W.; Callow, A.D. *Scanning Microscopy* **1990**, *4*, 341.
10. Chaikof, E.L.; Merrill, E.W.; Callow, A.D.; Connolly, R.J.; Verdon, S.L.; Ramberg, K. *J. Biomat. Mater. Res.* **1992**, *26*, 1163.
11. Sung, C.; Sobarzo, M.R.; Merrill, E.W. *Polymer* **1990**, *31*, 556.
12. Furusawa, K.; Shimura, Y.; Otobe, K.; Atsumi, K.; and Tsuda, K. *Kobanshi Ronbunshu* **1977**, *34*, 309.
13. Grainger, D.; Okano, T.; Kim, S.W. *Advances in Biomedical Polymers,* Plenum Press: New York, New York, 1987; pp. 229-247.
14. Bergstrom, K.; Holmberg, K.; Safrani, A.; Hoffman, A.S.; Edgell, M.J.; Kozlowski, A.; Hovanes, B.A.; Harris, J.M. *J. Biomat. Mater. Res.* **1992**, *26*, 779.
15. Bayer, E.; Rapp, W. In *Poly(Ethylene Glycol) Chemistry: Biotechnical and Biomedical Applications;* Harris, J.M., Ed.; Plenum Press: New York, New York, 1992; pp 325-344.
16. *Elastomers and Rubber Elasticity;* Mark J.E.; Lal, J., Eds.; ACS Symposium Series 193; American Chemical Society: Washington DC, 1982.
17. Desai, N.P.; Hubbell, J.A. *J. Biomed. Mater. Res.* **1991**, *25*, 829.
18. Shard, A.G.; Davies, M.C.; Tendler, S.J.B.; Nicholas, C.V.; Purbrick, M.D.; Watts, J.F. *Macromolecules* **1995**, *28*, 7855.
19. Drumheller, P.D.; Hubble, J.A. *J. Biomat. Mater. Res.* **1995**, *29*, 207.
20. Ferruti, P.; Penco, M.; D'Addato, P.; Ranucci, E.; Deghenghi, R. *Biomaterials* **1995**, *16*, 1423.
21. Ikeda, Y.; Kohjiya, S.; Takesako, S.; Yamashita, S. *Biomaterials* **1990**, *11*, 553.
22. Nagoaka, S.; Nakao, A. *Biomaterials* **1990**, *11*, 119.
23. Merrill, E.W.; Salzman, E.W. *Am. Soc. Artif. Intern. Organs* **1983**, *6*, 60.
24. Prime, K.L.; Whitesides, G.M. *J. Am. Chem. Soc.* **1993**, *15*, 10714.
25. Pale-Grosdemarge, C.; Simon, E.S.; Prime, K.L.; Whitesides, G.M. *J. Am. Chem. Soc.* **1991**, *113*, 12.
26. Osterberg, E.; Berstrom, K.; Holmberg, K.; Riggs, J.A.; Van Alstine, J.M.; Schuman, T.P.; Burns, N.L.; Harris, J.M. *Coll. Surf. A: Phys. Eng. Aspects* **1993**, *77*, 159.
27. Jeon, S.I.; Andrade, J.D.; deGennes, P.G. *J. Colloid Interface Sci* **1991**, *142*, 149.
28. Jeon, S.I.; Andrade, J.D. *J. Collard Interface Sci* **1991**, *142*, 159.
29. Auroy, P.; Auvray, L.; Léger, L. *Phys. Rev. Lett.* **1991**, *66*, 719.
30. Hommel, H.; Halli, A.; Touhami, A.; Legrand, A.P. *Colloids and Surfaces* **1996**, *111*, 67.

31. Gölander, C.-G.; Herron, J.N.; Lim, K.; Claesson, P.; Stenius, P.; Andrade, J.D. In *Poly(Ethylene Glycol) Chemistry: Biotechnical and Biomedical Applications;* Harris, J.M., Ed.; Plenum Press: New York, New York, 1992; pp 221-244.

32. Huang, S.-C.; Caldwell, K.D.; Lin, J.-N.; Wang, H.-K.; Herron, J.N. *Langmuir* **1996**, *12*, 4292.

33. Stenger, D.A.; George, J.H.; Dalcey, C.S.; Hickman, J.J.; Rudolph, A.S.; Nielsen, T.B.; McCort, S.M.; Calvert, J.M. *J. Am. Chem. Soc.* **1992**, *114*, 8435.

34. Nilsson, K.; and Mosback, K., *Biochem. Biophys. Res. Comm.,* **1981**, *102*, 449.

35. Nilsson, K., Mosback, K. *Methods Enzymology* **1984**, *104*, 56.

36. Sofia-Allgor, S.J., PhD. Thesis, Massachusetts Institute of Technology, 1996.

37. Gnanou, Y.; Lutz, P.; Rempp, P. *Makromol. Chem.* **1988**, *189*, 2885.

38. Lutz, P.; Rempp, P. *Makromol. Chem.* **1988**, *189*, 1051.

39. Bauer, B.J.; Fetter, L.J.; Graessley, W.W.; Hadjichristidis, N.; Quack, G. *Macromolecules* **1989**, *22*, 2337.

40. Daoud, M.; Cotton, J.P. *J. Phys.* (Paris, Fr.) **1982**, *43*, 531.

41. Douglas, J.F.; Roovers, J.; Freed, K.F. *Macromolecules* **1990**, *23*, 4168.

42. Laibinis, P.E.; Bain, C.D.; Whitesides, G.M. *J. Phys. Chem.* **1991**, *95*, 7017.

43. DeGennes, P.G. *Macromolecules* **1980**, *13*, 1069.

44. Herder, P.C.; Claesson, P.M.; Herder, C.E. *J. Colloid Interface Sci.* **1988**, *119*, 240.

45. Malmsten, M.; val Alstine, J.M. *J. Colloid Interface Sci.* **1996**, *177*, 502.

46. Flory, P. *Principles of Polymer Chemistry*; Cornell University Press: Ithaca, NY, 1953.

47. Kopf, A.; Baschnagel, J.; Wittmer, J.; Binder, K. *Macromolecules* **1996**, *29*, 1433.

48. Carignano, M.A.; Szleifer, I. *Macromolecules* **1995**, *28*, 3197.

49. Milner, S.T. *Science* **1991**, *251*, 905.

50. Antonsen, K.P.; Hoffman, A.S. In *Poly(Ethylene Glycol) Chemistry: Biotechnical and Biomedical Applications;* Harris, J.M., Ed.; Plenum Press: New York, New York, 1992; pp 15-27.

51. Abbott, N.L.; Blankschtein, D.; Hatton, T.A. *Macromolecules* **1991**, *24*, 4334.

52. Abbott, N.L.; Blankschtein, D.; Hatton, T.A. *Macromolecules* **1992**, *25*, 3917.

53. Abbott, N.L.; Blankschtein, D.; Hatton, T.A. *Macromolecules* **1992**, *25*, 3932.

54. Abbott, N.L.; Blankschtein, D.; Hatton, T.A. *Macromolecules* **1992**, *25*, 5192.

55. Szleifer, I. *Biophysical J.* **1997**, *72*, 595.

Chapter 23

Using Self-Assembled Monolayers That Present Oligo(ethylene glycol) Groups To Control the Interactions of Proteins with Surfaces

Milan Mrksich[1] and George M. Whitesides[2]

[1]Department of Chemistry, University of Chicago, 5735 South Ellis Avenue, Chicago, IL 60637
[2]Department of Chemistry, Harvard University, 12 Oxford Street, Cambridge, MA 02138

This chapter reviews the use of self-assembled monolayers of alkanethiolates on gold to control the interactions of proteins and cells with man-made materials. The work is based on the ability of monolayers that present oligo(ethylene glycol) groups to resist the non-specific adsorption of protein. The chapter describes the use of functionalized monolayers for the bio- and chemo-specific adsorption of proteins. The chapter concludes with a discussion of techniques that can pattern the formation of monolayers and that can prepare tailored substrates for the control of cell attachment.

The property of poly(ethylene glycol) (PEG) to resist the non-specific adsorption of protein has made this material the standard choice for applications requiring inert surfaces. There exist a variety of excellent strategies for tailoring the surfaces of materials with PEG. Most of the methods of using PEG are empirical, and a mechanistic understanding of the ability of PEG to resist adsorption is still incomplete. As a consequence, the structural parameters of PEG that make surfaces presenting it unable to adsorb proteins (that is, inert to adsorption) are still not well understood, and it is not yet routine to design new inert materials—or even simple derivatives of PEG—from basic principles.

We have used self-assembled monolayers (SAMs) of alkanethiolates on gold that present oligomers of the ethylene glycol group ($-EG_nOH$, n = 2-6, and $-EG_6OCH_3$) as model surfaces with which to study the properties of materials tailored with PEG (*1-4*). Several considerations make this class of SAMs the best that is currently available for fundamental studies of the relationships between the structure of a material and its interfacial properties. The structure of these interfaces is reasonably well-defined and stable over the intervals required for experiments involving protein adsorption and cell attachment. They can be systematically tailored using routine organic synthetic methods (*4-6*). Additional considerations that make this system particularly well-suited for studies of bio-interfacial phenomena include the use of surface plasmon resonance spectroscopy to measure the association of proteins with monolayers (*3*) and of microcontact printing to pattern the formation of monolayers (*7*).

This chapter reviews our work that has used monolayers presenting oligo(ethylene glycol) groups to control the interaction of proteins and cells with interfaces. The chapter begins with an introduction to SAMs, then discusses the properties of SAMs presenting oligo(ethylene glycol) groups, the interactions of proteins with functionalized SAMs, and methodologies that use techniques from microfabrication to create substrates that control the attachment of cells.

Self-Assembled Monolayers of Alkanethiolates on Gold.   Self-assembled monolayers of alkanethiolates on gold form upon the adsorption of long chain alkanethiols, RSH [R = $X(CH_2)_n$, n = 11-18] from solution (or vapor) to a gold surface:

$$RSH + Au(0)_n ----> RS^-Au(I) \cdot Au(0)_n + 1/2\ H_2(?)   \text{(eq 1)}$$

Extensive experimental work has shown that the sulfur atoms coordinate to the three-fold sites of the gold(111) surface and the close-packed alkyl chains are trans-extended and tilted approximately 30° from the normal to the surface (Figure 1). The terminal functional group X is presented at the surface and determines the properties of the interface. The properties of SAMs can be controlled further by formation of "mixed" SAMs from solutions of two or more alkanethiols.

Kinetics of Formation of Monolayers.   The mechanisms for the assembly of monolayers are complex and not completely understood. Several groups have studied the kinetics for assembly of alkanethiolates on gold (8-11). This work has used different methods, and although the data are not entirely consistent, most indicate that greater than 90% of the monolayer forms quickly—within minutes for mM solutions of thiol—and the remainder forms more slowly over hours. The kinetics for the initial, rapid assembly of the monolayer are probably dominated by the interaction between the thiolate and gold substrate and gives a monolayer that is locally ordered but contains defects. We presume that the second, slower phase of assembly involves the reordering of alkanethiolates on the surface and transfer of alkanethiol molecules from solution to the remaining vacant sites on the gold substrate.

For many terminal groups, the differences in properties of the monolayers formed under different conditions are minor; the contact angle of water on SAMs of octadecanethiolate, for example, is insensitive to the differences in structure of the phases formed in the terminal stages of assembly (9). For SAMs presenting other groups, however, the properties can change dramatically with increasing density of alkanethiolates. In a subsequent section we describe differences in the adsorption of protein to SAMs presenting oligo(ethylene glycol) groups that depend on the preparation of the monolayers.

Theory of the Mechanisms Underlying the Ability of PEG to Resist Adsorption of Protein.   In aqueous solution, poly(ethylene glycol) chains are solvated and disordered; measurements using NMR spectroscopy (12) and differential thermal analysis (13) indicate that as many as three water molecules are associated with each repeat unit. Further evidence for the large excluded volume of PEG comes from gel chromatography experiments that show PEGs are substantially larger than other polymers of similar molecular weight (14). De Gennes and Andrade have proposed that surfaces modified with long PEG chains resist the adsorption of protein by "steric stabilization" (15,16). Adsorption of protein to the surface causes the glycol chains to compress, with concomitant desolvation. The energetic penalty of transferring water to the bulk and the entropic penalty incurred upon compression of the layer both serve to resist protein adsorption.

Analytical Methods that Measure Adsorption.   Experimental studies of protein adsorption require analytical methods that can measure adsorption with high sensitivity; for example, 10% of a monolayer of a globular protein having a molecular weight of 30 kD corresponds to a density of ~0.3 ng/mm^2. It is also preferable that the techniques measure adsorption in real time to provide kinetic data and are non-invasive in that they do not affect or damage the layer of adsorbed protein. Several groups have used techniques based on ellipsometry (*1,2*), quartz crystal microbalance (*17*), surface acoustic waves (*17*), waveguide interferometry (*19*), and surface plasmon resonance (SPR) (*3*). Ellipsometry remains a convenient technique for measuring the amount of protein on substrates that had been removed from solution and dried; it is less convenient for *in situ* measurements. SPR fulfills all of these criteria and has the additional advantage that instruments are now available from commercial vendors. Because SPR itself uses thin films of gold, it is well-suited for characterizing adsorption on monolayers of alkanethiolates.

SAMs Presenting Oligo(ethylene glycol) Groups Resist the Non-specific Adsorption of Protein. We have used monolayers that present short oligomers of the ethylene glycol group to investigate the basis for PEG to resist the adsorption of protein (*1-3*). Our work has used mixed SAMs prepared from solutions containing a functionalized alkanethiol ($HS(CH_2)_{11}(OCH_2CH_2)_nOH$, n=2-6) and a methyl-terminated alkanethiol ($HS(CH_2)_{10}CH_3$). This system permits control over both the length and density of glycol chains at the surface (Figure 2). Because adsorption can depend dramatically on the structure of a protein (*19*), we have used a panel of representative proteins to characterize these monolayers. Both *ex situ* ellipsometry and *in situ* SPR show that SAMs presenting only oligo(ethylene glycol) groups resist almost entirely the adsorption of protein. Extensive work in our laboratory shows that these surfaces are very effective in resisting adsorption of proteins, and even resist adsorption from concentrated (1-10 mg/mL) solutions of mixtures of protein. Mixed SAMs presenting this group together with as much as 50% hydrophobic, methyl groups also resist the adsorption of protein; SAMs presenting methyl groups alone adsorb most proteins rapidly and irreversibly. The ability of the surface to resist adsorption increases with both the density and length of the oligomer.

Mechanisms of Inert SAMs. Our work shows that surfaces presenting densely packed short oligomers of the ethylene glycol group are highly effective at resisting the adsorption of protein. It is not clear that the mechanisms for this resistance of SAMs presenting short, oligo(ethylene glycol) chains are similar to those for high molecular weight PEG. The extensive solvation of PEG by water molecules is almost certainly critical to the properties of the bulk polymer, but SAMs presenting dense packed oligo(ethylene glycol) groups probably do not have sufficient volume to accommodate extensive solvation. Molecular modeling studies even suggest that the perfectly ordered SAMs cannot include any solvent (*20*). Because the glycol chains in the monolayers are each covalently tethered to the surface, these thin films should have conformational properties very different from those of the unconstrained bulk polymer.

Recent work from our laboratory, and that of Professor Michael Grunze at Heidelberg, have shown that the conditions used to prepare the SAMs are critical to the interfacial properties. SAMs prepared from solutions of a hexa(ethylene glycol)-terminated alkanethiol for periods of less than 12 hr—the usual conditions—resist the adsorption of protein. SAMs prepared from these same solutions, but allowing the equilibration of the structure of the SAMs to proceed for periods of 1-7 days in contact with the solution of thiol, are less effective in resisting the adsorption of protein; the amount of protein that adsorbs irreversibly to these surfaces increases with the time over which formation of the SAM is allowed to occur.

One explanation for these data is that the final stage of the assembly of alkanethiols substituted with oligo(ethylene glycol) groups onto a gold surface is slow, and that the interfacial properties depend strongly on the density of alkanethiolates in the monolayer. Immersion times of 12 hr may give SAMs that still have a substantial number of vacant coordination sites on the gold surface—we have no experimental data to suggest what this critical density of holes may be—and a surface that resists adsorption of protein. Longer immersion times may give SAMs having fewer vacant coordination sites and defects, and that adsorb protein. Other

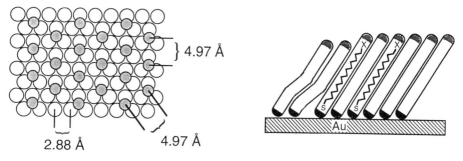

Figure 1. Representation of a self-assembled monolayer (SAM) of alkanethiolates on the surface of gold. (Left) Hexagonal coverage scheme of thiols coordinated to the gold (111) surface; the sulfur atoms (shaded circles) fill the hollow three-fold sites on the gold surface (open circles). (Right) The alkyl chains are close-packed and tilted approximately 30° from the normal to the surface. The properties of the SAM are controlled by changing the length of the alkyl chain and the terminal functional group X of the precursor alkanethiol. The missing row represents a common defect present in SAMs. The detailed structures of point and line defects have not been established.

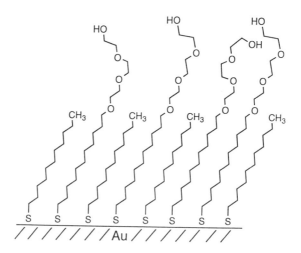

Figure 2. Representation of a mixed SAM terminated in methyl groups and tri(ethylene glycol) groups. We presume that the polymethylene chains are more ordered than the glycol groups. The density of the tri(ethylene glycol) groups at the surface is determined by the ratio of the two alkanethiols in the solution from which the SAM is formed.

factors that may be important in understanding the mechanisms for adsorption are the conformation of the oligo(ethylene glycol) groups in the monolayers and the role of other defects (for example, those produced by corrosion of the gold substrate). We are currently investigating this system in greater detail.

Design of Surfaces that Resist Adsorption. We sought to design a new material that shared the properties of PEG to resist the adsorption of protein. We chose the propylene sulfoxide group because oligomers of this group, like those of ethylene glycol, are hydrophilic, well-solvated by water and conformationally flexible. We prepared monolayers presenting tri(propylene sulfoxide) groups $(HS(CH_2)_{11}(SOCH_2CH_2CH_2)_3SOCH_3)$ (Figure 3). SPR showed that these SAMs resisted the non-specific adsorption of fibrinogen, RNAse A, and other proteins (*21*). These SAMs remained inert to adsorption even when mixed with as much as 50% methyl-terminated alkanethiolate. Although the different lengths of these two groups prevents a direct comparison, the data suggest that when presented at an interface they are similarly effective at resisting adsorption. The most important result from this work is the demonstration of a successful process—that is centered on the combination of SAMs and SPR—for the *de novo* design and testing of a new inert material. This study also suggests that PEG is not unique in its ability to serve as an inert surface, but that there will probably be many such polymers.

The inert surfaces provided by these monolayers serve as the basis for the design of biointerfaces having other properties. For example, ligands can be immobilized to these SAMs to create substrates that bind a specific receptor yet still resist the non-specific adsorption of other proteins. These inert monolayers also make possible a convenient and flexible methodology for creating patterned substrates that control the attachment of mammalian cells. The remainder of this chapter describes our work in these areas.

Bio-Specific Adsorption. The bio-specific adsorption of proteins to surfaces presenting appropriate ligands is important in drug screening, cell culture, biosensing, and other areas. These applications have also used empirical approaches and few studies have investigated fundamental aspects of biomolecular recognition at surfaces. The most serious problem encountered with surfaces designed for bio-specific adsorption is the non-specific adsorption of other proteins to the surface. The common strategy of coating a material with serum albumin, for example, suffers from poor reproducibility in the adsorption and from limited stability of the protein layer (*22*).

We have used SAMs presenting tri(ethylene glycol) groups and benzensulfonamide groups as model substrates with which to study the bio-specific adsorption of carbonic anhydrase (*23*). SPR showed that the protein bound reversibly to these SAMs and provided kinetic rate constants for association and dissociation (Figure 4). The binding of protein was bio-specific; addition of a soluble ligand of the CA to the protein-containing solution prior to the binding experiment inhibited adsorption of the protein to the surface. The amount of protein that bound at saturation increased with the density of the ligand on the surface; this density could be controlled by adjusting the ratio of the two alkanethiols in the solution from which the monolayer assembles. When a complex mixture containing nine proteins (2 mg/mL total concentration) was introduced into the flow cell, SPR recorded essentially no protein adsorption; however, when CA was present in this complex sample, SPR measured binding of the protein with no complications due to the other proteins (Figure 4). This system provides a convenient model for biophysical studies of biointerfacial recognition.

Chemo-Specific Adsorption. We have demonstrated a related immobilization strategy based on the well-known coordination of oligo(histidine) peptides by complexes of nickel (II) (*24*). Mixed SAMs presenting nitrilotriacetic acid (NTA)

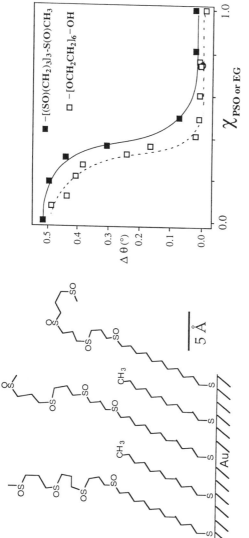

Figure 3. (Left) Representation of a mixed SAM terminated in methyl groups and tri(propylene sulfoxide) groups; the proportion of functionalized alkanethiolate is given by χ. The plot on the right compares the adsorption of fibrinogen on these SAMs with SAMs terminated in hexa(ethylene glycol) groups. The amount of protein that adsorbed to the monolayers was measured using surface plasmon resonance spectroscopy and is reported as a change in resonance angle ($\Delta\theta$) upon binding (3, 21).

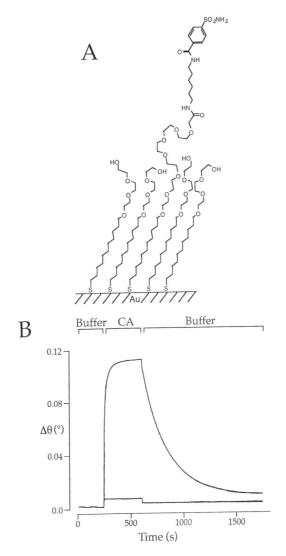

Figure 4. SPR was used to measure the rate and quantity of binding of carbonic anhydrase (CA) to a SAM terminated in EG$_3$ groups and benzenesulfonamide groups (A). The change in resonance angle ($\Delta\theta$) of light reflected from the SAM/gold is plotted against time; the time over which the solution of CA (5 μM) was allowed to flow through the cell is indicated at the top of the plot (B). The upper curve shows binding (and dissociation) of CA to a SAM containing ~5% of the ligand-terminated alkanethiolate. CA did not adsorb to a SAM presenting only ethylene glycol groups (lower curve). A response due to the change in index of refraction of the CA-containing solution was observed upon introduction of protein into the flow cell (evident in the lower curve). The difference between the measured response and this background signal represents binding of the CA to the SAM.

chelates of Ni(II) and tri(ethylene glycol) groups were used to capture histidine-tagged recombinant proteins from cell extracts (Figure 5). Only the his-tagged proteins adsorbed to the SAM; the other proteins in the sample did not interfere with the coordination nor did they adsorb to the SAM. This immobilized, his-tagged protein was stable but could be removed rapidly by adding imidazole as a competing ligand for the NTA group. This system has the additional advantages that the immobilized protein is presented in a single orientation and the density of protein can be controlled.

Using Microcontact Printing to Pattern Monolayers. Several groups have used photolithographic methods to pattern the formation of monolayers; these methods work well but the requirement for a lithography facility makes them inconvenient and inaccessible to many biological researchers. We have developed a new and convenient method for patterning SAMs of alkanethiolates on gold with features of sizes ranging down to 1 μm (7,25,26). Microcontact printing (μCP) uses an elastomeric stamp having on its surface a pattern of relief at the micron scale (Figure 6). This stamp can be coated with a solution of alkanethiol, dried, and brought into contact with a surface of gold to transfer the alkanethiol to discreet regions of the substrate. This process produces a pattern of SAM on the gold that is identical to the pattern of relief in the stamp. A different SAM can then be formed in the remaining regions of gold by immersing the substrate in a solution of the other alkanethiol. Conformal contact between the elastomeric stamp and surface allow surfaces that are rough (at the scale of 100 nm) to be patterned over areas several cm^2 in size with edge resolution of the features better than 50 nm. Multiple stamps can be cast from a single master and each stamp can be used hundreds of times. Microcontact printing has been used to pattern SAMs of alkylsiloxanes on oxide substrates (27) and can even form patterns on curved substrates (28).

Patterning the Adsorption of Protein. Microcontact printing can prepare substrates that adsorb protein in patterns. The method begins by contact printing a SAM of hexadecanethiolate on a gold substrate to give a pattern of hydrophobic, methyl-terminated SAM. Rinsing this substrate in a solution of oligo(ethylene glycol)-terminated alkanethiol renders the remaining regions of gold inert to protein adsorption. Immersion of the patterned substrate in a solution of protein results in the rapid and irreversible adsorption of protein to the hydrophobic, methyl-terminated regions of the monolayer (Figure 7). Scanning electron microscopy provides a convenient method for imaging the patterned protein (29). This method is experimentally simple and can pattern proteins at the micron scale. It has the limitation that it cannot pattern the adsorption of multiple proteins to a single substrate. Photolithographic methods that combine immobilization chemistries have been used to pattern the formation of multiple proteins on a single substrate (30).

Patterning the Attachment of Cells on Planar Substrates. This same methodology for patterning the adsorption of protein can be used to prepare substrates for patterning the attachment of mammalian cells (31,32). For attachment to surfaces, cells use membrane receptors to recognize immobilized ligands normally found in the extracellular matrix proteins. Consequently, surfaces presenting a pattern of a matrix protein will direct the attachment and spreading of cells, provided that the intervening regions of surface are inert to attachment. We have prepared substrates containing a pattern of adsorbed fibronectin (the most common matrix protein) and oligo(ethylene glycol) groups. Addition of a suspension of capillary endothelial cells to the substrate resulted in the attachment of cells only to the protein-coated regions (Figure 8). The spreading of the attached cells was also confined to the underlying pattern of protein (and SAM). This methodology was also used to pattern the attachment of *individual* hepatocytes (31).

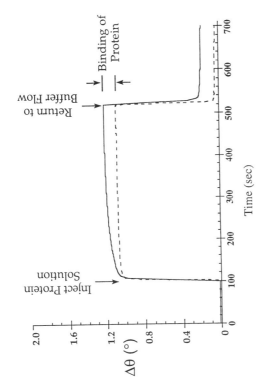

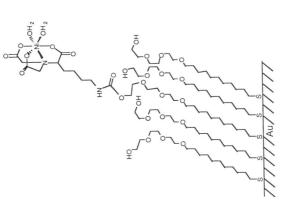

Figure 5. SPR was used to measure the rate and quantity of binding of a His-tagged T-cell receptor construct to a SAM terminated in EG_3 groups and Ni(II) complexes (24). (Left) The mixed SAM contains ~ 5% of the Ni(II)-functionalized alkanethiol. Imidazole rings of the His-tagged protein replace the water ligands of the Ni(II) complex. (Right) The change in resonance angle ($\Delta\theta$) of light reflected from the SAM/gold is plotted against time. A large response due to the change in index of refraction of the solution was observed upon introduction of protein into the flow cell (denoted by the dashed curve). The difference between the measured response and this background signal represents binding of the His-tagged protein to the SAM.

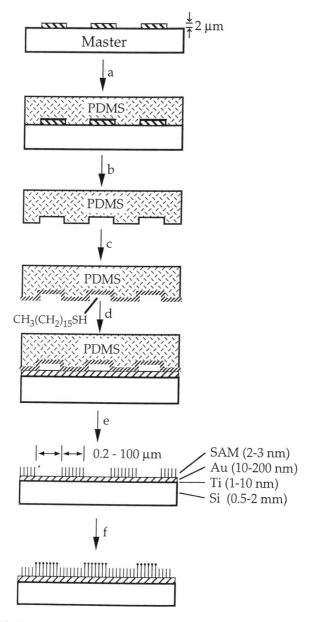

Figure 6. Microcontact printing starts with a master template containing a pattern of relief (a); this master can be fabricated by photolithography, or other methods. A polydimethylsiloxane (PDMS) stamp cast from this master (b) is "inked" with a solution of alkanethiol in ethanol (c) and used to transfer the alkanethiol to surface of gold (d); a SAM is formed only at those regions where the stamp contacts the surface (e). The bare regions of gold can then be derivatized with a different SAM by rinsing with a solution of a second alkanethiol (f).

-EG$_6$OH   -CH$_3$/Fibrinogen

100 μm

Figure 7. Scanning electron micrographs of fibrinogen adsorbed on a patterned SAM. A patterned hexadecanethiolate SAM on gold was formed by microcontact printing and the remainder of the surface was derivatized by exposure to a hexa(ethylene glycol)-terminated alkanethiol (HS(CH$_2$)$_{11}$(OCH$_2$CH$_2$)$_6$OH). The patterned substrate was immersed in an aqueous solution of fibrinogen (1 mg/mL) for 2 hr, removed from solution, rinsed with water, and dried. Fibrinogen adsorbed only to the methyl-terminated regions of the SAM, as illustrated by the dark regions in the SEM micrograph: secondary electron emission from the underlying gold is attenuated by the protein adlayer.

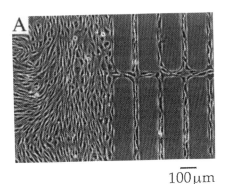
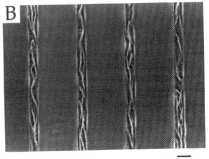

100 μm   50 μm

Figure 8. Control over the attachment of bovine capillary endothelial cells to planar substrates that were patterned into regions terminated in methyl groups and tri(ethylene glycol) groups using μCP. The substrates were coated with fibronectin prior to cell attachment; fibronectin adsorbed only to the regions of methyl-terminated SAM. (A) An optical micrograph showing attachment of endothelial cells to a non-patterned region (left) and to lines 30 μm in width. (B) A view at higher magnification of cells attached to the lines.

Controlling the Attachment of Cells on Contoured Substrates. We have combined this methodology with techniques for microfabrication to prepare contoured substrates that direct the attachment of cells (*33*). An elastomeric stamp was used to mold a film of polyurethane into alternating grooves and plateaus 50 μm in width. A thin, optically-transparent film of gold was evaporated onto this substrate on which monolayers could be formed. A flat PDMS stamp was used to form a SAM of hexadecanethiolate on the raised plateaus of the contoured surface by contact printing hexadecanethiol and a SAM terminated in tri(ethylene glycol) groups was subsequently formed on the bare gold remaining in the grooves by immersing the substrate in a solution of a second alkanethiol. Figure 9 shows that endothelial cells attached and spread only on the hydrophobic regions of the substrate that presented fibronectin.

Conclusions

The work described in this chapter presents a comprehensive methodology suitable for the study of biointerfacial phenomena. The flexibility offered by self-assembled monolayers to tailor the properties of an interface and present biologically relevant groups—including molecules, peptides and proteins—provides an opportunity to understand, in detail, the relationship between interfacial structure and properties. A range of analytical techniques, and surface plasmon resonance in particular, provide a methodology to understand the interactions and dynamics of surfaces with proteins and cells. Microcontact printing and related techniques for microfabrication make possible the design of a range of substrates with which to control and understand the biological responses to materials. This combination of techniques has already made possible new types of experiments relevant to biosurfaces and will certainly be important in work that follows.

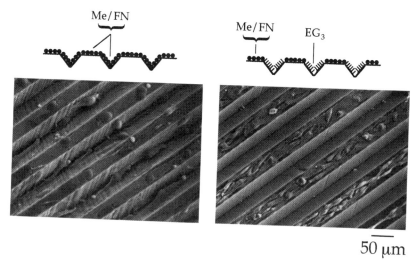

Figure 9. Control over the attachment of endothelial cells to contoured surfaces using self-assembled monolayers. The substrates are films of polyurethane (supported on glass slides) that were coated with gold and modified with SAMs of alkanethiolates terminated in methyl groups and tri(ethylene glycol) groups; the substrates were coated with fibronectin prior to cell attachment. (Left) Cells attached to both the ridges and grooves of substrates presenting fibronectin at all regions. (Right) Cells attached only to the ridges when the grooves were modified with a SAM presenting tri(ethylene glycol) groups.

Acknowledgments

This work was supported by the National Institutes of Health (GM 30367), the Office of Naval Research, and the Advanced Research Projects Agency.

Literature Cited

1. Prime, K.L.; Whitesides, G.M. *Science* **1991**, *252*, 1164.
2. Prime, K.L.; Whitesides, G.M. *J. Am. Chem. Soc.* **1993**, *115*, 10714.
3. Mrksich, M.; Sigal, G. B.; Whitesides, G.M. *Langmuir* **1995**, *11*, 4383.
4. Mrksich, M.; Whitesides, G.M. *Annu. Rev. Biophys. Biomol. Struct.* **1996**, *25*, 55.
5. Ulman, A. *Chem. Rev.* **1996**, *96*, 1533.
6. Dubois, L.H.; Nuzzo, R.G. *Ann. Rev. Phys. Chem.* **1992**, *43*, 437.
7. Mrksich, M.; Whitesides, G.M. *Trends in Biotechnol.* **1995**, *13*, 228.
8. Pan, W.; Durning, C.J.; Turro, N.J. *Langmuir* **1996**, *12*, 4469.
9. Bain, C.D.; Troughton, E.B.; Tao, Y.-T.; Evall, J.; Whitesides, G.M.; Nuzzo, R.G. *J. Am. Chem. Soc.* **1989**, *111*, 321.
10. Shimazu, K.; Yag, I.; Sato, Y.; Uosaki, K. *Langmuir* **1992**, *8*, 1385.
11. Schneider, T. W.; Buttry, D. A. *J. Am. Chem. Soc.* **1993**, *10*, 3315.
12. Breen, J.; Huis, D.; Bleijser, J. de; Leyte, J. C. *J. Chem. Soc. Faraday Trans. 1* **1988**, *84*, 293.
13. Vringer, T. de; Joosten, J. G. H.; Junginger, H. E. *Colloid Polym. Sci.* **1986**, *264*, 623.
14. Hellsing, K. *J. Chromatogr.* **1968**, *36*, 270.
15. Jeon, S.I.; Lee, J.H.; Andrade, J.D.; De Gennes, P.G. *J. Coll. Inter. Sci.* **1991**, *142*, 149.
16. Jeon, S.I.; Andrade, J.D. *J. Coll. Inter. Sci.* **1991**, *142*, 159.
17. Ward, M. D. and Buttry, D. A. *Science* **1990**, *249*, 1000.
18. Schlatter, D., Barner, R., Fattinger, Ch., Huber, W., Hubscher, J., Hurst, J., Koller, H., Mangold, C. and Muller, F. *Biosensors & Bioelectronics* **1993**, *8*, 109.
19. Ramsden, J.J.; Rousch, D.J.; Gill, D.S.; Kurrat, R.; Willson, R.C. *J. Am. Chem. Soc.* **1995**, *117*, 8511.
20. Chin, D. and Whitesides, G. M. unpublished.
21 Deng, L.; Mrksich, M.; Whitesides, G.M. *J. Am. Chem. Soc.* **1996**, *118*, 5136.
22. Bekos, E. J.; Ranieri, J. P.; Aebischer, P.; Gardella, J. A.; Bright, F. V. *Langmuir* **1995**, *11*, 984.
23. Mrksich, M.; Grunwell, J.R.; Whitesides, G.M. *J. Am. Chem. Soc.* **1995**, *117*, 12009.
24. Sigal, G.B.; Bamdad, C.; Barberis, A.; Strominger, J.; Whitesides, G.M. *Anal. Chem.* **1996**, *68*, 490.
25. Wilbur, J. L.; Kumar, A.; Biebuyck, H. A.; Kim, E.; Whitesides, G. M. *Nanotechnology* **1997**, in press.
26. Kumar, A.; Biebuyck, H. A.; Whitesides, G. M. *Langmuir* **1994**, *10*, 1498.
27. Xia, Y.; Mrksich, M.; Kim, E.; Whitesides, G. M. *J. Am. Chem. Soc.* **1995**, *117*, 9576.
28. Jackman, R. J.; Wilbur, J. L.; Whitesides, G. M. *Science* **1995**, *269*, 664.
29. Lopez, G. P.; Biebuyck, H. A.; Harter, R.; Kumar, A.; Whitesides, G. M. *J. Am. Chem. Soc.* **1993**, *115*, 10774.
30. Jacobs, J. W. and Fodor, S. P. A. *Trends in Biotechnol.* **1994**, *12*, 19.
31. Singhvi, R.; Kumar, A.; Lopez, G.P.; Stephanopoulos, G.N.; Wang, D.I.C.; Whitesides, G.M.; Ingber, D.E. *Science* **1994**, *264*, 696.
32. Lopez, G.P.; Albers, M.W.; Schreiber, S.L.; Carroll, R.; Peralta, E.; Whitesides, G.M. *J. Am. Chem. Soc.* **1993**, *115*, 5877.
33. Mrksich, M.; Chen, C. S.; Xia, Y.; Dike, L. E.; Ingber, D. E.; Whitesides, G. M. *Proc. Natl. Acad. Sci. USA* **1996**, *93*, 10775.

Chapter 24

Electrokinetic Analysis of Poly(ethylene glycol) Coating Chemistry

Kazunori Emoto[1], J. Milton Harris[1,3], and James M. Van Alstine[1,2]

[1]Department of Chemistry, The University of Alabama at Huntsville,
Huntsville, AL, 35899
[2]Departments of Chemical Engineering and Biotechnology, Royal
Institute of Technology, S–100 44, Stockholm, Sweden

Characterization of surface coatings formed with poly(ethylene glycol) (PEG) or related hydrophilic, neutral polymers can be achieved by means of free solution capillary electrophoresis. The method is nondestructive, relatively fast, mechanically simple, is performed in aqueous solution, and can be automated. Mathematical modeling provides estimates of surface density, pK of surface chemical groups and coating thickness for PEG coatings and for organosilane and other treatments used to activate surfaces prior to PEG grafting. This paper reviews recent studies by the authors related to coating quartz surfaces with PEG. Variables studied include surface activation, polymer molecular weight, linking chemistry, and hydrolytic and oxidative stability. The method improves understanding of PEG on surfaces, and provides methods for obtaining stable PEG coatings of controlled density. Chemistry for tethering molecules via PEG linkers is also discussed.

There is a recognized need for new analytical methods which allow rapid, nondestructive, analysis of polymer coatings under aqueous conditions (1,2). Such methods should augment results obtained by ellipsometry, x-ray photoelectron spectroscopy, Fourier transform infrared spectroscopy, and other surface chemical techniques. The results obtained should provide applications-related information such as grafting density, layer thickness, stability, and ability to "mask" underlying surface groups. Electrokinetic evaluation satisfies these criteria for many surfaces.

This chapter describes the use of free solution capillary electrophoresis for evaluating poly(ethylene glycol) (PEG) and related neutral polymer coatings on glass (2). The experimental method involves microscopic monitoring of the electrophoretic mobility of a charged particle in a coated quartz or glass capillary under an applied electric field. The influence of the coating on

[3]Corresponding author

particle mobility allows semiquantitative evaluation of various aspects of coating chemistry. However much of the information provided in the chapter also relates to polymer coatings on a variety of surfaces.

Although some theory is introduced in the chapter, the reader may wish to consult other references (*e. g., 3, 4*). More complete information related to the use of electrokinetic methods, including the use of automated apparatus with rectangular chambers, is available (*3-11*). It should be noted that electrokinetic characterization is also applicable to materials such as biological cells (*12,13*), polymer particles (*14-17*), and hydrogels (*18*).

A brief introduction to the technique of free solution capillary electrophoresis will be given followed by presentation of necessary theory and then a fuller discussion of results.

Free Solution Capillary Electrophoresis. A surface in contact with an aqueous solution generally possess acidic and basic groups whose surface density and pK influence the magnitude and nature of surface charge. Oppositely charged ions (counterions) associated with surface groups are unequally distributed near the surface, producing a zeta potential whose magnitude decreases as a function of distance from the surface. Because of the presence of water molecules attracted to the surface, the electrostatic potential measured electrikonetically is the potential at the "plane of shear" where attraction of water molecules to the surface is negligible. This potential reflects the charged nature of the surface and the medium composition (zeta potential varies inversely with bulk solution ionic strength). Free solution electrokinetic methods involve measurement of physical phenomena (*e. g.,* electroosmosis, electrophoretic mobility, streaming potential) associated with the zeta potential and thus provide insight into surface potentials and their related chemical groups.

Figure 1 schematically shows the apparatus we have used to measure electrophoretic mobility of particles and electroosmotic fluid mobility. When a potential is imposed on an aqueous solution in a capillary, migration of ions (and associated water molecules) occurs. The inequality in ion concentration near the surface results in net fluid flow at the wall ("electroosmosis") (*3*). In a closed cylindrical capillary, a compensatory "return" flow (Figure 2) will also exist. If a charged particle is suspended in the capillary, a parabolic mobility profile will result because of the imposition of electroosmotic and return flows on the inherent electrophoretic migration of the charged particle, Figure 3. Inherent particle mobility, which is a function only of particle surface charge, can be determined at the "stationary level" where the electroosmotic and return flows cancel (Figure 2). Inherent particle mobility is a parameter often used to characterize surfaces of colloidal particles (*5-8,12-20*).

Figure 3 was generated by use of the apparatus in Figure 1 to measure microsphere mobility at certain positions across a millimeter-diameter quartz capillary. Sulfated polystyrene microspheres were used as their surface charge is relatively constant from pH 3 to 11 and decreases reproducibly between pH 1 and 3 (*7*). The observed direction of electroosmotic flow

Modified Rank Electrophoresis Chamber

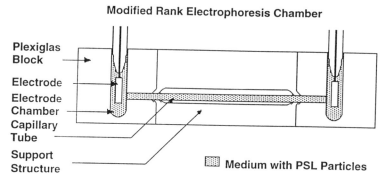

Plexiglas
Block

Electrode

Electrode
Chamber

Capillary
Tube

Support
Structure

▦ Medium with PSL Particles

Figure 1. Schematic of a Rank Mark I electrophoresis chamber modified for capillary chamber. Water bath, electrical connections and optics are not shown.

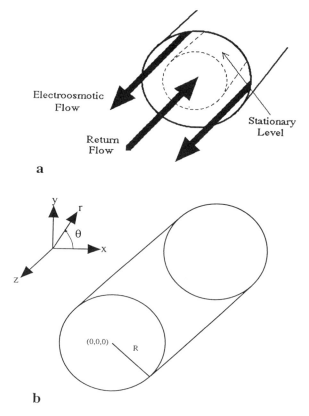

Figure 2. (a) Schematic of electroosmotic and compensatory flows, plus stationary level in closed capillary, and (b) coordinate system for describing the hydrodynamics of electroosmosis in the closed cylindrical chamber. The optical path is along the x-axis.

indicates the sign of charge on the capillary surface; by definition, positive flow indicates a negative surface and negative flow indicates a positive surface (*3*). The electroosmosis data of Figure 3 can be used to prepare a plot of electroosmotic mobility *versus* pH, Figure 4. Electroosmotic mobility is essentially the difference in particle mobility between the stationary level and the mobility in the center of the capillary (a more precise definition is given below). The data of Figure 4 were obtained by performing electroosmotic measurements on native and aminopropylsilane-modified quartz capillaries. For the native capillary, as pH increases surface silanol (SiOH) groups deprotonate to produce negatively charged siloxy (SiO$^-$) groups. The pH dependency is shown in Figure 4. Those who do not wish to delve too deeply into the theoretical section, which follows this introduction, can think of Figure 4 (and related figures) as a surface titration curve which reflects both surface charge type (+ or –) , as well as concentration and pK values related to surface charge groups.

As expected, grafting aminopropylsilane (APS) groups onto surface silanols (chemistry shown in Figure 5) significantly alters the electroosmosis versus pH "profile" by producing a net positive surface below about pH 7, Figure 4 (*10*). Figure 4 also shows that "hydrothermal treatment" (6 h, 60 $^\circ$C, 0.05 M Na phosphate, pH 10.3) removes some APS groups and renders a surface more like that of the native quartz capillary. The topic of silane modification is discussed in more detail below.

Electroosmotic flow can be reduced by adsorbing (*19*) or grafting (*20-22*) neutral, hydrophilic polymers onto a surface (some immobilization methods are shown in Figure 5). This reduction is primarily the result of the polymer coating creating a viscous layer that "masks" the surface by shifting the hydrodynamic plane of shear away from the surface to a position of significantly reduced zeta potential. These same coatings reduce particle mobility if the polymers are bound onto the particle (*21*). The magnitude of this effect is directly related to PEG molecular weight (*9,10,20-22*) (Figure 6) and grafting density (*10*) (see below).

PEG coatings also reduce nonspecific protein adsorption (*7,8*) (Figure 6) and improve wetting (*23,24*), and thus they are used for numerous biomaterials applications (*1,2*). These coatings improve the performance of fused silica capillaries used in analytical microcapillary electrophoresis (CE), (*20-27*), and they can be used to control various surface phenomena that adversely affect other separation methods (*28-31,34,35*).

Electrokinetic characterization is not limited to qualitative comparisons. Research by the authors and colleagues N. Burns and K. Holmberg at the Institute for Surface Chemistry, Stockholm (*9,11*) has built on the work of other groups (*15,32,33*) in applying electrokinetic theory to mathematically model electroosmosis *versus* pH data for various surfaces. This modeling can produce estimates of the surface density and pK associated with surface groups. The information obtained is not limited to charged groups. As described in more detail below, one can also obtain estimates of neutral

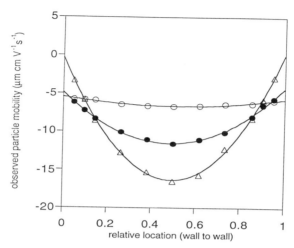

Figure 3. Measured mobility in 7.5 mM NaCl of a sulfated polystyrene latex particle sample, relative to location across the 1 mm diameter of a quartz cylindrical chamber. Profiles are associated with varying degrees of electroosmosis at pH 2 (o), pH 6 (●) and pH 11 (Δ). Error bars are approximated by symbol size.

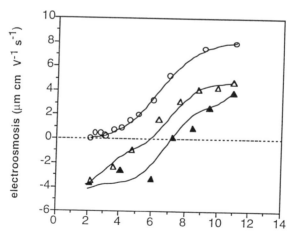

Figure 4. Electroosmosis profiles at 7.5 mM of (o) native, (▲) APS-modified, and (Δ) APS-modified then hydrothermally treated (6h, 60 °C, 0.05 M sodium phosphate pH 10.3) capillaries. Solid lines are modeling fits to experimental data shown by symbols.

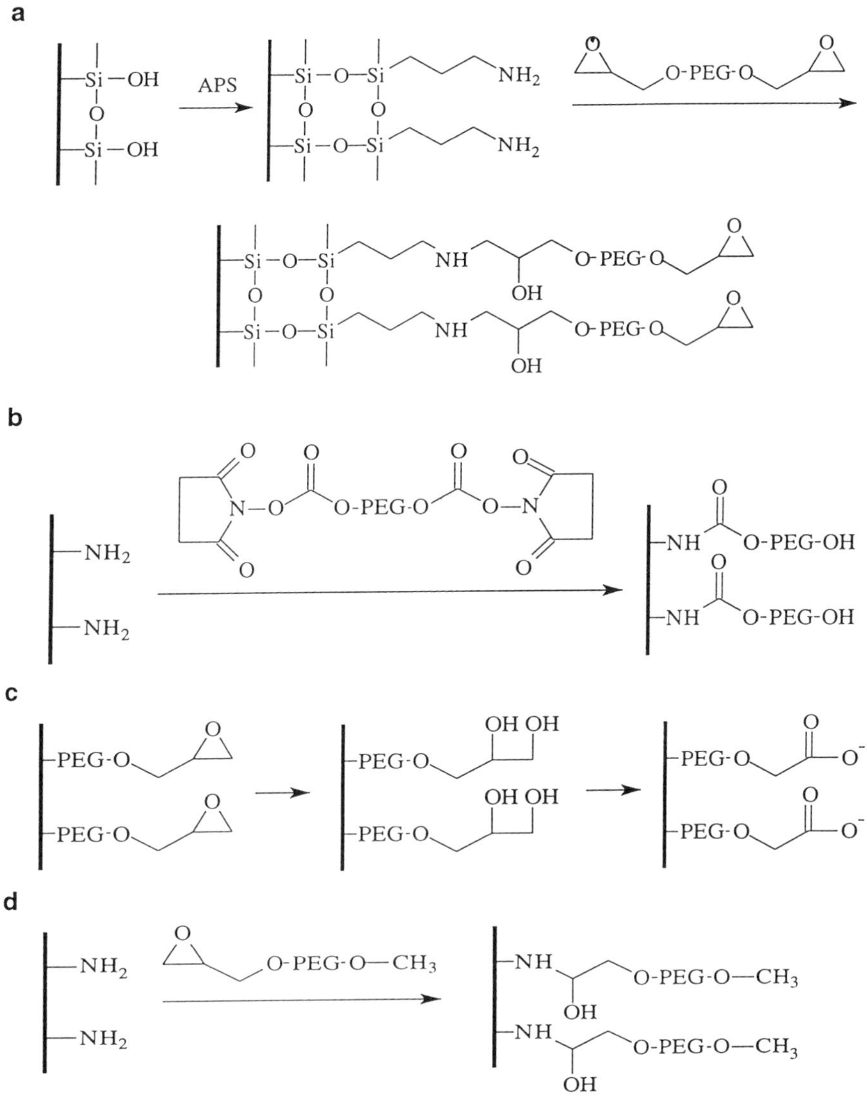

Figure 5. Reaction of (a) quartz with aminopropylsilane (APS) and APS-modified quartz with di-epoxide PEG (di-E-PEG), (b) APS-modified quartz with di-succinimidylcarbonate-PEG (di-SC-PEG), and hydrolysis of unreacted SC groups to hydroxyl groups, (c) hydrolysis and oxidation of unreacted epoxide groups to diol and carboxyl groups, and (d) methoxy-epoxide-PEG with APS-modified quartz.

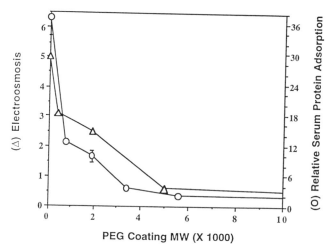

Figure 6. Effect of grafted-PEG molecular weight on electroosmosis in 7.5 mM buffer solution (△) and on relative total serum protein adsorption (O).

polymer coating thickness (*9-11,15,36*). In addition, when polymer grafting removes charged groups (*e. g.*, PEG-succinimidylcarbonate reacts with surface amino groups to form neutral urethanes, Figure 5b) polymer surface density can be estimated from the resulting change in surface charge.

The electrophoretic studies described in the following sections were performed with coatings on transparent capillaries. To extend the sample range for electroosmotic analysis the authors have recently developed a rectangular chamber with interchangeable walls. This new apparatus permits study of a wide range of non-transparent samples including slides that are also suitable for complementary XPS, ellipsometry, surface force, and FTIR analysis (*11*, see also Burns, N. B., et al. *J. Biomed. Materials Res.*, 1997, in press).

The following text covers the basic apparatus and theory related to surface characterization by electroosmosis measurement in cylindrical quartz capillaries. Some illustrative examples from our work with these capillaries modified with aminopropylsilane, polyethyleneimine, and PEG are presented. A major goal of this work is to investigate methods of grafting PEG to surfaces and improve our understanding of the properties of surface grafted PEG.

Practical and Theoretical Considerations

Experimental Considerations.   Electrokinetic analysis via electrophoretic mobility or electroosmosis measurement can be performed with various commercially available instruments. A simple apparatus (Figure 1) based on visual microscopic measurement of particle mobility is available from Rank Bros. (Cambridge, UK). Such an instrument has been used by the authors (*9,10,20-22*) and other researchers (*12,37*) and is described in detail elsewhere (*12,19,22*). It consists of a simple electrode assembly, a quartz capillary, and a microscope (with water immersion lens) to observe particle movement in the capillary (Figure 1). Total cost of such an apparatus is approximately $5,000. There are restrictions on the composition of capillaries in that they must be transparent, rigid and smooth, and the particles used must be spherical, not too dense and of uniform surface chemistry. The Rank instrument can be upgraded by attachment of a microvideo camera and video recorder that is connected to a computer with image processing software.

The electrophoresis capillary serves both as electrophoresis chamber and experimental substrate. The chamber is filled with a suspension of tracer particles (on the order of 1 μm) and fitted with electrodes. The authors typically use blank platinum electrodes although blank platinum, platinum black, palladium, or silver/silver chloride can be used (12). The chamber is immersed in a thermostatted water bath ($\pm$ 1 $^\circ$C), and electrophoresis of the particles is observed with a microscope at various locations across the diameter of the capillary. A stable DC power supply provides electric field strengths in the range of 10 to 150 V • cm^{-1} (typically 40 V is used for a 10 cm long capillary of 1 mm diameter) (*9,10,12*). As noted above, measurements involve determining the velocity of particles in a well characterized electrolyte solution exposed to an electric field of known strength.

Automated instruments are available for $50,000 to $100,000. These devices are produced by several manufacturers including Coulter Electronics (Hialeah, Florida) (31,36), Malvern Instruments (Malvern, Worcs., U. K.), Pen Kem Inc. (Bedford Hills, New York) (38) and Otsuka Electronics (Ft. Collins, Colorado) (18). These machines are typically designed to evaluate particle electrophoretic mobility and some modification may be required for electroosmosis measurements. The authors recently used a Coulter unit with a rectangular chamber to investigate use of PEG coatings for control of systematic errors (31). In most instruments fluid flow is followed optically via use of standard particles, however some applications may involve use of spectroscopically active molecules (25-27).

Basic Theory. Electroosmotic characterization involves measuring fluid flow at a surface in contact with an electrolyte subjected to an applied field. Particles suspended in the fluid medium are subjected to electrophoresis in a chamber, and the mobility of these particles is determined visually or optically. As described above, electroosmotic flow (Figure 2) influences the net mobility of a particle sample suspended in the chamber (Figure 3). Given an accurate description of chamber hydrodynamics, both particle electrophoretic mobility at the stationary level and electroosmosis at the chamber surface can be determined. This information relates directly to both particle and chamber surface charge (and surface chemical group) character . The hydrodynamics of fluid flow in microparticle electrophoresis chambers is described by solutions to the Navier-Stokes equation for steady laminar fluid flow with boundary values defined by the chamber geometry (for more details see 9,11). Though it is beyond the scope of this paper to derive such equations (3,4), it is useful to discuss relevant consequences of hydrodynamics and chamber geometry with respect to experimental practice. Measurement of electroosmosis in cylindrical electrophoresis chambers (Figure 2) is discussed below. More information (3,6,9,10), including that related to electrophoresis in rectangular chambers (11,12,31,37,38), can be found elsewhere.

Description of the hydrodynamics for a cylindrical chamber is relatively straightforward. Considering only electrostatic and fluid frictional forces acting upon the suspended particles, apparent particle mobility (U_0) at a given location (x) across the diameter of the capillary may be represented by:

$$U_0 = 2U_{eo}(x/r)^2 - U_{eo} + U_{el} \qquad [1]$$

where U_{eo} is wall electroosmotic fluid mobility, r is the capillary radius, and U_{el} is particle electrophoretic mobility (measured at the stationary level). U_{eo} is therefore taken as half the coefficient of the second order term of a second order curve fit of a plot of particle mobility versus x/r (relative location), Figure 3.

Electroosmosis is one of a number of electrokinetic phenomena related to surface charge and viscous forces acting upon the electrical double layer. According to classical electromagnetic theory and the Gouy-Chapman model of the electrical double layer (see *3*), for a 1:1 electrolyte fluid, electroosmotic mobility (U_{eo}) can be related to total surface charge (σ_0) (equation 1) by the equation:

$$\sigma_0 = (8c\varepsilon kT)^{1/2} \sinh\left(\frac{e\eta U_{eo}}{2\varepsilon kT}\right) \qquad [2]$$

given bulk electrolyte concentration (c), permittivity of the medium (ε), temperature (T), the Boltzman constant (k), coulombic charge (e), and viscosity (η) (*9,11*).

Electroosmosis is therefore an indirect measure of effective surface charge and can be used to monitor changes in charge properties upon surface modification. One way of determining the charge properties of a surface is to measure the pH dependence of electroosmotic fluid flow, via a "surface titration". Since the surface charge is a summation of protonated or deprotonated acid and base groups, it varies with pH. Healy, White *et al.* (*32,33*) described σ_0 for a uniformly charged planar surface as a function of discrete ionogenic sites, with independent dissociation constants (pK's) and surface densities.

$$\sigma_0 \quad = -e\sum_i [A_i^-] + e\sum_j [BH_j^+]$$

$$= \sum_i \left[\frac{-e\,Na_i}{1+10^{pKa_i - pH_s}}\right] + \sum_j \left[\frac{e\,Nb_j}{1+10^{pH_s - pKb_j}}\right] \qquad [3]$$

where $[A_i^-]$ and $[BH_j^+]$ are acid and base groups with pKa_i and pKb_j of density Na_i and Nb_j, respectively.

Data such as that in Figure 4 allows identification of surface acid and base groups by determination of their respective pH dependent equilibria (pK_a and pK_b), and surface densities (N_a and N_b, as groups per nm^2) can also be obtained. The methodology for achieving these determinations was recently presented in detail (*9-11, see also 15*) and involves site-dissociation modeling and exact solutions to the Poisson-Boltzman equation. Based on Gouy-Chapman theory; the electrostatic zeta potential (ζ) a distance (d) from a charged surface in a univalent electrolyte can be described in relation to the temperature, various known or determinable properties of the solution, and the surface potential (Ψ_0), which can be mathematically related to total surface charge (σ_0) as in equation 2.

Interfacial concentrations of ions such as hydronium ($[H^+]_S$) can be related to those in the bulk ($[H^+]$) by the Boltzman equation:

$$[H^+]_s = [H^+]\exp\left(\frac{-e\Psi_0}{kT}\right) \quad [4]$$

An expression relating electroosmosis to surface potential and ζ at the hydrodynamic plane of shear (*i. e.*, a defined d value) was provided by von Smoluchowski (see *3*):

$$U_{eo} = -\frac{\varepsilon\zeta}{\eta} \quad [5]$$

Analytical Modeling of Results. The pK and density of functional groups can be estimated by matching calculated and observed electroosmosis profiles, based on site-dissociation modeling and the above equations (*9*). In Figures 4 and 7 to 10 a best fit of calculated and observed electroosmosis was achieved for each surface over a wide range of pH. The solid curved lines in these figures are modeling profiles for experimental results indicated by symbols (whose error bars for individual experiments are typically within symbol dimensions). Each symbol represents a distinct parabolic distribution associated with a certain pH and ionic strength, as in Figure 3. The curve modeling represents effective parameters used in combination with Theorist , Matlab or similar software to calculate pH-dependent electroosmosis profiles which fit experimental data. Many years experience suggest that the results are reproducible even when operator, apparatus, and laboratory are varied.

Typical site-dissociation modeling results are provided in Tables I and II (*9,10*). A simple surface involving one acid group and one base group is defined by five parameters (pK_a, pK_b, N_{a}, N_{b} , plus d, the layer thickness), and a surface with three charged groups is defined by seven parameters. Assigning this number of parameters is difficult, especially when dealing with polymer-coated surfaces with rather linear profiles of electroosmosis versus pH. To this end, the parameter estimates arising from modeling should be considered to be approximate. In many cases, their accuracy is enhanced by employing some parameters (*e. g.*, pK's, or surface grafting densities) from the literature, or from complementary techniques (*9*)

It is undoubtedly possible to generate more accurate and complicated models for interpretation of electroosmosis. These may include, for example, including parameters related to a "Stern Layer" of solution ions reversibly bound at the surface (*3*). Such complications may offer little in regard to comparative analysis of polymer coated surfaces. The strength of the present approach is that it provides good qualitative estimates for parameters of obvious chemical and biochemical significance. Results are reproducible,

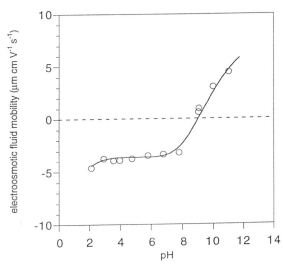

Figure 7. Electroosmosis profile for PEI-coated quartz capillaries in 7.5 mM NaCl.

Table I. Surface Density of Charged Groups and Shear Plane Distance Native and APS-Quartz Capillaries

Surface	Estim. Surface Group Density[a] [nm^{-2}]			Shear Plane
	SiOH$_i$ (pK$_a$ 3.6)	SiOH$_a$ (pK$_a$ 6.9)	NH$_2$ (pK$_b$ 10)	distance d (nm)
Native Quartz	0.03	5.00	0.00	0.9
Heated Quartz Control[b]	0.10	0.50	0.00	0.9
APS Quartz	0.07	0.20	0.20	1.3
APS-Quartz + HT[c]	0.18	0.23	0.20	1.3
	(pK$_a$ 4.1)	(pK$_a$ 8.0)	(pK$_b$ 10)	
Adsorbed PEG 25000[d]	0.10	0.50	0.00	1.1
	(pKa 2.2)		NH$_2$ (pK$_b$ 10)	
Adsorbed PEI[e]	0.90		1.00	0.9

[a]SiOH$_i$ represents isolated silanol groups, and SiOH$_a$ represents associated silanol groups. [b]Clean quartz capillary heated at 190 °C in vacuum as for APS coated capillary. [c]HT indicates hydrothermal treatment 6 h at pH 10.3, 60 °C, or pH 7.3, 95 °C. [d]Control adsorption of 5% (v/v) unfunctionalized PEG 25,000 for 6 h in 0.05 M Na phosphate, pH 7.3 at 60 °C on quartz. [e]PEI adsorbed for 2h (3% solution) in 0.05 Na carbonate, pH 9.5.

Table II.   Estimated Plane of Shear Distance for APS-Quartz Grafted with di-E-PEG 3400

Reaction Temperature (°C)	6h, 0.05M Na phosphate, pH 7.3	6h, 0.45M K$_2$SO$_4$, pH 6.9
20	N. D.	2.4
25	2.4	N. D.
30	N. D.	3.5
35	2.4	7.0
45	5.0	8.0
50	N. D.	8.0
60	7.0	N. D.
95	8.0	N. D.

useful and correlate well with related characteristics such as the ability of coatings to control protein adsorption (Figure 6, see also *7,8*).

Electroosmotic Analysis of Surface-Localized PEG. Theory suggests the manner in which electroosmosis measurements can provide information about non-charge-related structure near the solution-surface interface. Equation 2 includes a viscosity term, and although this term relates to bulk solution viscosity, it implies that a viscous entity near a surface will effectively reduce electroosmotic flow. This has been shown to be the case for several neutral polymer coatings. In principle, the limits of the thickness detection range for a surface-localized, neutral polymer layer corresponds to the hydrodynamic plane of shear value d where there is still appreciable zeta potential as evidenced by discernible electroosmosis (or electrophoretic mobility in the case of particles) (*5,6* see also *41*). A hydrodynamic thickness (d) of surface-localized, neutral polymer can be estimated according to classic electrokinetic theory based on characterizing the surface of interest before and after coating with polymer (*9*). Meaningful electrokinetic data can be obtained at physiological salt concentrations, but measurement sensitivity is enhanced by working at lower ionic strengths (*5,6,*).

The authors typically work with 1-10 mM NaCl solutions for which pH is adjusted by addition of NaOH or HCl of similar ionic strength (*6,9,20,22*). Choice of ions and careful monitoring of solution conductivity is required to meaningfully interpret and model the results. For a typical surface, such as quartz, these low salt concentrations provide sensitivity to polymer coating thicknesses (d) between 0 and •12 nm, a range of interest for many studies involving biomaterials (*6,9,11*). If the polymer layer is thicker than about 10 nm, electroosmosis goes to zero and it is possible only to note that the layer is thicker than that required to effectively eliminate electroosmosis.

Though analysis of PEG coatings presents a significant modeling challenge, meaningful data can be obtained (*9,10,11*). In many cases, estimates of surface density, pK, and coating thickness match those obtained by other methods such as ellipsometry, surface spectroscopy, radiolabel studies, and surface force analysis. This is particularly true when electrokinetic modeling is aided by inclusion of pK, surface density, or coating thickness data obtained via other analytical methods.

Experimental Results With PEG and Related Coatings

Various neutral polymers may function as useful surface coatings for biomaterials and other applications including separations technology. These polymers include linear and branched PEGs, ethylene oxide-propylene oxide copolymers, poly(vinyl alcohol)s, members the "dextran" polyglucose family, and ethoxylated celluloses such as ethylhydroxyethylcellulose (EHEC) (*2,6-8,19,20-31,34,39-40,44,45*). These polymers may be covalently grafted directly onto a native surface or onto a surface that has been "activated", for example via reaction with an organosilane reagent (Figure 5a). Another approach to surface activation is to adsorb a polymer that contains reactive groups. For example, branched poly(ethyleneimine) or PEI shows strong adsorption at glass, sulfated polystyrene and other negatively charged surfaces (*9,44,45,52,53*), and after adsorption this polymer presents primary amine groups which can be used to couple to groups on the neutral polymer. PEG-PEI conjugates can also be formed prior to adsorption.

In recent years PEG has emerged as the polymer of choice for many coating and tethering applications in biomaterials and biotechnology (*1,2*). The authors have long been involved in synthesis of new PEG derivatives (*7-10*), including those of use in surface modification (*20,21*). PEG is a useful test polymer as it is commercially available in linear or branched forms and fine-cut molecular weight fractions, variously functionalized at one or both terminal ends. Dextrans, which are reasonably linear, available in different molecular weights, and may also be grafted to surfaces in a manner similar to PEGs, often provide good comparison surfaces (*7,8,20,24,34,44*).

The following subsections review some of our research on the use of electroosmosis measurements to characterize: (a) the layer thickness and grafting density of PEG coatings of different molecular weights; (b) the effect on grafting density of PEG functionality and reaction conditions during PEG immobilization; (c) stability of PEG coatings, (d) susceptibility of PEG coatings to auto-oxidation, and (e) further modification, such as capping or tethering of other groups onto PEG layers. First we describe some of our work on quartz and quartz which has been activated by attachment of organosilanes and poly(ethyleneimine).

Quartz and Organosilane-Modified Quartz. Quartz and related materials such as glass are used in many biochemical, biotechnical and biomedical applications. Although much effort has been made to characterize such surfaces, their dynamic nature still presents many challenges. This dynamic nature includes alteration of surface charge and other properties with history and environment (*e. g.*, changes in pH) and the relative instability of most activated surfaces, particularly upon exposure to concentrated salt solutions or base. These concerns are of obvious biomedical significance (*2,10,28,29*).

Figure 4 presents the electroosmosis *versus* pH profile for a quartz capillary cleaned by sequential exposure to NaOH, HCl and H_2O_2 solutions. Similar results are seen following HF exposure (*9*) and for other "glass" surfaces (*20*), including the microcapillaries used in analytical microcapillary electrophoresis (*25*). At low pH there is little electroosmosis as surface silanol groups exist primarily in the protonated (SiOH) form. As pH increases, silanol groups deprotonate to the SiO⁻ form (siloxy) over a broad range of pH. Modeling results (Table I) indicate a range of surface silanol configurations which may be represented by two different types of silanols: "isolated" silanol groups of pK_a 3.6 and "associated" silanols (pK_a 6.9). The data shown in Table I, including total silanol surface density of •5 nm^{-2} and group heterogeneity, are commensurate with data and observations from other methods (*9,10*, see also *28,29,48*).

Exposure of the quartz surface to vacuum and heat results in formation of siloxane linkages between neighboring silanols, with commensurate reduction in the ratio of associated to isolated silanol groups (Table I). Heat and vacuum are also used to covalently graft aminopropylsilane (APS) or mercaptopropylsilane (MPS) groups to the quartz surface (Table I) (*9,10,28,49*).

The electroosmosis profile and matching modeling data for APS-modified quartz (APS-quartz) are also provided in Figure 4 and Table I. Corresponding information for MPS-quartz can be found in references *9* and *22*. Modeling results suggest that APS-quartz possesses approximately equal numbers of silanol and amino groups (0.27 nm^{-2} and 0.20 nm^{-2}, respectively), and the plane of shear is shifted outward by 0.4 nm. Not surprisingly, the APS-quartz surface exhibits a very different profile from native quartz (Figure 4). At low pH, cationic amino groups dominate the surface. These groups remain protonated and pH must increase to 7 before their contribution to surface charge is balanced by surface siloxy groups. As pH and siloxy group density increase the surface exhibits net negative charge (*i. e.*, positive electroosmosis).

The covalent nature of the APS coating of Table 1 is indicated by its stability over 6 h in pH 10.3 salt solution at 60 $^{\circ}$C (*10,49*). Such "hydrothermal" treatment, which mimics the reaction conditions sometimes used to graft PEGs, shifts the electroosmosis profile intermediate between that of native and APS-modified quartz. This treatment may not alter amino-group density, as evidenced by the similarity of electroosmosis at pH 2, where most silanols are protonated (Figure 4). This suggests that the hydrothermal treatment may increase silanol density via removal of amino groups ionically (not covalently) bound to surface silanols in a net neutral complex (Table I). Hydrothermal removal of these "physisorbed" (*28*) amino-groups does appear to increase the concentration of titratable surface silanol groups (for more discussion see reference *10*). The above surfaces were stored under nitrogen to minimize reaction of surface amino groups with atmospheric CO_2 (*50,51*).

Electroosmosis measurements such as those shown above have proven to be very helpful in defining grafting conditions that provide stable organosilane coatings to which PEG and other neutral polymers may be grafted (*9,10,20,22,24*).

Poly(ethyleneimine)-Modified Quartz. Various other coatings have been tested as "sublayers" to which PEG and other neutral polymer coatings can be applied. Much of this work has been done in conjunction with electrokinetic analysis, particularly in regard to the use of high molecular weight polyethyleneimine (PEI) (*9,22,31,52*). The PEI used in our studies (Polymin SN) (BASF) is a highly branched polymer of approximate weight average molecular weight of 2,000,000. The polymer contains primary, secondary and tertiary amino groups in the approximate molar ratio of 1:2:1. In addition to these strong basic groups (of similar pK_b), about one third of the amino groups may be associated with amido-bonded adipic acid residues. This is in addition to any acidic carbamate groups formed from reaction of the amino groups with atmospheric CO_2. Conductimetric titration suggests a basic group (pK_b •10) and an acidic group (pK_a • 3.8) that exhibits a broader titration curve and is less well defined (9).

PEI adsorbs strongly to various negatively-charged surfaces such as quartz (9,31,52,53), sulfated polystyrene latex (7,8), and oxidized, cast polystyrene and other "plastics" (44). Adsorbed PEI provides a dense layer of reactive amino groups to which various substances can be covalently grafted. Surfaces coated with PEG-linked to PEI exhibit greatly reduced protein adsorption (44,45) and have the potential to enhance the biocompatibility of surfaces in many non vivo applications (45). As noted by Gölander et al. (see 9,31,44) such polymers can be reacted with PEI before or after PEI localization at a surface. Reaction of PEG with PEI creates a water soluble reagent that can be localized at various surfaces by adsorption from a 1% solution. Although it is not irreversibly bound, such "one step" coatings are tenaciously held (9,31).

Figure 7 provides the electroosmosis versus pH profile for quartz that has been exposed to a dilute aqueous solution of PEI under the basic conditions which promote good bonding between the polycationic polymer and the negatively charged surface. The results can be modeled as one acidic and one basic group (Table I) (9). It is of interest to note that the basic group (pK_b = 10) has a surface density of one amino group per nm^2, which is considerably greater than the amino density on most APS-modified quartz or on surfaces modified via radio-frequency plasma discharge with diaminocyclohexane (54). It is also interesting to note that PEI binding appears to have little effect on shear plane distances (Table I). This result suggests that PEI lies flat on the surface (9), an interpretation which is supported by recent surface force analysis (52).

Effect of PEG Functional Group and Molecular Weight on PEG-Modified Quartz. The authors have coated amino-modified quartz and other surfaces with PEGs functionalized with electrophilic groups such as cyanuric chloride (CC-PEG) (20), succinimidylcarbonate (SC-PEG) (9,22), epoxide (E-PEG) (10), aldehyde and various other moieties (55,56) developed by various groups (57,58) including the authors (e. g., 47). Unmodified quartz has also been derivatized with silane-derivatized PEG. In all cases a significant reduction in electroosmosis suggests that surface masking can be obtained with moderately dense layers of PEGs in the molecular weight range 3,000 to 20,000. The surface density of PEG grafting may be limited by PEG-PEG interactions (see below), and the grafting chemistry and functional groups involved may be of secondary importance in many coating applications. Factors such as reaction conditions (e. g., reagent costs, organic versus aqueous solvents, heated *versus* room temperature, neutral *versus* basic pH) may be more important, especially when coating delicate materials. Sublayer stability, linkage hydrolytic stability, behavior of unreacted functional groups, and PEG oxidative stability are also of concern.

Early studies involving CC-PEG and dextran polymers of different molecular weight (MW) grafted to glass or polystyrene latex particles and APS-modified quartz capillaries suggested that electroosmosis and particle electrophoretic mobility both decrease with increasing polymer molecular weight (Figure 6, *22*). A major problem with these studies was lack of control over polymer grafting density (molecules per nm^2). Radiolabel studies suggest the electroosmosis data in Figure 6 is derived from coatings of •0.1 PEG nm^{-2} (*22*).

Later studies led to improved methods for obtaining coatings of similar grafting density for both dextran (*23*) and PEG (*9*). Figure 8 illustrates the electroosmosis versus pH profile for APS-quartz grafted from anhydrous toluene with SC-PEGs of number average molecular weight (M_n) of 500 to 35,000. The grafting reaction converts cationic amino groups to neutral urethanes (Figure 5b), thus allowing estimation of grafting density (difficult to obtain by other indirect methods) (e.g., *56*). These estimates are consistent with equal grafting density for the different molecular weights. As expected (*20, 22*) electroosmosis decreases as PEG MW increases, reaching •0 for the PEG 35,000 coating (*9*). The layer thickness associated with the PEG 8,000 coating can be modeled (curved lines in Figure 8) at •5 nm, whereas that associated with the PEG 35,000 coating cannot be estimated (because electroosmosis has gone to zero), but must be > 10 nm (see *9*). Such results correlate with data of Hlady et al. (*59*) on the ability of PEG coatings of similar MW and grafting chemistry to inhibit serum protein adsorption on APS-quartz (Figure 6, see also references *7* and *8*).

Effect of Reaction Conditions on PEG Grafting Density. An obvious strength of electroomosis determination is that it can provide qualitative information on polymer coating thickness and grafting density. The relative contribution of polymer size and grafting density to the ability of PEG coatings to mask the surface and reduce nonspecific protein and cell adsorption is of concern in biomaterials research, as is the development of methods to generate and analyze coatings of varied grafting density (*2,39,40,45,56-61*). Here we describe our work on generating PEG coatings of varied grafting density. These studies have focused primarily on grafting epoxide-activated PEG (E-PEG) to the relatively stable and well characterized APS-modified quartz surface (Figure 5a) (*10,55*). Epoxide-derivatized PEG (E-PEG) was studied because it reacts in aqueous solution and because the epoxy group is relatively stable in water; epoxide groups on difunctionalized-PEG that do not participate in surface grafting are thus available for subsequent reaction to tether proteins and other groups to the PEG-coated surface, Figure 5a. An additional feature of PEG-epoxide is that grafting at APS surfaces occurs without loss of charge since a secondary amine is produced. In the following studies PEG molecular weight (M_n) was kept constant at 3400 so that alteration in coating-associated electroosmosis could be related to grafting density.

Initial E-PEG grafting reactions involved coating an APS-modified quartz surface (which had been stored under nitrogen) with a 5-10 % solution for 6 h at 60 °C in 0.05 M Na phosphate at pH 9 (9,10). This results in a coating whose electroosmosis profile is similar to that for SC-PEG 3400 in Figure 8. Additional studies (not shown) indicated that alteration of pH (7 to 10.3) and reaction time (6 h to 24 h) had little effect on electroomosis (10). However, increasing reaction temperature was found to reduce electroomosis, Figure 9. Our interpretation of these results is that the temperature increase leads to increased grafting density and an associated increase in thickness of the polymer layer.   The modeling data presented in Table II support this interpretation.

The increased packing density with increased coating temperature appears to result from reduction of PEG inter-chain and intra-chain repulsion. This phenomenon (discussed more fully in 10,39,40) relates to other research that suggests that polymer-polymer attraction increases as temperature is raised towards the lower critical solution temperature (LCST; approximately 100 °C under the above conditions). As the LCST is approached, there is a decrease in PEG molecular dimensions, PEG-PEG inter-chain interactions and PEG solubility (i. e., surface affinity increases). All of these effects can lead to increased grafting density.

Our hypothesis that the PEG layer thickness increases because of an increase in packing density is consistent with other work. Even at relatively low surface densities (e. g., < 0.1 nm^{-2}) various analyses suggest that unfavorable inter-chain interactions cause elongation of polymer coils normal to the surface (2,39,40). Malmsten and Van Alstine (39) recently reported ellipsometric data which suggests that at the above grafting density (on various modified silicas) terminally localized PEG 6000 molecules form a dense layer which is several times thicker than the random coil diameter of the polymer molecule in solution.

PEG-PEG interactions (and solution solubility) are also influenced by salt concentration and lyotropic nature. As a result, greater grafting density is expected from reactions carried out at higher temperatures, or in sulfate- as opposed to phosphate-containing solutions (for more information see 10 and 56). Figure 9 and Table II illustrate the manner in which use of potassium sulfate instead of sodium phosphate buffer allows a dense coating to be achieved even at 35 °C. It is thus possible to obtain coatings of variable density at fairly benign pH and temperature (60).

PEG Coating Stability and Auto-Oxidation. Electroosmosis studies have also aided analysis of stability of PEG coatings. Exposure to oxygen and base were considered because of the known ready auto-oxidation of

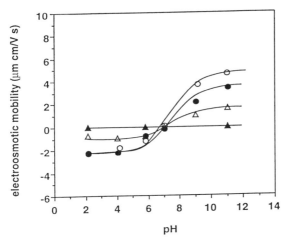

Figure 8. Effect of PEG molecular weight on electroosmosis in 7.5 mM NaCl for APS-modified quartz grafted with PEG of M_n 500 (o), 3,400 (●), 8,000 (Δ) and 35,000 (▲). Corresponding modeling indicates grafted layer thickness of 1.6, 2.4, 5.0 and >10 nm, respectively.

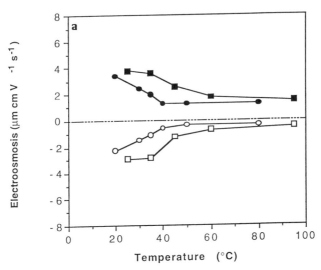

Figure 9. Electroosmosis measured in 7.5 mM NaCl at pH 2 (open symbols) or 11 (filled symbols) for APS-quartz capillaries grafted with di-E-PEG 3400 at various temperatures either in 0.05 M sodium phosphate, pH 7.3 (squares) or in 0.45 M potassium sulfate, pH 6.9 (circles).

ethers in general and PEG in particular (2) and because of the known instability of quartz coatings under aqueous conditions and high pH.

Figure 10 presents the electroosmosis profile obtained for a capillary coated with PEG by grafting di-E-PEG 3400 to APS-quartz followed by storage in 0.5M Na phosphate buffer of pH 11 for up to three weeks. The observation of electroosmosis becoming increasingly positive with exposure time is consistent with formation of a carboxyl group at the exposed PEG terminus, Figure 5c. This change occurs even if various antioxidants are added to control oxidative chain cleavage (55). Consistent with this hypothesis, coatings prepared with methoxy-PEG-epoxide (mE-PEG 5000) (Figure 5d) are stable to basic storage (Figure 11). Di-SPA-PEG 3400 coatings made with difunctional succinimidyl esters of PEG propionic acid (whose unreacted functional group readily hydrolyzes to carboxylic acid) exhibit almost identical post-reaction electroosmosis profiles to the 3 week di-E-PEG profiles in Figure 11. Site dissociation modeling suggests that the fresh m-E-PEG, di-E-PEG and di-SPA-PEG coatings are of similar grafting density ($\bullet$ 0.1 $\bullet$ nm^{-2}) and thickness ($\bullet$ 8 nm), with the fresh di-SPA-PEG, and pH-11 stored di-E-PEG coatings exhibiting surface carboxyl groups at matching grafting density and pKa.

It appears that at these grafting densities a significant number of difunctional PEG molecules are only linked to the surface via one end group (55). Under conditions that provide for reasonably dense grafting density, PEG-PEG inter-chain repulsion leads to elongatation of the PEG molecular coils normal to the surface (2,39,40). This should favor subsequent reactions by PEG surfaces as it promotes exposure of unreacted terminal groups.

PEG Coated Surfaces for Controlled Tethering. "Tethering" refers to covalently immobilizing a molecule such as an affinity ligand, chromophore reporter group, or enzyme at the free end of a surface-bound polymer "tether". Tethering to PEG-coated surfaces may reduce problems associated with nonspecific surface adsorption and allow the tethered material more molecular freedom. Similarly, surface-associated phenomena that may compromise activity of the tethered molecule will be reduced (2,30). In the present context "controlled tethering" refers to controlling the density of tethered molecules, as well as the polymer density and other properties of the underlying surface.

As shown in the preceding section, exposure to high pH for extended periods will not harm silane grafting chemistry, and PEG autooxidative chain cleavage is miminal under the usual coating conditions. At high grafting densities, difunctional PEGs are expected to react with the surface via only one functional group. The remaining functional group is therefore available for tethering, although in time epoxide and other functional groups may hydrolyze or oxidize to generate carboxyl or other undesired groups.

To test for the possibility that the terminal epoxide group from the above di-E-PEG coatings has hydrolyzed or oxidized, we reacted a fresh di-E-PEG capillary (stored for a day in water) with butylamine and then measured the

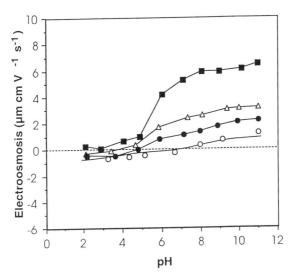

Figure 10. Electroosmosis in 7.5 mM NaCl for APS-quartz capillaries coated with di-E-PEG 3400 and then stored at pH 11 for 0 h (O), 12 h (●), 6 days (Δ) and 3 weeks (■).

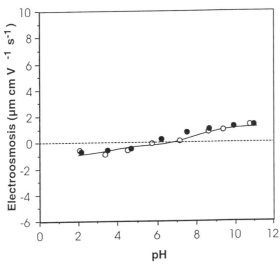

Figure 11. Electroosmosis profiles for APS-modified quartz capillaries coated with methoxy-epoxide-PEG 3400 (Figure 5d) before (open circles) and after (filled circles) storage in 0.5 M Na phosphate pH 11, at 22 °C for 3 weeks.

electroosmosis characteristics of this capillary. The resulting electroosmosis versus pH plot (not shown) showed substantial positive charge at lower pHs, consistent with formation of the expected secodary amine (55). A control capillary prepared with m-E-PEG was unchanged by reaction with butylamine. Current work is concerned with use of these PEG coatings to tether proteins to surfaces at different PEG and protein surface density.

Concluding Remarks

In summary, we have demonstrated that determination of electroosmosis is a relatively simple surface analysis technique capable of providing significant information regarding PEG and related coatings. Coatings studied in this work included aminopropylsilane, PEI and PEG. The silane and PEI "sublayers" are stable under a range of conditions. Relatively simple aqueous reaction conditions were developed for varying and determining the density and thickness of PEG coatings. It is of interest that increasing PEG packing density on a surface leads to increased thickness of the PEG layer. PEG immobilized on glass via several linkages was found to be stable to hydrolysis and oxidation for a period of several weeks. The utility of PEG coatings for tethering other molecules at varying densities has been shown.

Acknowledgments

Various aspects of this work were supported by the National Science Foundation, National Aeronautics and Space Administration, National Institutes of Health, Stiftelsen Wenner-Gren and Naturvetenskapliga Forskningsrådet.

Literature Cited

1. *Dynamic Aspects of Polymer Surfaces*; Andrade, J. D., Ed.; Plenum Press: New York, NY, 1988.
2. *Poly(ethylene glycol) Chemistry: Biotechnical and Biomedical Applications*; Harris, J. M., Ed.; Plenum Press: New York, NY, 1992.
3. Hunter, R. S.; *Zeta Potential in Colloid Science - Principles and Applications*; Academic Press: London, 1981.
4. Russel, W. B.; Saville, D. A.; Schowalter W. R. *Colloidal Dispersions*; Cambridge University Press: London, 1989; pp 211-257.
5. Lyklema, J. In *Surface and Interfacial Aspects of Biomedical Polymers, Surface Chemistry and Physics*; Andrade, J. D., Ed.; Plenum Press: New York, NY, 1985, Vol. 1; pp 293-336.
6. Koopal, L. K.; Hlady, V.; Lyklema, J. *J. Colloid Interf. Sci.* **1988**, *121*, 49-62.
7. Österberg, E.; Bergström, K.; Holmberg, K.; Van Alstine, J. M.; Riggs, J. A.; Burns, N. L.; Harris, J. M. *Colloids and Surfaces A* **1993**, *77*, 159-169.

8. Österberg, E.; Bergström, K.; Holmberg, K.; Schuman, T. P.; Riggs, J. A.; Burns, N. L.; Van Alstine, J. M.; Harris, J. M. *J. Biomed. Materials Res.* **1995**, 29, 741-747.
9. Burns, N. L.; Van Alstine, J. M.; Harris, J. M. *Langmuir* **1995**, *11*, 2768-277.
10. Emoto, K.; Harris, J. M.; Van Alstine, J. M. *Analytical Chem.* **1996**, *68*, 3751-3761.
11. Burns, N. L. *J. Colloid Interf. Sci.* **1996**, *183*, 249-259.
12. Seaman, G. V. F. In *The Red Blood Cell,* Surgenor, D. M., Ed.; Academic Press: New York, N.Y., 1975, 2nd Ed., Vol. 1.; pp 1135-1229.
13. Bauer, J. *J. Chromatogr.* **1988**, *418* , 359-383.
14. Bastos-González, D.; Hidalgo-Alvarez, R.; De Las Nieves, F. J. *J. Colloid Interf. Sci.* **1996**, *177*, 372-379.
15. Marlow, B. J.; Rowell, R. L. *Langmuir* **1991**, *8,* 2970-2980.
16. Bale, M.D.; Mosher, D. F.; Wolfarht, L.; Sutton, R. C. *J. Colloid Interface Sci.* **1988**, *125*, 516-525.
17. Norde W.; J. Lyklema, J. In *Surface and Interfacial Aspects of Biomedical Polymers, Protein Adsorption*; Andrade, J. D., Ed.; Plenum Press: New York, NY, 1985, Vol. 2; pp 293-336.
18. Makino, K; Suzuki, K.; Sakurai, Y.; Okano, T.; Oshima, H. *J. Colloid Interf. Sci.* **1995**, *174*, 400-404.
19. Nordt, F. J. ; Knox, R. J.; Seaman, G. V. F. In *Hydrogels for Medical and Related Applications*; Andrade, J. D. Ed.; ACS Symposium Series No. 31, American Chemical Society: Washington, D. C., 1976; pp 225-240.
20. Herren, B. J.; Shafer, S. G.; Van Alstine, J. M.; Harris, J. M., Snyder, R. S. *J. Colloid Interf. Sci.* **1987**, *115* , 46-55.
21. Van Alstine, J. M., U.S. Patent 4,690,749.
22. Van Alstine, J. M.; Burns, N. L.; Riggs, J. A.; Holmberg, K.; Harris, J. M.*Colloids and Surfaces A* **1993**, *77*, 149-158.
23. Boyce, J. F.; Hovanes, B. A.; Harris, J. M.; Van Alstine, J. M.; Brooks, D. E. *J. Colloid Interf. Sci.* **1992**, *149*, 153-158.
24. Harris, J. M., Brooks, D. E. Boyce, J. F.; Snyder, R. S.; Van Alstine, J. M. In *Dynamic Aspects of Polymer Surfaces*; Andrade J. D. Ed.; Plenum Press: New York, NY, 1987; pp 111-118.
25. Grossman, P. D. "In *Capillary Electrophoresis: Theory and Practice*; Grossman P. D.; Colburn, J. C. Eds.; Academic Press: New York, NY, 1992; pp 3-43.
26. O'Neill, K.; Shao, X; Zhao, Z.; Malik, A.; Lee, M. L. *Analytical Biochem.* **1994**, *222*, 185-189.
27. Gilges, M.; Kleemiss, M. H.; Schomburg, G. *Analytical Chem.* **1994**, *66,* 2038-2046.
28. Xu, B.; Vermeullen, N. P. E. *J. Chromatogr.* **1988**, *445*, 1-28.
29. Anderson, D. J. *Analytical Chem.* **1995**, *65*, 475R-486R.
30. Thomas, V.; Bergstrom, K.; Quash, G.; Holmberg, K. *Colloids and Surfaces A* **1993**, *77*, 125-139.

31. Knox, R. J.; Burns; N. L., Van Alstine, J. M.; Harris, J. M.; Seaman, G. V. F. *Lanmguir* **1997**, submitted.
32. Healy, T. W.; White, L. R., *Adv. Colloid Interf. Sci.* **1978**, *9*, 303-307.
33. Scales, P. J.: Grieser, F.; Healy, T. W.; White, L. R.; Chan, D. Y. C. *Langmuir* **1992**, *8*, 965-974 .
34. Van Alstine, J. M.; Burns, N. L.; Riggs, J. A.; Holmberg; K.; Harris, J. A. In *Carbohydrates and Carbohydrate Polymers*; Yalpani, M. Ed.; ATL Press: New York, NY, 1993; pp 298-309.
35. Van Alstine, J. M. U. S. Patent 5,108,568.
36. Dunstan, D. *J. Chem. Soc. Faraday Trans.* **1994**, *90*, 1261-1263.
37. Debacher , N.; Ottewill, R. H. *Colloids and Surfaces* **1992**, *65*, 51-69.
38. Doren, A.; Lemaitre, J.; Rouxhet, P. G. *J. Colloid Interf. Sci.* **1989**, *130*, 146-156.
39. Malmsten, M.; Van Alstine, J. M. *J. Colloid Interf. Sci.* **1996**, *177*, 502-512.
40. Hommel, H.; Halli, A.; Touhami, A.; Legrand, A. P. *Colloids and Surfaces A* **1996**, *111*, 67-74.
41. Sharp, K. A.; Brooks, D. E. *Biophys. J.* **1985**, *147*, 563-566.
42. Eremenko, B. V.; Platonov, B.É.; Uskov, I. A.; Lyubchenko, I. N. *Kolloidnyi Zhurnal* **1974**, *36*, 240-244.
43. Churaev, N. V.; Sergeeva, I. P.; Sobolev, V. D. *J.Colloid Interface Sci.* **1995**, *169*, 300-305.
44. Brink, C.; Österberg, E.; Homberg, K.; Tiberg, F. *Colloids and Surfaces* **1992**, *66*, 149-156.
45. Malmsten, M.; Lassen, B.; Van Alstine, J. M.; Nilsson, U. R. *J. Colloid and Interf. Sci.* **1996**, *178*, 123-134 .
46. Brooks, D. E. *J. Colloid Interface Sci.* **1973**, *43*, 714-726.
47. Harris, J. M.; Struck, E. C.; Shanon, S.; Paley, S. M.; Yalpani, M.; Van Alstine, J. M.; Brooks, D. E. *J. Polymer Sci., Polymer Chem. Edn.* **1984**, *22*, 341-352.
48. Köhler , J.; Kirkland, J. J. *J. Chromatogr.* **1987**, *385*, 125-150.
49. Vandenburg, E. T.; Bertilsson, L.; Liedberg, B.; Uvdal, K.; Erlandsson, R.; Elwing, H.; Lundström, L. *J. Colloid and Interf. Sci.* **1991**, *147*, 103-118.
50. Battjes, K.P.; Barolo, A. M.; Dreyfuss, P. In *Silanes and Other Coupling Agents*; Mittal, K. L. Ed; VSP; Zeist, The Netherlands, 1992; pp 199-213.
51. Burns, N. L.; Holmberg, K.; Brink, C. *J. Colloid and Interf. Sci.* **1996**, *178*, 116-122.
52. Claesson, P. M., Blomberg, E.; Paulson, O.; Malmsten, M. *Colloids and Surfaces A* **1996**, *112*, 131-139.
53. Towns, J. K.; Regnier, R. *J. Chromatography* **1990**, *616*, 69-78.
54. Lassen, B.; Malmsten, M. *J . Materials Sci.: Materials in Med.* **1994**, 5, 662-665.
55. Emoto, K.; Van Alstine, J. M.; Harris, J. M. *Langmuir* **1997**, submitted.
56. Lin, Y. S.; Hlady, V.; Gölander, C.-G. *Colloids and Surfaces B* **1994**, *3*, 49-62.

57. Zalipsky, S.; Seltzer, R.; Menon-Rudolph, S. *Biotechnol. and Appl. Biochem.* **1992**, *15*, 100-114.
58. Zalipsky, S. *Advanced Drug Deliv. Rev.* **1995**, 16, 157-182.
59. Hlady, V.; Van Wagenen, R. A.; Andrade, J. D. In *Surface and Interfacial Aspects of Biomedical Polymers, Protein Adsorption*; Andrade, J. D. Ed.; Plenum Press: New York, NY, 1985, Vol. 2; pp 109-123.
60. Emoto, K.; Harris, J. M.; Van Alstine, J. M. *Anal. Chemistry,* **1997**, submitted, .
61. Prime, K. L; Whitesides, G. M. *J. Am. Chem. Soc.* **1993**, *115*, 10714-10721.

Chapter 25

Surface Modifications with Adsorbed Poly(ethylene oxide)-Based Block Coploymers

Physical Characteristics and Biological Use

Karin D. Caldwell

Center for Biopolymers at Interfaces, Department of Bioengineering, University of Utah, Salt Lake City, UT 84112

Pluronic surfactants with varying PEO block lengths are found to adsorb to polystyrene nanoparticles. Stable adsorption complexes can be formed with similar packing densities, allowing an evaluation of chain length on the physical and biological properties of the coating. The ad-layer thickness at maximum surface coverage increases monotonically with increasing chain length, and the chain dynamics parallels this increase. Various complexes have been examined for their ability to repel fibrinogen, a capability that likewise increases with increasing chain length up to 129 EO units (F108), in analogy with previous observations on covalently modified surfaces. At maximum repulsion efficiency, the fibrinogen uptake is two orders of magnitude below that of the bare surface. By end-group activating the PEO chains, proteins and other ligands can be tethered to the surface, where they remain structurally intact. Surfaces modified in this manner have shown to support the growth of anchor dependent cells.

Ever since the middle of the 1960's, biochemists have known to precipitate selected proteins from complex mixtures by controlled additions of polyethylene glycol (PEG, PEO). This convenient procedure led to careful and systematic studies (*1-5*) which established that the PEG molecule was preferentially solvated in aqueous media, effectively excluding other solutes from its solvation sphere in proportion to their size. Specifically, Ingham was able to show (*3*) that for a given protein, in his case human serum albumin (HSA), precipitation from solutions of a fixed protein concentration was accomplished at ever smaller weight concentrations of polymer the higher the PEG molecular weight. This exclusion effect was shown to reach its maximum for molecular weights around 6000 Da. Subsequently, he expanded the study to include proteins of a variety of sizes and concluded that there was a general increase in the exclusion efficiency with an increase in protein size. Although later investigations (*6,7*) have confirmed that PEG interacts only weakly with most proteins, it was clear already from the early studies (*1,2*) that PEG was unique among water soluble polymers in its ability to exclude, concentrate, and under special conditions precipitate proteins, even those of limited stability, without inflicting losses in their biological activity.

Due to its ability to sterically exclude other macromolecules and particles, PEG has in recent years come to be used extensively as a stabilizer of colloidal systems and as a protective coating of surfaces intended for contact with protein containing fluids. Since its general inertness precludes direct adsorption as a means of forming a stable complex with the surface, the most common strategy for achieving a PEG-based protection has involved a covalent attachment of the linear PEG (or polyethylene oxide, PEO) chains to the surface. Surfaces modified in this manner were carefully examined by several groups *(8-10)* to determine what constructs might be most effective in suppressing the adsorption of proteins and cells. Of prime importance was the establishment of a relationship between repulsion efficiency and PEO chain length. Here, the pioneering work by Nagaoka *(8)* indicated that the positive effects of increasing the PEO molecular weight reached a plateau around 5,000 Da, a result that has since been confirmed by many others *(9,10)*. It is interesting to note that these findings suggest the surface-attached PEO to behave in a manner similar to PEO in solution, as discussed above, even though the presence of the substrate must place significant constraints on both structure and dynamics of the polymer chain.

The covalent surface modification with PEO is attractive from the point of view of generating a stable product whose properties are readily evaluated. However, for steric reasons, it is less clear whether the substitution of chains of different lengths proceeds to one and the same degree and, therefore, whether the observed plateau is a true phenomenon or an artifact due to incomplete close-packing of the longer chains and a resulting exposure of certain bare areas to contacting proteins. Such areas would likely display residual reactive groups, previously introduced to the surface in large excess to allow a desirable degree of substitution to occur even for long chains with their low concentration of active terminal groups. These concerns have suggested a different approach to surface modification with PEO, namely one based on adsorption.

While PEO itself is slightly hydrophobic, as suggested above, and therefore showing weak tendencies to adsorb to hydrophobic surfaces in aqueous environments, these adsorption complexes are relatively unstable, and the PEO is easily displaced by proteins and other, more strongly adsorbing, compounds. However, the adsorption of PEO containing block copolymers is by now a well known route to surface protection. By linking PEO chains of different molecular weights to hydrophobic anchor blocks of different lengths and composition, one can obtain surface coatings that vary extensively, both in stability and repulsion efficiency *(11-16)*. Of particular interest were the early observations by Illum and others *(17-20)* regarding the reduced macrophage uptake and the *in vivo* tolerance of polystyrene latex particles coated with polymeric triblock surfactants of the poloxamer type, i.e. compounds of the general composition $(EO)_m$-$(PO)_n$-$(EO)_m$ whose polypropylene oxide (PPO) center block is highly hydrophobic and serves to anchor the PEO chains to the hydrophobic substrate. From this work it has become clear that adsorptive coatings may be generated that have some highly interesting and useful protein repellent properties. Since the adsorption is likely to be strongly regulated by the length of the PPO block, the possibility exists of generating surfaces with one and the same molar concentration of PPO, and hence of PEO, regardless of the lengths of the latter chains. Poloxamer-coated surfaces might therefore provide some valuable insights into the physical characteristics of an optimally repulsive PEO surface layer.

The work by Nagaoka *(8)* had not only shown an increased repulsion efficiency of PEO chains of increasing length, but had also demonstrated that the longer chains were considerably more mobile than the shorter ones. This observation led to the formulation of a mechanism for the repulsion which has its base in the large configurational entropy conferred on the interface by the chain dynamics. In order to adsorb to the surface, an approaching macromolecule or particle would have to constrain this motion by reducing the space available for the polymer chains, and thereby reduce the entropy of the system. In the absence of a strong attraction such a reduction would be prohibitive, and the surface would therefore be protected in proportion both to the dynamics of the polymer chains and the size of the approaching particle. This notion was explored in some theoretical detail by Andrade and coworkers (*21,22*), and more recently by Torchilin et al. (*23*) who showed that the flexible PEO molecule is more effectively protecting a liposome surface from protein fouling than a, comparatively more rigid, dextran molecule of similar molar mass.

In contrast to the notion that molecular dynamics play a significant role in protecting surfaces from protein adsorption stands the notion that the key to surface protection is the reduction of interfacial energy. This can effectively be accomplished by covering the surface with a perfectly close-packed layer of hydroxy-terminated alkanes, exemplified by the self-assembled monolayers (SAM) formed on a gold substrate if the alkane contains a thiol as its second terminus. Work by Whitesides and others (*24-26*) clearly demonstrates that hydroxylated or EO-substituted surfaces are less prone to protein uptake than surfaces covered by alkanes terminated in hydrophobic groups. However, comparing protein uptakes between laboratories can easily lead to erroneous conclusions, and it is therefore difficult to promote one mechanism over another. From a practical standpoint, however, the versatility of the adsorptive coatings outweighs that of the SAM:s which require highly specific substrates for their formation. Despite much work in the area of surface protection, many questions still remain unanswered. For some time to come, surface chemists will be attempting to sort out questions concerning the chemical and physical nature of the most effective surface coatings. Specifically, the importance of interfacial chain mobility, close-packing, and layer thickness will continue to intrigue the workers in this field.

Analytical Strategies

Even though many blood contacting surfaces in need of protection are of a flat geometry, the flat surface is undesirable from an analytical perspective, in that it generally presents an inconvenient surface-to volume ratio and leads to the adsorption of only minute amounts of surfactant or protein. Although situations present themselves in which the quantification of flat surface deposits is unavoidable, demanding labeling with radioactive isotopes for the required sensitivity, our own work has by and large relied on adsorption to uniformly sized polystyrene (PS) nanoparticles. By means of an example, a sample of 200 nm PS latex particles presents just above 71 cm^2 of surface area per mg of solids. Despite the fact that these particles carry a substantial amount of surface charge, displaying zeta potentials of the order of - 50 mV, their uptake of the non-ionic poloxamers is in relatively good agreement with that seen on neat polystyrene discs (*27*).

The composition of ad-layers that form on particulate substrates can be found either through *direct analysis*, carried out after centrifugation and wash procedures have eliminated all loosely adsorbed material. The residual adsorption complex is therefore similar to that observed in an actual use situation. The alternative approach involves an *indirect analysis* based on the composition of the supernatant after removal of the particulates. This mode of analysis tends to over-estimate the surface uptake, as it will include even loosely associated components.

One group of analytical techniques that has been relied upon to perform a direct determination of the surface concentration of adsorbed surfactant involves labeling, either with ^{125}I or with a pyridyl disulfide group that can be quantified spectroscopically following reductive cleavage of the disulfide bond with release of the thiopyridone (*28*). Another has its base in the field-flow fractionation (FFF) strategy, whereby bare

or coated particles are forced to migrate through a thin separation channel (100-250 μm in thickness) under the constant influence of an externally applied field (*29*). If the field is a sedimentation field, a particle's movement in the field is reflective of its mass. Even subtle differences in particle mass, such as those deriving from the adsorption of a surface layer, result in different positions within the sharply pointed parabolic flow profile formed by the mobile phase, and hence of different transport velocities through the analytical cell. While in transit through the channel the particle is continuously washed by the mobile phase, and as it elutes its residence time is recorded as an exact measure of its mass. By comparing elution positions of the bare and coated particles, one can directly calculate the surface concentration per particle, without introducing either labeling artifacts or errors due to particle losses during wash.

The uptake of protein, either on surfactant-coated or bare particles, is regularly determined by amino acid analysis (*30*) of the coated particles. At times, the uptake is also evaluated following radioiodination of the proteins to be adsorbed.

Indirect methods for assessing surfactant uptake include the colorimetric Baleux method (*31*), in addition to the thiopyridone quantification mentioned above, while indirect methods for protein quantification involve either radiolabelling, amino acid analysis, or the colorimetric micro-BCA technique (*32*).

The ad-layer thickness is determined from the difference in size between bare and coated particles. This determination is done either by flow FFF (*33*), a technique which yields the particle's diffusion coefficient from its retention in a hydraulic field, or by photon correlation spectroscopy (PCS)(*34*), preferably on particles having undergone prior fractionation by FFF with removal of aggregates that invariably form as the large surfactant molecules bridge pairs of particles during adsorption.

The mobility of the PEO chains, attached to the particle surface via the adsorbed PPO-block, can be followed by electron spin resonance spectroscopy, following the introduction of a spin label, such as the proxyl radical, into the terminal hydroxyls of the PEO (*35*). The slower the movements of these labeled ends, the longer are the correlation times, τ, observed in the measurement.

Model Systems

Surfactants. The poloxamer surfactants used in this study were of the Pluronic type produced by the BASF Corporation and kindly donated to us. Although these products exist in a variety of block lengths, we selected four with reasonably comparable PPO blocks, but with PEO blocks of widely varying lengths, as seen in Table 1. Of the four, Pluronic P105 is highly prone to micellization in aqueous solution, while the other three are not.

Table I. Physical properties of selected Pluronics studied here. MW represents the molecular weight of Pluronic, and m and n represent the number of monomer units of polyethylene oxide and polypropylene oxide respectively.

Pluronics	MW	m	n
F108	14,600	129	56
F88	11,400	104	39
F68	8,400	76	30
P105	6,500	37	56

Adsorption Conditions. In the original phase of this study, the adsorption was always carried out from surfactant solutions whose concentrations were 4% by weight, using adsorption times of 24 hours or longer. Both in terms of time and concentration these conditions were excessive, as can be seen from Figures 1 and 2. Figure 1 displays two adsorption isotherms recorded for Pluronic F108 (36). The one labeled a) is obtained from the product, as delivered by the manufacturer. The non-ideal features of this isotherm suggested the presence of some impurity. Indeed, from size exclusion chromatography it became clear that the product consisted of two components, one about half the size of the other. Chemically, the two appeared identical, to judge from their IR spectra. The larger of these components was collected after fractionation and used to yield the isotherm labeled b). Both isotherms coincide at the plateau, which is reached at solution concentrations of between 0.01 and 0.02% using adsorption times of 24 hours. It is therefore reasonable to assume that adsorptions at the significantly higher concentration of 4% (initially) produced surfaces that were exclusively populated by the larger component.

The rate of adsorption of Pluronic F108 was analyzed by two direct methods (37). The first of these was the above mentioned spectroscopic technique and involved the adsorption of a pyridyl disulfide labeled surfactant. In this case, samples were harvested at specified times, beginning with a five minutes exposure, and were then extensively washed on a Centricon filter. The wash procedure took a minimum of 15 minutes, so the early time points are extremely uncertain. The second method was based on quantification by sedimentation FFF. Also here, the first time points are weak, since a sample needs to relax into its equilibrium position in the separator in the absence of flow, before the analysis can begin. This relaxation process requires 20 minutes under the chosen conditions. Nevertheless, as seen from Figure 2 both methods indicate that the plateau value is reached after about five hours. The adsorption process is clearly bi-phasic, with a rapid first phase in which the surface is populated to about 75%, and a slow second phase in which molecular rearrangements occur that permit a better close-packing on the particle surface.

In on-going studies, we are using the FFF analysis method in its electrical mode to investigate the rapid phase of the kinetics. During electrical FFF (elFFF) the relaxation is sufficiently fast to allow a continuous separation process without the type of flow interruptions required for the corresponding sedimentation analysis. The elFFF method retains particles in accordance with their charge, and due to the high zeta potential of the bare particles, they are significantly retained even at weak fields, as seen in Figure 3 (Y-S Gao, personal communication). Upon adsorption of the uncharged surfactant the particle's charge is screened which results in a weaker retention, as seen in the figure. It is interesting to note that even at maximum surface coverage there is still a residual charge on the particle, presumably reflective of areas in between PPO blocks that, for steric reasons, remain uncovered. Since, upon injection in the elFFF channel, the charged particle is immediately separated from the uncharged surfactant, retention in this system is a direct reflection of the degree of surface coverage obtained at a given reaction time. Sampling of the particle-surfactant mixture can at present take place in 10 seconds (37), which allows a closer look at the first rapid phase of the adsorption kinetics (see Figure 4). It is remarkable that complexes are formed in such short times which are capable of withstanding the extensive washing that takes place during the subsequent separation process.

Stability of the Adsorption Complex. Due to the absence of any covalent linkages between surfactant and substrate, the adsorption complex is inherently reversible in nature, although the activation energy for release appears to be large. Since one of the practical uses of surfactant coated surfaces is to render them non-adsorptive to proteins, bacteria, and other cells, it is important to establish their stability not only in a pure aqueous environment, but also in the presence of proteins that might interact more strongly with the particle and thereby displace the surfactant. Figure 5 shows the retention of radiolabeled Pluronic F108 by a particulate substrate following washing and suspension in three different media, namely phosphate buffered saline (PBS), PBS with a 0.8% content of human serum albumin, and human whole plasma

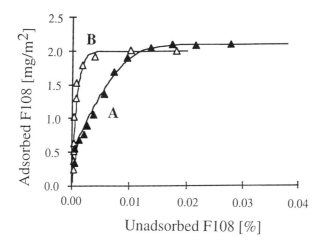

Figure 1. Adsorption isotherms of Pluronic F108 surfactant on PS 252 nm latex.
A. Surfactant as delivered; B. High molecular weight fraction collected after GPC
on a Superose 12 column from Pharmacia

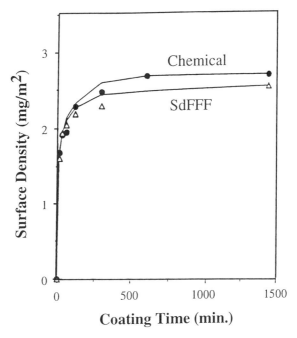

Figure 2. Adsorption kinetics of Pluronic F108 on PS 261 nm latex. The
surfactant concentration is 4% by weight. The two analytical techniques are both
"direct methods" The upper trace represents results from a spectrophotometric
quantitation after filtration, and the lower represents data from sedimentation FFF.

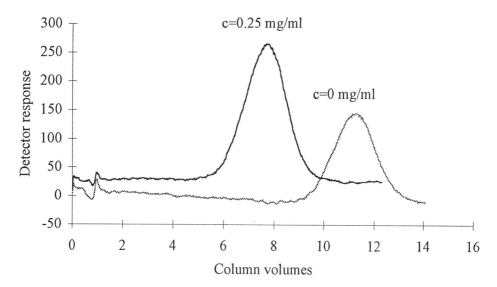

Figure 4. Retention shifts in elFFF as a function of surfactant adsorption time. The concentration of Pluronic F108 was 1% by weight.

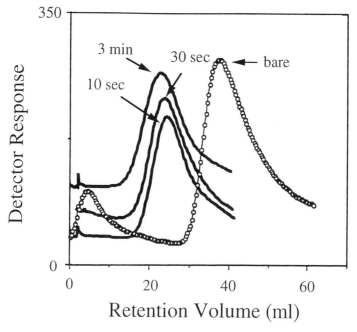

Figure 3. Retention in elFFF of bare and Pluronic F108 coated PS 165 nm latex. The applied voltage was 1.89 V across a 127μm thick separation channel. The carrier fluid was DI water.

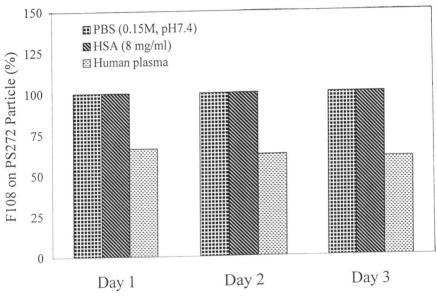

Figure 5. Stability of Pluronic F108 coated PS latex particles, suspended in various media. The surfactant was end-group labeled with a ^{125}I-Bolton Hunter reagent. (Adapted from ref. *33*).

(*33*). In the first two media there is no evidence of any surfactant loss during three days of observation. The third environment produces a rapid and significant loss of surfactant. The exact nature of the displacing entity is not clear. However, SDS gel electrophoresis has indicated a strong presence of some low molecular weight compounds which we have suspected to be apo-lipoproteins. It is interesting to note that the careful analytical work by Blunk (*38*) identifies several apo-lipoproteins as being preferentially adsorbed on latex particles coated with a comparable poloxamer.

Even after equilibration with human plasma, there is no change in the hydrodynamic size of Pluronic F108 coated PS latex particles (*33*), as illustrated by the flowFFF (flFFF) fractograms of Figure 6. As mentioned earlier, the flFFF technique separates samples based on their diffusivity, or hydrodynamic size. The figure shows elution positions for bare and F108-coated latex particles, as well as for the same coated particles equilibrated in plasma diluted to different degrees. While there is no change in size of the coated particles, the corresponding bare particles have experienced extensive aggregation, as seen in the figure. A highly significant proof of the retention of a protective layer of Pluronic F108 under *in vivo* conditions was reported by Tan et al. (*39*), who found the half life of clearance of coated PS latex particles, 73 nm in diameter, to increase to 13 hours, compared to less than 30 minutes for the bare particles. Although surfaces adsorption-coated in this manner are less stable than those covalently modified, the stability appears adequate for many short-term uses. In case a more extended exposure to plasma is needed, the Pluronic surfactants can be covalently attached to their substrate by means of the gamma-radiation procedure used for sterilization, as shown by Amiji and Park (*40*).

Ad-layer Characteristics

Surface Density. The four Pluronic compounds of Table 1 were adsorbed to an array of particles, ranging in size from 69 nm to 394 nm, using the 4% concentration/24 hour adsorption time protocol discussed above. In this process a certain curvature effect was noted (*29*), in that any one surfactant appeared to adsorb with a lesser concentration on the smaller particles. Curvature effects in ad-layer thickness had previously been observed by Baker and Berg (*34*). In comparing surface concentrations of the different surfactants, it is therefore necessary to use carrier particles of one and the same size. As seen in Figure 7, the four surfactants adsorb with comparable surface concentration on a 69 nm PS latex. Similarities in surface concentrations were also seen for the other particle sizes under investigation (*29*). Although the PEO blocks vary in length by more than a factor three, this is not reflected in the close-packing on the surface, which instead appears to be governed primarily by the length of the anchor block. The fortuitous similarity in surface concentration allows a direct evaluation of the effect of PEO chain length on protein repulsion, to be discussed below. The mentioned curvature effect in surface density was found through direct analysis of the amount of surfactant present on the particles after adsorption under the above conditions. Attempts at demonstrating this effect using an indirect analysis protocol were unsuccessful (*36*).

Ad-layer Thickness. In accord with observations by Baker and Berg (*34*) we found the ad-layer thickness to vary with substrate size in such a way that the PEO-chains on the smaller particles form a more collapsed layer than those attached to the larger spheres. Given the observed differences in surface concentration, this could be explained by a transition from a "mushroom"-like conformation at the low surface concentration to a more "brush"-like conformation at the higher concentrations, to use the de Gennes nomenclature (*41*). The layer thicknesses measured for the different surfactants on the 69 nm substrate are shown in Figure 8, which also includes a trace outlining the diameters which PEO-blocks of various lengths would assume if free in solution. The latter data derive from observations by Bhat and Timasheff (*42*), and the

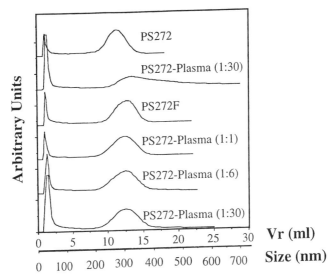

Figure 6. FlFFF fractograms of PS 272 nm after various treatments: Counted from the top, the first represents untreated particles, the second represents particles exposed to human plasma with a particle:protein weight ratio of 1:30, the third represents F108 coated particles, the fourth, fifth, and sixth represent coated particles after 4 h of contact with plasma at the specified particle:protein weight ratios. (Reproduced with permission from ref. *33*. Copyright 1996, Elsevier Science B.V..)

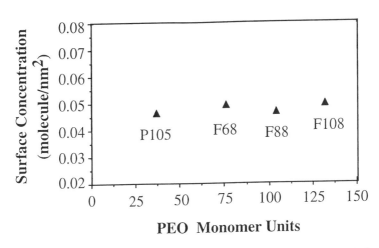

Figure 7. Surface Concentration of Pluronic surfactants of various PEO chain lengths, adsorbed to PS 69 nm latex. (Reproduced with permission from ref. *50*. Copyright 1996, American Chemical Society.)

close agreement between the two traces suggests that the PEO-chains on this particle are weakly, if at all, constrained by the presence of the substrate.

Ad-layer Dynamics. There is a clear increase in chain dynamics with increasing PEO block length, to judge from the ESR-derived relaxation times of the proxyl-labeled Pluronic surfactants(*35*) displayed in Figure 9. In a separate analysis, the dynamics of Pluronic F108 adsorbed on the PS 69 nm latex was shown to be only slightly dampened compared to that observed for the compound in solution, while the dynamics of the molecule on a 272 nm substrate was significantly reduced. Again, these data seem to be in conformity with the observed variations in layer thickness and in surface concentration.

In view of our interest in the effect of chain dynamics on protein repulsion, an attempt was made to generate less than completely close-packed Pluronic F108 coatings on the surfaces of moderately large (252 nm) particles. The intent was to escape the "brush" regime and produce a more "mushroom"-like conformation of the PEO-chains with a dynamic behavior similar to what had been observed previously on the small (69 nm) particles. This attempt failed (*43*), as the chains appeared virtually immobilized at low surface concentration and only gradually increased their motion as the concentration increased. Whether the immobilization is due to a general hydrophobic interaction between PEO and the PS surface, manifested at the low osmotic pressure in the forming film, or whether it is due to a specific interaction between the surface and the terminally linked proxyl radical is as yet unclear.

Surface Protection by Adsorbed Surfactants

Effects on Fibrinogen. The plasma protein fibrinogen is notorious for its tendency to strongly adsorb to surfaces of different kinds. Its presence on a blood contacting material has multiple adverse effects, including the activation of platelets, and thus the clotting cascade, as well as the recruitment of neutrophils and activation of the cellular immune response. In evaluating the protective effects of a surface coating, its ability to reduce fibrinogen uptake is therefore highly relevant. The different Pluronic surfactants listed in Table 1 were all somewhat effective in this regard, and the efficiency increased starkly with increasing PEO chain length, as seen in Figure 10 (*33*). Although we have not examined blocks with more than 129 EO units, corresponding to a molar mass of around 5,600 Da, the notion of an optimal PEO molecular weight somewhere in the vicinity of 5,000 Da, put forth by Nagaoka (*8*) for covalently attached chains, appears to hold well also for the adsorbed triblocks of this study. It should be noted that the Pluronic F108 at full close packing reduces the fibrinogen uptake by about two orders of magnitude. This reduction is fully implemented only at a completely covered surface, as seen in Figure 11 (*43*).

At maximum surface coverage with Pluronic F108, the two latex samples with diameters of 69 and 272 nm respectively show significantly different fibrinogen uptake (*33*), with the smaller particle taking up only half the amount compared to that adsorbed to the larger particles (0.021 ± 0.003 mg/m^2 vs. 0.038 ± 0.002 mg/m^2). This difference occurs despite the higher surface density of PEO and the thicker polymer layer that forms on the larger particles. However, due to the higher PEO close-packing on the 272 nm particle, the chain dynamics are much reduced, a fact that possibly explains its lower fibrinogen repulsion efficiency. It is interesting to note that Blunk (*44*) in his studies of uptake from whole plasma observes a similar trend of weaker protein depositions on the smaller coated particles.

Effects on Other Proteins and Particles. The efficient surface protection against fibrinogen uptake, demonstrated by a fully F108 covered surface, is not universal in nature and coated particles (272 nm) suspended in human plasma adsorb around 10% of the protein load taken up by uncoated particles under identical

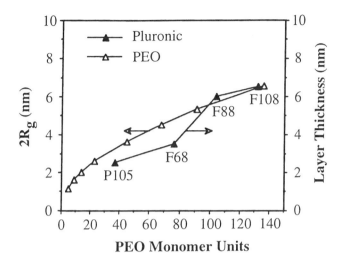

Figure 8. Thickness of the various Pluronic ad-layers on PS 69 nm latex. Included for comparison are the solution dimensions of PEO molecules of molar masses in the range of the PEO blocks of the various surfactants. (Reproduced with permission from ref. *50*. Copyright 1996, American Chemical Society.)

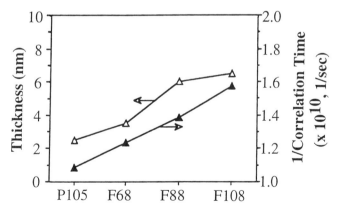

Figure 9. Comparison of ad-layer thickness and chain mobility. (Reproduced with permission from ref. *50*. Copyright 1996, American Chemical Society.)

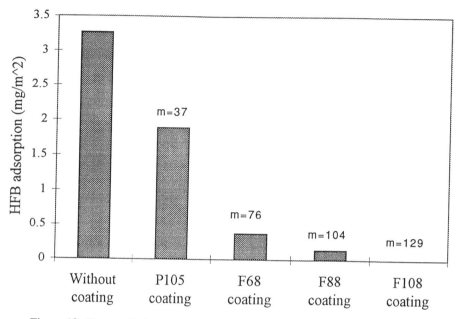

Figure 10. Human fibrinogen adsorption on PS 272 nm latex coated with different Pluronic surfactants; *m* represents the number of EO units per flanking block in each surfactant. (Adapted from ref. 33.)

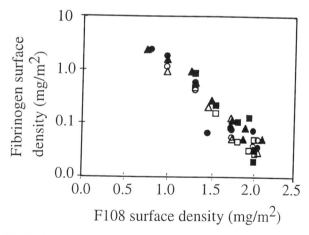

Figure 11. Reduction in Fibrinogen uptake (log scale) by PS 252 nm latex particles, coated to different degrees with Pluronic F108.

conditions (*33*). While Norman et al. (*45*) report finding all plasma proteins in extracts from Pluronic coated particles after plasma exposure and subsequent wash, the studies by Blunk referred to above (*38*) see an enrichment of proteins of the lower molar masses. The latter finding, which supports some of our own observations, is more in line with the protein exclusion behavior displayed by the PEO homopolymer in solution, as discussed above (5). In order to further examine the question of protein size and its effect on the exclusion efficiency of an F108 surface, given amounts of coated particles (261 nm) were added to three protein solutions containing equimolar

(2.9 pM) concentrations of the following proteins: β-casein (24,100 Da), bovine serum albumin (67,000 Da), and human immunoglobulin G (155,000 Da). After overnight incubation, the particles were carefully washed and subjected to amino acid analysis, which showed the protein contents per gram of washed particles to be 6.06 mg, 2.35 mg, and 1.06 mg, respectively. Within this limited sample, the trend of an increased repulsion efficiency with increasing protein size is clearly demonstrated..

The repulsion efficiency is particularly clear for coated surfaces that come in contact with bacteria and other cells. The effective suppression of bacterial adhesion to surfaces coated with Pluronic surfactants was reported by Bridgett et al. (*46*), who noted a two orders of magnitude reduction in the number of staphylococcal cells that adhered to a PS substrate following surfactant treatment, compared to the number adhering to the bare substrate. Our own work is fully supporting this observation, as demonstrated by an unpublished study which compared bacterial adhesion to coated and uncoated polycarbonate and polyacetal ingots.

Similarly, the inoculation with human fibroblasts was performed in tissue culture dishes that had had prior exposure to drops of 1% Pluronic F108 in discrete patterns (*27*). After rapid spray rinsing with DI water, the dishes had been filled with growth medium containing 10% fetal calf serum and inoculated with the cells. Although, to begin with, there were cells equally distributed over the entire dish surfaces, there soon began to develop images of the drop-patterns, as cells migrated away from the precoated regions and anchored themselves to the Pluronic-free areas of the dishes. The patterns persisted for over two weeks of incubation, reflecting the stable nature of this adsorption complex even in the presence of substantial levels of serum proteins.

Specific Ligand Attachment to Pluronic F108 Coated Surfaces

Rationale. The generally high level of protein repulsion accomplished by coatings of PEO-containing surfactants makes these compounds interesting as potential surface passivation agents, e.g. in immunodiagnostic and other analytical applications where accuracy is compromised by non-specific protein adsorption. Yet, by surrounding an active attachment site with a PEO-based steric exclusion barrier, one may well impede the specific binding reaction which the surface is designed to execute. An attractive way around this problem is to activate the terminal hydroxyls of the PEO blocks and prepare them for covalent attachment of a desired ligand, such as an antibody or its F_{ab}-fragment. By lifting the ligand above the steric barrier, one might well accomplish an accessible presentation of a specific binding entity, while shielding the underlying surface from non-specific fouling. An end-group activation of PEO to some desired degree, before or after covalent attachment to a surface, has been performed by others (*47,48*). The aim of our approach (*49*) has been to introduce a leaving group that is stable in aqueous environments and hence allows ample time for surface coating, but that can be readily replaced by a desired ligand under gentle conditions of temperature and pH. One such leaving group is the pyridyl disulfide group, which is inherently hydrolysis resistant but can be easily cleaved off by contact with free thiol. The resulting thiopyridone is easily quantified spectrophotometrically to assess coupling yield. Surfactants substituted in this manner are found to be stable for years in dry form. Furthermore, they adsorb with identical binding strength to that found for underivatized Pluronic. This gives the possibility to mix derivatized and underivatized surfactant in some desired proportion, and subsequently perform the adsorption reaction to generate a surface with a desired ligand density (50).

Surface preparation for Tissue Culture. Although originally intended for immunological use, the pyridyl-disulfide activated Pluronic F108 (PDS-F108) has proven useful in several other applications. Of particular interest has been the grafting of thiolated cell adhesion peptides to tissue culture dishes precoated with the modified surfactant (27). As discussed above, there is no evidence of cell attachment and growth on surfaces coated with the unperturbed F108. However, upon attachment of a thiolated RGD-peptide to PDS-F108 coated surfaces, cells readily attach and spread, as seen in Figure 12. The ability to easily vary the ligand density in a controlled manner has in our laboratories been the impetus for studying the effects of attachment site density on cell morphology and differentiation. From a practical point, the ease of coating even oddly shaped objects with the activated Pluronic and subsequently, at the convenient moment, link in the appropriate adhesion peptides or proteins followed by a seeding with a desired type of cells, makes this approach attractive for future tissue engineering applications.

Structural Aspects of Protein Immobilization. Immunodiagnostics, immobilized enzymes, and other protein-based solid phase systems often rely on simple adsorption as a practical method of immobilizing the macromolecule. The preparation of tissue culture ware by coating with extracellular matrix proteins (ECM:s) is likewise a common practice in cell biology today. However, it is a well known fact that most proteins undergo some level of conformational rearrangement upon adsorption, just as they may do when covalently attached to a surface through some multi-point fixation approach. In our laboratory, we have employed the differential scanning calorimetric (DSC) technique to survey differences in the enthalpy of denaturation between proteins in free and adsorbed form and have in some cases found some startling losses of structure in the surface-bound molecules (51). For applications such as tissue engineering, where cellular phenotype will closely reflect both the chemistry and morphology of the cell's environment, such conformational shifts may be highly undesirable and even destructive. In ongoing work, we have used the calorimeter to compare structural integrities of a number of proteins in free and adsorbed form, as well as upon attachment to the substrate through the intermediary of adsorbed PDS-F108. In the latter case, the polymer coated surfaces were first subjected to a reductive cleavage with DTT to leave them in a thiolated form, while the proteins were thiolated by means of the difunctional SPDP reagent (28), leaving the pyridyl disulfide groups in place on the newly introduced protein thiols. Suspension of the thiol-Pluronic coated particles in solutions of the activated proteins led to their attachment via the PEO tether in a form that was easily compared with the protein in a direct adsorption complex with the bare particles. Figure 13 shows a series of thermograms of bare particles, the same particles containing an adsorbed layer of the adhesion protein fibronectin (FN), and the particles with FN linked to their surfaces via the PEO tether (52). While the adsorbed protein shows no thermal transitions in the observed temperature range, the tethered protein displays all structural features characteristic of the macromolecule in solution (not shown). The strategy of precoating surfaces with the activated Pluronic, which has high storage stability, and subsequently linking in even such fragile substances as FN with retained structural integrity at the time of need, is thus a simple and effective way to produce biologically active surfaces.

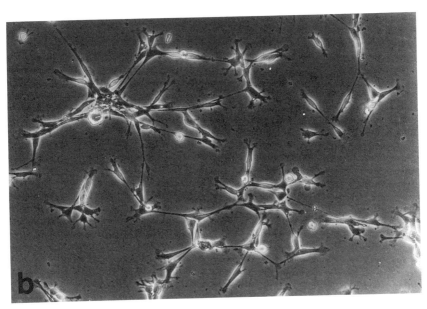

Figure 12. Behavior of NIH 3T3 fibroblasts in medium containing 10% calf serum. The substrate is neat PS, coated with pyridyl disulfide derivatized Pluronic F108. In A, the GRDSY adhesion peptide was added during the coating to test for non-specific adsorption; in B, the GRDSY peptide had been covalently linked to the F108 prior to seeding.

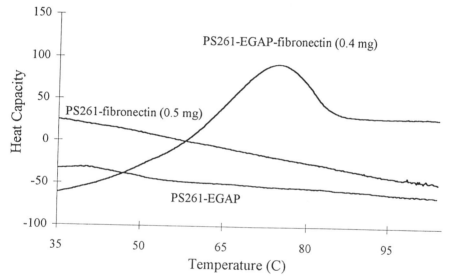

Figure 13. Differential scanning calorimetry of porcine fibronectin, covalently attached to the end-group activated Pluronic F108 (EGAP) previously adsorbed to PS 261 nm latex. Also shown are thermograms for directly adsorbed fibronectin, and EGAP coated PS 261, respectively.

Conclusions

The powerful protection against surface fouling afforded by the covalent attachment of PEO chains can for many hydrophobic surfaces be accomplished in a simple, rapid, and stable manner by coating the surfaces with polymeric surfactants of the block copolymer type. Suitable copolymers for this purpose contain one or more PEO blocks that are linked to a hydrophobic block whose size is of importance not only for complex stability but also for the molecules' ability to close-pack on the surface. By creating surfaces with nearly equal close-packing regardless of PEO chain length, it has been possible to verify previous findings of an increasing protein and particle repulsion efficiency the larger the PEO block, up to a level of 129 EO units. Although we have not probed the effects of even longer chains, they are likely to be small, at least for proteins the size of fibrinogen and beyond for which the 129 EO chain length at maximum close packing results in a 100-fold reduction in adsorption compared to the bare substrate.

 Comparing uptake data between different studies is fraught with problems, since the employed adsorption conditions are never identical. Yet, it is worth noting that the fibrinogen uptakes on Pluronic F108 coated surfaces, repeatedly found by different investigators in our laboratories and verified by several techniques, are more than an order of magnitude lower than those reported e.g. by Gombotz et al. (10) for a PEO (Mw 20,000 Da) derivatized PET film, i.e. 0.02 mg/m^2 vs. about 0.8 mg/m^2. Given the relatively high uptake numbers, this covalently attached coating must for steric reasons have left significant surface areas unprotected. In an entirely different approach to surface protection, investigators have designed surfaces for complete compositional uniformity and tested them for protein uptake. Among reported numbers for mass uptake, Margel et al. (26) find the fibrinogen adsorption on self-assembled monolayers with exposed hydroxylic end-groups to equal 25.7 pmol/cm^2 (85 mg/m^2), a value much larger even than the 3.5 mg/m^2 that we record in adsorption to bare PS surfaces from solutions containing fibrinogen as the sole protein (0.5 mg/mL, - a concentration comparable to that used by Margel et al.). It would indeed be desirable to reconcile these widely differing numbers in one single adsorption experiment.

 The approaches taken to reduce protein adsorption at solid-liquid interfaces range from attempts to create as small an interfacial tension as possible, best done by decorating the solid surface with hydroxyls able to provide the same hydrogen bonding opportunities as those present in the liquid, to the creation of a highly dynamic surface which, even though the interfacial tension may be relatively substantial, gives rise to an entropic penalty for surface access. The data summarized above would tend to support the latter hypothesis, in view of the high correlation between PEO chain mobility and fibrinogen repulsion, and between PEO repulsion efficiency and protein size.

 By end-group derivatizing its PEO blocks, and thereby introducing an active yet hydrolysis resistant structure that can readily be released in favor of attaching a specific ligand, the Pluronic surfactants have been given expanded use as versatile coupling intermediaries. Because of the ease of mixing an activated surfactant with its unmodified analog prior to coating a given surface, one has a straightforward means of controlling the density of attachment sites. This flexibility may prove to be a great asset in future studies of the optimal design of biologically active surfaces, whether for analytical, diagnostic, or therapeutic use.

Acknowledgments

Partial support for this work derived from NIH grant GM 38008-05, with additional support provided by the Center for Biopolymers at Interfaces. Both are gratefully acknowledged. I am particularly indebted to Dr. Julia S. Tan for originally steering me into this area of investigation.

Literature Cited

1. Laurent, T.C. *Biochem. J.* **1963**, *89*, 253.
2. Polson, A.; Potgieter, G.M.; Largier, J.F.; Mears, G.E.F.; Joubert, F.J. *Biochim. Biophys. Acta* **1964**, *82*, 463.
3. Ingham, K.C. *Arch. Biochem. Biophys.* **1978**, *186*, 106.
4. Atha, D.H.; Ingham, K.C. *J. Biol. Chem.* **1981**, *256*, 12108.
5. Middaugh, C.R.; Tisel, W.A.; Haire, R.N.; Rosenberg, A. *J. Biol. Chem.* **1979**, *254*, 367.
6. Lee, L. L-Y.; Lee, J.C. *Biochemistry* **1987**, *26*, 7813.
7. Lee, J.C.; Lee, L.L.Y. *J. Biol. Chem.* **1981**, *256*, 625.
8. Nagaoka, S. *Trans. Am. Soc. Intern. Organs* **1988**, *10*, 76.
9. Gölander, C.-G.; Herron, J.N.; Lim, K.; Claesson, P.; Stenius, P.; Andrade J.D.in *Poly(ethylene glycol) Chemistry*; J.M. Harris, Ed.; Plenum Press, New York, NY, 1992; pp. 221-245.
10. Gombotz, W.R.; Guanghui, W.; Horbett, T.A.; Hoffman, A.S. *J. Biomed. Matls. Res.* **1993**, *27*, 861.
11. Amiji, M.; Park, K. *Biomaterials* **1992**, *13*, 682.
12. Grainger, D.W.; Okano, T.; Kim, S.W. *J. Coll. Interf. Sci.* **1989**, *132*, 161.
13. Lee, J.H.; Kopeckova, P.; Kopecek, J.; Andrade, J.D. *Biomaterials* **1990**, *11*, 455.
14. Horbett, T.A.; Weathersby, R.K. *J. Biomed. Matls. Res.* **1981**, *15*, 403.
15. Lee, J.; Martic, P.A.; Tan, J.S. *J. Coll. Interf. Sci.* **1989**, *131*, 252.
16. Müller, R.H. in "Colloidal Carriers for Controlled Drug Delivery and Targeting", CRC Press, Boca Raton, FL,1991.
17. Illum, L.; Jacobsen, L.O.; Müller, E.M.; Davis, S.S. *Biomaterials* **1987**, *8*, 113.
18. Davis, S.S.; Illum, L. *Biomaterials* **1988**, *9*, 111.
19. Illum, L.; Davis, S.S.; Müller, R.H.; Mak, E.; West, P. *Life Sci.* **1987**, *40*, 367.
20. Harper, G.R.; Davies, M.C.; Davis, S.S.; Tadros, Th.F.; Taylor, D.C.; Irving, M.P.; Waters, J.A., *Biomaterials* **1991**, *12*, 695.
21. Jeon, S.I.; Lee, J.H.; Andrade, J.D.; de Gennes, P.G. *J. Coll. Interf. Sci.* **1991**, *142*, 149.
22. Jeon, S.I. ; Andrade, J.D. *J. Coll. Interf. Sci.* **1991**, *142*, 159.
23. Torchilin, V.P.; Omelyanenko, V.G.; Papisov, M.I.; Bogdanov, A.A.; Trubetskoy, V.S.; Herron, J.N.; Gentry, C.A. *Biochim. Biophys. Acta* **1994**, *1195* , 11.
24. Prime, K.L.; Whitesides, G.M. *Science* **1991**, *252*, 1164.
25. Mrksich, M.; Whitsides, G.M. *Ann. Rev. Biophys. Biomol. Struct.* **1996**, *25*, 55.
26. Margel, S.; Vogler, E.A.; Firment, L.; Watt, T., Haynie, S.; Sogah, D.Y. *J. Biomed. Matls. Res.* **1993**, *27*, 1463.
27. Neff, J.A.; Caldwell, K.D.; Tresco, P.A. "Use of PEO Containing Surfactants for Immobilization of Cell Adhesion Peptides. I. Control of Surface Ligand Density", *Manuscript in preparation.*
28. Carlsson, J.; Drevin,H.; Axén, R. *Biochem. J.* **1978**, *173*, 723.
29. Li, J.-T.; Caldwell, K.D. *Langmuir* **1991**, *7*, 2034.
30. Yan, G.; Nyquist,G.; Caldwell, K.D.; Payor, R.; McCraw, E.C. *Inv. Ophthalmol. Vis. Sci.* **1993**, *34*, 1804.
31. Baleux, M.B.; *C. R. Acad. Sci. Paris* **1972**, *274*, 1617.

32. Smith, P.K.; Krohn, R.I.; Hermanson, G.T.; Mallia, A.K.; Gartner, F.H.; Provenzano, M.D.; Fujimino, E.K.; Goeke, N.M.; Olsen, B.J.; Klenk, D.C. *Anal. Biochem.* **1985**, *150*, 76.
33. Li, J.-T.; Caldwell, K.D. *Colloids and Surfaces B: Biointerfaces* **1996**, *7*, 9.
34. Baker, J.A.; Berg, J.C. *Langmuir* **1988**, *4*, 1055.
35. Li, J.-T.; Rapoport, N,; Caldwell, K.D. *Langmuir 10*, 4475.
36. Bohner M.; Ring, T.A.,; Caldwell, K.D. "Effects of Particle Size and Copolymer Polydispersity on the Adsorption of a PEO/PPO/PEO Copolymer on PS Latex Particles", *manuscript in preparation*.
37. Chen, Q., "Adsorption Kinetics of a Polymeric Surfactant to Polystyrene Latex", M.S. Thesis, University of Utah (1996)
38. Blunk, T. "Plasmaproteinadsorption auf kolloidalen Arzneistoffträgern", PhD. Dissertation, Christian-Albrechts-Universität zu Kiel (1994), Chapter 7.
39. Tan, J.S.; Butterfield, D.E.; Voycheck, C.L.; Caldwell, K.D.; Li, J.-T., *Biomaterials* **1993**, *14*, 823.
40. Amiji, M.; Park, K. *J. Coll. Interf. Sci.* **1993**, *155*, 251.
41. deGennes, P.G. *Macromolecules* **1980**, *13*, 1069.
42. Bhat, R.; Timasheff, S.N. *Protein Science* **1992**, *1*, 1133.
43. Bohner, M.; Ring, T.A.; Rapoport, N.; Caldwell, K.D. "Fibrinogen Uptake on PS Latex Particles Coated with Various Amounts of a PEO/PPO/PEO Triblock Copolymer", *manuscript in preparation*.
44. Blunk, T. "Plasmaproteinadsorption auf kolloidalen Arzneistoffträgern", PhD. Dissertation, Christian-Albrechts-Universität zu Kiel (1994), Chapter 9.
45. Norman, M.E.; Williams, P.; Illum, L. *J. Biomed. Matls. Res.* **1993**, *27*, 861.
46. Bridgett, M.J.; Davies, M.C.; Denyer, S.P. *Biomaterials* **1992**, *13*, 411.
47. Harris, J.M.; Sedaghat-Herati, M.R.; Sather, P.J.; Brooks, D.E.; Fyles, T.M. "Synthesis of New Poly(ethylene glycol) Derivatives", In *Poly(ethylene glycol) Chemistry: Biotechnical and Biomedical Applications*, J.M. Harris, Ed., Plenum Press, New York (1992) pp. 371-381.
48. Holmberg, K.; Bergström, K.; Stark, M.-B. "Immobilization of Proteins via PEG Chains", In *Poly(ethylene glycol) Chemistry: Biotechnical and Biomedical Applications*, J.M. Harris, Ed., Plenum Press, New York (1992) pp. 303-324.
49. Li, J.-T.; Carlsson, J.; Lin, J.-N.; Caldwell, K.D. *Bioconj. Chem.* **1996**, *7*, 592.
50. Li, J.-T.; Carlsson, J.; Huang, S.-C.; Caldwell, K.D. "Adsorption of Poly(ethylene oxide)-Containing Block Copolymers: A Route to Protein Resistance", In *Hydrophilic Polymers: Performance with Environmental Acceptability*, J.E. Glass, ed., ACS Advances in Chemistry 248, ACS Press, Washington (1996), pp.61-78.
51. Yan, G.; Li, J.-T.; Huang, S.-C.; Caldwell, K.D., in *Proteins at Interfaces II*, T.A. Horbett ; J.L Brash, Eds., ACS Symposium Series 602, American Chemical Society, Washington, D.C. (1995)pp. 256-268.
52. Huang, S.-C.; Tresco, P.A.; Caldwell, K.D. *Manusript in preparation*

Chapter 26

In vivo Stability of Poly(ethylene glycol)-Collagen Composites

Woonza Rhee[1], Joel Rosenblatt[2], Marisol Castro[1], Jacqueline Schroeder[1], Prema R. Rao[1], Carol F. H. Harner[1], and Richard A. Berg[1]

[1]Collagen Corporation, Palo Alto, CA 94303
[2]Johnson & Johnson Corporate Biomaterials Center, Somerville, NJ 08876

Collagen materials were crosslinked by two different difunctional poly ethylene glycols (PEGs) for a variety of applications. Crosslinking was mediated by carbamation of primary amines on collagen with N-hydroxysuccinimide ester end-groups on the PEG crosslinkers. One crosslinker, disuccinimidyl glutarate PEG (di-SG-PEG), contained internal ester linkages which were capable of hydrolyzing and hence produced degradable networks. The other crosslinker, disuccinimidyl propionate PEG (di-SE-PEG), contained hydrolytically stable internal ether linkages and hence produced a more durable network. In vitro experiments verified that di-SE-PEG networks were more hydrolytically stable. *In vivo* experiments (rat subcutaneous model) revealed similar degradation rates for both networks indicating that other degradation mechanisms (probably enzymatic) also influence degradation *in vivo*. Histologic analysis revealed a mild inflammatory response to the implants accompanied by fibroblast ingrowth. A second *in vivo* study (bovine collagen hypersensitized guinea pigs), examined the immunogenicity of di-SG-PEG crosslinked materials and revealed that the PEG crosslinked implants showed lower antibody titers compared to non PEG-crosslinked controls. Another *in vivo* study compared the osteoconductive properties of di-SG-PEG crosslinked fibrillar collagen with uncrosslinked fibrillar collagen in a rat parietal defect model. Both of the collagen implant materials displayed elevated levels of new bone formation relative the non implanted sites with the PEG crosslinked implant displaying a slightly higher density of new bone forming cells. In an ophthalmic application, methylated collagen was crosslinked by either di-SG- PEG or di-SE-PEG crosslinkers to form clear lenticules. In de-epithelialized primate corneas, di-SG-PEG and crosslinked implants resorbed within 3 months whereas the di-SE-PEG crosslinked lenticules exhibited enhanced stability.

Polyethylene glycols (PEGs) have been extensively used to modify the properties and behavior of proteins in medical and biotechnology applications. The benefits of PEG modification revolve around its ability to increase the solubility of proteins and to alter their biological properties *(1, 4)*. Other advantages of PEG in medical applications are its biologically inert property and reduced antigenicity of PEG-proteins *(5, 6)*. Extensive reviews of PEG modification are available *(7)*.

Previous work on PEG modified collagen has involved conjugating monomethoxy-PEG (mPEG) to collagen. In order to covalently bind PEG to collagen, hydroxyl end groups of PEG must first be activated with a suitable reagent *(8)*. N-hydroxysuccinimide (NHS) ester modifications of PEG are particularly useful since they react under mild conditions within a relatively short period of time (30 min, pH < 7.8, 25^0C) *(9)*. Certain applications require that a collagen implant reside in the treated tissues for several months. These applications require collagen materials with very low immunogenicity. Zyderm® collagen implants (fibrillar bovine atelocollagen) and Zyplast® collagen implants (glutaraldehyde crosslinked bovine atelocollagen) have substantially less immunogenecity in humans than native bovine collagen since the telopeptide regions of the collagen have been enzymatically cleaved. Monomethyl-PEG (5000 daltons) modification of atelocollagen was shown to further reduce the antigenicity of bovine collagen *(10)*.

More recent work on PEG-collagen has focused on the use of difunctionally activated PEG capable of crosslinking collagen. This chapter describes the properties of collagen crosslinked by two different difunctionally activated PEGs. One was developed for applications where bio-degradability of implant materials is desirable. The other was developed for a more permanent implant. Both PEG-crosslinkers are activated using NHS. The degradable crosslinker (denoted di-SG-PEG) is formed from a PEG-glutaric acid intermediate and contains hydrolytically cleavable ester linkages *(11)*. The other crosslinker (denoted di-SE-PEG) is formed from a PEG-propionic acid intermediate and contains no cleavable ester linkages. The novel feature of PEG crosslinked collagen is undoubtedly the hydrophilic nature of PEG which incorporates more water molecules into a collagen network. Here we describe a range of biomaterials that can be obtained from PEG crosslinking of collagen and review the properties and behavior of PEG crosslinked collagen biomaterials in orthopedic, opthalmic and soft tissue augmentation applications.

Cross-linkers Used for Collagen Modification. PEG-N-succinimidyl glutarate and PEG-N-succinimidyl propionate were prepared as described by A. Abuchowski et. al. with minor modification *(12)*. PEG was converted to the active ester in two chemical steps. In the first step, PEG-diacid was prepared from the PEG (3400 daltons). The second step was dicyclohexylcarbodiimide-mediated condensation of the PEG-diacid with N-hydroxysuccinimide. The resultant PEG esters were characterized by UV (Ultraviolet Visible), FT-IR (Fourier Transform Infrared), NMR (Nuclear Magnetic Resonance), and HPLC (High Performance Liquid Chromatography). Recrystalization and extensive washing with anhydrous solvent were followed by high vacuum evaporation of the final product. This procedure eliminated any trace of organic solvents with high boiling points.

The succinimidyl ester of PEG yields stable products upon reaction with amines in the collagen molecule. There are 92 primary amino groups in a collagen molecule including three (3) amines in the N-terminal. Therefore, theoretically, 92 moles of PEG ester react with one mole of collagen molecule. A amide linkage is formed when the crosslinking agent reacts with primary amines, releasing N-hydroxysuccinimide (NHS). The reaction of di-SG-PEG and di-SE-PEG with collagen are shown below (Schemes 1 and 2).

Stability of Collagen Gels In-vitro Model.   The stability of collagen implants determines its usefullnes for various applications *(25, 26)*. The degradation of a collagen polymer matrix may involve three possible reactions.   One is the hydrolysis reaction of ester if introduced into the PEG. The second reaction is enzymatic degradation of the polyamide linkage. The third is proteolytic cleavage of collagen. *In vitro* hydrolysis of crosslinked collagen gels was performed to compare the effect of two cross-linkers, one containing an ester and one containing an ether linkage introduced at the end of the PEG. Depending on the structure of the linkages connecting PEG chains, hydrolytically degradable bonds may be introduced into the collagen-polymer matrix.

PEG-N-succinimidyl glutarate (di-SG-PEG, 3772 daltons, Union Carbide, Danbury, CT) contains the ester linkage between the polymer and the glutaric ester residue which   is susceptible to hydrolytic cleavage. However, PEG-N-succinimidyl propionate (di-SE-PEG, 3400 daltons, Shearwater Polymers, Huntsville, AL) does not possess this ester linkage and has ether linkage only. Therefore, a collagen gel crosslinked by these two reagents is expected to undergo a different hydrolysis *in vitro*. K. Ulbrich and his co-workers studied the relationship between the structure of PEG-oligopeptide (amino acid) congugates and their chymotrypsin catalyzed hydrolysis *(27)*. They found that the polymer chain is apparently cleaved rapidly by hydrolysis of ester bond between PEG and oligopeptide.

Zyderm® collagen implant is a 35 mg/ml dispersion of highly purified fibrillar collagen in physiologic buffer. Zyplast® collagen implant is a 35 mg/ml particulate suspension of glutaraldehyde crosslinked collagen in a physiologic buffer. When Zyderm is mixed with SG- or SE-PEG at 20^0C and neutral pH it forms a solid gel (denoted di-SG-PEG-ZD, di-SE-PEG-ZD respectively) in less than 10 minutes.   This is undesirable in applications (such as soft tissue augmentation) where the preferred mode of collagen implantation is by injection since the working time for injectability is limited.   When Zyplast was premixed with Zyderm and then crosslinked with difunctional PEG, the Zyplast acts as a plasticizing agent (since it is already precrosslinked) and increases the time required for gel formation.   Zyderm, however, is capable of attaining higher crosslink densities (since there is no precrosslinking) which may be advantageous in other applications. Therefore the stability of both Zyderm and a Zyplast/Zyderm mixture (denoted Z/Z) crosslinked by di-SG and di-SE PEG (denoted di-SG-PEG-

Scheme 1.
Crosslinking collagen with di-SG-PEG

$(COCH_2)_2N-O-CO-(CH_2)_3-CO-O-(CH_2CH_2O)_{75}-CO-(CH_2)_3-CO-O-N(COCH_2)_2$

(di-SG-PEG)

+   2 Collagen

$|$
$\downarrow$

collagen-HN-CO-$(CH_2)_3$-CO-O-$(CH_2CH_2O)_{75}$-CO-$(CH_2)_3$-CO-NH-collagen

+ 2 HO-N$(COCH_2)_2$

Scheme 2.
Crosslinking collagen with di-SE-PEG

$(COCH_2)_2N-O-CO-(CH_2)_2-O-(CH_2CH_2O)_{75}-(CH_2)_2-CO-O-N(COCH_2)_2$

(di-SE-PEG)

+   2 Collagen

$|$
$\downarrow$

collagen-HN-CO-$(CH_2)_2$-O-$(CH_2CH_2O)_{75}$-$(CH_2)_2$-CO-NH -collagen

+   2 HO-N$(COCH_2)_2$

Z/Z, di-SE-PEG-Z/Z respectively) were studied. Gel disks were aseptically prepared by mixing di-SG- and di-SE-PEG with Zyderm and Z/Z (final crosslinker concentration < 1% (w/v). The disks were incubated in sterile saline for up to 40 Days. Disks from each preparation were sampled on selective days and compressed to failure at a constant rate (2 mm/min) on an Instron testing instrument (Canton, MA). Gel strength was recorded as the average peak force sustained by disks.

As Z/Z contains less Zyderm per volume, the gel strength of Z/Z-PEG gel is significantly lower than that of Zyderm-PEG gel. The results indicate that Z/Z gel crosslinked with di-SG-PEG was found to be hydrolytically unstable under physiological conditions in an *in vitro* assay (Figure 1). Apparently, the conjugate is cleaved by hydrolysis of the ester bond between PEG and glutaric acid. Gels of collagen crosslinked with di-SE-PEG were stable to hydrolysis under identical conditions.

In vivo Biocompatibility of Collagen Gels

Fibrillar collagen was crosslinked using di-SG-PEG and di-SE-PEG to form collagen-PEG composite gels. These gels are typically evaluated by implantation in experimental animals. The models discussed here have been useful to study biocompatibility and have been used as predictors of preclinical safety.

Rat Subcutaneous Model. This study investigated the biocompatibility and long term persistence in the rat subcutaneous model of fibrillar collagen implants crosslinked *in situ* using succinimidyl activated difunctional crosslinkers. The crosslinkers include di-SG-PEG, di-SE-PEG, and DSS (disuccinimidyl suberate, Pierce, Rockford, IL), the latter being a nonPEG crosslinker having difunctional succinimidyl esters. The hydrophilic nature of PEG is expected to increase water content in collagen-polymer gel, and is expected to form a more biocompatible collagen matrix than the hydrophobic, much smaller nature of DSS. Collagen was a mixture of commercial Zyplast® and Zyderm® collagen and each crosslinker concentration was < 1% (w/v). The fibrillar collagen was mixed with solid crosslinkers by syringe to syringe exchange immediately prior to injection.

In this study, the wet weight recovery and histological appearance of retrieved implants were studied at intervals up to 90 Days after injection. The mechanical strength of tissue adherence of the implants was examined using a mechanical jig to pull the implants from the subcanteous wall. Forty-eight female Sprague-Dawley rats received two subcanteous injections and twelve animals were euthanized on Day 7, Day 14, Day 28, and Day 90 following implantation. The implant sites of eight animals were collected and analyzed by standard histological sectioning and staining with hematoxylin. Eight pieces of skin attached with implants were collected for the mechanical test to measure the anchoring force of the implant and adjacent tissue.

For each of the crosslinkers, a moderate inflammatory response occurred in the early time points, then progressively decreased in time and was absent by Day

90. Ingrowth of fibroblasts into the implant was present in the majority of implants crosslinked with di-SG-PEG and di-SE-PEG, however, it was not seen in implants crosslinked with DSS. All of the implants developed extensive calcification by Day 90 as is often the case in collagen implants in rodents. The long term persistence of crosslinked collagen materials were studied using the methods of wet weight recovery and biochemical testing. Wet weight for recovered Z/Z without crosslinker and Z/Z with crosslinkers was determined at 7, 14, 28, and 90 Days.

In vivo wet weight recovery showed no significant differences among Z/Z collagen materials crosslinked with or without di-SE-PEG and di-SG-PEG (Figure 2). It did not correlate with the results obtained from the *in vitro* stability study. *In vivo* degradation of collagen matrix appeared to be more dominated by enzymatic degradation of the protein rather than by the hydrolytic degradation of the ester bond in this animal model. In contrast, DSS crosslinked Z/Z demonstrated a slight increase in the percent weight recovery of the implant and was $99\pm 20\%$ of initial implant weight on Day 90.

In situ crosslinking potentially enhances the persistence of an implant in two ways. One is that the final crosslinked implant is a cohesive solid as opposed to Zyderm® or Zyplast® collagen which are softer particle-aggregates. Elastic moduli of PEG-crosslinked Zyderm® collagen implants range from 5-10 x 10^4 dyne/cm^2 *(13, 14)*. The second potential enhancement in persistence derives from the ability of the implant to become crosslinked and covalently bound to collagen fibers in the host dermis. The force required to dislodge the implant crosslinked materials is 4-7 newtons compared to 0 newtons for non-*in situ* crosslinked collagen material. The implants with crosslinker seemed to adhere to dermis with significantly higher force than did the controls (Figure 3). This indicates that the collagen molecules are not only crosslinked to form a gel but are also attached to the surrounding tissues through chemical bonding.

Immunogenicity Study in a Sensitized Guinea Pig Model. Removal of telopeptides from bovine dermal collagen in the manufacture of Zyderm® and Zyplast® collagen implants reduces the immunogenicity of these products. Extensive data from over a decade of use clinically indicate their high biocompatability and low immunogenicity. Nevertheless, the low level of immunogenicity manifests itself in approximately 3-5% of the population as localized hypersensitivity that resolves with time and is accompanied by antibodies specific to bovine collagen *(15)*.

To evaluate (1) whether di-SG-PEG crosslinking Z/Z reduces collagen immunogenicity and (2) whether sensitization to PEG-collagen results in hypersensitivity to PEG, we used a sensitized guinea pig model similar to that described previously *(10)*. This hypersensitivity model employs adjuvant to amplify immunoreactivity and allows subsequent assessment of humoral and cellular skin responses.

Three groups of guinea pigs were sensitized with subcutaneous immunizations of di-SG-PEG crosslinked Z/Z (denoted di-SG-PEG-Z/Z), Zyderm,

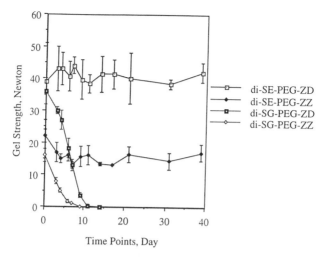

Figure 1. Effect of di-SE-PEG and di-SG-PEG cross-linkers on stability of collagen gels *in-vitro* model. Zyderm® collagen (ZD) and a mixture of Zyderm® and Zyplast® collagen (ZZ) were crosslinked with < 1% (w/v) of di-SE-PEG and di-SG-PEG to produce di-SE-PEG-ZD (□), di-SE-PEG-ZZ (◆), di-SG-PEG-ZD (■), di-SG-PEG-ZZ (◇). ZZ gel crosslinked with di-SG-PEG is more rapidly cleaved by hydrolysis of ester bond than ZZ gel crosslinked with di-SE-PEG.

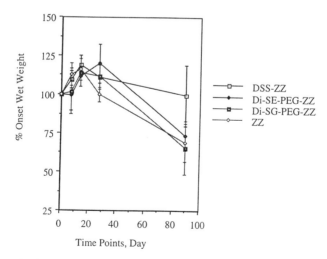

Figure 2. Result of wet weight recovery at each time point in Rat Subcutaneous model. A mixture of Zyderm® and Zyplast® collagen (ZZ) were crosslinked with < 1% (w/v) of disuccinimidyl suberate (DSS), di-SE-PEG, and di-SG-PEG to produce DSS-ZZ (□), Di-SE-PEG-ZZ (◆), di-SG-PEG-ZZ (■), and ZZ (◇) as a control. *In vivo* wet weight recovery showed no significant differences among Z/Z collagen materials crosslinked with di-SE-PEG and di-SG-PEG.

or Zyplast in Complete Freund's Adjuvant on day 0 and Incomplete Freund's Adjuvant on day 14. Each sensitized group was used to compare (1) the humoral antibody response to bovine collagen following sensitization with di-SG-PEG-Z/Z, Zyderm, or Zyplast at days 28, 31, and 59; (2) the gross delayed type hypersensitivity (DTH) response to specific intradermal challenge on day 28 (n = 6 per group) and to additional cross-challenge with all collagen implant formulations and di-SG-PEG on day 56; and (3) the cellular, immune and inflammatory response observed histologically at each site explanted 72 hrs post-DTH challenge.

Table 1 summarizes the humoral response evaluated. Guinea pigs immunized with di-SG-PEG-Z/Z developed lower levels of antibody titers than did those immunized with either Zyderm® or Zyplast® collagen at the one month sensitization time points. In fact, only three of the six di-SG-PEG-Z/Z guinea pigs tested at day 28 had developed positive titer of 160 prior to DTH challenge; three gave negative or equivocal results with titers of 20, 40, and 80 respectively. Guinea pigs sensitized with either Zyderm® or Zyplast® collagen showed no significant difference in their humoral immune responses at any time point tested; and all showed positive antibody titers at the one-month and two-month sensitization time points. At day 31, 72 hours post-DTH, all guinea pigs sensitized with di-SG-PEG-Z/Z had become positive for anti-bovine collagen antibodies; however, titers remained significantly lower than those of the other sensitized groups. By day 59 of sensitization, 31 Days following the first DTH challenge, no significant difference existed between any of the groups.

Skin challenge sites were monitored for delayed type hypersensitivity (DTH); and responses were grossly observed and measured for wheal diameter with calibrated calipers through 72 hours post-challenge. All sites were then explanted and stained with H&E for histological evaluation of inflammatory cellular response.

All three formulations effectively sensitized guinea pigs to respond to all three collagen implants, and all positive skin-test sites demonstrated the profile of lymphohistiocytic cell infiltration characteristic of DTH in this animal model. However, sensitization with di-SG-PEG-Z/Z resulted in a markedly lower short-term response to challenge; and DTH challenge with di-SG-PEG-Z/Z resulted in a markedly lower response in all sensitized groups. This is evident in both the gross observation of the DTH response and in the histological evaluation of the cellular inflammatory response in explants 72 hours post-DTH (Figures 4a and 4b). Furthermore, di-SG-PEG alone elicited no skin-response in any group (4c); and histological evaluation showed no inflammatory response at 3 days (4c) and 31 Days post-intradermal injection. Sensitization with di-SG-PEG-Z/Z resulted in sensitization to the PEG-collagen, but no hypersensitivity developed to the PEG crosslinker.

The results of this study suggest that di-SG-PEG crosslinking reduces the immunogenicity of bovine collagen implants. When tested as a sensitizing agent in guinea pigs, it results in lower humoral antibody titers than Zyderm® or Zyplast® collagen. When used as a dermal implant following various sensitization regimens, it results in a lower cellular DTH response than does Zyderm® collagen.

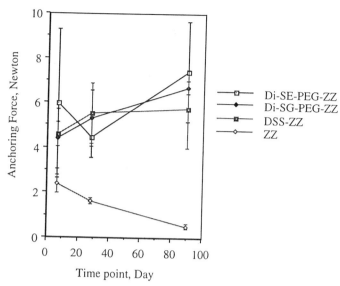

Figure 3. Mechanical test of anchoring force in Rat Subcutaneous model. A mixture of Zyderm® and Zyplast® collagen (ZZ) were crosslinked with < 1% (w/v) of di-SG-PEG, di-SE-PEG and DSS to produce Di-SE-PEG-ZZ (□), Di-SG-PEG-ZZ (◆), DSS-ZZ (■), and ZZ (◇) as a control. The implants with crosslinker seemed to adhere to dermis with higher force than did the controls without any crosslinker.

Table 1: Anti-Collagen Antibody Titers Guinea Pigs Immunized with di-SG-PEG-Z/Z, Zyderm®, and Zyplast® collagen implants. Note the lower antibody titers in di SG-PEG-immunized guinea pigs.

	Antibody Titers to Bovine Collagen as Assayed by ELISA		
	Day 28 (pre-DTH)	Day 31 (72 hrs post-1st DTH)	Day 59 (72 hrs post-2nd DTH)
Groups Sensitized with: Di-SG-PEG-Z/Z (n=6)	Mean (±SEM) 130 (±45)	Mean (±SEM) 933 (±515)	Mean (±SEM) 2560+
Zyderm (n=5)	1920 (±405)	2560+	2560+
Zyplast (n=5)	1296 (±526)	2560+	2560+

a

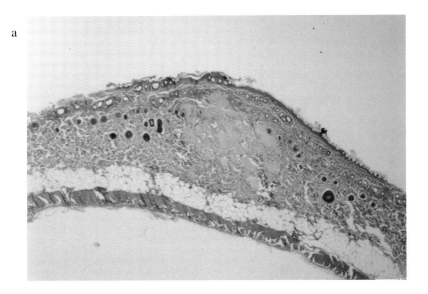

b

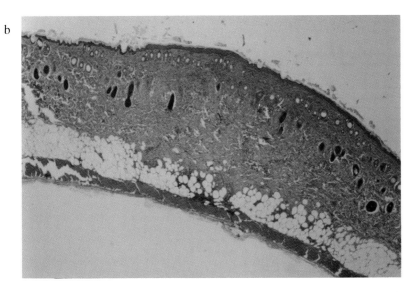

Figures 4-a, 4-b, and 4-c. Histologic appearance of 72 hrs intradermal challenge sites (a) di-SG-PEG-Z/Z, (b) Zyderm®, and (c) di-SG-PEG; stained with Hematoxylin & Eosin; (a-b) 5x and (c) 10x magnified. Sites were explanted from di-SG-PEG-Z/Z-immunized guinea pig. Note the mild inflammatory response around the di-SG-PEG-Z/Z implant in dermis, the moderate inflammatory response in the dermis and subcutis associated with the Zyderm® collagen implant, and the absence of inflammatory cells at the di-SG-PEG intradermal challenge site.

Continued on next page.

c

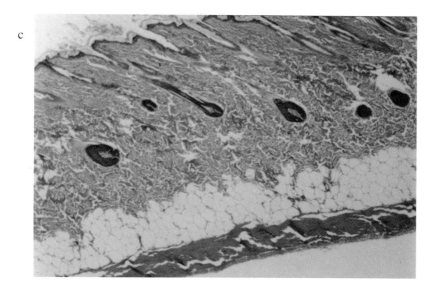

Figure 4. *Continued.*

A decrease in the specific antibody reactivity, and a lower incidence of skin hypersensitivity reactions have been historically observed and reported for Zyplast® collagen than Zyderm® collagen *(16)*. This *in vivo* study provides an evaluation of di-SG-PEG-Z/Z, and suggests that this new collagen implant formulation may demonstrate a similar reduction in immunogenicity in patients.

Orthopedic Application. Fibrillar collagen is known to be osteoconductive *(17, 18)*. Injectible bone repair formulations are attractive because material flows into the defect site and fills the empty space ensuring good contact with the margins of the host bone. Fibrillar collagen alone or mixed with the multifunctional polyethylene glycol crosslinker was injected into a bone defect and shown to promote bone growth. The di-SG-PEG crosslinked the fibrillar collagen in the host forming a firm osteoconductive gel which stays in place.

Fibrillar collagen (FC) and crosslinked fibrillar collagen were injected into bilateral rat parietal defects. Histological evaluations were made from routine Trichrome stained sections prepared from the various bone defect sites. Bone cell concentrations (osteocytes, osteoblasts and osteoclasts as-well-as "primitive mesenchymal" cells) were determined by counting the number of the various cellular populations present within a 100 x 100 μ m^2 area measured at 400 x magnification. Statistical analysis was evaluated using ANOVA. Differences were considered significant if the p value was equal or less than 0.05.

Representative histological specimens at Day 28 are shown in Figure 5. Figures 5a, 5b, and 5c show control sites with no implant, FC, and FC crosslinked with di-SG-PEG at Day 28, respectively. Primitive mesenchymal cells are abundant throughout the FC and FC crosslinked implant areas at Day 14 and their concentrations are significantly increased compared to the control sites as shown in Figure 6. The defect sites containing both FC and FC crosslinked with di-SG-PEG have an increase in new bone area compared to the control sites (Figure 7). New bone formation occurs when the conductive implant material has direct contact with the host bone. If there is not a close approximation of implant to bone, a loose connective tissue matrix is found at the interface between the implant and the host bone. Both FC and crosslinked FC flow into the defect site and have good initial apposition to the host bone providing an ideal osteoconductive environment.

Ophthalmic Application. Synthetic epikeratoplasty involves removal of the corneal epithelial layer and placement of a clear lenticule directly on the cornea to correct refractive errors. Three novel biomaterials were evaluated as in-situ polymerizable lenticules for synthetic epikeratoplasty in primates.

Methylated collagen was prepared by reacting dehydrated Zyderm® collagen with acidified anhydrous methanol *(19, 20)*. At neutral pH, methylated collagen is

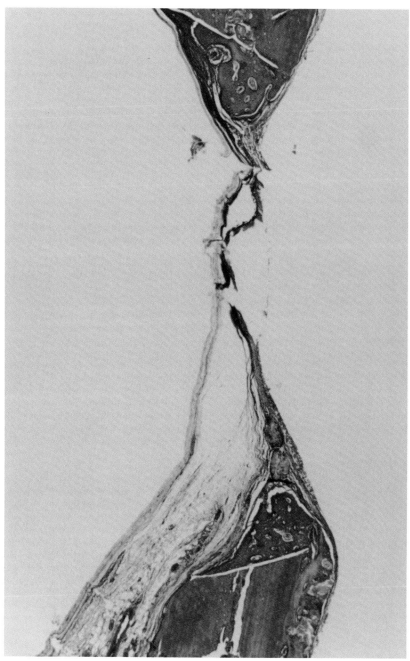

a

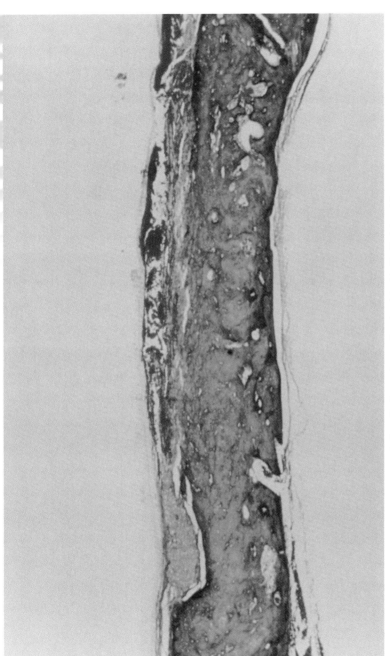

b

Figure 5-a, 5-b, and 5-c. Defect site stained with H & E (a) with no implant at 28 Days, (b) crosslinked with Fibrillar collagen (FC) at 28 Days, and (c) crosslinked with di-SG-PEG at 28 Days.

Continued on next page.

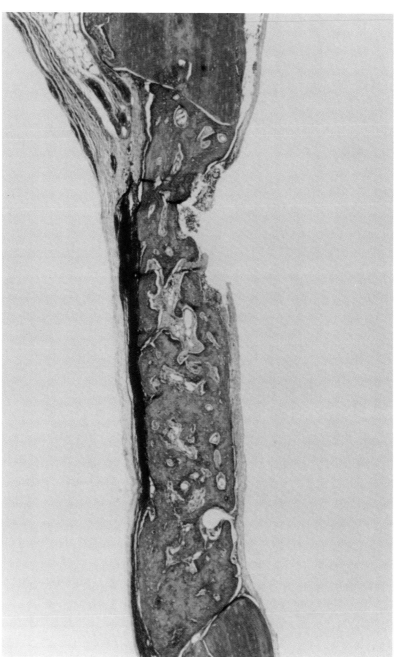

Figure 5. *Continued.*

c

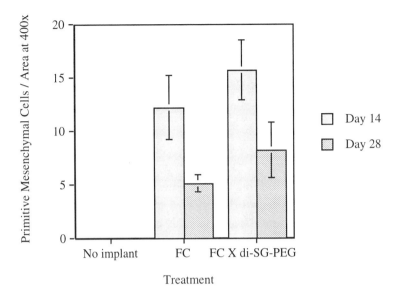

Figure 6. Histomorphometric results showing primitive mesenchymal cell concentrations within a 100 × 100 μm square area measured at 400× magnification at days 14 and 28. Shaded bars represent day 14 and hatched bars representday 28. Fibrillar collagen (FC) and FC cross-linked have significantly more cells as compared to the control, *p* values of 0.0047 and 0.0001, respectively.

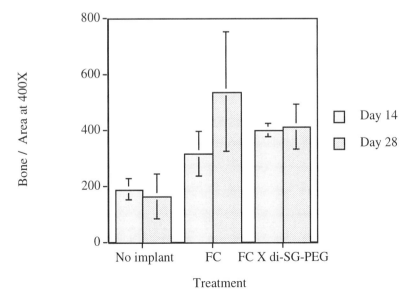

Figure 7. Histomorphometric results showing amount of bone within the defect site at days 14 and 28. Shaded bars represent day 14 and hatched bars representday 28. Fibrillar collagen (FC) and FC cross-linked have significantly more cells as compared to the control, *p* values of 0.003 and 0.0001, respectively.

Table 2: *In vitro* data for optimal formulations.

Test	Set 1	Set 2	Set 3
Clarity	Clear	Clear	Clear
Polymerization time, start	10 seconds	30 seconds	3 minutes
Polymerization time, finish	3 minutes	3 minutes	6 minutes
Attachment to cornea	Firm	Firm	Firm
Sensitivity to trypsin	15 %	10 %	4 %
Sensitivity to b. collagenase	31 %	36 %	9 %
Permeability: inulin, cm/sec	$(2.82 \pm 0.54) \times 10^{-7}$	$(33.7 \pm 6.68) \times 10^{-8}$	$(28.7 \pm 2.09) \times 10^{-8}$
Permeability: dextran, cm/sec	$(1.60 \pm 0.57) \times 10^{-8}$	$(1.90 \pm 0.76) \times 10^{-8}$	$(1.4 \pm 0.50) \times 10^{-8}$
Direct contact, MRC5	confluent at day 6	confluent at day 6	confluent at day 3
Direct contact, keratinocyte	confluent at day 6	confluent at day 6	confluent at day 3
Air exposed, keratinocyte	good morphology	multilayered cells	multilayered cells

afibrillar and therefore optically clear. Di-SG-PEG (and di-SE-PEG) were dissolved in 200 μl of PBS (phosphate buffer in saline) and mixed with 0.9 ml methylated collagen. The collagen and cross-linker used for the three formulations is outlined below.

Material	pH of collagen	Cross-linker
methylated collagen	7.3	Di-SG-PEG
methylated collagen	7.3	Di-SE-PEG
methylated collagen	< 7.3	Di-SE-PEG

To test attachment to the cornea, a viscoelastic mixture of crosslinker and methylated collagen was delivered to de-epithelialized rabbit cornea and polymerized *in-situ*. The degree of attachment was qualitatively evaluated while attempting to "peel" the lenticule from the cornea. All three optimal formulations attached firmly to the cornea.

Biomaterial discs were tested for enzymatic sensitivity by incubation in trypsin (Worthington Enzymes, 20 μg trypsin per mg collagen dry weight) at 37 ^{0}C for 13 - 20 h. In a separate experiment, discs were incubated at 37 ^{0}C in bacterial collagenase (Merck, 44 μg collagenase per mg dry weight collagen) for 16 - 20 h. In both cases, hydroxyproline content of the soluble fraction determined the percent of disc sensitive to enzymatic degradation. Methylated collagen crosslinked with di-SE-PEG at a pH < 7.3 was the most stable against trypsin and collagenase treatment.

Permeability to Inulin and dextran was tested by polymerizing the biomaterials (n = 5 - 8) on Millicell® - PCF filter inserts (Millipore 12 mm diameter, 12 μm pore size). Inulin (mw = 5,200, molecular radius = 14 Å) and dextran (mw = 70,000, molecular radius = 38 Å) fluxes across the collagen sheets were measured according to previously described procedures *(21)*. Permeability (Ktrans) values were calculated according to method of Maffly *(22)* and reported in units of cm/sec. All three formulations displayed similar permeability profiles to the model solutes.

Biocompatibility was tested in two cell culture models: 1) direct contact (using human lung MRC5 and mouse keratinocytes) to determine cytotoxicity and 2) air exposed model (using keratinocytes) to determine capability to support multilayered cells in air exposed conditions were employed. Controls were culture dish plastic and a Type IV collagen formulation known to support re-epithelialization in primates *(23)*. All three formulations were non-cytotoxic and supported cell growth and proliferation.

The ideal biomaterial should be clear, remain attached to cornea, permeable, permeable to nutrients, capable of promoting re-epithelialization and stable up to three (3) years *in vivo*. The animal model used in this study was developed by K. Thompson *(23)*. Fifteen primates (Macaca mulatta) were anesthetized and the central zone of both corneas was gently marked using an 8.5 mm trephine. Epithelium within the trephined area was gently removed using a Gill knife. Methylated collagen (pH ≤ 7.3) was mixed with di-SG-PEG or di-SE-PEG to produce three formulations as described above. Each formulation was delivered immediately after mixing onto de-epithelialized cornea (n = 5 per formulation) and molded, *in situ*, with the aid of a polysulfone mold to form the lenticule within 10 minutes. The contralateral eye served as a sham control. No patching or tarsorrhaphy was utilized. Epithelial migration over the lenticule was photographically recorded after fluorescein staining and the extent of re-epithelialization determined by digitizing the photographs. Translation of a two-dimensional photo to a three-dimensional cornea sphere was taken into account *(24)*. The primate study evaluated the lenticules' for 1) attachment to cornea, 2) re-epithelialization and 3) stability.

All three lenticule formulations attached firmly to the de-epithelialized cornea. The three formulations supported re-epithelialization. Control corneas re-epithelialized by Day 3 post-op. At Day 7, lenticules consisting of methylated collagen crosslinked with di-SG-PEG were 94% re-epithelialized, lenticules consisting of methylated collagen crosslinked with di-SE-PEG were 85% re-epithelialized and lenticules consisting of methylated collagen crosslinked with di-SE-PEG, pH < 7.3 were 93% re-epithelialized. Lenticule thickness varied from 200 - 400 µm. Thicker lenticules did not attain 100% re-epithelialization due, possibly, to mechanical debridement by the lids. Some neovascularization and edema were observed in these corneas. Epithelia appeared normal under the slit lamp. Epithelial cells were well attached and did not slough despite manual manipulation during fluorescein staining. No ocular irritation was observed clinically or behaviorally.

Lenticules of methylated collagen crosslinked with di-SG-PEG resorbed within 3 months post-op; histological analyses of one primate at ~ 2 months post-op suggested resorption at the epithelial/lenticule interface. Of the five primates with lenticules of methylated collagen crosslinked with di-SE-PEG, one primate was still on study at 14 months, one primate was euthanized at 6 months and three primates had their lenticules removed at 4 months post-op due to problems in maintaining re-epithelialization. One primate with lenticules of methylated collagen crosslinked with di-SE-PEG at pH < 7.3 was still on study at 14 months with a good re-epithelialization profile. The lenticules on the other four primates were removed at 4 - 5 months post-op due to incomplete re-epithelialization. It was speculated that re-epithelialization was mechanically hindered due to excessively thick (350 - 400 µm) lenticules. To test this hypothesis, one primate was implanted with a thin (~ 100 µm) lenticule of methylated collagen crosslinked with di-SE-PEG at pH < 7.3. This primate is still on study at 7 months post-op with a good re-epithelialization profile.

Summary. The studies reported here indicate that various collagen formulations were crosslinked by two different difunctional PEGs for a variety of applications. The results of *in vivo* experiments (rat subcutaneous model, rat parietal defect model, and de-epithelialized primate cornea model) show PEG-collagen gels are biocompatible and stable. An immunogenicity study in a sensitized guinea pig model revealed that the PEG crosslinked collagen implants showed lower antibody titers than collagen formulations themselves. These materials have wide potential in several engineered tissue applications.

Literature Cited

1. Harris, J. M. *J. Macromol. Chem. Phys.* **1985**, C25, 325.
2. Weiner B. Z.; Zillha, A. *J. Med. Chem.* **1973**, 16, 573.
3. Katre, N. V.; Knauf, M. J.; Laird, W. J. *Proc. Natl. Acad. Sci. U.S.A.* **1987**, 84, 1487.
4. Abuchowski, A.; van Es,T.; Palczuk, N.C.; Davis, F.F. *J. Biol. Chem.* **1977**, 252, 3578.
5. Golander, C.; Herron, J.; Lim, K.; Claesson, P.; Stenins, P.; Andrade, J. D. In *Poly (Ethylene Glycol) Chemistry, Biotechnical and Biomedical Applications*; Harris, J.M., Ed., Plenum: New York, New York, 1992, pp 221-245.
6. Abuchowski, A.; McCoy, J. R; Palczuk, N. C.; van Es, T.; Davis, F. F. *J. Biol. Chem.* **1977**, 252, 3582.
7. *Poly (Ethylene Glycol) Chemistry In Biotechnical and Biomedical Applications;* Harris J. M., Ed.; Plenum: New York, New York, 1992.
8. Rhee, W.; Wallace, D. G.; Michaels, A.; Burns, R.; Fries, L.; Delustro, F.; Bentz, H., Collagen-Polymer Conjugates, *U.S. Patent 5,162,430,* **1992**.
9. Zalipsky, S.; Seltzer, R.; Menon-Rudolph, S. *Biotechno. Appl. Biochem.* **1992**, 15, 100-114.
10. Rhee, W.; Carlino, J.; Chu, S.; Higley, H. In *Poly(Ethylene Glycol) Chemistry: Biotechnical and Biomedical Applications*; Harris J. M., Ed.; Plenum: New York, New York, 1992, pp 183-198.
11. Rhee, W.; Wallace, D. G.; Michaels, A.; Burns, R.; Fries, L.; Delustro, F.; Bentz, H. Collagen-Polymer Conjugates, *U.S. Patent 5,328,955,* **1994**.
12. Abuchowski, A.; Kazo, G. M.; Verhoest, C. R., Jr.; van Es, T.; Kafkewitz, D.; Nucci, M. L.; Viau, A. T.; Davis, F. F. Cancer Biochem. Biophys., **1984**, 7, 175-186.
13. Rosenblatt, J.; Rhee, W.; Berg, R. Mechanical Properties of PEG Crosslinked Collagen Hydrogels, *The 21st Annual Meeting of the Society for Biomaterials*, San Francisco, CA, 1995, March 18-22, 241.
14. Rosenblatt, J.; Rhee, W.; Berg, R. A New Injectable, In situ Crosslinkable Collagen Biomaterial, *The Fifth World Biomaterials Congress*, Toronto, Canada, 1996, May 29-June 2, I-344.
15. Keefe, J.; Wauk, L.; Chu, S.; DeLustro, F. *Clin. Mat.* **1992**, 9, 155-162.
16. DeLustro, F.; MacKinnon, V.; Swanson, N. *J. Dermatol. Surg. Oncol.* **1988**, 14 (Suppl):49-55.
17. Deporter, D.; Komori, N.; Howley, T.; Shiga, A.; Ghent, A.; Hansel, P.; Parisien, K. *Calcif. Tissue Int.* **1988**, 42:321-325 .

18. Schroeder-Tefft, J.; Bentz, H.; Wilson, F. D. Osteogenic Effects Produced by Collagen Matrices Evaluated in The Rat Parietal Bone Defect Model, *The Fifth World Biomaterials Congress*, Toronto, Canada, 1996, May 29-June 2, I-981.

19. Miyata, T.; Rubin, A.; Stenzel, K.; Dunn, M. *U.S. Patent 4,164,559*, **1979**.

20. Rauterberg, J.; Kuhn K. *Hoppe-Seyler's Z Physiol Chem* **1968**, 349, 611.

21. Geroski D. H.; Hadley, A. *Current Eye Research* **1992**, 11, 61.

22. Maffly, R.; Hays, R.; Lamdin, E. *J. Clin. Invest.* **1960**, 39, 630.

23. Thompson, K.; Hanna, K.; Gipson, I.; Gravagna, P.; Waring III, G.; Johnson-Wint, B. *Cornea* **1993**, 12(1): 35-45.

24. Crosson, C. E.; Klyce, S. D.; Beuerman, R. W. *Invest. Ophthalmol. Vis. Sci.* **1986**, 27, 464.

25. Kligman, A. M.; Armstrong, R.C. *J. Dermatol. Surg. Oncol.*, **1986**, 12: 351-357.

26. Stegman, S. J.; Chu, S.; Bensch, K.; Armstrong, R.C. *Arch. Dermatol.*, **1987**, 123: 1644-1649.

27. Ulbrich, K.; Strohalm, J.; Kopecek, J. *Makromol. Chem.* **1986**, 187, 1131-1144.

Chapter 27

Temperature-Induced Gelation Pluronic-g-Poly(acrylic acid) Graft Copolymers for Prolonged Drug Delivery to the Eye

Guohua Chen[1], Allan S. Hoffman[1,3], Bhagwati Kabra[2], and Kiran Randeri[2]

[1]Center for Bioengineering, University of Washington, Box 352255, Seattle, WA 98195
[2]Alcon Laboratories, Inc., Forth Worth, TX 78134

Copolymers of PEO/PPO/PEO triblock polyethers (Pluronic® polyols) grafted onto a bioadhesive polymer backbone, polyacrylic acid (PAAc), exhibit unique, thermally-induced solution-to-gel phase transitions when their aqueous solutions are warmed. The bioadhesive, thermally-gelling properties of these graft copolymers makes them useful for prolonged delivery of drugs to mucosal surfaces, such as the surface of the eye. *In vitro* drug release from the graft copolymer matrices is prolonged significantly at 34 °C in PBS compared to homo-PAAc or physical mixtures of PAAc and Pluronic® polyols, especially when Pluronic® polyols with relatively long PPO segments are grafted.

Mucoadhesive polymers have attracted large interest as drug carriers for controlled drug delivery in the last decade. (*1*) They are said to promote the delivery of a drug at a particular site by adhering to the mucous layer at that site where delivery of the drug is desired. Mucosal surfaces are present on buccal, gastric, enteric, nasal, ophthalmic, and genito-urinary tissues. Polyacrylic acid (PAAc) is well-known as a bioadhesive polymer and widely used in the development of controlled-release bioadhesive drug delivery systems. (*2*) PAAc is also a pH-sensitive polymer, and the carboxyl groups are ionized at physiological condition (pH 7.4). Thus, drug may be released too rapidly from PAAc carriers due to the high swelling (or fast dissolution) of the ionized PAAc. Therefore, drug formulations containing this polymer have been modified in order to slow down the drug release rate. Environmentally-induced gelation of the drug formulation, especially for topical delivery, has been the most popular approach.

The various approaches taken to develop formulations that gel in contact with mucosa include gelation due to (a) an increase in temperature (*3,4*), (b) an increase in pH (*5,6*), (c) interaction with specific ions (*7*), or (c) a change from alcoholic to aqueous solvent conditions (*8*). The temperature-induced gelling systems are based on formulations containing copolymers of PEO/PPO/PEO triblock polyethers (Pluronic® polyols) or ethyl(hydroxyethyl)cellulose (EHEC) (*4*), but these polymers are not very bioadhesive. (*3*) Formulations containing PAAc (or poly[methacrylic acid], [PMAAc]) are bioadhesive and also are pH-sensitive. pH-Induced gelation of such systems is

[3]Corresponding author

based on ionization and swelling of the unionized matrix, passing through a gel state when going from the unionized -COOH state at pH 5-6 to the ionized -COO⁻ state at 7.4. (5) However, these systems depend on delivery in an acidic condition, which is not always desirable, due to the harshness of low phs on some mucosal surfaces, such as the eye. Other pH-sensitive, bioadhesive formulations form gels at lower phs due to complexation of PAAc with PEO (poly[ethylene oxide]). (6) However, these complexes dissociate and therefore should rapidly release drug when the PAAc ionizes at pH 7.4. More recently, injectable gelling systems have been developed that are based on solutions of PEO co-dissolved with PAAc in alcohol-water mixtures. (8) These solutions form gels when placed in aqueous environments; however, the use of alcohol restricts their use to injectable systems. Topical application of this system could be problematical since some mucosal surfaces would be severely irritated by the alcohol.

Our goal has been to design and synthesize new polymeric compositions and structures for ophthalmic drug delivery that (a) are bioadhesive in contact with the surface of the eye, (b) exhibit thermally-induced transitions from low viscosity liquids to gels upon contact with the eye surface, which is at 34 °C, (c) do not irritate the eye and (d) do not interfere with vision. The first two actions should combine to prolong the residence time of the formulation on the eye surface and also to reduce the rate of drug release from the formulation. Such polymeric compositions could also be generally useful as carriers for drug delivery to mucosal surfaces other than the eye, such as to the nasal, buccal, vaginal or oral mucosa. In order to achieve these goals, our approach has been to synthesize unique copolymer compositions and structures that combine bioadhesive polymers (eg, PAAc) with temperature-sensitive polymers in the same polymer molecule.

There has also been great interest recently in polymers that are responsive to external stimuli such as temperature (9-12), electric fields (13,14) and pH (15-17). These polymers have been widely applied to the fields of drug delivery, diagnostics, separations and robotics. (10) In attempting to synthesize copolymers with both bioadhesive and thermally-induced gelation properties, we first synthesized random copolymers of AAc and N-isopropylacrylamide (NIPAAm). (18, 19) As noted above, PAAc is bioadhesive, and poly(NIPAAm) (or PNIPAAm) exhibits a sharp phase transition at 32 °C called the lower critical solution temperature (LCST). These random copolymers did not provide both properties of bioadhesivity and temperature sensitivity because the bioadhesive property of PAAc depends on having long sequences of PAAc, while the LCST transition of PNIPAAm also depends on the presence of reasonably long sequences of homo-PNIPAAm.

In order to provide both properties in one polymer structure, we graft copolymerized PNIPAAm (19) or its random copolymer with butyl methacrylate (BMA) (19, 20) onto PAAc backbones, and investigated their properties as matrices for diffusion and erosion-controlled, mucosal drug delivery. (Figure 1) We found that at pH 7.4, where the carboxyl groups of PAAc are ionized, the graft copolymers exhibited LCSTs similar to those of homo-PNIPAAm or its copolymer with BMA. We also found that the graft copolymer vehicles prolonged the duration of drug release compared with either NIPAAm-AAc random copolymers of similar compositions, or homo-PAAc by itself. (21) However, these graft copolymers did not form high viscosity gels when warmed above 34 °C. In addition, PNIPAAm is not on the FDA list of "generally regarded as safe" (GRAS) polymers.

Therefore, we decided to synthesize similar graft copolymers using thermally-sensitive "GRAS" polymers that are known to form gels at low concentrations. (21) Although poly(ethylene oxide) (PEO) is a well-known GRAS polymer, its solutions are only thermally-sensitive at higher temperatures (eg, ~80°C) and graft copolymers of PEO onto PAAc do not form gels at body temperatures. (22) However, Pluronic® polyols {triblock copolymers of PEO and poly(propylene oxide) (PPO), or PEO-PPO-PEO} are also GRAS polymers, and solutions of Pluronic® polyols do form gels in the

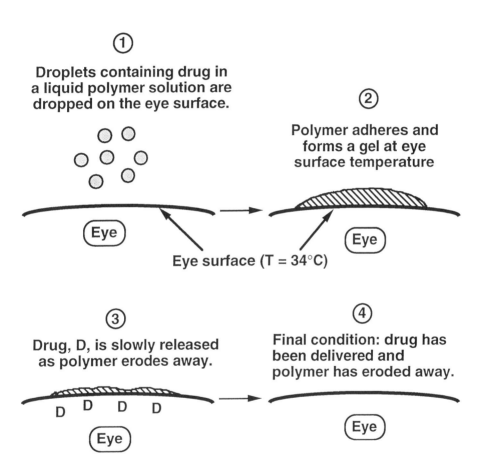

Figure 1. Temperature-sensitive drug delivery system for ophthalmic drug delivery.

temperature range between ambient and body temperatures. *(21)* Therefore, we grafted Pluronic® polyols onto PAAc backbones *(23,24)*, and report here on the synthesis and unusual thermally-induced gelation properties of Pluronic®-g-PAAc copolymers, and their drug release behavior.

Experimental

Materials. Polyacrylic acid (PAAc, Mv = 250,000, Aldrich, Milwaukee, USA) dicyclohexylcarbodiimide (DCC, Pierce, Rockford, USA), 4-nitrophenyl chloroformate (Sigma, St. Louis, USA) and diaminoethylene (Aldrich, Milwaukee, USA) were used without further purification. Block copolymers of polyethylene oxide - polypropylene oxide - polyethylene oxide (PEO-PPO-PEO) were kindly provided by BASF Corp. (Wyandotte, USA) under the tradename Pluronic®. The structures and properties of the Pluronic® polyols used in this study are summarized in Table I. *(24)* An ophthalmic drug, timolol hydrogen maleate (Timolol), was provided by Alcon Laboratories, Inc., Fort Worth, TX, USA. All other reagents used were of analytical grade.

Table I Structure and selected properties of Pluronic®* used in this study

The structure of Pluronic® surfactant

$$HO\text{---}[CH_2CH_2O]_x\text{---}\begin{matrix}CH_3\\|\\[CH_2CHO]_y\end{matrix}\text{---}[CH_2CH_2O]_x\text{---}H$$

| EO | PO | EO |

Pluronic®	EO-PO-EO	Total M.W. (Avg.)	Cloud point in water (°C) (wt. %)	
			1%	10%
L-61	4/30/4	2,000	24	17
L-81	7/38/7	2,750	20	16
L-92	10/50/10	3,650	26	16
L-122	13/69/13	5,000	19	13

* Data from BASF brochure

Synthesis of Amino-terminated PEO/PPO/PEO (Pluronic®).

The triblock copolymers with a reactive amino-terminal were prepared by a two step reaction. First, 4 mmoles of Pluronic® were reacted with 5 mmoles of 4-nitrophenyl chloroformate in methylene chloride in the presence of triethyleneamine at room temperature for 4 hours to yield a 4-nitrophenyl formate-derivatized intermediate (Figure 2). This intermediate was recovered by extraction three times using petroleum ether (35 - 60 °C), resulting in a yield of 70 - 80% by weight of product. In the second step, 2 mmoles of the intermediate were reacted with 6 mmoles of diaminoethylene in methylene chloride at room temperature overnight. After reaction, the mixture was extracted three times with petroleum ether, then dialyzed against distilled water using a membrane with a molecular weight cut-off of 1000 or 3,500 (depending on the molecular weight of Pluronic® used)

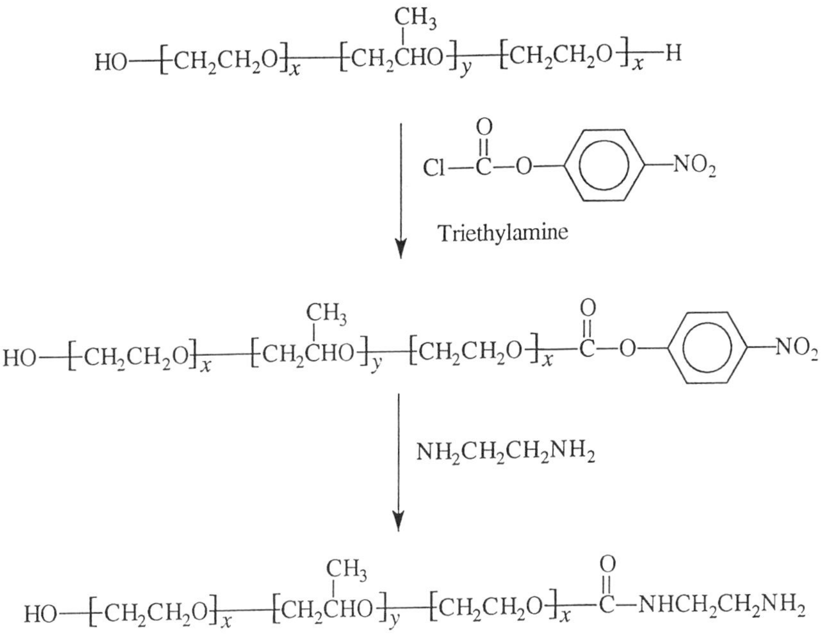

Figure 2. Schematic synthesis of amino-terminated Pluronic® derivitive.

for three days at 4 °C, and then lyophilized to result in the product, the amino-terminated Pluronic® with a yield of 85 to 90% by weight. Functionality of the amino-terminated derivative was determined by back titration. The derivative was dissolved in an excess of 0.01 N HCl to neutralize the amino-terminal groups and the excess of HCl was back-titrated with 0.01 N NaOH.

Synthesis of Graft Copolymers of Pluronic®-g-PAAc. Graft copolymers of Pluronic®-g-PAAc were prepared by coupling the reactive amino-terminals of the derivatized block copolymers of PEO-PPO-PEO (Pluronic®) onto the PAAc backbone. Specifically, reaction between the amino group of the amino-terminated Pluronic® derivative and carboxyl groups of PAAc in the presence of dicyclohexyl carbodiimide (DCC) resulted in amide bond formation (Figure 3). The reaction was carried out in methanol at room temperature for 24 hours, with a mole ratio of the block copolymer of PEO/PPO/PEO to DCC of 2:1. The resulting graft copolymers were recovered from the reaction solution by precipitation into tetrahydrofuran (THF) or diethylether.

Characterization of Graft Copolymers of Pluronic®-g-PAAc.

Gel Permeation Chromatography (GPC). The graft copolymers were characterized by GPC (Waters 501 equipped with Waters 500 Å, 10^3 Å and 10^4 Å Ultrastygel® columns, and Waters 410 differential refractometer and 486 tunable absorbance detectors), dimethyl formamide (DMF) as mobile phase with an elution rate of 0.7 ml/min, at 40 °C.

Composition. The composition of the graft copolymers were determined by back titration. An autotitrator (Model Mettler DL25) was used to determine the composition of graft copolymers. 0.1 g of the graft copolymers of Pluronic®-g-PAAc was dissolved in an excess of 0.1 N NaOH to neutralize the COOH groups and the excess of NaOH was back-titrated with 0.1 N HCl.

Temperature-sensitive Behavior. The temperature-sensitive behavior of the Pluronic®-g-PAAc graft copolymers solutions was determined by measuring the solution viscosity and elastic modulus as a function of temperature. 2.5 wt% of aqueous graft copolymer solutions at pH 7.2 were used for viscosity measurements at various temperatures ranging from 25 °C to 40 °C using a Brookfield DV III RV viscometer fitted with a CP-52 spindle. The same solutions were used for elasticity measurements at various temperatures using Bohlin Rheometer fitted with a CP 4/40 spindle under constant oscillatory shear stress of 1 Pa. The temperature was controlled by circulating water from a constant temperature water bath through the jacket of the viscometer cup. About 0.6 ml of the solution was placed in the viscometer cup. The viscosity measurements were carried out at 0.1 rpm (shear rate 0.2/second) for 3 minutes at a fixed temperature. The viscosity value at the end of 3 minutes was recorded. The phase transition temperature of the graft coplymer was defined as 25% of maximum viscosity.

Drug Loading and Release. Drug solutions were prepared in methanol, containing 10% polymer with 5% (with respect to polymer) of the timolol hydrogen maleate salt (Timolol). One drop of each solution from a syringe was placed on a clean cover glass, air dried overnight, and then dried completely under vacuum for 24 hrs at room temperature. The glass disc, coated with the drug-loaded polymer film with thickness of ca. 150 mm and diameter of 0.54 - 0.58 mm, was hung in a vessel containing well-stirred PBS buffer (pH 7.4) at 34 °C for drug release measurements. The amount of drug released from the polymer matrix was measured periodically at 294 nm by circulating the release medium (PBS buffer, pH 7.4) through a UV cuvette (Figure 4).

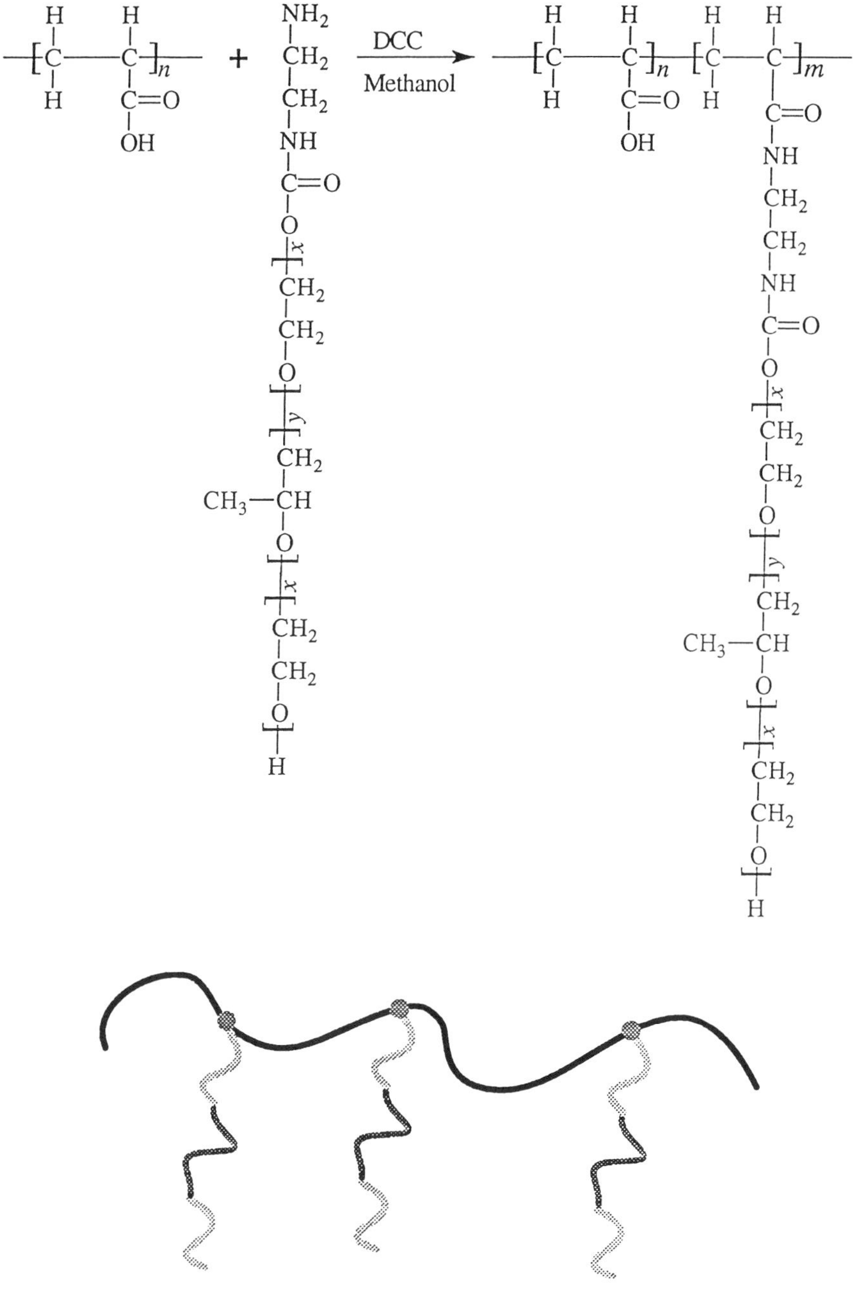

Figure 3. Schematic synthesis of graft copolymers of Pluronic®-g-PAAc.

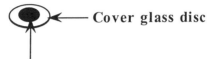 Cover glass disc

Film cast from [Polymer (P) + drug (D)]
solution in methanol

[P]: 10 wt%; [D]/[P]: 5 wt%

Figure 4b. Apparatus for measurement of drug release rates from eroding polymers.

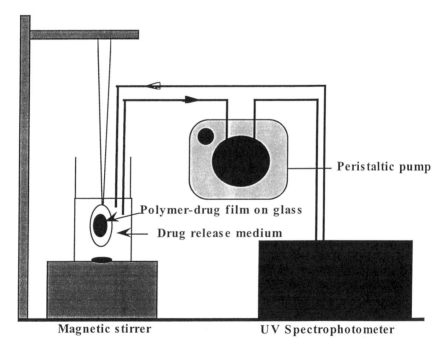

Figure 4a. Films of polymer + drug cast on glass discs (thickness of films, ca. 150μ; diameter of films, ca. 0.54 - 0.58 mm; weight of each film, ca. 5 mg).

Results and Discussion

Pluronic® Polyols. Pluronic® polyols were chosen as the temperature-sensitive polymers to be grafted to the PAAc backbone because some of them are approved for use by the FDA as food additives. The Pluronic® polyols used in this study have cloud points (CP) (which indicate the temperature-induced phase transition) between 19 - 26 °C for 1 wt% solutions in water. (Table I)

Amination of Pluronic® Polyols at One End Only. In order to graft the Pluronic® onto the backbone of PAAc, a reactive amino group is required at one end of the Pluronic® PEG blocks. Thus, amino-terminated Pluronic® derivatives at one end only were prepared by adjusting the stoichiometry (Figure 2). In the first step, it is important to control the stoichiometry of the reaction so that only one end hydroxyl group in the Pluronic® PEG blocks is activated. In the second step, an access amount of diaminoethylene is needed to ensure that all of the activated Pluronic® end groups are converted to amino groups. Using this method, we were able to derivatize the Pluronic® to an average amino functionality of f = 0.91 ± 0.1, as determined by back-titration. Statistically speaking, it might be possible to have some f = 0 or f = 2 derivatives existing in the product, but no attempt was made to separate them from the product (ion exchange chromatography could be used with a pH gradient). In fact, any f = 0 Pluronic® fraction was probably isolated during precipitation of the graft copolymer. In addition, reaction of any doubly-derivatized Pluronic® fraction with PAAc should result in a crosslinked network; however, the graft copolymers we synthesized were completely soluble, indicating that the amounts of doubly derivatized Pluronic® polyols in the product were negligible (see section below).

Synthesis and Isolation of the Graft Copolymers. The synthesis of graft copolymers of Pluronic®-g-PAAc, should produce a mixture of graft copolymer, unreacted Pluronic® and ungrafted PAAc. After the Pluronic® L-61, or L-92, or L-122 polyols were grafted onto PAAc they were recovered from the reaction solutions by precipitation into THF or diethylether, both of which are solvents for Pluronic®. Any unreacted Pluronic® will therefore remain in the supernatant. However, since homoPAAc also precipitates in these solvents, we can't separate any ungrafted PAAc from the graft copolymer product. On the other hand, the chance for a PAAc chain to have absolutely no Pluronic® grafted to it is very small, especially for high degrees of grafting. The graft copolymer compositions that were synthesized are summarized in Table II.

Formation of the graft copolymer was verified by gel permeation chromatography (GPC). As illustrated in Figure 5, the Pluronic® L-122 peak at 37.78 min can hardly be seen in the graft copolymers, but it is clearly seen in the physical mixture of PAAc with L-122 (30% L-122), indicating that almost no ungrafted Pluronic® was physically-entrapped in the graft copolymers. In addition, the graft copolymers elute later than the "parent" homoPAAc, indicating that they have smaller hydrodynamic radii than PAAc, probably due to associations between the grafted Pluronic® chains.

Temperature-induced Gelation of Graft Copolymer Solutions. In our previous papers (*19,20,23*), we have reported that at pH 7.4 the solutions of graft copolymers of polyNIPAAm-g-PAAc or copoly(NIPAAm-BMA)-g-PAAc exhibited temperature-induced precipitation behavior, as indicated by a clouding of the solutions, despite the ionization of the backbone PAAc. The graft copolymer phase transitions also occurred at temperatures similar to the LCSTs of the grafted side chains. In contrast, solutions of Pluronic®-g-PAAc graft copolymers showed unique temperature-induced gelation behavior. Figure 6 shows that the viscosities of the graft copolymers increase with temperature, an opposite trend to viscosity/temperature behavior of most polymer

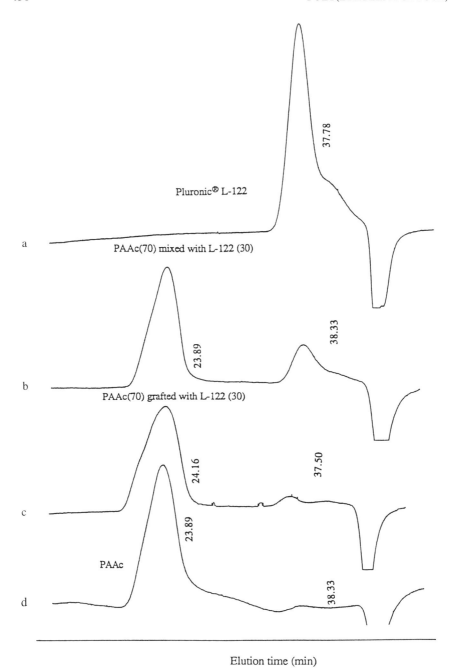

Figure 5. GPC diagrams of a) free Pluronic® L-122; b) physical mixture of Pluronic® L-122 with PAAc (30/70 by weight); c) graft copolymer of Pluronic® L-122-g-PAAc (30/70 by weight); d) parent homopolymer of PAAc. Eluent, DMF; temperature, 40 °C; elution rate, 0.7 ml/min.

solutions. It can be seen in Figure 6 that viscosities also increase with concentration, but this is expected. The viscosities of the graft copolymer solutions increase dramatically just above the cloud point of the free Pluronic®, and a gel forms. In contrast, equivalent concentration solutions of the same Pluronic® by itself, or physically mixed with PAAc, become cloudy and precipitate as the Pluronic® phase transition temperature is reached. The shear modulus of the system also increases with temperature, while the loss modulus shows a maximum around 28 - 32 °C, corresponding to the solution-to-gel phase transition (Figure 7).

Table II. Preparation of graft copolymers of Pluronic®-g-PAAc*

Code	Pluronic®	$W_{Pluronic®}/W_{PAAc}$ in feed	Yield (wt%)	Pluronic® in copolymers** (wt%)
L61-1	L-61	10/90	79	15
L61-2	L-61	20/80	55	24
L61-3	L-61	30/70	86	34
L92-3	L-92	30/70	94	32
L122-1	L-122	10/90	80	15
L122-2	L-122	20/80	80	18
L122-3	L-122	30/70	68	21
L122-4	L-122	40/60	78	27
L122-5	L-122	50/50	65	41

*Solvent: methanol, 100 mL; reaction temperature and time: room temperature, 24 h; Mole ratios of DCC to Pluronic®, 2 : 1. The polymers were recovered by precipitation into THF or diethylether.
**The composition of Pluronics® in the graft copolymers were determined by back titration.

It is worth noting that solutions of the graft copolymers have a higher phase transition temperature than do those of the pure Pluronic®, since association of the grafted Pluronic® chains–especially when grafted to the ionized PAAc backbone–should be less efficient than that of free Pluronic® chains. (This also makes the Pluronic®-g-PAAc graft copolymers more readily soluble when continuous dilution occurs, as seen in the drug release studies discussed below).

The mechanism of the temperature-induced gelation is related to the cloud point phenomenon of Pluronic® polyols, and can be explained by the hydrophobic interactions of the polypropylene oxide (PPO) segments among the grafted Pluronic® chains. When the temperature is raised above the phase transition temperature (cloud point) of the Pluronic®, the PPO segments among the grafted chains tend to aggregate together. Since the Pluronic® chains are covalently bonded to the PAAc backbone, those assemblies of PPO segments act as physical crosslinks, leading to formation of a physically-crosslinked hydrogel. In contrast, Pluronic® by itself, or physically mixed with PAAc, both form precipitates and do not gel as temperature increases. These contrasting systems are sketched in Figure 8.

The hydrophobic aggregation of the PPO segments of the grafted chains is relatively "fragile", as evidenced by the fact that the viscosity increase with temperature can be reduced or even reversed by increase of shear rate or temperature. As can be seen in Figure 9, the viscosity vs. temperature curve reaches a maximum and then decreases with further increase of temperature. It can be seen that the maximum viscosity and the temperature at which maximum viscosity is observed decreases with increase of shear rate. This suggests that the aggregation of the PPO segments on the grafted Pluronic® chains has to compete with the increase in molecular motions at higher temperatures and/or mechanical shear. Hence, at first the viscosity increases with temperature, as

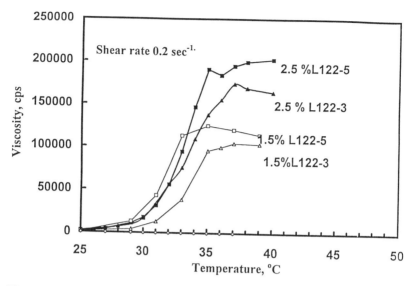

Figure 6. Viscosity of graft copolymer solution (1.5 and 2.5 wt%) of Pluronic® L-122-g-PAAc (L122-3 and L122-5) at pH 7.2 vs. temperature.

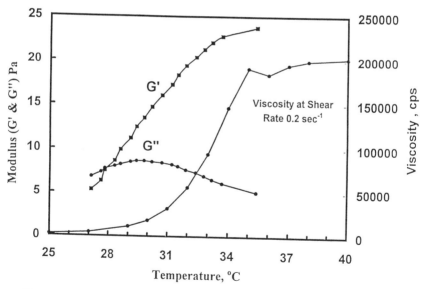

Figure 7. Elasticity and viscosity of Pluronic®-g-PAAc (L122-5) solution vs. temperature.

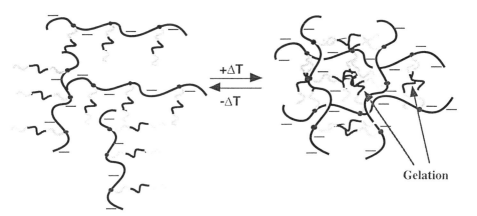

Figure 8a. Sketch for the thermally-sensitive behavior of the graft copolymers of Pluronic®-g-PAAc at pH 7.4.

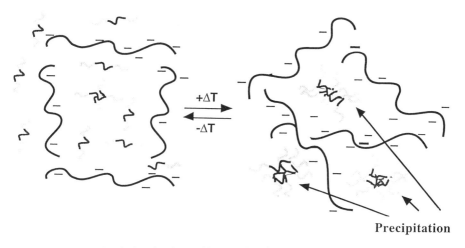

Figure 8b. Sketch for the thermally-sensitive behavior of the physical mixtures of Pluronic® with PAAc at pH 7.4.

hydrophobic aggregation of PPO segments on the Pluronic® grafted chains is the dominant action, but as temperature continues to increase, the molecular motions resulting from thermal energy dominate. Addition of mechanical energy in the form of shear also helps to break up the PPO aggregates. These effects are reversible, and the viscosity is regained instantaneously upon removal of shear.

Bioadhesive properties of the graft copolymers. The bioadhesive properties of the graft copolymers were evaluated (by N. Peppas and C. Brazel) by comparing crosslinked hydrogels of Pluronic®-g-PAAc and PAAc. It was found that the graft copolymer hydrogels showed good bioadhesive properties, similar to PAAc gels. The detailed synthesis of the grafted copolymer hydrogels and their bioadhesive characteristics will be reported in a separate publication.

Drug release from graft copolymer gels. Timolol hydrogen maleate (Timolol), an ophthalmic drug, was used as a model drug to evaluate the graft copolymers of Pluronic®-g-PAAc as drug carriers. Timolol was loaded to the graft copolymer matrix by dissolving both polymer and the drug in a neutral solvent (methanol) and a film was cast on a glass disc from the copolymer-drug mixture. The dried coated glass discs were suspended in PBS buffer (pH 7.4) at 34 °C, and the amount of the drug released from the film was measured as a function of time. Figure 10 illustrates the drug release profiles from copolymers having varying Pluronic® L-61 graft levels. For comparison, the drug release profiles from homoPAAc as well as a physical mixture of 80% homoPAAc and 20% L-61, are also presented in Figure 10. It can be seen that the graft copolymers show significantly prolonged drug release profiles compared to homoPAAc alone or the physical mixture of PAAc with Pluronic® L-61. Complete drug release from the graft copolymers with 20% or more Pluronic®L-61 content was delayed to over 1 hour compared to 15 min from the homoPAAc formulation or 25 min from the physical mixture formulation.

The prolonged drug release profile is due to the gelation of the graft copolymer under the conditions of the drug release experiment (pH 7.4 and 34 °C). The gelation not only prolongs the drug diffusion rate out of the polymeric matrix, but it also slows the dissolution rate of the graft copolymer. In contrast, the drug formulation with the homoPAAc swells and dissolves very rapidly, leading to fast drug release. Furthermore, the Pluronic® in the physical mixture with PAAc phase separates at the drug release conditions. As a result, the highly soluble drug is also released rapidly due to the fast swelling, separation and dissolution of PAAc from the mixture; this is similar to the very rapid drug release from homoPAAc. However, the presence of the precipitated Pluronic® phase in the mixture still prolongs slightly the overall release rate profile compared to the homoPAAc formulation.

Figure 11 compares the drug release profiles for 30% grafts of three different Pluronic® polyols. The times for complete drug release from those three different polymer matrices are prolonged in the order of L-122 > L-92 > L-61, which is the same order as increasing lengths of the PPO segments in those Pluronic® polyols. This indicates that the length of the hydrophobic segments in the grafted Pluronics® chains plays an important role in prolonging the drug release from these interesting graft copolymer carriers.

Figure 12 illustrates the drug release profiles from the graft copolymers having Pluronic® L-122 as the grafted chains, the triblock polyol with the longest PPO segment studied. It can be seen that drug release was more than 4 hours from carriers having 30% or more Pluronic® L-122 grafted to PAAc. This indicates that longer hydrophobic PPO segments in the grafted chains probably enhance the self-association of the grafts and viscosity buildup, thus most effectively prolonging the drug release from the polymer matrix.

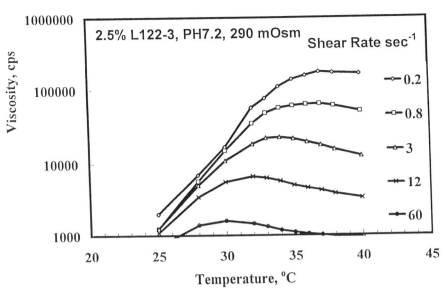

Figure 9. Viscosity of Pluronic®-g-PAAc solution vs. temperature. Effect of shear rate on the maximum viscosity.

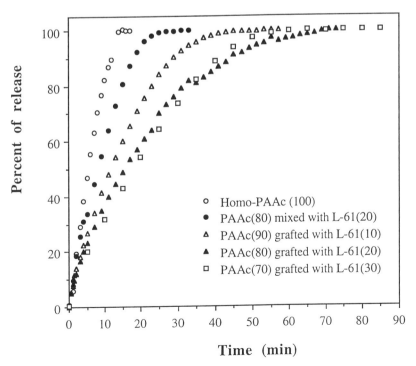

Figure 10. Percent of Timolol release from the polymer matrices as a function of time.

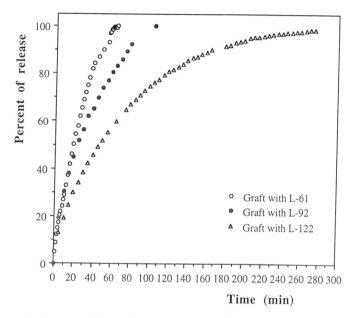

Figure 11. Kinetics of Timolol release from copolymer matrices with different Pluronics® in the graft chains.

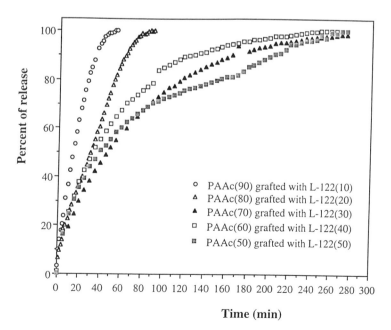

Figure 12. Percent of Timolol release from the Pluronic® L-122-g-PAAc graft copolymer matrices as a function of time.

In conclusion, the major compositional variables in graft copolymers of triblock PEO-PPO-PEO polymers grafted to PAAc are the PEO/PPO ratio in the graft, and the grafting level. Controlling these two variables permits one to "molecularly engineer" a wide range of properties of such graft copolymers for use as drug carriers.

Conclusions

Graft copolymers of thermally-sensitive PEO/PPO/PEO (Pluronic®) copolymers grafted to PAAc show a unique temperature-induced gelation. When these copolymers are used as drug delivery carriers, drug release rates are significantly reduced, and complete drug release from the matrices is significantly prolonged to over four hours, compared to ca. 15 min from homoPAAc. Bioadhesive properties of hydrogels prepared from these graft copolymers are excellent; these results will be published elsewhere.

Literature Cited

1. Gu, J. M.; Robinson, J. R.; Leung, S. S. *Crit. Rev. Therap. Drug Carrier Sys.* **1988,** *5,* 21.
2. Park, H.; Robinson, J. R. *Pharm. Res.* **1987,** *4* , 457.
3. Schmolka, I. R. *Ann. New York Acad. Sci.* **1994,** *720,* 92.
4. Lindell, K.; Engstrom, S. *Int. J. Pharm.* **1993,** *95,* 219.
5. Leung, S. H.; Irons, B. K.; Robinson, J. R. *J. Biomater. Sci. Polym. Ed.* **1993,** *4,* 483.
6. Klier, J.; Scranton, A. B.; Peppas, N. A., *Macromol.* **1990,** *23,* 4944.
7. Rozier, A.; et. al. *Int. J. Pharm.* **1989,** *57,* 163.
8. Haglund, B. O.; Joshi, R.; Himmelstein, K.J. *J. Contr. Rel.* **1996,** *41,* 229.
9. Tanaka, T. *Polymer* **1979,** *20,* 1404.
10. Hoffman, A. S. *J. Contr. Rel.* **1987,** *6* , 297.
11. Bae, Y. H.; Okano, T.; Hsu, R.; Kim, S. W. *Makromol. Chem., Rapid Commun.* **1987,** *8,* 481.
12. Chen, G. H.; Hoffman, A. S. *Bioconj. Chem.* **1993,** *4,* 509.
13. Kwon, I. K.; Bae, Y. H.; Kim, S. W. *Nature (London)* **1991,** *354,* 291.
14. Kishi, R.; Hara, M.; Sawahata, K.; Osada, Y. in *Polymer Gels*, DeRossi, D. et. al., Eds., Plenum, New York, **1991,** pp.205.
15. Siegel, R. A.; Firestone, B. A. *Macromol.* **1988,** *21,* 3254.
16. Peppas, N. A.; Buri, P. A. *J. Contr. Rel.* **1985,** *2,* 257.
17. Park, K.; Robinson, J. R. *J. Contr. Rel.* **1985,** *2,* 47.
18. Dong, L. C.; Hoffman, A. S. *J. Contr. Rel.* **1991,** *15,* 141.
19. Chen, G. H.; Hoffman, A. S. *Nature (London)* **1995,** *373,* 49.
20. Chen, G. H.; Hoffman, A. S. *Macromol. Rapid Commun.* **1995,** *16,* 175.
21. BASF brochure on Pluronic® polyols
22. Hourdet, D.; L'Alloret, F.; Audebert, R. *Polym. Prepr.* **1993,** *34,* 972.
23. Hoffman, A. S.; Chen, G. H.; Kaang, S.; Priebe, D.T. *Proc. Intern. Symp.Contr. Rel. Bioact. Mater.* **1995,** *22,* 159.
24. Chen, G. H.; Hoffman, A. S.; Ron, E. *Proc. Intern. Symp.Contr. Rel. Bioact. Mater.* **1995,** *22,* 167.

Chapter 28

Novel Degradable Poly(ethylene glycol) Esters for Drug Delivery

Xuan Zhao and J. Milton Harris[1]

Department of Chemistry and Materials Science Program, The University of Alabama at Huntsville, Huntsville, AL 35899

In this chapter we describe two applications of new hydrolytically-degradable PEG derivatives. The first involves synthesis of ester-containing PEGs for preparation of soluble PEG-drug conjugates in which the PEG is attached to the drug via a hydrolytically degradable linkage. The second application involves use of these and related PEGs to prepare degradable PEG hydrogels. Degradation rates can be controlled by variation in molecular structure of the esters and in the degree of branching of the PEGs.

Polyethylene glycol (PEG) is a biocompatible, nontoxic, nonimmunogenic and water-soluble polymer of much use in biomaterials, biotechnology and medicine (1). In the area of drug delivery, PEG derivatives have been widely used in covalent attachment ("PEGylation") to proteins to reduce immunogenicity, proteolysis and kidney clearance and to enhance solubility ($1,2$). Similarly, PEG has been attached to low molecular weight (MW) drugs to enhance solubility, reduce toxicity and alter biodistribution (2). A closely related application is synthesis of crosslinked PEG networks or hydrogels for use in drug delivery ($3,4$) since the same chemistry used in design of degradable, soluble drug carriers can also be used in design of degradable gels.

In this chapter we review our recent work on preparation and characterization of hydrolytically degradable PEGs designed for use as soluble drug carriers, and we present preliminary studies of hydrogels prepared by use of similar degradable PEGs.

As noted above, conjugation with PEG gives a protein or peptide an impressive set of new properties, but PEGylated proteins also usually exhibit a significant decrease in activity, even though the PEG attachment site may be removed from the protein active site (5). Protein PEGylation is generally achieved by formation of a stable urethane, amide or amine linkage between an amino group

---

[1]Corresponding author.

on the protein and an active carbonate, ester, aldehyde or tresylate on PEG (*1,2,6*). In some instances it could be advantageous to slowly detach the PEG from the protein, thus releasing free protein or lightly PEGylated protein into circulation. If the release rate is controlled, the protein activity level in serum could be maintained, even as the total amount of protein in serum decreases by clearance.

Examples of reversible PEGylation of proteins have been described previously. Glass and colleagues coupled 4-phenoxy-3,5-dinitrobenzoyl-polyethylene glycol to several small peptides and to the B-chain of bovine insulin by thiol (cysteine) displacement of the 4-phenoxy group (*7*). The resulting conjugate could be "de-PEGylated" by exposure to thiols at pH 7. These workers suggested use of this chemistry in peptide and protein purification. Reversible PEGylation of protein thiol groups was later described by Woghiren (*8*) and Musu (*9*) by use of 2- or 4-dithiopyridine PEG. PEG was cleaved from these PEG-protein conjugates by mild reduction.

Garman has adopted the known reversibility of dimethylmaleic anhydride with lysine to PEGylation by attaching this anhydride to PEG and then coupling the new PEG derivative to plasminogen activator proteins (*10*). The resulting conjugate gave complete hydrolytic cleavage of the PEG after about two days in buffer, in the process restoring active protein. Similarly, PEGylation with the well known succinimidyl succinate of PEG is susceptible to hydrolytic cleavage through the existing ester linkage (*2*), and Vestling and coworkers have taken advantage of this hydrolytic instability to tag lysines with succinate groups and thus determine the position of PEGylation by mass spectrometry (*11*).

Similar to the work on protein PEGylation for pharmaceutical applications, there has also been much work on PEGylation of low MW drug molecules. Unlike protein PEGylation, however, it has long been appreciated that reversible linkages between low MW drug and PEG are desirable since many PEG conjugates with low MW drugs are inactive (*2,12*). Hydrolytically degradable ester linkages are readily provided by coupling low MW drugs containing carboxyl groups to the terminal hydroxyl of PEG (*2,13*). Similarly, Ouchi et al., coupled 5-fluorouracil to PEG through hydrolyzable ester linkages (*14*), and Pendri et al. (*12*) and Desai et al. (*15*) have prepared PEG-taxol conjugates by coupling PEG acids directly to the hydroxyl groups of the drug. Nichifor et al. have prepared PEG-fluorouracil conjugates connected by an enzymatically-degradable peptide linkage (*16*), and Veronese and coworkers have coupled doxorubicin to PEG through amino acid linkers (*17*).

We describe below the preparation of a new family of hydrolytically degradable PEGs which are useful as soluble drug carriers for release of proteins and low MW drugs. These same PEGs can be used to prepare degradable PEG hydrogels. PEG-based hydrogels have been prepared by others for use in drug delivery, enzyme immobilization and wound covering (*3,4*). These gels have been prepared in several fundamentally different ways. The simplest route is to crosslink PEG by exposure to ionizing radiation (*18*). Graham and his coworkers pioneered use of coupling reactions between a difunctional PEG, a triol and a multifunctional crosslinking agent (*3*). For example, such a gel can be prepared by coupling linear PEG diol, 1,2,6-hexane triol and diphenylmethane-4,4'-diisocyanate. These gels can be made degradable by inclusion of copolymers containing ester linkages (*3*). Suitably activated PEGs can be crosslinked by a variety of crosslinkers. E.g., Fortier and coworkers have crosslinked PEG p-nitrophenylcarbonate with bovine serum albumin (*19*). Similarly, Nathan et al., have crosslinked PEG-lysine copolymers by activating the pendant carboxyl groups of this polymer with hexanediamine (*20*), and Llanos and Sefton have crosslinked PEG with poly(vinyl alcohol) (*21*).

Hubbell and coworkers have extensively utilized PEG gels prepared by use of PEG acrylates having lactide/glycolide linkages between the PEG and the acrylate group (*4*). Acrylate polymerization provides a crosslinked network, and the lactide/glycolide linkages provide degradability. Russell and coworkers have extended this concept by use of PEGs with an acrylate at one terminus and an enzyme at the other (*22*); the resulting gels have potential for use as industrial catalysts. Similarly, the PEG-lysine copolymers of Nathan et al., can have acrylate pendant groups that can be polymerized to make hydrolytically degradable PEG gels for drug delivery (*20*).

In another synthetic approach, PEG gels can also be formed by photochemical [2 + 2] coupling of vinyl PEGs; this reaction converts the vinyl groups to cyclobutane crosslinks (*23,24*). These reactions have the advantage of being reversible upon irradiation at a different wavelength (*24*).

Degradable hydrogels have two major advantages for drug delivery. First, surgical removal of the drug-depleted device is not necessary, and second, drug release kinetics can be controlled to approach desired zero order release. However, the degradation products of the gel must be biocompatible and nontoxic. The biophysical and biochemical properties of PEG make it an ideal material for application *in vivo*. In most existing degradable PEG hydrogels, small organic crosslinkers or copolymers have been used. The appearance of these non-PEG components in degradation products is a possible toxicity concern.

In the following discussion we describe a family of new degradable PEG hydrogels containing ester linkages that are formed by reaction of PEG carboxylic acids with hydroxyl-terminated PEGs. Upon hydrolysis of these linkages, the hydrogels degrade into only PEG and PEG acids. All final degradation products are less than 15 kD in molecular weight and thus can be readily cleared by the body (*25*). Degradation rates of the gels can be fine-tuned by slight variation in molecular structure of the esters and in degree of branching of the PEGs.

In the course of preparing these ester-linked PEG gels, it became apparent that loading large drug molecules into the gels was difficult. Thus we devised a second approach that utilizes mild reactions for gel formation in the presence of the drug molecule. The same degradable PEGs used for preparation of hydrolyzable, soluble drug carriers are used in this second approach to gel formation.

A discussion of these new degradable PEG gels and soluble PEG-drug conjugates follows.

Soluble, Degradable Drug Carriers Based on Double-Ester PEGs

To prepare these drug carriers we synthesized a new family of degradable PEG esters. These PEG derivatives were prepared by synthesizing an ester from a PEG carboxylic acid and a small hydroxy-acid. The resulting acid can then be activated (for example by reaction with N-hydroxylsuccinimide, NHS) for subsequent reaction with an amine.

$$PEG\text{-}CO_2H + HO\text{-}R\text{-}CO_2H \longrightarrow PEG\text{-}CO_2\text{-}R\text{-}CO_2H \tag{1}$$

$$PEG\text{-}CO_2\text{-}R\text{-}CO_2H + NHS \longrightarrow PEG\text{-}CO_2\text{-}R\text{-}CO_2\text{-}NHS \tag{2}$$

double-ester PEG

$$PEG\text{-}CO_2\text{-}R\text{-}CO_2\text{-}NHS + R'\text{-}NH_2 \longrightarrow PEG\text{-}CO_2\text{-}R\text{-}CONH\text{-}R' \tag{3}$$

$$PEG\text{-}CO_2\text{-}R\text{-}CONH\text{-}R' + H_2O \longrightarrow PEG\text{-}CO_2\text{-}H + HO\text{-}R\text{-}CONH\text{-}R' \tag{4}$$

These new "double-ester" PEGs are similar to the well known PEG succinimidyl succinate and succinimidyl glutarate, in that hydrolysis of the PEG-drug conjugate leaves a "tag" attached to the drug, equation 4. An advantage of the new double-ester PEGs is that their hydrolysis rates can be precisely controlled by variation of the PEG acid and the hydroxy acid. Also the hydroxy acid can be chosen to be a natural metabolite.

In the current work we describe the preparation and characterization of four of the double-ester, methoxy-PEGs **1**, utilizing the natural metabolites, glycolic acid (GA) and 2-hydroxybutyric acid (HBA) and two PEG carboxylic acids, carboxymethylated PEG (CM) and the propionic acid derivative of PEG (PA). Hence there are four combinations, labeled as CM-GA (**1a**), PA-GA (**1b**), CM-HBA (**1c**), and PA-HBA (**1d**).

$$\overset{\text{NHS}}{mPEG\text{-}O(CH_2)_n\text{-}CO_2\text{-}H + HO\text{-}R\text{-}CO_2\text{-}H \longrightarrow mPEG\text{-}O(CH_2)_n\text{-}CO_2\text{-}R\text{-}CO_2\text{-}NHS}$$

$$\mathbf{1}$$

1a $n = 1$, $R = \text{-}CH_2\text{-}$ ("CM-GA")

1b $n = 2$, $R = \text{-}CH_2\text{-}$ ("PA-GA")

1c $n = 1$, $R = \text{-}CH(CH_3)CH_2\text{-}$ ("CM-HBA")

1d $n = 2$, $R = \text{-}CH(CH_3)CH_2\text{-}$ ("PA-HBA")

Typically these new PEGs would be used to couple to amines through an amide linkage, eq. 3, although they could be used to couple to alcohols through an ester linkage. A possible disadvantage of this double-ester approach is that a tag remains on the drug after hydrolysis, if coupling is done via an amide linkage, eq. 4.

The synthetic route used for preparation of the desired compounds is illustrated in Scheme 1 for the combinations of mPEG acids ($n = 1$ and 2) with glycolic acid and 3-hydroxybutyric acid. Yields were typically greater than 85% with end group conversions of about 90% (Submitted for publication).

Hydrolysis Kinetics of Conjugates of 1 and mPEG Amine. To determine the hydrolysis rates of the ester bonds in conjugates of **1** with amines, soluble dimers of **1** and mPEG amine were prepared and subjected to hydrolysis. Hydrolysis was followed by HPLC as the dimer degraded into mPEG monomers, Scheme 2 (HBA conjugates are used for illustration). The kinetic results for these experiments are presented in Table I.

Scheme 1

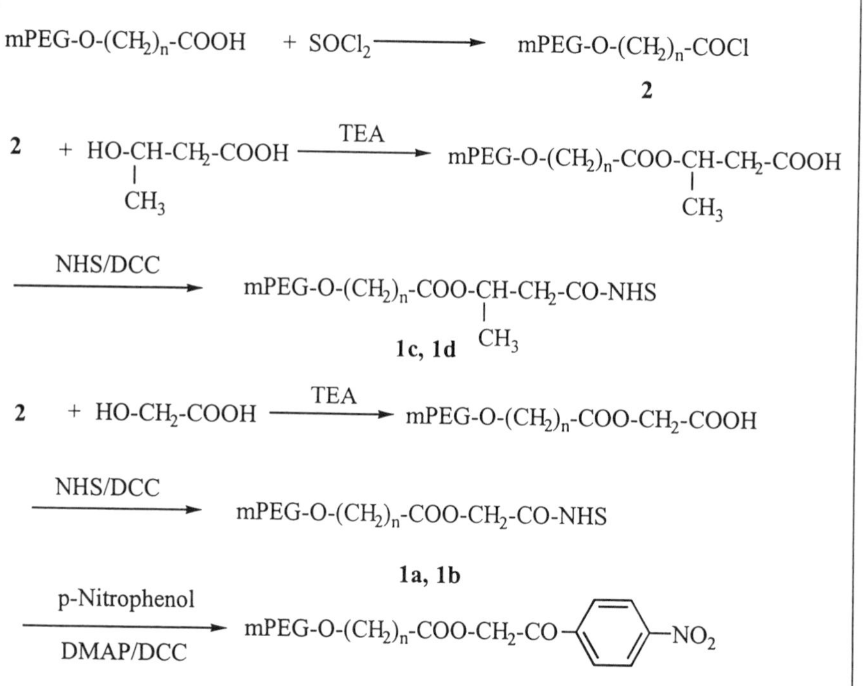

$$mPEG\text{-}O\text{-}(CH_2)_n\text{-}COOH \quad + \quad SOCl_2 \longrightarrow \quad mPEG\text{-}O\text{-}(CH_2)_n\text{-}COCl$$

$$\textbf{2}$$

$$\textbf{2} \quad + \quad \underset{\underset{CH_3}{|}}{HO\text{-}CH\text{-}CH_2\text{-}COOH} \xrightarrow{\ TEA\ } \quad mPEG\text{-}O\text{-}(CH_2)_n\text{-}COO\text{-}\underset{\underset{CH_3}{|}}{CH}\text{-}CH_2\text{-}COOH$$

$$\xrightarrow{\ NHS/DCC\ } \quad mPEG\text{-}O\text{-}(CH_2)_n\text{-}COO\text{-}\underset{\underset{CH_3}{|}}{CH}\text{-}CH_2\text{-}CO\text{-}NHS$$

$$\textbf{1c, 1d}$$

$$\textbf{2} \quad + \quad HO\text{-}CH_2\text{-}COOH \xrightarrow{\ TEA\ } \quad mPEG\text{-}O\text{-}(CH_2)_n\text{-}COO\text{-}CH_2\text{-}COOH$$

$$\xrightarrow{\ NHS/DCC\ } \quad mPEG\text{-}O\text{-}(CH_2)_n\text{-}COO\text{-}CH_2\text{-}CO\text{-}NHS$$

$$\textbf{1a, 1b}$$

$$\xrightarrow[DMAP/DCC]{\ p\text{-Nitrophenol}\ } \quad mPEG\text{-}O\text{-}(CH_2)_n\text{-}COO\text{-}CH_2\text{-}CO\text{-}\!\!\left\langle\!\!\bigcirc\!\!\right\rangle\!\!-\!NO_2$$

n=1 : PEG carboxymethyl acid (CM PEG)
n=2 : PEG propionic acid (PA PEG)
DCC: Dicyclohexylcarbodiimide
NHS: N-hydroxysuccinimide
DMAP: 4-Dimethylaminopyridine

As can be seen from Table I, the esters of glycolic acid are more susceptible to hydrolysis than those of hydroxybutyric acid, and CM esters are more susceptible to hydrolysis than PA esters. Overall, the half lives of the esters at 37 °C and pH 7 range from 14 hours for the conjugate of CM-GA and mPEG amine to more than 100 days for the conjugate of PA-HBA and mPEG amine.

Hydrolysis kinetics of active esters 1. The new PEG-ester acids have been activated as N-hydroxysuccinimide (or NHS) esters to provide an activated form, **1**, which can easily react with proteins or peptides. The hydrolysis rates of these NHS esters at pH 8.1 (typical PEGylation condition) were determined by UV measurement of released NHS, Table II. From experience we have found that NHS esters that hydrolyze with half lives shorter than about 5 minutes are too reactive for aqueous PEGylation in that much of the active ester is lost before reactants can be mixed. Thus, both CM-GA and PA-GA, with hydrolysis half lives of a few seconds, are too reactive as NHS esters for aqueous PEGylation. To improve the properties of these compounds, we also prepared p-nitrophenyl active esters, which have half lives of 7 minutes at pH 8.1.

Scheme 2

$$\text{mPEG-O-(CH}_2)_n\text{-COO-CH-CH}_2\text{-CO-NH-PEGm} \xrightarrow{\text{pH, T}}$$
$$\underset{\displaystyle CH_3}{|}$$

$$\text{mPEG-O-(CH}_2)_n\text{-COOH} \quad + \quad \text{mPEG-NH-CO-CH}_2\text{-CH-OH}$$
$$\underset{\displaystyle CH_3}{|}$$

Table I. Hydrolysis Half Lives (days, unless noted otherwise) of the Ester Linkages Formed Between 1 and mPEG Amine ($\pm$ 10%).

	Double-Ester PEG used						
	CM-GA	PA-GA		CM-HBA		PA-HBA	
pH	7.0	7.0	8.1	7.0	8.1	7.0	8.1
23 °C	3.2	43	6.5	-	15	-	120
37 °C	14 h	7.6	-	14	-	112	-
50 °C	4 h	2.2	-	5	-	58	-

Table II. Hydrolysis Half Lives of Succinimidyl Active Esters (R = NHS) and p-nitrophenyl Active Esters (R = NP) of PEG-ester Acids at pH 8.1 and Room Temperature.

R	CM-GA-R	M-PA-GA-R	M-CM-HBA-R	M-PA-HBA-R
NHS	11 s	11 s	12 min	12 min
NP	7 min	7 min	-	-

Hydrolytic Cleavage of PEG-Subtilisin Conjugate. To further assess the rates at which the new PEG derivatives cleave from a conjugate, we have coupled the CM-HBA derivative, **1c** (MW 3000), to subtilisin and followed hydrolysis at pH 8 and room temperature. The hydrolysis reaction of the PEG-subtilisin conjugate was monitored by use of MALDI-MS to reveal the amount of conjugate versus the amount of non-PEGylated subtilisin. As shown in Figure 1, the reaction product is composed of a mixture of non-PEGylated subtilisin (MW 27,500) and subtilisin with one PEG (MW 30,500), two PEGs (MW 33,500) and three PEGs (MW 36,600) attached. After one week exposure to pH 8 buffer, hydrolysis of the ester linkages reduces the mixture to non-PEGylated subtilisin and subtilisin with one PEG attached. Exposure for an additional week shows that the conjugate is greater than 90% cleaved.

Reference to Table I shows that the CM-HBA conjugate with mPEG amine reacts more slowly under these conditions (half life of two weeks) than this CM-HBA conjugate with subtilisin. Possibly the more rapid hydrolysis of the PEG-protein conjugate is the result of enzymatic catalysis, or it may reveal more subtle neighboring group effects. In any event, it is apparent that the hydrolysis rates of the conjugates of **1** and mPEG amine (Table I) are useful only as an approximation of the expected hydrolysis rates of conjugates of **1** with other molecules.

Summary of results for soluble, degradable PEGs. The preceding discussion documents the preparation and characterization of a new family of hydrolytically degradable PEG esters, having hydrolysis half lives ranging from 14 hours to more than 100 days at 37 °C and pH 7. These new PEGs should be very useful for PEGylation of small molecule drugs to enhance solubility and reduce toxicity. Similarly, protein PEGylation should also be of interest. A possible disadvantage of these derivatives is that hydrolytic cleavage of the PEG from the protein or drug results in attachment of a hydroxy-acid tag to the protein or drug. Since the tags are small (104 Da for HBA and 76 Da for GA) and nontoxic molecules, it is expected that these tags will not present problems for most proteins and many low MW drugs.

Degradable PEG Hydrogels

Our work on PEG hydrogels has concentrated on two basic types of gels. The first type ("one-step gel") is prepared by reaction of PEG carboxylic acids with PEG alcohols; hydrolytic degradability is provided by the newly formed ester linkages. Control of hydrolysis rates is provided by variation of the carboxylic acid (e.g., from carboxymethylated PEG to propionic acid PEG). The second type of PEG gel ("two-step gel") is prepared by first synthesizing a double-ester PEG (as described above), and then reacting this intermediate with a PEG amine. In this case, degradation rate is controlled by varying the PEG carboxylic acid and the hydroxy acid used in preparing the double-ester PEG. The amide link formed from reaction of the double-ester PEG with PEG amine is stable to hydrolysis. For both types of gel, increasing the extent of branching in the PEG precursors decreases degradation rate of the gel.

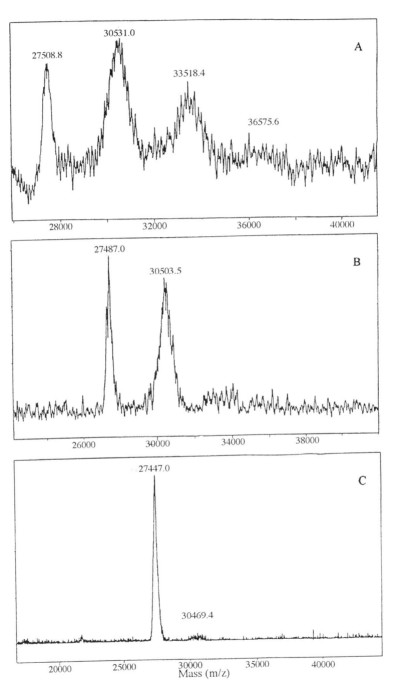

Figure 1. MALDI-MS spectrum of the mPEG-HBA subtilisin conjugate at different times (t) after preparation (pH 8 and room temperature). A. t = 1 day. B. t = 8 days. C. t = 14 days.

The following discussion provides a brief review of our work with the two types of PEG gels.

One-Step PEG Hydrogels. The one-step PEG hydrogels, represented schematically in Figure 2a, were prepared in a condensation reaction as shown in Scheme 3. Note that the R group in this Scheme is a branched core moiety, generally a polyol that is ethoxylated to form a branched PEG. E.g., for 4-arm PEG this core moiety is simply C, and the polyol which is ethoxylated is pentaerythritol. The PEG carboxylic acids of this Scheme were prepared using linear PEG (MW 2000 to 20,000) and the branched PEG alcohols were 4-arm or 8-arm (MW 2000 to 15,000). For the one step gel the weak linkage W in Figure 2a is the ester linkage $-O-(CH_2)_n-CO_2-CH_2CH_2O-$, where n = 1 (carboxymethylated PEG) and n = 2 (PEG propionic acid).

Scheme 3

m $HOCO-(CH_2)_n-O-PEG-O-(CH_2)_n-COOH$ +   2 $R-(PEG-OH)_m$

$$\xrightarrow[130\ ^{\circ}C/vacuum]{Sn(Oct)_2} \quad PEG\ gel$$

$n=1$: PEG carboxymethyl acid (CM PEG)
$n=2$: PEG propionic acid (PA PEG)
m : 4,8 (number of branches)
Sn(Oct)$_2$:stannous-2-ethylhexanoate

The gels were characterized in terms of gel conversion, swelling degree and water content. All gels made by this method have a high gel conversion with greater than 92% of PEG reactants incorporated into the gel. Also these gels have high water content in the equilibrium state of >80%. A slight temperature dependence of equilibrium swelling was also observed.

Scheme 4

mPEG-O-(CH$_2$)$_n$-COO-PEGm $\xrightarrow{\text{pH, T}}$ mPEG-O-(CH$_2$)$_n$-COOH + mPEG-OH

3 mPEG-dimer

To determine hydrolysis rates of the ester linkages, soluble dimers of mPEG acids and mPEG were prepared as in Scheme 3. Hydrolysis was followed by GPC as the dimer degraded into mPEG monomer, Scheme 4. Hydrolysis half lives of the ester linkages are listed in Table III.

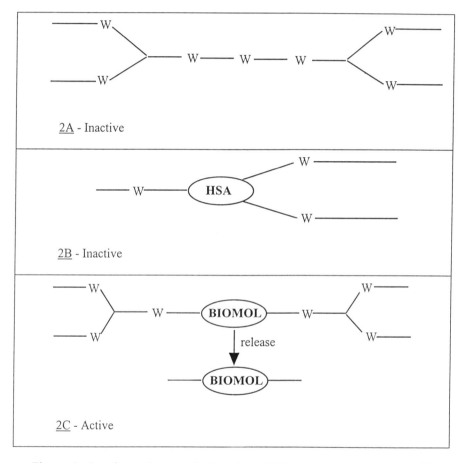

Figure 2. Inactive gel networks based on PEG (represented as —). W represents hydrolytically weak linkages. BIOMOL represents the active molecule to be delivered. Note that after release BIOMOL has PEG attached.

As can be seen from the Table , there is a large difference between hydrolysis rates of the PA ester bond and the CM ester bond. At the same temperature and pH, the CM ester hydrolyzes 7-10 times faster than the PA ester. The pH dependence of hydrolysis is also substantial; gels are most stable at pH 5.5, and hydrolysis rate increases with pH.

Table III. Hydrolysis Half Lives (days, ± 10%) of mPEG Dimer 3.

	PA ester linkage[a]			CM ester linkage[b]		
pKa of acid	4.45±0.1			3.67±0.05		
pH	5.5	7.0	8.1	5.5	7.0	8.1
23 °C	> 500	250	37	>100	30	5
37 °C	-	43	-	-	4	-
50 °C	-	15	-	-	1.5	-

a PA is propionic acid ester, **3**, n = 2.
b CM is carboxymethylated PEG ester, **3**, n = 1.

Degradation of the gels was investigated by following weight loss with time. Two distinct time periods were observed during degradation. One can be referred to as the "stable period" during which the gel maintains its shape and mass loss is gradual. The other is the "unstable period" during which the gel starts to lose shape, becomes viscous, and mass loss is much faster. Table IV gives the degradation times of the gels. Note that the degree of branching in the PEG alcohol used in preparing the gel (4-arm or 8-arm) also influences the degradation process.

Table IV. Degradation Half Lives of Ester 3 and Dissolution Times of One-step Hydrogels at pH=7.0 and 37°C.

Gel type	$t_{1/2}$ (days)	Complete Dissolution of the Gel (days)	
		4-arm PEG	8-arm PEG
PA[a]	43	25-30	60-70
CM[b]	4	2-3	5-7

a PA is propionic acid ester, **3**, n=2.
b CM is carboxymethylated PEG ester, **3**, n=1.

Release of macromolecules, such as lysozyme and PEG dyes, from the one-step gels has been studied by use of UV spectroscopy to monitor release of the macromolecule. The gels are first loaded by immersing dry gel in an aqueous solution of the macromolecule for several days. Typically the macromolecule concentration in the gel after this time is found to be only 25-50% of the solution concentration. An example of release of mPEG-disperse-yellow-9 from one-step gel is shown in Figure 3, series 1.

In summary, a family of degradable, one-step PEG hydrogels, crosslinked with ester linkages, was synthesized and characterized. The hydrolysis rates of the esters can be controlled by the choice of the PEG carboxylic acid used in preparing the gels and by the choice of the degree of branching of the PEG alcohol used in preparing the gels. The hydrogels were tested for delivery of macromolecular drugs and show potential for controlled release of protein drugs.

Two-Step PEG Hydrogels. Our second route for preparing hydrogels utilizes a two-step process in which a double-ester PEG (described above) is first synthesized and then coupled to an amine crosslinker. Our work with these materials is still in its early stages, so the following material is primarily presented to give an indication of the potential of these interesting new materials. The major advantage of the two-step hydrogels is that they can be prepared under mild conditions that are not harmful to most biologically active molecules. Therefore these gels can be prepared with the drug present, thus greatly minimizing problems with loading drugs into preformed gels (such as those described in the previous section); with this two-step approach the drug is loaded instantly upon gel formation, and there is essentially no loss of drug since the entire volume of drug solution can be incorporated into the gel. If the drug possesses an amino group, then the drug may become incorporated into the degradable matrix, so that its release is controlled by hydrolysis of the ester linkage to the drug as well as by diffusion from the matrix and degradation of the matrix. This two-step approach also provides more control over degradation rates since the hydroxy-acid can be varied as well as the PEG acid and the degree of branching, as with the one-step gel.

PEG hydrogels can be prepared from the double-ester PEGs by reacting a PEG containing two or more double-ester termini with a branched PEG amine. The first such example of a two-step gel was prepared by coupling linear, difunctional, double-ester PEG of MW 2000 with four-arm PEG amine of MW 2000.

$$PEG\text{-}[O(CH_2)_n\text{-}CO_2\text{-}R\text{-}CO_2\text{-}NHS]_2 + C(CH_2O\text{-}PEG\text{-}NH_2)_4 \longrightarrow gel$$

The resulting gels can be described schematically as shown in Figure 2a where $W = -O(CH_2)_n\text{-}CO_2\text{-}R\text{-}CONH\text{-}$. As with the soluble double-esters described above, the hydroxy acids used (R groups in above equation) are glycolic acid and 3-hydroxybutyric acid, and n is 1 and 2. The gels prepared by this route are optically clear materials of similar "firm jelly-like" consistency to the one-step gels. Early studies indicate that the gels degrade with kinetics similar to that found for the soluble double-esters, Table 1; i.e, with degradation lives at 37 °C and pH 7.0 of several hours for the CM-GA linkage, to a half week for the PA-GA linkage, to more than one week for CM-HBA, to very stable for the PA-HBA linkage. As with the one-step gels, the degradation timess can be extended by increasing the degree of

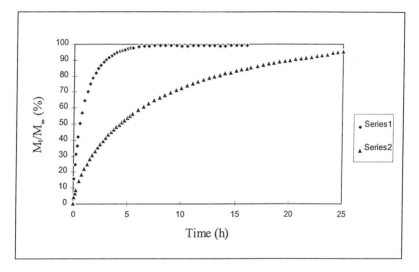

Figure 3. Release of PEG-dye 10k from PEG hydrogels. In series 1, PEG-dye is loaded into the gel (CM PEG 2k + 4 arm PEG 2k) by diffusion, while in series 2, the gel (CM-HBA 2k + 4 arm PEG amine 2k) is formed in presence of the PEG-dye.

branching in the PEG amine; preliminary experiments have been done with 8-arm PEG amine.

To characterize the release characteristics of these gels we have formed the gels in the presence of PEG-disperse yellow 9. As can be seen from Figure 3, release of PEG-dye is slower for the two-step gel. Presumably this results because the dye is loaded throughout the gel, not just in a layer near the outer surface, as is the case with the diffusion-loaded, one-step gel. Also it is noteworthy that more dye is loaded into the two-step gel (i.e., $M_{\infty}^2 > M_{\infty}^1$); the amount of dye (or drug) loaded with the two-step gel is controlled simply by the concentration of dye in the initial solution used to form the gel.

The double-ester PEGs can also be used to prepare gels by reaction with protein crosslinkers rather than branched PEG amines. For example, gels are readily formed by coupling a double-ester PEG with human serum albumin (HSA), Figure 2b. This work is similar to that of Fortier and coworkers who have coupled PEG p-nitrophenylcarbonate to BSA to form nondegradable gels for drug delivery (*19*). Use of the double-ester PEGs provides control over degradation rates of the gel.

An additional advantage of the two-step gels is that a protein drug can be incorporated into the gel in the formation step. The protein could be the sole crosslinker, or it could be used in conjunction with a branched PEG amine. Also the protein could be PEGylated previously by reaction with a conventional mPEG derivative. With this approach, release of the protein drug would be controlled by hydrolysis of the ester linkage between the gel and the protein, by diffusion from the gel, and by degradation of the gel, Figure 2c.

In summary, a family degradable, two-step PEG hydrogels, crosslinked with ester linkages, was synthesized from double-ester PEGs and branched PEG amines or HSA. The hydrolysis rates of the esters can be controlled by choice of the PEG carboxylic acid and the hydroxy acid used in preparing the double-ester PEGs and by choice of the degree of branching of the amine crosslinker. Future experiments will test the utility of these hydrogels for drug delivery.

Acknowledgment. The authors gratefully acknowledge the partial financial support of this work by the National Science Foundation, the technical assistance of Michael Roberts in obtaining MALDI spectra and the cooperation of Research Genetics, Inc. in providing access to the MALDI spectrometer.

Literature Cited

1. *Poly(Ethylene Glycol) Chemistry*; Harris, J. M., Ed.; Plenum Press: New York, NY, 1992.
2. Zalipsky, S. *Adv. Drug Del. Rev.* **1995**, *16*, 157-182.
3. Graham, N.B.In *Poly(Ethylene Glycol) Chemistry*; Harris, J. M., Ed.; Plenum Press: New York, NY, 1992; Chap. 17.
4. Sawhney, A.S.; Pathak, C.P.; Hubbell, J.A. *Macromolecules* **1993**, *26*, 581-587.

5. Clark, R.; Olson, K.; Fuh, G.; Marian, M.; Mortensen, D.; Teshima, G.; Chang, S.; Chu, H.; Mukku, V.; Canova-Davis, E.; Somers, T.; Cronin, M.; Winkler, M.; Wells, J.A. *J. Bio. Chem.* **1996**, *271*, 21969-21977.

6. Zalipsky, S.; Lee, C. In *Poly(Ethylene Glycol) Chemistry*; Harris, J.M., Ed.; Plenum Press: New York, NY, 1992; Chap. 21.

7. Glass, J. D.; Silver, L.; Sondheimer, J.; Pande, C. S.; Coderre, J. *Biopolymers* **1979**, *18*, 383-392.

8. Woghiren, C.; Sharma, B.; Stein, S. *Bioconjugate Chem.* **1993**, *4*, 314-318.

9. Musu, T.; Azarkan, M.; Brygier, J.; Paul, C.; Vincentelli, J.; Baeyens-Volant, D.; Guermant, C.; Nijs, M.; Looze, Y. *Applied Biochem. Biotech.* **1996**, *56*, 243-263.

10. Garman, A. J.; Kalindjian, S. G. *FEBS Lett.* **1987**, 223.

11. Vestling, M. M.; Murphy, C. M.; Keller, D. A.; Fenselau, C.; Dedinas, J.; Ladd, D. L.; Olsen, M. A. *Drug Metabolism & Disposition*, **1993**, *21*, 911-916.

12. Pendri, A.; Gilbert, C. W.; Soundararajan, S.; Bolikal, D.; Shorr, R. G. L.; Greenwald, R. B. *J. Bioact. Compat. Polym.* **1996**, *11*, 122-134.

13. Zalipsky, S.; Gilon, C.; Zilkha, A. *Eur. Poly. J.* **1983**, *19*, 1177-1183.

14. Ouchi, T.; Yuyama, H.; Vogl. O. *J. Macromol. Sci. Chem.* **1987**, *A24*, 1011-1032.

15. Desai, N. P.; Soon-Shiong, P.; Sandford, P. A. International Patent PCT 93/24476.

16. Nichifor, M.; Schacht, E.H.; Seymour, L.W. *J. Controlled Release* **1996**, *39*, 79-92.

17. Caliceti, P., Monfardini, C., Sartore, L., Schiavon, O., Baccichetti, F., Carlassare, F. and Veronese, F. M. *Il Farmaco* **1993**, *48*, 919-932.

18. Royappa, A. T.; Lopina, S. T.; Cima, L. G. *Mat. Res. Soc. Symp. Proc.* **1994**, *331*, 245-250.

19. Gayet, J.-C.; Fortier, G.; *J. Controlled Release* **1996**, *38*, 177-184.

20. Nathan, A.; Bolikal, D.; Vvavahare, N.; Zalipsky, S.; Kohn, J. *Macromolecules* **1992**, *25*, 4476-4484.

21. Llanos, G. R.; Sefton, M. V. *J. Biomed. Mat. Res.*, **1993**, *27*, 1383-1391.

22. Yang, Z.; Mesiano, A. J.; Venkatasubramanian, S.; Gross, S. H.; Harris, J. M.; Russell, A. J. *J. Amer. Chem. Soc.* **1995**, *117*, 4843-4850.

23. Matsuda, T., Nakayama, Y. *J. Polym. Sci., Part A*, **1992**, *30*, 2451-2457.

24. Andreopoulos, F. M.; Deible, C. R.; Stauffer, M. T.; Weber, S. G.; Wagner, W. R.; Beckman, E. J.; Russell, A. J. *J. Amer. Chem. Soc.* **1996**, *118*, 6235-6240.

25. Yamaoka; T.; Yasuhiko, T.; Ikada, Y. *J. Pharm. Sci.* **1994**, *83*, 601-606.

Author Index

Affiliation Index

Subject Index

474

Bestsellers from ACS Books

The ACS Style Guide: A Manual for Authors and Editors (2nd Edition)
Edited by Janet S. Dodd
470 pp; clothbound ISBN 0–8412–3461–2; paperback ISBN 0–8412–3462–0

Writing the Laboratory Notebook
By Howard M. Kanare
145 pp; clothbound ISBN 0–8412–0906–5; paperback ISBN 0–8412–0933–2

Career Transitions for Chemists
By Dorothy P. Rodmann, Donald D. Bly, Frederick H. Owens, and Anne-Claire Anderson
240 pp; clothbound ISBN 0–8412–3052–8; paperback ISBN 0–8412–3038–2

Chemical Activities (student and teacher editions)
By Christie L. Borgford and Lee R. Summerlin
330 pp; spiralbound ISBN 0–8412–1417–4; teacher edition, ISBN 0–8412–1416–6

Chemical Demonstrations: A Sourcebook for Teachers, Volumes 1 and 2, Second Edition
Volume 1 by Lee R. Summerlin and James L. Ealy, Jr.
198 pp; spiralbound ISBN 0–8412–1481–6
Volume 2 by Lee R. Summerlin, Christie L. Borgford, and Julie B. Ealy
234 pp; spiralbound ISBN 0–8412–1535–9

From Caveman to Chemist
By Hugh W. Salzberg
300 pp; clothbound ISBN 0–8412–1786–6; paperback ISBN 0–8412–1787–4

The Internet: A Guide for Chemists
Edited by Steven M. Bachrach
360 pp; clothbound ISBN 0–8412–3223–7; paperback ISBN 0–8412–3224–5

Laboratory Waste Management: A Guidebook
ACS Task Force on Laboratory Waste Management
250 pp; clothbound ISBN 0–8412–2735–7; paperback ISBN 0–8412–2849–3

Reagent Chemicals, Eighth Edition
700 pp; clothbound ISBN 0–8412–2502–8

Good Laboratory Practice Standards: Applications for Field and Laboratory Studies
Edited by Willa Y. Garner, Maureen S. Barge, and James P. Ussary
571 pp; clothbound ISBN 0–8412–2192–8

For further information contact:

American Chemical Society
1155 Sixteenth Street, NW ◆ Washington, DC 20036
Telephone 800–227–9919 ◆ 202–776–8100 (outside U.S.)

The ACS Publications Catalog is available on the Internet at
http://pubs.acs.org/books

Highlights from ACS Books